An Introduction to Turbulent Flow

Most natural and industrial flows are turbulent in whole or in part. The atmosphere and oceans, automobile and aircraft engines, all provide examples of this ubiquitous phenomenon. In recent years, turbulence has become a very lively area of scientific research and application, attracting many newcomers who need a basic introduction to the subject.

An Introduction to Turbulent Flow offers a solid grounding in the subject of turbulence, developing both physical insight and the mathematical framework needed to express the theory. It begins with a review of the physical nature of turbulence, statistical tools, and space and time scales of turbulence. Basic theory is presented next, illustrated by examples of simple turbulent flows and developed through classical models of jets, wakes, and boundary layers. A deeper understanding of turbulence dynamics is provided by spectral analysis and its applications. The final chapter introduces the numerical simulation of turbulent flows.

This well-balanced text will interest graduate students in mechanical, aerospace, chemical, and civil engineering, as well as in applied mathematics and the physical sciences. It will also be a useful reference for practicing engineers and scientists.

An Introduction to Turbulent Flow

JEAN MATHIEU
JULIAN SCOTT

Laboratoire de Mécanique des Fluides et d'Acoustique,
Ecole Centrale de Lyon, France

CAMBRIDGE
UNIVERSITY PRESS

CAMBRIDGE UNIVERSITY PRESS
Cambridge, New York, Melbourne, Madrid, Cape Town,
Singapore, São Paulo, Delhi, Tokyo, Mexico City

Cambridge University Press
The Edinburgh Building, Cambridge CB2 8RU, UK

Published in the United States of America by Cambridge University Press, New York

www.cambridge.org
Information on this title: www.cambridge.org/9780521775380

First published 2000

A catalogue record for this publication is available from the British Library

Library of Congress Cataloguing in Publication data
Mathieu, Jean, 1924–
An introduction to turbulent flow / Jean Mathieu, Julian Scott.
p. cm.
Includes bibliographical references.
1. Turbulence I. Scott, Julian, 1954–. II. Title.
QA913.M39 2000 99-16742
532′.0527 – dc21 CIP

ISBN 978-0-521-57066-4 Hardback
ISBN 978-0-521-77538-0 Paperback

Contents

Preface and Roadmap

The theory of turbulence has seen considerable progress in the last fifty years, although these advances are, at first sight, less marked than might have been hoped at the outset. Prior to that period, Reynolds, Taylor, Prandtl, Von Karman, and Kolmogorov, to name just some of the eminent workers, had laid the foundations of the subject and a number of excellent books and surveys appeared in the 1950s. Batchelor's book, *The Theory of Homogeneous Turbulence*, was first published in 1953 and mainly concerned the development of spectral methods, quickly gaining a following among existing and aspiring specialists in turbulence. Shortly afterwards (1956), another classic appeared in the shape of Townsend's *The Structure of Turbulent Shear Flow*, whose approach was based more on physical reasoning and placed more emphasis on the more realistic, but less theoretically tractable case of inhomogeneous turbulent flows.

Many other excellent works have appeared since then, among them the books given in the General References list following this preface. In each of them, the authors view turbulence from their own perspective, different from that of other authors, for, owing to the nature of the subject, an author's personal viewpoint colors the presentation perhaps more than in most areas of science. The book by Tennekes and Lumley is somewhat different from most of the others because it gives considerable weight to physical ideas, order-of-magnitude discussions, and other means for gaining intuition about turbulent flows, while the authors remind us that "several dozen introductory texts in general fluid dynamics exist, but the gap between monographs and advanced texts in turbulence is wide." The book gives the reader excellent food for thought before tackling more specialized ones, such as Monin and Yaglom's two-volume work, which remains the undisputed (if slightly dated) bible for the subject.

The present book does not seek to replace these existing works, but to complement them, giving the reader a solid grounding in the theory of turbulence and developing both physical insight and the mathematical framework needed to express the theory. Turbulence is treated using one-point statistics in the early chapters of the book, while multipoint statistics, analyzed by spectral methods, are introduced later, the two approaches being complementary and mutually illuminating. Employing a one-point formulation, Chapter 4 gives an introduction, illustrated by simple examples, to the basic statistical measures of turbulence, such as mean velocity, turbulent energy, and Reynolds stresses, as well as their governing equations, while Chapter 5 describes classical theories of the main classes of turbulent flows encountered in practice, namely jets, wakes, and boundary layers. Among the advantages of one-point statistical methods are that the mathematical armory required is less extensive and the range of allowed flows wider than those, namely homogeneous flows, whose multipoint statistical properties are tractable by spectral methods. Furthermore, one-point techniques are directly related to those presently used in most industrial modeling. Multipoint methods, on the other hand, are mathematically more demanding, but undoubtedly more powerful for those flows to which they can currently be applied. The multipoint approach, implemented using spectral techniques, is introduced in Chapter 6, allowing quantitative expression of the notions, already discussed in qualitative terms in Chapter 3, of a continuum of turbulent scales, the Richardson cascade of energy to smaller scales, and the resulting dissipation of energy by viscosity at the smallest scales. These are among the most important physical mechanisms at work in turbulence and form the main focus of Chapter 7, exploiting the work of Chapter 6 to introduce Kolmogorov's theory of the small scales, one of the cornerstones of the subject. The book is brought to a close in Chapter 8 with an overview of the principal techniques (DNS, LES, and one-point statistical models) for the numerical simulation of turbulent flows, critically examined in the light of the theory developed in earlier chapters, techniques that are of considerable practical importance.

This book is intended as an introduction to the theory of turbulence, with a second volume due to appear that builds on the material developed here. The reader will no doubt also want to refer to other, more specialized, books and articles for further study. The references given in each chapter and in the General References following this preface form a guide to further reading. Throughout the book, citations from the General References will be identified by "G" in front of the date (e.g., Monin and Yaglom (G 1975) will be found in the general reference list, rather than the list at the end of the individual chapter).

In conclusion, a short description of the origins of this book may be of interest. The idea was born following a doctoral course of lectures on turbulence given by one of the authors at Stanford University at the invitation of Professor Moin. Using the lecture course as a basis and adding much new material, the authors have remodeled it, to such an extent that, while mainly adhering to the overall block structure of the course, most parts have undergone a complete transformation in the book. In their original form, the Stanford lectures were inspired by a third-year undergraduate course at the Ecole Centrale de Lyon, developed by Professor Comte-Bellot, whose consequent contribution to the book we gratefully acknowledge. Our gratitude also to Professors Gence, Perkins, and Sabelnikov for their expertise in the area of probabilistic and stochastic methods in turbulence, which make their appearance in the second volume of this book in the contexts of scalar mixing by turbulence and the theory of small-scale intermittency. We owe a similar debt to our colleagues Drs Bertoglio and Cambon for help in developing the chapters on spectral methods and modeling, as well as to Dr Ladhari, who generously provided some of the figures. Last, but not least, many, many thanks to Drs Isnard and Nicolleau, whose remarks and corrections to the manuscript greatly improved the result, and to Sarah and Francine, who did an outstanding job of typing in a language other than their own.

Finally, our gratitude for the understanding, patience, and professionalism of the editor, Florence Padgett, and her colleagues at CUP, who encouraged us despite the lateness and excessive length of the manuscript and did a magnificent job of turning our poor efforts into the book you have before you. Faced with the product of some seven years of work, we ask the reader's forbearance and comprehension of the many difficulties that beset anyone trying to put together a book of this type, imperfect though the result is.

Ecully, March 1999

General References

Batchelor, G. K., 1953. *The Theory of Homogeneous Turbulence.* Cambridge University Press, Cambridge.

Batchelor, G. K., 1967. *An Introduction to Fluid Dynamics.* Cambridge University Press, Cambridge.

Bradshaw, P., 1971. *An Introduction to Turbulence and Its Measurement.* Pergamon Press, New York.

Corrsin, S., 1963. Turbulence: experimental methods. *Handbuch der Physik, Fluid Dynamics II*, eds. Flugge & Truesdell, pp. 524–90. Springer, Berlin.

Favre, A., Kovasznay, L. S. G., Dumas, R., Caviglio, J., Coantic, M., 1976. *La Turbulence en Mécanique des Fluides.* Gauthier-Villars, Paris.

Frisch, U., 1995. *Turbulence.* Cambridge University Press, Cambridge.

Hinze, J. O., 1975. *Turbulence: an Introduction to Its Mechanisms and Theory.* McGraw-Hill, New York.

Landahl, M. T., Mollo-Christensen, E., 1986. *Turbulence and Random Processes in Fluid Mechanics.* Cambridge University Press, Cambridge.

Landau, L. D., Lifshitz, E. M., 1987. *Fluid Mechanics.* Pergamon Press, Oxford.

Lesieur, M., 1990. *Turbulence in Fluids. Stochastic and Numerical Modelling.* Kluwer, Dordrecht.

Leslie, D. C., 1973. *Developments in the Theory of Turbulence.* Clarendon Press, Oxford.

McComb, W. D., 1990. *The Physics of Fluid Turbulence.* Oxford Science Publications, Oxford.

Monin, A. S., Yaglom, A. M., 1971 (vol. 1), 1975 (vol. 2). *Statistical Fluid Mechanics.* MIT Press, Cambridge, MA.

Tennekes, I. I., Lumley, J. L., 1972. *A First Course in Turbulence.* MIT Press, Cambridge, MA.

Townsend, A. A., 1976. *The Structure of Turbulent Shear Flow*, 2nd edition. Cambridge University Press, Cambridge.

Tritton, D. J., 1988. *Physical Fluid Dynamics.* Clarendon Press, Oxford.

Van Dyke, M., 1982. *An Album of Fluid Motion.* Parabolic Press, Stanford, CA.

CHAPTER ONE

An Introduction to Turbulence

This book, intended as an introductory course in turbulence, presumes the reader to be well acquainted with basic fluid mechanics and to have a reasonable level of mathematical sophistication (the General References following the Preface include some textbooks on fluid dynamics to which the reader may want to refer if necessary). In this chapter we take a critical look at the concepts underlying a description of turbulence and lay the foundations for the chapters to follow.

Most flows occurring in nature and in industrial applications are said to be turbulent. Everyday life gives us an intuitive understanding of turbulence in fluids and thus the concept is often accepted without further discussion. Among other properties, typical turbulent flows have apparently random velocity fluctuations with a wide range of different length and time scales. Thus, for instance, a graph of turbulent velocity as a function of time at a fixed point in space shows random variations. The graph has a "furry" appearance and successively higher magnifications of the "fur" show smaller fluctuations of shorter time scales, until finally the shortest is reached and the velocity is revealed as a smooth function. The same is true if the velocity is plotted as a function of one of the spatial coordinates at fixed time. However, these characteristics of turbulence do not provide a rigorous definition allowing one to objectively distinguish turbulence from a complicated laminar flow. In fact, it is difficult to give a watertight definition of turbulence, although it is easy to give specific examples in which turbulence is present in at least part of the flow:

- the rapid flow of fluid around a bluff body or airfoil,
- the majority of terrestrial atmospheric and oceanic currents,
- the motion of the atmospheres of the sun and planets,
- the flow inside most industrial plant.

It would certainly help in the systematic study of turbulence if there were a complete and succinct definition of a turbulent flow, but no fully satisfactory one has been found to date and so, rather than attempt a strict definition, we will describe specific examples and some general properties of turbulence in the course of this chapter. We begin by considering how turbulence arises in the first place, via the instability of laminar flows.

The term laminar is used to describe a flow which is not turbulent. Typically, one has in mind simple flows, such as are described in basic textbooks on fluid mechanics. For instance, consider the case of incompressible flow in a long circular pipe driven by a constant pressure difference between the ends, an example that was studied experimentally by Reynolds in his path-breaking work on turbulence towards the end of the nineteenth century. In the laminar regime, the flow far from the ends of the pipe takes on an asymptotic limiting form, known as Poiseuille flow, which is steady and axisymmetric. The velocity is parallel to the pipe axis with a parabolic distribution as a function of distance from the pipe axis, while the pressure is a linear function of streamwise distance. This simple flow is always a mathematical possibility in an infinite pipe, since it is an exact solution of the Navier–Stokes equations, but, in reality, it is subject to perturbations coming from the inlet,[1] even if these are small. Viscosity tends to damp out flow perturbations as they are convected downstream and, in the laminar regime, the perturbations are indeed attenuated, resulting in Poiseuille flow asymptotically at large downstream distances. However, once a certain flow rate is exceeded, the perturbations no longer decay and may instead be amplified, a phenomenon known as instability. In the case of the pipe, instability leads to the breakdown of Poiseuille flow to turbulence and the resulting turbulent flow is no longer approximately steady, parallel, or parabolic. The change from laminar to turbulent flow due to instability of the laminar state is referred to as transition.

The parameter that, along with the amplitude and type of perturbations, determines the onset of pipe turbulence is the Reynolds number, UD/ν, where U is the averaged flow speed (volumetric flow rate divided by cross-sectional area), D the pipe diameter, and ν the kinematic viscosity of the fluid. Viscosity provides a dissipative mechanism that attempts to damp out perturbations, but its effects diminish as the Reynolds number rises. In consequence, the tendency to instability increases with the Reynolds number. For sufficiently large perturbations, the flow becomes unstable as soon as the Reynolds number exceeds a critical value of around 2,000. The flow then enters a transitional regime in which sporadic bursts of turbulence alternate with laminar flow. As the Reynolds number is further increased, turbulence becomes less intermittent and eventually occurs continuously, yielding the fully developed turbulent regime. However, if the amplitude of perturbations is small enough, Poiseuille flow remains stable at Reynolds numbers above 2,000, and, in fact, is stable to *infinitesimal* perturbations at all Reynolds numbers. Thus, Poiseuille flow becomes sensitive to perturbations at Reynolds numbers higher than 2,000, but the appearance of instability depends on the amplitude and type of the flow perturbation if the perturbation is sufficiently small. Experimentally, taking great care to suppress perturbations, the Reynolds number at which turbulence arises has been increased by more than an order of magnitude above 2,000, but the flow is increasingly sensitive to perturbations the higher the Reynolds number and eventually becomes turbulent in any case. This illustrates the fact that transition from laminar to turbulent flow need not be simply a function of the base flow considered, but also of the amplitude and type of perturbations.

Another example of a flow instability is provided by an infinite, circular cylinder placed in uniform, steady flow perpendicular to the axis of the cylinder. The Reynolds number is defined using the uniform flow velocity (far from the cylinder) and cylinder diameter as scales. At Reynolds numbers below the critical value for

[1] Perturbations may also arise from imperfections and vibrations of the pipe walls.

instability, the flow is steady and symmetric under reflection in a plane through the cylinder axis parallel to the uniform flow. As the Reynolds number is increased prior to instability, two symmetrically placed, attached eddies develop behind the cylinder, whose extent grows with Reynolds number (see, e.g., Batchelor (G 1967), figure 4.12.1). Boundary layers form on the cylinder surface, extending from the forward stagnation point to symmetrically placed separation points on the sides of the cylinder, from where the shear layers resulting from boundary-layer separation skirt the recirculating eddies and feed the wake with vorticity. Once the critical Reynolds number of around forty-five is exceeded, although the above steady flow remains a possible solution of the equations of motion, it is unstable even to infinitesimal perturbations and hence not observable in practice. However, in contrast with pipe flow, the result of this instability is not a step change in the qualitative character of the flow, as in transition to turbulence, but another laminar flow whose spatial structure is rather similar to the one described above and in which the cylinder wake undergoes time-periodic, wavelike oscillations. Also unlike pipe flow, the onset of the instability can be predicted using the theory of infinitesimal perturbations, in which the equations governing the flow perturbation (derived from the Navier–Stokes equations) are linearized. In consequence, the critical Reynolds number does not depend on the amplitude of the perturbation and is a definite number, calculable from linear theory. As the Reynolds number is further increased, the new, time-periodic, flow develops into the well-known and beautifully regular Von Karman vortex street shown in Figure 1.1, in which the vortex wake results from oscillatory shedding of vorticity by unsteady separation of the boundary layers on the two sides of the cylinder. As the Reynolds number is increased again, turbulence is usually found to begin in the far wake, no doubt due to an instability of the vortex street. The zone of turbulence approaches the cylinder at still higher Reynolds numbers, while, above a Reynolds number in the thousands, the entire wake of the cylinder has become turbulent. The flow is then aperiodic, although it retains remnants of the cyclic vortex shedding process, and apparently organized spatial structures can be discerned within the turbulent wake up to much higher Reynolds numbers (see Figure 1.2). A region of fine-grained turbulence is present immediately behind the cylinder, giving way to more organized turbulent vortices further downstream. Outside the wake, the flow is still laminar, showing that turbulence can coexist with laminar flow, and furthermore, that the two may alternate, as bulges of wake turbulence are convected past a given point in space. The cylinder boundary layer stays laminar up to a Reynolds number of about 10^5–10^6, at which point the attempt by the laminar boundary layer to separate, which suc-

Figure 1.1. The Von Karman vortex street behind a cylinder placed in a uniform flow. (Courtesy of Sadotoshi Taneda.)

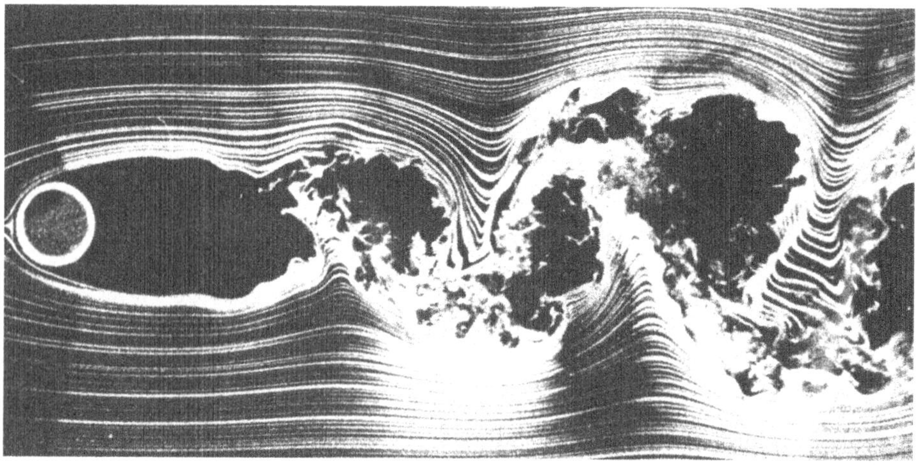

(a)

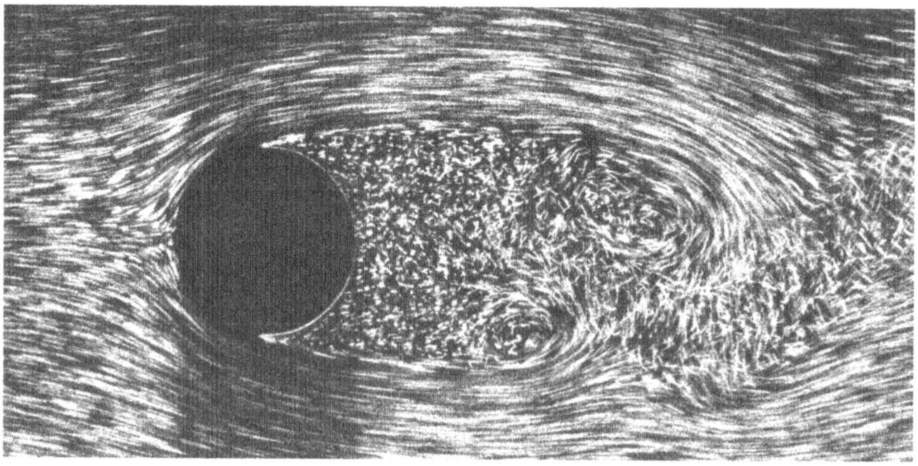

(b)

Figure 1.2. Visualizations of a turbulent cylinder wake; (b) shows a close-up. ((a) Courtesy of Thomas Corke and Hassan Nagib; (b) ONERA photograph, Werlé and Gallon (1972), reproduced with permission.)

ceeds in shedding the layer into the wake at lower Reynolds numbers, instead triggers transition of the layer, leading to a turbulent boundary layer that remains attached to the surface and extends considerably further along the cylinder surface until it too separates. As the Reynolds number is increased still more, boundary-layer transition takes place further and further upstream.

The above example shows that not all instabilities of laminar flow lead to transition. Furthermore, transition may never occur in some parts of a flow, while in others turbulent and laminar flow alternate even at very large Reynolds numbers. Among those regions that eventually become turbulent with increasing Reynolds number, transition may take place at different stages, depending upon which region is considered. Thus, in the example, transition first occurs in the far wake, spreading

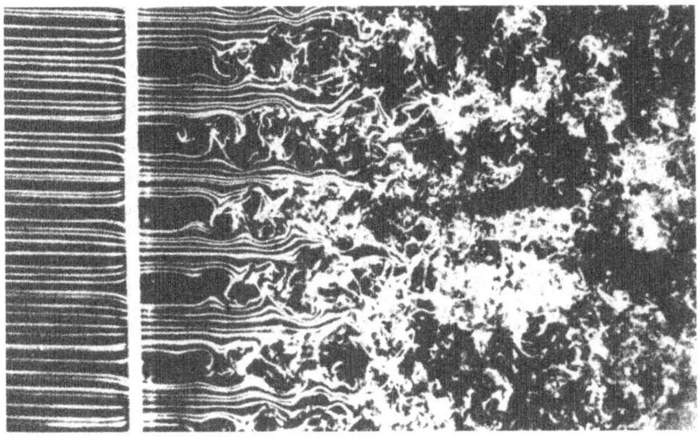

Figure 1.3. Grid-generated turbulence. (Courtesy of Thomas Corke and Hassan Nagib.)

progressively to the near wake, then in the boundary layer, progressively moving upstream. As an aside, the wakes of cylinders (not necessarily circular) are often experimentally exploited to generate grid turbulence. A grid of rods is placed in a uniform flow at high Reynolds number, leading to a double system of horizontally and vertically aligned wakes, whose complicated interactions fill the fluid far downstream of the grid with turbulence. The result is illustrated in Figure 1.3 and exhibits a clearly random character. Far downstream of the grid, the resulting flow is found to be a good approximation to statistically homogeneous[2] turbulence without mean shear, a theoretical ideal to which we will often return later in this book. In so doing, we hope and expect that at least some of the properties of the homogeneous case extend to more general turbulent flows.

Boundary-layer transition, which, as described above, occurs as the final step in the development of the cylinder flow with increasing Reynolds number, is an important topic in its own right, of which the simplest example consists of a semi-infinite flat plate placed in an infinite, steady, uniform flow aligned with the plate and perpendicular to its upstream edge. In this case, one refers to streamwise, normal, and spanwise directions. The Reynolds number that is found to characterize the instability is formed from the uniform velocity outside the layer and a boundary-layer thickness scale, usually taken to be the displacement thickness of the boundary layer. Sufficiently far downstream from the leading edge of the plate, the basic, steady laminar flow whose instability is considered is described by the Blasius similarity solution of the boundary-layer equations, which indicates a growth of layer thickness proportional to the square-root of distance from the leading edge. Thus, *the Reynolds number, and hence the tendency to instability, increases with streamwise distance*. Classical linear theory, which uses a parallel-flow approximation, predicts instability once the Reynolds number exceeds a value of about 500, in agreement with careful experiments in which upstream and other perturbations are kept to low enough levels. The result of the instability is Tollmien–Schlichting waves, which propagate and grow in amplitude downstream of the location for onset of instability and whose wavecrests are aligned in the spanwise direction. The com-

[2] Homogeneity means that the statistical properties of the turbulent velocity fluctuations are independent of spatial position. We will discuss this and other symmetries in more detail in the next chapter.

bined flow, consisting of the steady Blasius solution and the unsteady waves, itself becomes unstable once the amplitude of the waves has reached a sufficient, rather small, amplitude. This secondary instability is to perturbations which vary with spanwise position across the flow, and their growth produces a complicated, aperiodic three-dimensional flow, which is not yet turbulent (the interested reader may refer to the review by Kachanov (1994)). The remainder of the boundary-layer transition process is less well understood, but is observed to eventually give rise to sporadic turbulent spots, which appear at apparently random locations and spread out as they are convected downstream. The spots become more and more frequent with increasing downstream distance until they fill the boundary layer with turbulence (see Figure 1.4, which shows a fully turbulent boundary layer visualized by smoke injection).

The above description applies to sufficiently small perturbations, but is not what is observed when the perturbation amplitude is larger than a certain quite small threshold. For larger perturbations, transition is found to occur *upstream* of the location predicted by classical theory and appears to depend on the amplitude and type of perturbations, as for the case of the pipe discussed earlier. Such "bypass" transition has not yet been as fully elucidated as that based on Tollmien–Schlichting waves described above, but a likely candidate is the amplification of upstream perturbations by a linear mechanism of algebraic growth (see the brief review by Henningson and Reddy (1994)). This mechanism is active at lower Reynolds numbers than the Tollmien–Schlichting instability and is not captured by the classical theory because that concerns itself only with exponential growth or decay of the perturbations. If the amplitude of perturbations incident on the boundary layer from upstream is sufficient, algebraic amplification may generate large enough perturbations that secondary instability and breakdown to turbulent spots can occur upstream of the classical location for instability, inducing bypass transition. Theory

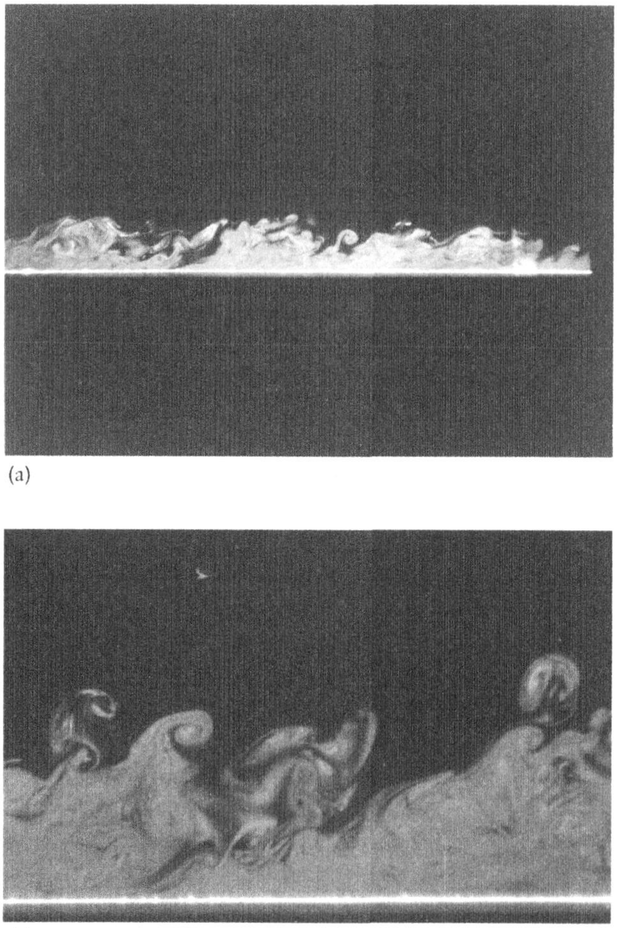

(a)

(b)

Figure 1.4. Smoke visualization of a fully developed turbulent boundary layer; (b) shows a close-up. (Courtesy of F. Ladhari; see Ladhari (1983) for experimental details.)

predicts that algebraic growth is accompanied by elongation of the perturbations in the streamwise direction, leading to perturbations whose spatial structures are aligned with the flow, rather than the spanwise direction, as for Tollmien–Schlichting waves. Clearly, the physical mechanisms responsible for the initial growth of perturbations (and no doubt those implicated in secondary instability) differ in the two cases, illustrating the possibility of multiple paths to transition through different sequences of laminar flow instabilities depending on the amplitude (and type) of perturbations.

We could continue to give other examples of laminar flow instabilities and transition, but the main points should by now be clear. Turbulence is thought to arise via the instability of laminar flow as the Reynolds number[3] is increased, though not all instabilities lead to transition and both the onset of and route taken to transition may vary depending on the amplitude and nature of the perturbations. Research on the transition problem, which presents an important theoretical challenge, continues apace, but we now turn to the question of the description of turbulence once it has arisen, which will occupy us for the remainder of this book. A change of pace is also in order as we review the equations which govern the flow, partly as a reminder to the reader and partly to define the notation used.

Turbulent flow appears random in time and space and is not experimentally reproducible in detail. It might be thought that precise mathematical analysis of such flows would be irrelevant or even impossible; however, this point of view proves too pessimistic. We have no reason to believe that the basic dynamical equations governing laminar flows somehow do not apply to turbulent ones, and experience suggests that turbulence is governed by the same fundamental equations as laminar flows. The source of disorder in turbulent flow is to thus be looked for as a consequence of the equations rather than in a breakdown of the quantitative model.

In this book, we will restrict attention to the flow of incompressible, Newtonian fluids, whose viscosity and density are taken constant, and which are not acted on by body forces. Of course, in reality these assumptions will not be respected exactly. They are commonly used approximations, which simplify the flow description, but whose validity depends on the flow and fluid considered, as indeed does the continuum approximation that underlies the entire description. With these approximations, the velocity and pressure are described by the celebrated Navier–Stokes equation

$$\frac{\partial U_i}{\partial t} + \underbrace{U_j \frac{\partial U_i}{\partial x_j}}_{\substack{\text{Nonlinear} \\ \text{convective term}}} = -\frac{1}{\rho}\frac{\partial P}{\partial x_i} + \underbrace{\nu \frac{\partial^2 U_i}{\partial x_j \partial x_j}}_{\substack{\text{Viscous} \\ \text{term}}} \tag{1.1}$$

and the incompressibility condition

$$\frac{\partial U_i}{\partial x_i} = 0 \tag{1.2}$$

[3] In the above examples there is one dimensionless parameter – a Reynolds number – describing the base flow, but it should be borne in mind that several such parameters are needed in general.

which are often referred to collectively as the Navier–Stokes equations, and which should be supplemented by appropriate boundary and initial conditions to form a complete description of the fluid velocity and pressure fields, $U_i(\mathbf{x}, t)$ and $P(\mathbf{x}, t)$. In the above equations, x_i are Cartesian spatial coordinates and ρ, ν are the constant density and kinematic viscosity of the fluid. Here and throughout this book, we denote vectors by bold symbols and use subscript notation for the Cartesian components of vectors and tensors, with subscripts taking the values 1, 2, or 3. The summation convention is employed, that is, a single term containing one or more repeated subscripts represents an implied sum over all three values of each repeated subscript. The nonlinear term in (1.1) is due to momentum convection by the flow and plays a very important role in the generation and maintenance of turbulence. Viscosity dissipates kinetic energy of the flow as thermal energy of the incompressible fluid at the rate

$$\Delta = \frac{1}{2}\nu\left(\frac{\partial U_i}{\partial x_j} + \frac{\partial U_j}{\partial x_i}\right)\left(\frac{\partial U_i}{\partial x_j} + \frac{\partial U_j}{\partial x_i}\right) \tag{1.3}$$

per unit mass of fluid, where there is an implied sum over i and j.

Thanks to their mathematical simplicity, many classical solutions of equations (1.1) and (1.2) are special flows for which the nonlinear convective term is zero or negligible. These include parallel flows, for example, Poiseuille flow and that generated by in-plane oscillations of an infinite flat plate, and flows having very low Reynolds number, often referred to as Stokes or creeping flows. Important though such examples are, they are not representative of more general flows, which can have important effects of nonlinearity. The very simplicity of such solutions means that they tend to occupy pride of place in most basic courses on fluid mechanics and it is rarely pointed out that, in this respect, they are rather special cases. This might lead a naive student to the conclusion that the nonlinear terms are of secondary importance, whereas nonlinearity is essential to the dynamics of turbulent flow and its importance is stressed here.

The momentum and continuity equations take the particularly simple forms (1.1) and (1.2), with constant ρ and ν, under the assumptions made above. However, although we do not consider the possibility in this book, these equations may be complicated by effects we have excluded, such as variable density, compressibility, body forces, or the non-Newtonian nature of the fluid, and one may also need to take the thermodynamics of the fluid into account via an energy equation and equation of state, together with yet further equations if the fluid is a mixture or has other internal degrees of freedom (see Bird, Warren, and Lightfoot (1960) for further information). Even under the simplifying assumptions made in this book, which allow us to describe the velocity and pressure fields using (1.1) and (1.2) alone, one may still be interested in the heat transfer properties of the flow, which require consideration of thermodynamics. We will briefly discuss the turbulent mixing and dispersion of scalar fields such as fluid temperature later in this chapter.

The Navier–Stokes equations, (1.1) and (1.2), should be supplemented by appropriate boundary and initial conditions. For instance, the velocity is prescribed according to the no-slip condition at solid boundaries and at infinity for a body placed in an infinite flow. In such cases, the only parameter which the fluid brings to the party is the kinematic viscosity, ν, since one can absorb the density into the pressure field by working with the quantity P/ρ. To the extent that the above mathematical model accurately describes many real flows, both turbulent and laminar, the precise nature of the fluid, for example, liquid or gas, is thus of secondary importance to the flow. In fact, since only nondimensional parameters can be of fundamental significance, ν, which is dimensional, should be combined with geometrical and other parameters arising from the boundary and initial conditions to form, for example, a Reynolds number, as we saw in the examples of flow instabilities given earlier. More complicated boundary conditions can appear, for instance, those involving the stress at the interface between two fluids, and which can introduce other fluid properties, such as surface tension.

In general, initial conditions on the velocity field are also needed for (1.1) and (1.2). Mathematically, the solution of these equations with prescribed initial and boundary conditions on the velocity is unique.[4] However, turbulent flow is not experimentally reproducible in detail and a different flow occurs each time the same experiment is performed. The source of this lack of reproducibility is extreme sensitivity of the flow to small changes in the initial and/or boundary conditions,[5] which the experimentalist cannot, of course, control with infinite precision. In theoretical analyses of turbulence one usually imagines an experiment performed many times under nominally the same conditions, leading to an ensemble of different realizations of the turbulent flow. Ensemble averages and probability distributions can then be defined by considering a large (in principle infinite) number of different realizations, thus introducing a statistical description of turbulence. For instance, the mean flow velocity is defined as the average of the velocity over a large number of realizations, while the departure of the velocity in a given realization from the mean is known as the velocity fluctuation of that realization and is usually identified with the turbulence itself. Flow statistics are assumed to be reproducible, even though individual realizations are not, which is the fundamental reason for adopting a statistical description. This is not to say that the dynamics of individual realizations are unimportant, and indeed the basic governing equations of fluid motion, (1.1) and (1.2), *describe single realizations rather than the statistics of the ensemble.* One approach to a theoretical description of turbulence consists of first determining the properties of individual realizations, usually via brute-force numerical simulation of the governing equations, and then to calculate the statistical properties of the flow from these. A second way forward is to work with equations for the statistics themselves, but, as we shall see in Chapter 4, these equations are not closed, a fundamental problem that is usually resolved by the expedient of introducing additional heuristic approximations, known as closure models or hypotheses. Independently of the approach used, the ultimate aim of most studies of turbulent flows is to describe their statistics.

In many cases, the flow is found to eventually forget its initial conditions and to become steady, where, in the case of turbulent flow, steady means that its statistical properties are independent of time, *not* that the flow is steady in any one realization.

[4] However, existence, which is generally believed to hold, has, to the authors' knowledge, yet to be proven.

[5] Sensitivity of turbulent flow to these conditions is no doubt related to the flow instabilities discussed earlier.

The statistical properties of steady turbulence are taken to be independent of the initial conditions, which the flow forgets, and, as for turbulent flows in general, insensitive to small changes in the boundary conditions. Thus, whereas one cannot predict the flow in any given realization, it is a reasonable goal to describe the statistical properties of the ensemble, both experimentally and theoretically. For statistically steady turbulence, one may also define averages and probabilities by sampling at a large number of times over a long enough period (in principle infinite), which forms the basis for most experimental measurements of flow statistics. It is usually assumed that the results of performing repeated experiments (ensemble) and time sampling a single flow lead to the same results for steady flows. This question, among others, is addressed in Chapter 2, which is devoted to statistical methods, in part to remind the reader of some of the analytical techniques involved, but also to set them in the context of the theory of turbulence for the remainder of the book.

1.1 The Physical Nature of Turbulence

Since the subject matter of this book is turbulent flows, we consider it important that the reader have some understanding of the properties of turbulence from the start. It is thus the intention of this section to provoke and enlarge on the knowledge of the reader. In lieu of a precise definition of turbulence, we shall discuss some important characteristics of turbulent flows. Some of the concepts are probably unfamiliar to the reader and we do not want to interrupt the presentation by fully defining all terms here. Detailed discussion will be given in later chapters. In particular, the first two of the properties listed below, namely apparent randomness and the possession of a wide continuum of space and time scales at high Reynolds number, are probably the most important characteristics of turbulence and form the subject matter of the next two chapters of this book.

TURBULENCE IS A RANDOM PROCESS

Turbulent flow is time and space dependent with a very large number of spatial degrees of freedom. The random nature of the flow is easily observed using, for instance, hot-wire or laser anemometry. Such probes produce signals which show random fluctuations and, at high turbulent Reynolds number (defined below), have "furry" graphs as functions of time. As discussed above, although turbulence is unpredictable in detail, its statistical properties are supposed reproducible and it is fruitful to consider averages and probability distributions of flow quantities, as we shall do throughout this book.

TURBULENCE CONTAINS A WIDE RANGE OF DIFFERENT SCALES

The "fur" apparent in velocity measurements at high Reynolds number (see Figure 3.1a) reflects the existence of a continuum of different space and time scales of the flow. The large scales are evident in the overall fluctuations of a graph of velocity versus spatial position or time, whereas small ones are apparent through the fine-scale fur. The dynamics of turbulence involve all scales, as we shall see in the course of this book. Different scales coexist and are superimposed in the flow, with smaller ones living inside larger ones. The continuum of scales forms one of the main topics of Chapter 3 and, like the statistical description, recurs throughout this book.

Measurements of turbulent fluctuations from two spatially separated probes seem unrelated at first sight, but statistical analysis reveals that they are correlated. The degree of statistical correlation can be measured using the velocity correlations, which are averages of the products of velocity fluctuations at the two probes and are found to fall off[6] as the distance between the probes increases, indicating that the velocities become less and less closely related the more they are separated. At small separations, the velocities at the two probes have essentially the same values and their correlations are large, being close to the value at zero separation, whereas, as the probe separation increases, the difference of velocity fluctuations between the two probes grows and the correlations are less. The characteristic length scale, L, for decorrelation, or correlation length, provides an important measure of the distance over which the velocity fluctuations differ significantly and hence of the size of the *large scales* of turbulence. A number of precise definitions of correlation length can be given, but the correlations do not drop off suddenly at a definite separation, and what we have in mind here is the order of magnitude of the separation required for significant decorrelation.

In addition to correlation lengths, one may define temporal correlations by averaging products of velocity fluctuations at a single point and two distinct times, leading to correlation times that characterize the time delay needed for significant temporal decorrelation. More generally, correlations involving separations in both space and time can be studied.

A measure of the smallness of viscous effects on the *large scales* of turbulence is provided by the turbulent Reynolds number, $\mathrm{Re}_L = u'L/\nu$, where u' characterizes the overall turbulent velocity fluctuations (u' is usually taken as the root-mean-squared velocity fluctuation of one of the components of velocity) and L is a length scale representing the order of size of the large scales (e.g., a correlation length). The quantities u' and L, and hence the turbulent Reynolds number, Re_L, are characteristic of the large scales of turbulence. The Reynolds number, Re_L, is typically high for turbulent flows, implying little direct effect of viscosity on the large scales, although, as we shall see, this does not mean that viscosity can be neglected. Indeed, viscous effects play an important role in the dynamics of the smallest scales.

As noted above, a graph of velocity as a function of position is furry when the large scales of a high-Reynolds-number flow are considered. Applying higher and higher magnifications, one eventually finds a scale at which the velocity is revealed to be a smooth function, defining the smallest scales of the flow. Smoothness reflects viscous action and thus the size of the smallest scales depends on the viscosity, decreasing relative to the large scales as the Reynolds number is increased. The size of the large scales is typically fixed by the overall geometry of the flow, for example, the width of a turbulent jet, whereas that of the smallest ones adjusts itself according to the viscosity.

[6] Not necessarily monotonically, however.

TURBULENCE HAS SMALL-SCALE RANDOM VORTICITY

Turbulent flow is rotational, that is, it contains vorticity, and not just any vorticity. Laminar flows can, of course, possess vorticity, but, a characteristic of high-Reynolds-number turbulence is that the vorticity has intense, *small-scale*, random variations in both space and time. The magnitude of these vorticity fluctuations is much larger than the mean vorticity and they are randomly orientated in direction. The vorticity is defined as the curl of the velocity, $\mathbf{\Omega} = \nabla \times \mathbf{U}$, and thus involves its spatial derivatives. Taking the derivative brings out the fine-scale fur in the velocity field and, as a result, the spatial scale for vorticity fluctuations is the smallest in the continuum of turbulent scales. That is, velocity derivatives are dominated by the smallest scales of turbulence, at which the velocity field appears as smooth (and, in particular, differentiable) when examined at a sequence of higher and higher magnifications. This scale is known as the Kolmogorov length scale, and viscosity, which has little influence on larger scales at high Re_L, is important at these, the smallest.

Notice that, in such descriptions of the different length scales of turbulence, we are only really concerned with orders of magnitude. Thus, the correlation length, which gives the order of spatial separation over which velocities decorrelate, and the Kolmogorov length, which describes the scale at which there are significant viscous effects, are both order-of-magnitude quantities, forming the upper and lower ends of the continuum of turbulent scales.

The reader will recall (see, e.g., Batchelor (G 1967), section 2.3) that the vorticity can be physically interpreted by expanding the instantaneous velocity field as a spatial Taylor's series up to linear terms about some point in the flow. In this way, the instantaneous velocity inside a small particle can be written as the sum of the velocity of its centroid, simple strainings along three perpendicular axes, and rotation at angular velocity $\mathbf{\Omega}/2$. Thus, the vorticity represents rotation of small fluid particles about their centroid. An evolution equation for the vorticity can be derived by taking the curl of (1.1) and using (1.2), leading to

$$\frac{\partial \Omega_i}{\partial t} + \underbrace{U_j \frac{\partial \Omega_i}{\partial x_j}}_{\text{Convection}} = \underbrace{\Omega_j \frac{\partial U_i}{\partial x_j}}_{\text{Stretching}} + \underbrace{\nu \frac{\partial^2 \Omega_i}{\partial x_j \partial x_j}}_{\text{Viscous diffusion}} \tag{1.4}$$

which describes the time evolution of the vorticity vector, $\mathbf{\Omega}$. Both the terms labeled convection and stretching arise from the convection term in (1.1). If these terms are neglected, we have the simple diffusion equation, hence the labeling of the viscous term. On the other hand, *in the absence of viscosity*, the convection term in (1.4) causes vorticity to move at the fluid velocity, although the stretching term is responsible for changes in both the magnitude and direction of the vorticity vector as it is convected. Vortex lines, defined as everywhere tangential to the vorticity vector, are carried by the flow, that is, a given vortex line always consists of the same fluid particles. As the flow convects the vortex line, it generally changes in orientation and the vorticity vector obviously does likewise. Stretching of a vortex line by the flow causes the vorticity to be amplified,[7] its magnitude increasing in

[7] The importance of vortex stretching as a means of amplifying vorticity appeared early in the development of fluid dynamics in the work of Kelvin, Lagrange, Helmholtz, and Cauchy on inviscid rotational flows.

proportion to the distance between two infinitesimally separated fluid particles among those making up the line. Convection and stretching are inviscid mechanisms and, with nonzero viscosity, vorticity diffuses due to the viscous term in (1.4).

Qualitatively, vortex convection and stretching may be thought of as the mechanisms by which the intense, fine-scale vorticity fluctuations of high-Reynolds-number turbulence are generated and maintained. In this view, the intensity of vorticity fluctuations is due to amplification by stretching, whereas their fine scale reflects transverse reduction in size of a blob of vorticity that is drawn out along the direction of stretching. These are inviscid mechanisms, stemming from the nonlinear convective term in (1.1), and are continuously going on inside high-Reynolds-number turbulence. However, the viscous diffusive term in (1.4) is called into play at sufficiently small scales because the viscous term in (1.4) contains second derivatives and thus increases in relative importance as the scale considered decreases. Since viscous diffusion tends to cause vorticity to spread, it counteracts both the amplifying and scale-reducing effects of vortex stretching. Thus, viscous diffusion of vorticity places a lower limit (the Kolmogorov scale) on the size of vortical structures attainable by stretching, resulting in the velocity field being smooth when viewed at such scales, and an upper limit on the amplification of vorticity fluctuations. Because viscosity is called into play, there is associated dissipation of mechanical energy to heat at the smallest scales of turbulence.

We will return to the subject of vorticity in turbulent flows in Chapter 4.

TURBULENCE ARISES AT HIGH REYNOLDS NUMBERS

We saw earlier that transition to turbulence occurs due to instability of laminar flow at large Reynolds number. With rising Reynolds number, the nonlinear convective term in the Navier–Stokes equation assumes increasing importance compared with the viscous term, and the tendency to instability, which is damped by viscosity, increases. Thus, a large Reynolds number is a prerequisite for the production of turbulence.

Once a turbulent flow is established at high Reynolds number, flow instabilities are responsible for continued generation of turbulence, producing large-scale eddies, which are themselves unstable, giving rise to smaller ones, and so on, until viscosity becomes important at the smallest scales, as described above. This cascade process, by which small scales are produced from larger ones, is going on incessantly inside high-Reynolds-number turbulence and extracts energy from large scales, passing it down via smaller and smaller ones until it is dissipated by viscous action at the smallest scales. The continuum of spatial scales generated by the energy cascade, between the large ones, over which fluid velocities are correlated, and the much smaller viscous scales of the vorticity fluctuations, at which energy is dissipated as heat, gets wider as the turbulent Reynolds number increases. The size of the large scales is generally determined by the environment of the flow, with the correlation length scaling on, but being somewhat smaller than, for instance, the width of a turbulent wake or jet. On the other hand, the smallest, dissipative scales change in size with the fluid viscosity, becoming smaller at larger Reynolds numbers.

If the turbulent Reynolds number drops too low, the cascade can no longer operate. There are then only the large scales of turbulence, whose velocity fluctuations decay due to direct viscous damping, rather than indirect loss of energy via the cascade, followed by dissipation at the Kolmogorov scale. This is the eventual fate of turbulence left to decay without an energy source to continuously replenish the large scales. Its Reynolds number falls, and with it the range of turbulent scales, until, eventually, the turbulence decays passively under viscosity, at which point it is essentially dead. There appears to be little to distinguish a random laminar flow from such an end result of turbulent decay and, in studies of turbulent flows, one is usually more interested in the behavior of high-Reynolds-number turbulence. We will concentrate on the case of high Reynolds number throughout much of this book.

TURBULENCE DISSIPATES ENERGY

Since a high-Reynolds-number cascade is essentially inviscid, it conserves mechanical energy. Thus, as smaller scales are formed, they can be thought of as sapping some of the mechanical energy of their larger parents and transmitting it to their own offspring. As a result, the cascade generating the smaller scales is associated with a mean flux of energy from large to small turbulent scales, a flux which is controlled by the dynamics of the large scales, with viscous dissipation as heat at the smallest scales. Because the small scales are only responsible for fine wiggles of velocity graphs, it is the large scales that contribute most to the overall turbulent fluctuations of velocity. That is, the large scales contain the major part of the turbulent kinetic energy. They continuously feed energy via the cascade to the smallest eddies, where it is dissipated. Since the intermediate scales contain only a small fraction of the overall turbulent kinetic energy, they should not be thought of as hoarding the energy that is passed to them by the large scales, but rather as quickly passing it on via the cascade to yet smaller scales.

As a result of the cascade, turbulent flows dissipate energy rapidly as the viscous stresses act on the fine scales. For example, under otherwise similar conditions, the energy lost in a circular pipe is roughly one hundred or more times larger if the flow is turbulent than if it is laminar. Statistically steady turbulence thus requires a continuous supply of energy. Overall, a steady turbulent flow can be considered as driven by the existence of a large-scale source of energy, namely extraction from the mean flow by instability, the cascade of energy through smaller and smaller scales, and the dissipation of energy at the smallest scales. If there is no energy supply to maintain the flow, turbulence decays and eventually ceases to be active because the Reynolds number is no longer large enough.

For homogeneous turbulence, quantification of the above qualitative ideas of different scales of turbulence, their contributions to the turbulent energy, and the cascade of energy from large to small scales is provided by spectral theory, developed in Chapters 6 and 7.

TURBULENCE IS A CONTINUUM PHENOMENON

The smallest scales associated with turbulence are those dictated by viscosity. These scales are typically many orders of magnitude larger than the molecular free paths.[8] Thus, turbulent flows can be described by a continuum approximation, such as the Navier–Stokes equations.

[8] Exceptions may occur for gases and plasmas if the turbulent Mach number is large enough. We do not consider such cases in this book.

TURBULENCE IS INTRINSICALLY THREE-DIMENSIONAL

For a strictly two-dimensional flow, that is, one which is independent of x_3, say, and has $U_3 = 0$, vorticity is directed in the x_3-direction and the vortex stretching term in (1.4) is zero. Thus, stretching does not act in two-dimensional flow and, in the absence of viscous diffusion, the vorticity is passively convected, unchanged by the flow. The high-Reynolds-number cascade of energy to smaller scales cannot take place and turbulence, as we understand the term in this book, is not initiated. While it is true that such flows can exhibit complicated flow structures with a degree of randomness, they do not possess the ubiquitous fine scales associated with three-dimensional turbulence at high Reynolds numbers, and correlations tend to be present throughout the flow. Numerical simulations of high-Reynolds-number two-dimensional flows with random initial conditions suggest that small-scale structures coalesce to form larger ones. This contrasts with the three-dimensional case, for which large structures produce smaller ones, leading to intense random vorticity and energy dissipation on a small scale at which viscosity intervenes. It should be made clear that turbulent flow can nonetheless be two-dimensional in a statistical sense (statistical properties independent of x_3 with zero mean-velocity component in that direction), but that the turbulence in any given realization is three dimensional. Furthermore, turbulent flows may show larger correlation lengths in some direction, thus resembling the two-dimensional case, as they tend to when subjected to strong rotation, for instance.

In summary, although many respected authors have talked of two-dimensional turbulence, in our view turbulent flow proper can only occur if the random fluctuations are three dimensional. We do not intend to suggest by these remarks that the study of random two-dimensional flow is unimportant, indeed it has stimulated much interesting work, but to point out that the physical mechanisms that govern such flows are sufficiently different from what is usually meant by turbulence that they should be placed in a different category. We will briefly return to this subject in Chapter 4.

THE LARGE SCALES OF TURBULENCE ARE INSENSITIVE TO VISCOSITY AT HIGH ENOUGH REYNOLDS NUMBER

If the turbulent Reynolds number is high enough, the dynamics of the large scales are essentially inviscid and hence insensitive to the precise value of the large Reynolds number. While the size of the smallest scales adjusts to changes in the fluid viscosity so as to dissipate energy at an appropriate rate, controlled by the large scales, the smallest scales are thought to have little direct effect on the large scales, which interact mainly with scales immediately below them in the cascade. Thus, those properties of turbulence which are determined by the large scales should be largely unaffected by changes in the viscosity. This leads to the conjecture that, for instance, the mean energy dissipation rate, mean velocity and root-mean-squared velocity fluctuations of a turbulent flow approach limiting values as the Reynolds number tends to infinity (or at least that they ought to vary much more slowly with changes in the Reynolds number). One would, for example, expect the mean velocity and root-mean-squared turbulent fluctuations of a turbulent jet or wake to approach limiting profiles at high Reynolds numbers. Furthermore, although viscous energy dissipation is dominated by the smallest scales, the average rate of energy dissipation reflects the mean rate of energy supply from the large scales and is consequently also believed to tend to a limit as the viscosity goes to zero.

Insensitivity of the large-scale properties of turbulence to changes in viscosity at large Reynolds number is borne out experimentally. Wall-bounded flows are more subtle than free-shear ones, such as jets or wakes, owing to the existence of thin viscous layers at solid boundaries, within which the turbulent Reynolds number takes moderate values even when the overall Reynolds number is very large. Thus, viscosity affects the large scales within the viscous layers and, even outside, their existence may mean that (at least some) large-scale properties of wall-bounded flows do not approach limiting values, but rather continue to be weakly (perhaps logarithmically) dependent on the viscosity. Viscous layers are discussed in more detail in Chapters 4 and 5 and should not be confused with turbulent boundary layers, of which they form a comparatively thin sublayer at the wall.

1.2 Some Practical Consequences: Energy Loss, Drag, and Dispersion

We have already noted the greater energy dissipation of turbulent flow and so we will concentrate on the related issue of drag and on the dispersion of heat and material. The existence of turbulence results in more rapid dispersal of momentum, heat, and material compared with laminar flows, leading to substantially different properties for turbulent flows (see, e.g., Figure 1.5a). Increased momentum transfer is usually undesirable because it gives rise to greater skin friction and hence drag. Figure 1.5b shows the drag coefficient of a flat plate for the idealized cases in which the boundary layer is either laminar or turbulent over the whole plate, illustrating the greater skin friction of the turbulent regime. In practice, below a certain value (typically between 10^5 and 10^7) of the Reynolds number, Re, based on the plate length, the boundary layer is indeed entirely laminar and the lower line in the figure applies. However, it becomes turbulent over the downstream part of the plate at higher Reynolds numbers. The layer is then laminar upstream and turbulent downstream of a transition zone, leading to a value of c_f intermediate between the two shown in the figure. Thus, as the Reynolds number increases, the downstream part of the boundary layer becomes turbulent and the resulting measured curve of c_f as a function of Re departs from the lower line in the figure, rising towards the upper line, to which it asymptotes at large enough Re.

Unlike increased drag due to momentum dispersal, the turbulent dispersion of heat or material is often a positive and crucial aspect of many practical flows. Without it, the reactants in a chemical reactor would take a very long time to become sufficiently mixed to undergo a reaction and pollution released into the atmosphere would remain undiluted by the mixing process. In any case, turbulence is an unavoidable feature of most industrial flows, because high throughputs imply large Reynolds numbers.

DISPERSAL OF MOMENTUM

At high Reynolds number, close to a solid boundary, a boundary layer is formed. Visualization of turbulent boundary layers shows the existence of complicated structures of all scales within the layer and a convoluted boundary separating the external laminar flow from the turbulent flow inside the layer (see Figure 1.4). Irregular bulges form on the boundary and engulf and entrain fluid from outside, taking it into the layer. Measurements of instantaneous velocities within the layer show large gradients and strong fluctuations. The skin friction coefficient is much higher for a turbulent than for a laminar boundary layer (see Figure 1.5b), hence the tendency to increased drag due to turbulence. The greater skin friction is a reflection of faster dispersal of momentum as the turbulent layer entrains fluid from outside, slowing it down as it is engulfed by the layer and becomes turbulent itself. The entrainment process is far more efficient at spreading a boundary layer than the viscous molecular diffusion that acts in a laminar boundary layer.

Let us first consider a flat plate. A laminar boundary layer on the plate evolves with downstream distance, x, in response to two processes. Firstly, fluid particles are convected by the flow; secondly, viscous diffusion causes the layer to grow in thickness. If the external flow velocity is U_e, a characteristic time for convection to the position x is

$$T_c \approx \frac{x}{U_e} \tag{1.5}$$

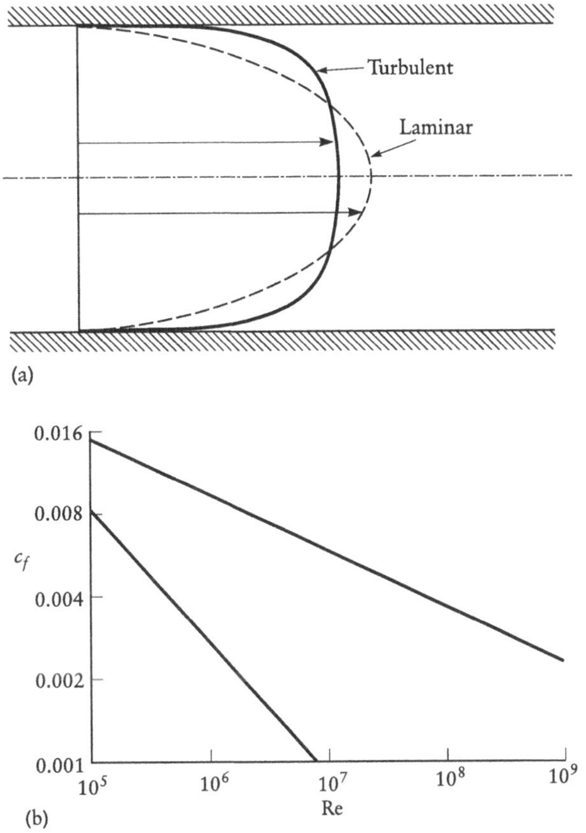

Figure 1.5. Turbulent vs. laminar flows: (a) effects of turbulence on the mean-velocity profile in a pipe; (b) the drag coefficient of a flat-plate boundary layer. Both are reflections of more efficient momentum transfer in turbulent flow. In (b), $\mathrm{Re} = U_e L/\nu$ is the Reynolds number based on the length of the plate, L, and U_e is the free-stream velocity, while $c_f = 2F_D/(\rho U_e^2 L)$ defines the drag coefficient, F_D being the drag force.

whereas for viscous diffusion to produce a layer of thickness δ requires a time of order

$$T_d \approx \frac{\delta^2}{\nu} \tag{1.6}$$

where ν is the kinematic viscosity. Setting $T_c \approx T_d$ gives us the thickness

$$\delta \approx \left(\frac{\nu}{U_e}\right)^{1/2} x^{1/2} \tag{1.7}$$

at downstream distance x. We observe that $\delta \propto x^{1/2}$ (so that the laminar layer spreads at a steadily decreasing rate) and that this expression can also be written

$$\delta \approx \frac{x}{\mathrm{Re}_x^{1/2}} \tag{1.8}$$

where

$$\mathrm{Re}_x = \frac{U_e x}{\nu} \tag{1.9}$$

is the Reynolds number based on x and U_e. Thus, a large Reynolds number implies a layer thickness much less than x, growing like $x^{1/2}$.

The spreading of a turbulent boundary layer is not controlled by viscous diffusion, but by entrainment. A measure of entrainment is the velocity u' of turbulent fluctuations within the layer. Supposing that the layer grows at a constant velocity u', we have a characteristic time for entrainment to thickness δ of

$$T_e \approx \frac{\delta}{u'} \tag{1.10}$$

Setting $T_c \approx T_e$ as before we have

$$\delta \approx \frac{u'}{U_e} x \tag{1.11}$$

which makes $\delta \propto x$. The linear growth with streamwise distance means that, at large x, the turbulent layer will be much thicker than the laminar one, which only grows like $x^{1/2}$. Figure 1.6 sketches the growth of both laminar and turbulent boundary layers on a flat plate. We should observe that the above argument assumes constant u', whereas there is in fact a slow decrease of turbulent velocities with increasing x. As u' falls the turbulent layer spreads at a gradually decreasing rate, but still much more rapidly than the equivalent laminar layer. In consequence, a turbulent layer is considerably thicker than its laminar counterpart at the large values of Re_x which are typically required for turbulence to occur. Overall, the boundary layer on a long flat plate evolves as follows (recall the discussion towards the beginning of the chapter). As the layer thickens, the effective Reynolds number increases. A length of laminar flow near the leading edge of the plate is followed by a range over which transition takes place and the layer becomes turbulent. Its thickness is subsequently larger and grows more rapidly than would a corresponding laminar layer. Such behavior is typical of streamlined bodies for which there is no separation, at high enough Reynolds numbers that transition to turbulence occurs in the boundary layer.

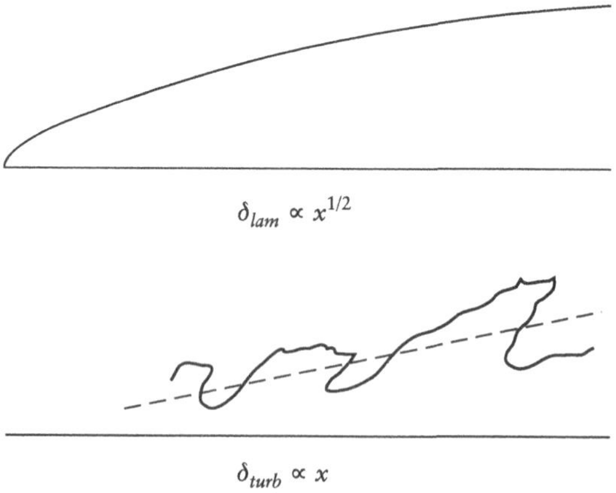

Figure 1.6. Sketches of the thickening of flat-plate laminar and turbulent boundary layers. For clarity, the scale has been expanded normal to the plate: boundary layers are very thin at the high Reynolds numbers presumed here. For instance, the angle of growth of a turbulent boundary layer is typically of order 1°.

For flow past more general bodies boundary-layer separation may occur (see Figure 1.7). This results from a sufficiently large adverse pressure gradient in the flow external to the layer, which decelerates the fluid near the surface to rest, producing a separation point to which fluid converges from both sides and at which the boundary layer leaves the surface. Laminar layers are less resistant to separation than turbulent ones. This is essentially because a turbulent layer disperses momentum faster than a laminar one and hence the effective viscosity is higher. Thus, an adverse pressure gradient finds it harder to produce the reversed flow shown in Figure 1.7 and which characterizes separation.

The total drag on a body can be considered as the sum of a form drag (due to pressure forces) and skin drag (due to viscous friction). Skin drag is increased by turbulence, but form drag can be decreased because separation is delayed. The total drag on a bluff body may often be decreased by provoking transition earlier than it would occur naturally. In Figure 1.8a, which shows a sphere, the boundary layer is laminar up to the separation point and skin friction is small. Separation occurs rapidly where the pressure gradient becomes adverse and consequently the form drag is high. A trip wire on the surface of the sphere provokes early transition in Figure 1.8b and the flow remains attached much longer. The form drag is reduced, but the skin drag increased. The total drag is reduced.

A jet flow is shown in Figure 1.9. This flow can be roughly divided into three regions as a function of downstream distance. Close to the outlet of the jet, fluid is only weakly entrained from outside. There, the entrainment process depends strongly on the state of the boundary layer coming from inside the pipe. Next, the outer parts of the jet, which are essentially shear layers, begin to roll up due to Kelvin–Helmholtz instability, producing ordered structures similar to vortex rings and external fluid is entrained more strongly. Finally, the ordered vortical structures break down and the flow becomes complicated and turbulent throughout. A wide range of eddy sizes are then apparent and the jet rapidly entrains external fluid. The

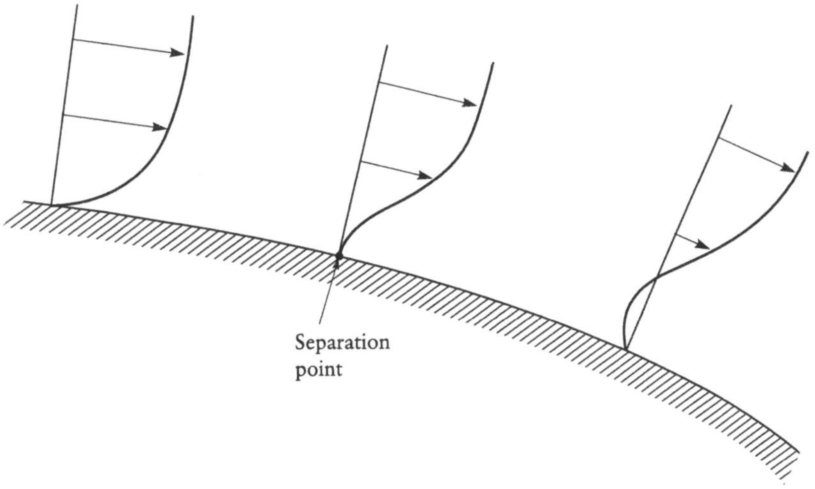

Figure 1.7. Sketches of the velocity profiles at different positions near boundary-layer separation on a bluff body.

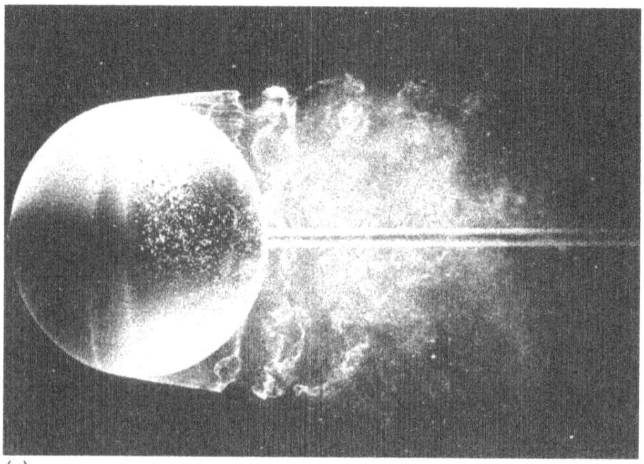

(a)

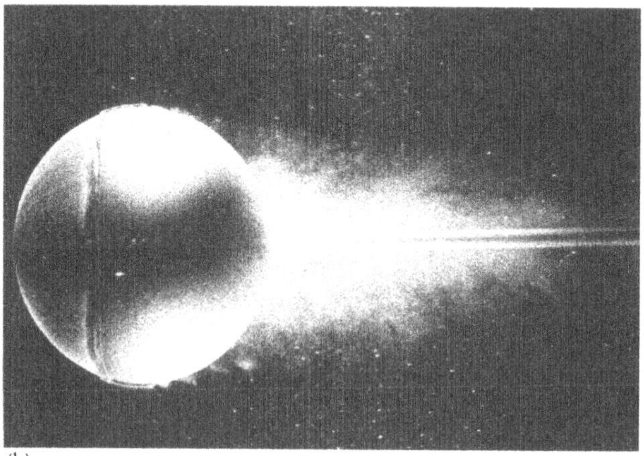

(b)

Figure 1.8. A solid sphere placed in uniform flow: (a) the boundary layer is laminar and separates to form the wake behind the sphere; (b) a trip wire has triggered boundary-layer transition upstream of where flow separation occurs in Figure 1.8a and hence delayed separation. (ONERA photograph, Werlé (1980), reproduced with permission.)

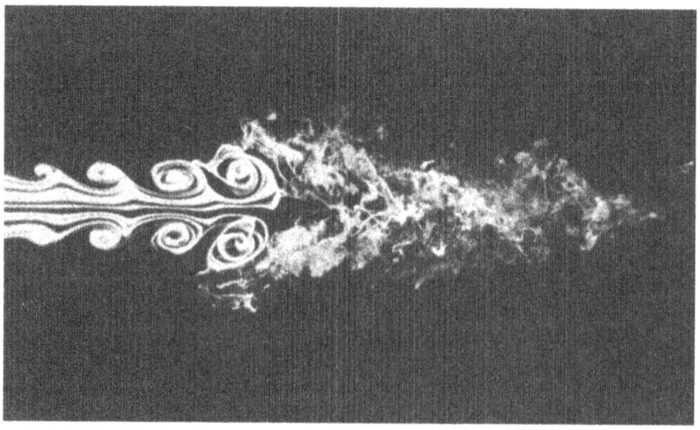

Figure 1.9. Visualization of a jet. (Courtesy of Robert Drubka and Hassan Nagib.)

location of breakdown to turbulence tends to move upstream with increasing jet Reynolds number and is also sensitive to the level and type of perturbations (recall the discussion of transition at the beginning of this chapter).

Since turbulence is a result of the dynamics of the flow, if it can be described as having an objective, this would be the dispersal of momentum rather than material. The dispersion of fluid momentum is essential at high Reynolds number because viscosity can no longer relieve the high strain rates sufficiently quickly as to avoid instabilities and overturning of the larger structures. Such overturning transfers momentum quickly and helps to reduce the high shear which produces the instability, and hence the turbulence, in the first place. As described earlier, it also leads to a cascade of energy and vorticity to smaller scales and hence the characteristic intense, small-scale vorticity and viscous energy dissipation.

DISPERSAL OF MATERIAL AND HEAT

A by-product of turbulence is the mixing of material, which, as noted above, is vital to many industrial processes and to the dilution of pollutants in the environment.

As an example of the potentially unexpected effects of pollutant dispersal in the atmosphere, consider the situation shown in Figure 1.10a. A stably stratified layer exists above an unstable layer near the surface. The flow

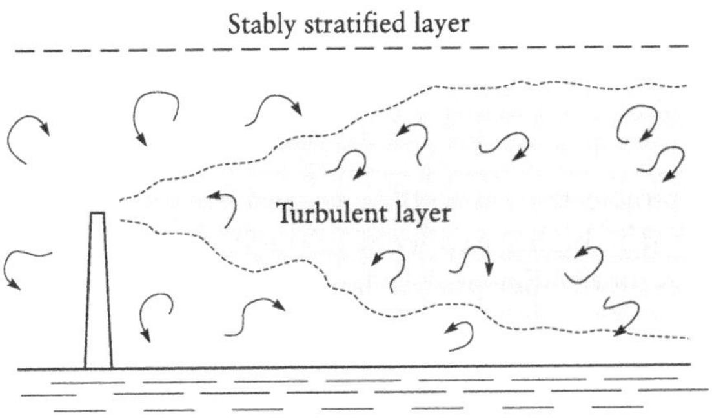

(a)

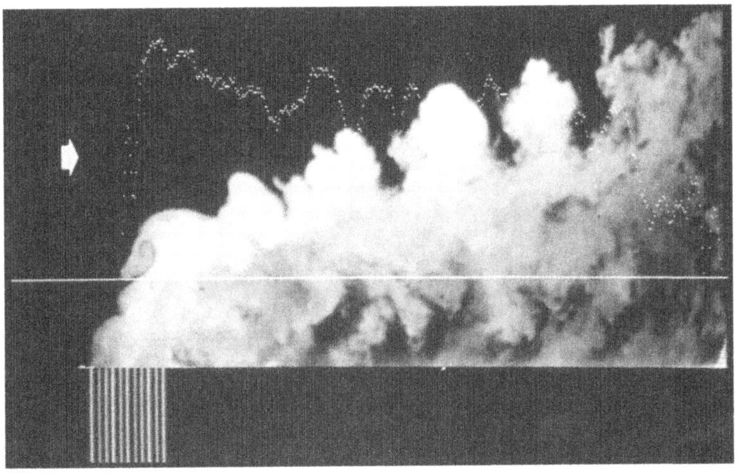

(b)

Figure 1.10. (a) Sketch of pollutant release, illustrating trapping by a stably stratified layer some distance above the ground; (b) measured and visualized concentrations of scalar released by a jet into a cross-stream. The dotted line represents the concentration level along the continuous line. ((b) Courtesy of F. Ladhari; see Ladhari (1984) for experimental details.)

in the lower layer is turbulent and there is rapid dispersion of pollutants from a source in that layer. However, little of the pollution gets carried into the upper layer, because there is no turbulence there. As a result the pollution is spread rapidly in the lower layer, but is trapped there.

Atmospheric turbulence tends to dilute pollutant releases. A laboratory simulation which illustrates this effect is shown in Figure 1.10b. A round jet, simulating a pollutant release, exhausts vertically into a cross-stream. The instantaneous distribution of pollutant concentration along the continuous horizontal line is represented by the dotted line.

Heat transfer is perhaps even more important than the dispersion of material in industrial applications. The enhancement of heat transfer by turbulence is due to the same effects of convective transport by turbulent flows that result in mixing of materials and dispersal of momentum. For example, the following empirical formulae for heat transfer from a flat plate in flow indicate the greater transfer which occurs at high Reynolds number in turbulent flows:

laminar $\quad \mathrm{Nu}_x \propto \mathrm{Re}_x^{1/2}\mathrm{Pr}^{1/3}$ (1.12)

turbulent $\quad \mathrm{Nu}_x \propto \mathrm{Re}_x^{4/5}\mathrm{Pr}^{1/3}$ (1.13)

Here, Nu_x is the Nusselt number, characterizing the efficiency of heat transfer, and Pr is the Prandtl number of the fluid. The higher exponent of the Reynolds number indicates that turbulent heat transfer is more efficient than laminar transfer when Re_x is large.

As remarked earlier in the chapter, the detailed analysis of heat transfer requires consideration of the energy equation even in cases in which the flow itself may be described by the incompressible Navier–Stokes equations (1.1) and (1.2), as supposed in this book. In many flows, the equation for the internal energy of the fluid may be approximated by (see Landau and Lifshitz (G[9] 1987); Bird et al. (1960)) the convection–diffusion equation

$$\frac{\partial T}{\partial t} + \underbrace{U_i \frac{\partial T}{\partial x_i}}_{\text{Convection}} = \underbrace{\kappa \frac{\partial^2 T}{\partial x_i \partial x_i}}_{\text{Diffusion}} \tag{1.14}$$

for the fluid temperature T, where κ is a constant thermal diffusivity. Equation (1.14) is often employed, together with appropriate thermal boundary conditions, in heat transfer calculations, but, since it involves a number of approximations of the original energy equation, its validity should be checked in particular cases, as indeed should that of (1.1), (1.2). In this formulation, the flow is governed by (1.1) and (1.2), independent of the temperature field, but the velocity appears in the evolution equation, (1.14), for the temperature. Thus, there is coupling between the flow and heat transfer problems in one direction only (temperature acts as a passive scalar). According to (1.14) without heat conduction ($\kappa = 0$), the temperature of the fluid is convected unchanged following a fluid particle, while conduction introduces thermal diffusion, tending to produce a uniform temperature. There is a certain qualitative similarity with the convection and viscous diffusion of momentum; however, the analogy is far from exact because, comparing the evolution equations, (1.1) and (1.14), for the velocity and temperature fields, it is evident that the former contains a pressure gradient term and describes a vector quantity, whereas the latter concerns the evolution of a scalar. Furthermore, the associated diffusivities, ν and κ, are generally not the same, with their ratio, $\mathrm{Pr} = \nu/\kappa$, defining the Prandtl number. In problems such as that of the heat transfer of a flat plate considered above, there is a thermal layer at the surface, in which, like the velocity, the temperature adjusts from its value at the surface to that of the external stream. For gases, the Prandtl number is typically of order 1, making the thickness of the thermal layer comparable with that of the boundary layer of the velocity field. More generally, the combined effects of turbulent convective mixing and thermal diffusion by conduction, both represented by terms in (1.14), try to make the temperature field uniform, while nonuniformities of temperature may, for instance, be introduced by heating the boundaries, via the

[9] Recall that, as explained in the preface, "G" with a date refers to the "General References" following the preface.

initial conditions on the temperature, or by volumetric heat sources, such as chemical reactions, which would add a source term to the right-hand side of (1.14).

We can illustrate the enhancement of dispersion of heat by flow via a simple calculation. Consider a room of size L. Under pure molecular diffusion, with diffusion coefficient κ, a time of order L^2/κ is required for dispersal, as can be seen from equation (1.14), dropping the convective term and performing order-of-magnitude estimates of the derivatives. On the other hand, the time taken for turbulence of fluctuating velocity u' to disperse heat through the room may be estimated as of order L/u'. The ratio of the two times is Lu'/κ, a quantity referred to as the Peclet number and which is analogous to the Reynolds number, with the scalar diffusivity, κ, instead of the viscosity, ν. If we take $L = 5$ m and $u' = 0.05$ m/s, the Peclet number is about 10^4, that is, it takes around ten thousand times longer for molecular diffusion than for turbulent convection to transfer heat through a room-sized region. Using the above parameters, the time taken for pure conduction is of order 10^6 s, while that for turbulent convection is about 100 s.

Evidently, the enhancement of heat transfer by the flow is due to the large Peclet number. At high Peclet numbers, the convective term in (1.14) dominates the diffusive one, at least for the large scales of the temperature field, which means that the overall heat transfer is mainly due to convection by the flow. At the same time as heat is dispersed in this way, convective stirring by turbulence tends to generate smaller and smaller scales of the temperature field, a process similar to the energy cascade in the velocity field, bringing conduction into play at the smallest scales and dissipating thermal nonuniformities in a fashion analogous to the dissipation of mechanical energy by viscosity. Thus, there are simultaneous processes of dispersal and mixing of temperature in turbulent flows, accompanied by thermal diffusion at the smallest scales.

Equation (1.14) describes the convection and diffusion of scalar quantities other than the temperature, such as the concentration of dissolved substances in a liquid, provided T is replaced by the given scalar and κ by its diffusivity. Thus, one may use (1.14) to study the mixing, dispersion, and diffusion of passive scalars in general, for instance the problems of pollutant dispersal referred to above. Various techniques may be used to analyze the scalar field, including the spectral methods that will be developed for the velocity field in Chapters 6 and 7. However, detailed discussion of the properties of passive scalars convected by turbulence lies beyond the scope of this introductory volume.

1.3 Remarks on Mathematical Chaos

It has been proposed that there may be some connection between turbulence and mathematical "chaos," whose onset marks a definite and well-defined change in the properties of systems of differential equations in which it appears. Chaotic systems exhibit a form of deterministic randomness that is reminiscent of turbulent flows. Furthermore, nonlinearity is crucial to the appearance of chaos and is also important in generating and maintaining turbulence. The mathematical theory of chaos may in the future contribute to our understanding of turbulent flow. Even at this stage in their development, the concepts are interesting and provocative when viewed from the standpoint of turbulent flows, although, to our knowledge, they have yet to contribute new and concrete results.

Many attempts have been made to analyze the mathematical properties of the Navier–Stokes equations, in particular the existence and uniqueness of their solutions. Pioneering work in this area is described by Ladyzhenskaya (1969). We take the existence and uniqueness of solutions to the Navier–Stokes equations with prescribed boundary *and* initial data for granted, even though it has never been proved. If, however, one specifies the *boundary conditions alone*, the solution need not settle down to a unique limiting flow as $t \to \infty$. Assuming that the boundary conditions are steady, at sufficiently low Reynolds numbers the asymptotic flow at large times is also steady and uniquely determined by the boundary conditions: it is then said to be globally stable. Statistically steady turbulence is at the opposite extreme and is highly nonunique, producing a different flow each time the experiment is performed. Starting from two slightly different initial conditions, the two flows diverge rapidly and quickly bear no resemblance to one another. However, although steady turbulent flow is nonunique in detail, its *statistical* properties (probabilities and averages) do seem to be uniquely determined by the boundary conditions alone, that is, the flow forgets its initial conditions and always settles into the same statistical state. Thus, even though the Navier–Stokes equations provide a deterministic model of turbulent fluid flow (because exactly specified boundary *and* initial conditions lead to a unique flow), at later times the resulting solution may still be extremely sensitive to the precise conditions used. In this case, although mathematically deterministic, the detailed flow can appear experimentally unrepeatable and unpredictable, as is found to be the case for turbulence. These are among the fundamental reasons for using statistical methods, as noted earlier in the chapter.

Formally, sensitivity to initial conditions can be expressed using the concept of "state space." At any given time, the flow can be represented by its velocity components U_i at each point in space. If the number of points of physical space were finite (which it is not, of course) we could use the corresponding velocities U_i as coordinates in a state space. A point in state space would then represent the flow. This idea may be taken over to the infinite number of points of physical space that are actually present. Thus, a point in (infinite-dimensional) state space represents an entire flow at a given time. If the velocity is specified at time t, that is, the point in state space representing the flow is given, then the flow is uniquely determined thereafter by the Navier–Stokes equations (and the given boundary conditions). The point in state space can be thought of as moving along a trajectory which is fixed by the initial position of the point. This is the point of view adopted in the mathematical theory of dynamical systems. The main difficulty in applying that theory to flows is that almost all of the theory is restricted to finite-dimensional state spaces and, indeed, practical calculations have only been carried out for rather low-dimensional spaces, whereas for the flows we have in mind the number of effective degrees of freedom is very large and the dimension of the state space infinite. However, we provisionally adopt the idea of thinking in terms of state space to try and explain the possible relevance of mathematical chaos to turbulence.

The sensitivity of a flow to the initial conditions can be expressed in state space as follows. Two points representing different flows that are initially close together move rapidly apart with time (here we have not defined distance in state space, but we have in mind some suitable norm). That the points are initially close together expresses the idea that the initial flows should be nearly the same. That they move apart rapidly indicates that the slight differences in initial conditions lead to large differences later. In finite-dimensional systems, the separation of two points with infinitesimal separation is an exponential function of time at large times. The coefficient (or growth rate) of the exponential is known as the Liapounov exponent and is positive for systems in which initially close points of state space move apart. The Liapounov exponent is necessarily positive for chaotic systems, representing exponential divergence of neighboring points, but this is not a sufficient condition for mathematical chaos.

Dynamical systems with dissipation (like viscosity in fluid flow) usually have subsets of state space called attractors (see, e.g., Bergé et al. (1984)). An attractor is such that points of state space that lie within another subset, called the domain of attraction, are drawn towards it, approaching the attractor ever more closely as $t \to \infty$. The point representing the system is still free to wander around the attractor at large times. The importance of the attractors of a system is that the ultimate behavior, as $t \to \infty$, can be characterized by their study.

The interest in these concepts has increased sharply since it was discovered that even relatively simple nonlinear systems could have remarkably complicated attractors. The classic example is the Lorenz system of three equations in three unknowns, which Lorenz derived as a simple model of thermal convection. Results of numerical integration of the Lorenz system are shown in Figure 1.11 and illustrate mathematical chaos, which has now been identified in many low-dimensional dynamical systems, including ones that represent physical systems. The attractor is indicated by the "A" in the upper part of the figure and has an overall "butterfly" shape. Within the butterfly, however, the attractor has a complicated structure known as a fractal (for fractional dimension). Fractals cannot be represented by conventional smooth surfaces and have intricate fine structure of ever decreasing scales. Many fractals exhibit self-similarity at small scales, that is, structures repeatedly recur with smaller and smaller sizes.[10] Such attractors are called strange and the motion within the attractor is complicated, has an element of randomness, and may be described by statistical methods.

These ideas are suggestive when we come to consider turbulence and have stimulated a considerable amount of research (see, e.g., Lumley (1990)). The sensitivity to initial conditions and deterministic randomness are indeed characteristic of turbulence. Might it be that, if one could find the correct mathematical description, turbulence would appear as a strange attractor of the flow? Statistical properties of the turbulence might then be deduced from the properties of the attractor. As pointed out by Lesieur (G 1990), there is no a priori contradiction in philosophy between chaos theory and the point of view usually adopted in the theory of turbulence. However, as previously stated, turbulence has a very large number of degrees of freedom and an infinite-dimensional state space. It shows randomness in both space and time, whereas the finite-dimensional chaos of dynamical systems theory has only

[10] Fractals will briefly crop up again in Chapter 7. There they will represent objects in physical space, rather than in state space. The two concepts are quite different and should not be confused, despite the similarity in mathematical description.

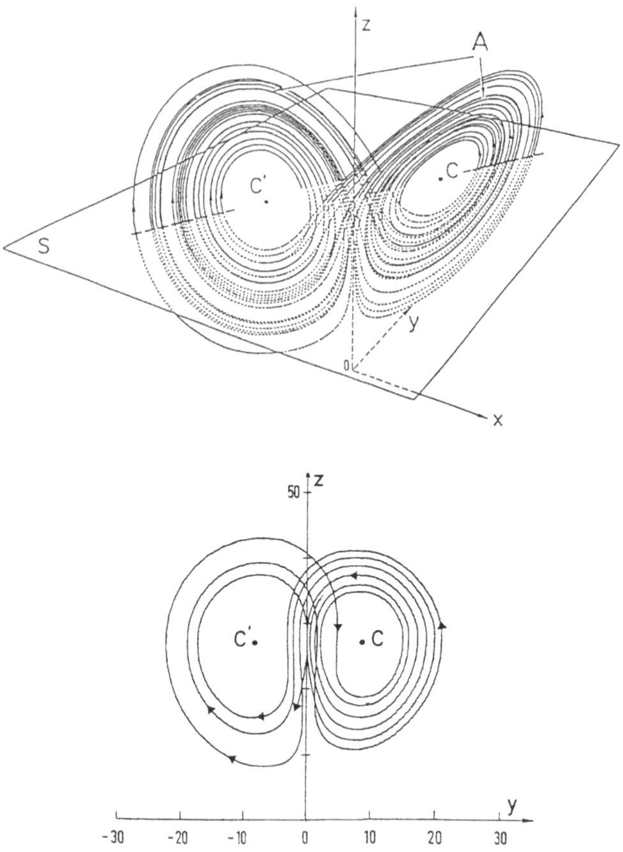

Figure 1.11. Illustrations of the Lorenz attractor. (Bergé et al. (1984), reproduced with permission.)

a rather limited number of degrees of freedom and is thus chaotic in time only. The theory of mathematical chaos is tantalizing, but it is as yet unclear whether it will prove useful in the study of turbulence, and, to the authors' knowledge, no significant new results have so far been obtained from its application to turbulence. For these reasons we will not discuss chaos theory further in this book. Nonetheless, it may be instructive to give a simple example of one of the difficulties in putative chaos theory modeling of decaying, rather than steady turbulence.

Consider viscous fluid inside a fixed container. The fluid is vigorously stirred, leading to what would usually be called turbulence, and then left to itself. Any initial flow can be shown to eventually approach the state of rest, which is therefore the only attractor, whose domain of attraction is the whole state space. Clearly this global attractor is not chaotic, but despite this, the flow can be turbulent. Turbulence is not described by the attractor because the flow first becomes turbulent and then tends towards the attractor (state of rest) by turbulent decay. In such cases, the flow might be assumed to approach a subset of phase space or "turbulent slow manifold," similar to an attractor, but evolving with time due to statistical unsteadiness of the turbulence. However, giving precise meaning to such a concept, let alone constructing a useful theory from it, would appear to be far from obvious.

1.4 Miscellany

An important and relatively recent method for the study of turbulence is by direct numerical solution of the Navier–Stokes equations. Such solutions have become possible at moderately high Reynolds numbers in the last decade owing to the rapid development of both the speed and memory of computers. Direct simulation allows one to conduct *numerical experiments*, often permitting better control of the flow parameters and the "measurement" of variables within flows that are hard to envisage in the laboratory. Furthermore, flow "codes" can be designed to be more flexible, allowing one to simulate quite a wide class of flows with minimal changes. As for laboratory measurements, it is important to design numerical experiments to answer real questions and to analyze the results intelligently, so that the end product is not simply a set of numbers or pretty pictures in color, but increased understanding of the flow. Direct simulation does not obviate the continuing need for laboratory experiments, still less for theoretical analysis, but provides a complementary and increasingly powerful set of tools for studying turbulence as computer technology advances. We shall occasionally refer to results obtained using direct simulation, but such results should not be accepted uncritically, any more than one would regard an experimental measurement or theoretical result as definitive without considering the errors and limitations implicit in the method used. Turbulence, being unsteady and having structures of very diverse spatial scales, is a particularly difficult case for numerical simulation and one often has few cross-checks on the validity of the results. No doubt computers will continue to evolve rapidly and this will open up more flows to direct simulation, although such progress is likely to be rather slow, since the computer time involved appears to increase rapidly with the Reynolds number used. We shall have more to say about this and other essentially numerical prediction methods, such as large-eddy simulation, in the final chapter of this book.

A classical method for theoretical description of turbulence is the decomposition of the flow into Fourier components, known as spectral analysis. Spatial Fourier analysis allows one to give precise meaning to the important notion of different scales of turbulence and their nonlinear interactions (e.g., the cascade) and will be developed in detail for homogeneous turbulence in Chapters 6 and 7. We illustrate the procedure using the simple nonlinear equation

$$\frac{\partial U}{\partial t} + U\frac{\partial U}{\partial x} = 0 \tag{1.15}$$

Let us suppose that initially, at time $t = t_0$, the "velocity" U has the sinusoidal form

$$U(x, t_0) = B\cos kx \tag{1.16}$$

For small $t - t_0$, the solution can be found as follows:

$$\begin{aligned} U(x,t) &= U(x,t_0) + (t-t_0)\left.\frac{\partial U}{\partial t}\right|_{t_0} + \ldots \\ &= U(x,t_0) - (t-t_0)\left[U\frac{\partial U}{\partial x}\right]_{t_0} + \ldots \\ &= B\cos kx + (t-t_0)B^2 k \sin kx \cos kx + \ldots \end{aligned} \tag{1.17}$$

which shows the development of a term in $\sin 2kx$. This is the first result of nonlinearity.

If we go to a slightly more complicated example which models a two-dimensional field:

$$\frac{\partial U}{\partial t} + U\frac{\partial U}{\partial x} + V\frac{\partial U}{\partial y} = 0 \tag{1.18}$$

$$\frac{\partial V}{\partial t} + U\frac{\partial V}{\partial x} + V\frac{\partial V}{\partial y} = 0 \tag{1.19}$$

with

$$U(x, t_0) = B\cos kx \cos ly \tag{1.20}$$

$$V(x, t_0) = B'\cos k'x \cos l'y \tag{1.21}$$

then for small $t - t_0$

$$\begin{aligned} U(x, t) = B\cos kx\cos ly + (t - t_0)(B^2 k \sin kx \cos kx \cos^2 ly \\ + BB'l \cos kx \cos k'x \sin ly \cos l'y) + \ldots \end{aligned} \tag{1.22}$$

$$\begin{aligned} V(x, t) = B'\cos k'x \cos l'y + (t - t_0)(BB'k' \sin k'x \cos kx \cos ly \cos l'y \\ + B'^2 l' \cos^2 k'x \sin l'y \cos l'y) + \ldots \end{aligned} \tag{1.23}$$

of which the terms in $t - t_0$ can be expressed as a sum of terms, each of which is a product of sines or cosines, such as,

$$\sin(k + k')x \cos(l + l')y \tag{1.24}$$

The wavenumbers of this term are $k + k'$ and $l + l'$, whereas those of the original disturbance were k, l and k', l'. In general, one finds that quadratic nonlinear interactions generate sums and differences of the original wavenumbers. Thus nonlinearity has the effect of generating different wavenumbers and so leading to disturbances of both shorter and longer wavelengths than those present initially. From these examples, we see that the effect of nonlinearity is to couple together the Fourier components. Thus, if a turbulent flow is decomposed into such modes, they interact with one another, and given sufficient time all modes will have acted on all other modes. Such modal interactions are an expression of the fundamental difficulties resulting from nonlinearity.

The spatial Fourier analysis considered above turns out to be especially useful in the study of homogeneous turbulence. Likewise, it can be convenient to describe statistically steady turbulence in terms of its frequency spectrum, that is, to decompose the flow into Fourier components in time. The spectrum, either spatial or temporal, describes the contribution of each component to the total energy of the flow. The Fourier modes are supplied with, or yield energy by, nonlinear interactions with other modes, and also lose energy by viscous dissipation. The evolution of each mode depends on its nonlinear interactions with all other modes, as will be seen in detail in Chapter 6, which introduces Fourier analysis of turbulence.

In the presence of a mean flow, it is usual to carry out statistical splitting of the flow field, the total field being expressed as the sum of a mean flow and a fluctuation, generally identified with the turbulence. This splitting can be usefully carried out whether or not the turbulence is homogeneous. There is generally strong coupling of the mean and fluctuating components due to nonlinearity, with mean-flow gradients appearing in the equations for the turbulent fluctuations, while inhomogeneity of the turbulence occurs in the mean-flow equations via a forcing term, known as the Reynolds stress term, as we will see in Chapter 4. Observe that it is the nonlinear term in the Navier–Stokes equation that produces coupling of the mean flow and fluctuations, as it does for Fourier components in spectral analysis of homogeneous turbulence.

A special class of mean flows can coexist with homogeneous turbulence, namely those that have velocity fields that are linear functions of the spatial coordinates. This allows one to carry out statistical splitting and spatial Fourier analysis. Homogeneity makes the Reynolds stress term zero, so there is no forcing of the mean flow by the turbulent fluctuations, and one may study the distortion of turbulence by a prescribed mean flow.

1.5 Conclusions

This chapter has presented an introduction to turbulence, which we hope will prove useful to the beginner and of interest to specialists in fluid dynamics. The number and diversity of disciplines in which turbulence is important are impressively large: combustion, chemical engineering, multiphase flows, meteorology, stellar dynamics, pollutant dispersal, and so on – indeed, almost anywhere that fluids flow.

A brief summary of the remainder of the book may be found in the preface, while the next chapter gives an overview of those statistical methods we will need later in the book and is intended as a review of a subject whose elements are already familiar to the reader.

References

Bergé, P., Pomeau, Y., Vidal, C., 1984. *Ordre dans le Chaos: vers une Approche Déterministe de la Turbulence*. Hermann, Paris. (English translation by Tuckerman, L., 1986. *Order within Chaos: Towards a Deterministic Approach to Turbulence*. Wiley, New York.)

Bird, R. B., Warren, E. S., Lightfoot, E. N., 1960. *Transport Phenomena*. Wiley, New York.

Henningson, D. S., Reddy, S. C., 1994. On the role of linear mechanisms in transition to turbulence. *Phys. Fluids*, **6**, 1396–8.

Kachanov, Y. S., 1994. Physical mechanisms of laminar-boundary-layer transition. *Ann. Rev. Fluid Mech.*, **26**, 411–82.

Ladhari, F., 1983. Application of image processing analysis to the study of the turbulent boundary layer structure. *3rd Int. Symposium on Flow Visualisation*, University of Michigan, Ann Arbor, Sept. 6–9, 1983.

Ladhari, F., 1984. Hot wire measurements and visual observations in a hot buoyant jet in a turbulent cross flow. *12th All-Union Heat and Mass Transfer Conference*, Minsk, May 21–23, 1984.

Ladyzhenskaya, O. A., 1969. *The Mathematical Theory of Viscous Incompressible Fluid*. Gordon and Breach, New York.

Lumley, J. L., 1990. *Whither Turbulence? Turbulence at the Crossroads*. Lecture Notes in Physics, **357**, ed. J. L. Lumley. Springer, Berlin.

Werlé, H., 1980. Transition et décollement: visualisations en tunnel hydrodynamique de l'ONERA. *Rech. Aérosp.*, **1980-5**, 331–45.

Werlé, H., Gallon, M., 1972. Contrôle d'écoulements par jet transversal. *Aéronaut. Astronaut.*, **34**, 21–33.

CHAPTER TWO

Statistical Tools

The apparently random character of turbulent flows strongly suggests that statistical methods will be fruitful. In this chapter, we discuss some statistical techniques, in preparation for later chapters. The reader is presumed to have a modest background in probability theory and this chapter summarizes the main elements, while placing statistical ideas in the context of turbulent flows (those whose knowledge of the basic theory is rusty or nonexistent should perhaps keep one of the many textbooks on the mathematics of probability and statistics to hand; Lumley (1970) and Monin and Yaglom (G 1971) describe the theory as applied to turbulence).

As discussed in Chapter 1, despite being generally considered as deterministic, turbulent flows are highly nonunique in practice. Thus, if an experiment is repeatedly carried out, a different velocity field is obtained in each realization, even if the experimental conditions are nominally the same (i.e., the experimenter endeavors to reproduce the same flow). This is because the detailed behavior of the flow in any one realization is extremely sensitive to small changes in the initial or boundary conditions, which the experimenter cannot control to infinite precision. This type of problem is ideally suited to statistical methods. Indeed, it was the nonrepeatability of an experiment such as tossing a coin that led to the idea of probabilities in the first place. By looking at the statistical properties of an ensemble of different flow realizations, all obtained using the same nominal conditions, one hopes to extract useful quantities, namely probabilities and averages, which depend only on parameters that the experimenter controls. To take an example, the detailed behavior of the turbulent wake of a sphere placed in a uniform flow may vary with tiny perturbations in the incoming stream and small vibrations of the sphere, to name just two possible extraneous factors, but one hopes and expects that, for instance, the average velocity is well defined. Of course, the flow statistics depend on the gross experimental conditions used, for example, a turbulent boundary layer can have quite different thickness and drag with suction than without, but we expect them not to change significantly with small variations in those conditions.

Given an ensemble of different flow realizations, we can define the associated probabilities of flow variables[1] taking on particular values, or more precisely, ranges of values, since they are generally continuous variables. Thus, we imagine an experiment performed very many times under nominally the same conditions, each time producing a different realization of the flow, and use the frequency with which a

[1] Flow variables range from simple ones, such as the pressure at a single point and time, to more complicated ones that are tensorial or obtained from the flow at many points and times.

given flow variable falls into a given range of values as a definition of the probability for that range.

Mean (or average) values are of particular importance in the theory of turbulence. We can define means by taking the average over an increasingly large number of realizations of the flow under the same nominal conditions. The mean value can also be calculated in the usual way from a knowledge of the probabilities of the given quantity as a sum (or integral for continuous variables) over all possible values, weighted by their probabilities. In particular, the mean flow is defined as having the average values of the velocity and pressure (together with the density and perhaps other fluid properties if the fluid is compressible, a case which is not considered in this book). The departure of any given realization from the mean can be calculated by subtracting the mean flow and is conventionally identified with turbulence. That is, the total flow is split into a mean part and a fluctuating component, whose average is zero and which is usually thought of as representing the turbulence. For instance, the mean values of the squared fluctuating velocities are often used to characterize the intensity of turbulence. We shall come across numerous other important quantities defined by averages during the course of this book.

2.1 Probabilities and Averaging

Consider any flow variable U, which might represent a velocity component or the pressure at a given position and time, or a more complicated quantity derived from different points and times. U will take on different values in different experiments. One may average the quantity U by summing over experimental realizations, dividing by the number of realizations, and letting that number go to infinity. Thus we obtain the mean of U, variously denoted as $\overline{U}$ or $\langle U \rangle$ (or sometimes $E(U)$, then called the expectation of U). The probability distribution function, $P(U)$, can also be defined so that

$$\int_{U_-}^{U_+} P(U)\, dU \tag{2.1}$$

gives the proportion of realizations in the ensemble for which U takes on values in the range $U_- < U < U_+$, that is, the probability that it falls in that range. The idea here is that $P(U)dU$ gives the probability of U lying between U and $U + dU$ and that we should sum up such elementary contributions over the range U_- to U_+ to determine the overall probability that $U_- < U < U_+$. The probability of a given event is thus the proportion of the ensemble for which it occurs and (2.1) gives the probability that $U_- < U < U_+$.

Since U must always take on some value,

$$\int_{-\infty}^{+\infty} P(U)\, dU = 1 \tag{2.2}$$

is an identity which the distribution function, $P(U)$, must always satisfy. The mean of U can also be calculated from $P(U)$ via

$$\overline{U} = \int_{-\infty}^{+\infty} UP(U)\, dU \tag{2.3}$$

which expresses the fact that the average can be computed by taking the proportion of the sample in the range U to $U + dU$, which is $P(U)dU$, multiply it by U and sum over all possible values of U. We can likewise determine the mean of any function of U from

$$\overline{f(U)} = \int_{-\infty}^{+\infty} f(U)P(U)\, dU \tag{2.4}$$

Given $U_- < U_+$, if we take $f(U)$ as the window function

$$f(U) = 1 \quad \text{if} \quad U_- < U < U_+ \tag{2.5}$$

$$f(U) = 0 \quad \text{if} \quad U \leq U_- \text{ or } U \geq U_+$$

the average value of $f(U)$ is

$$\overline{f(U)} = \int_{U_-}^{U_+} P(U)dU \tag{2.6}$$

that is, the probability that U falls in the range $U_- < U < U_+$, from which $P(U)$ can be obtained by differentiation with respect to U_- or U_+. Thus, knowledge of the probability distribution function allows calculation of average values from (2.4), while, working in the other direction, averaging can be used to derive the distribution function. Because mean values are often easier to determine experimentally, averaging of window functions has formed the basis of a popular experimental technique for the measurement of probability distributions. In this method, an experimental signal, $U(t)$, is sent through an electronic device whose output is 1 for a certain range of values and 0 outside that range, followed by averaging to obtain the probability that $U(t)$ lies in the given range. Experimentally, time averaging is usually employed for steady flows, whereas we have defined both probabilities and mean values via an ensemble of experiments. Conditions under which the two approaches give the same results will be discussed in the next section.

An obvious property of averaging is that it is a linear operation, that is, if λ is any constant

$$\overline{\lambda U} = \lambda \overline{U} \tag{2.7}$$

and, if V is any flow quantity

$$\overline{U + V} = \overline{U} + \overline{V} \tag{2.8}$$

Since flow quantities are usually functions of spatial location and time (e.g., a velocity component), governed by differential equations, we will frequently need to take averages of their derivatives. Linearity allows one to write

$$\overline{\frac{U(x+h) - U(x)}{h}} = \frac{\overline{U}(x+h) - \overline{U}(x)}{h} \tag{2.9}$$

and taking the limit as $h \to 0$ we have

$$\overline{\frac{\partial U}{\partial x}} = \frac{\partial \overline{U}}{\partial x} \tag{2.10}$$

which indicates that one can take averages inside derivatives, an operation we will often perform in later chapters without explicitly noting the fact. Note that this holds for both space and time derivatives and that similar results apply to integrals over either space or time. Mathematically, we might say that averaging commutes with differentiation and integration.

Most quantities occurring in the theory of turbulence are continuous variables, that is, they take on a continuum of values and the value in one experiment is never exactly the same as that in any other. This behavior may be contrasted with that of discrete variables (e.g., the number of times the velocity exceeds a certain critical value in a given time range), which can only take on values within a certain countable set of numbers and thus tend to repeat themselves. The distribution function of a discrete variable consists of Dirac functions, whose amplitudes give the associated probabilities. For instance, the trivial case in which the variable U takes the same value U_0 in all realizations is described by $P(U) = \delta(U - U_0)$, where the amplitude is 1 since that gives the probability that $U = U_0$. One can also consider variables which are part continuous and part discrete, whose distribution functions consist of Dirac functions embedded within a non-Dirac continuum. Notice that, although the probability distributions of continuous variables derived from turbulence are usually smooth functions, in general the probability distribution of a continuous variable may be a discontinuous function. For example, a uniformly distributed random variable in the range $0 < U < 1$ has $P(U) = 1$ in that range, and zero outside. However, we do not intend to open the door on the various mathematical pathologies which might arise and, at worst, the distribution functions we have in mind for continuous variables show isolated jump discontinuities of the type illustrated by this example.

Examples of continuous $U(t)$ and their probability distributions are shown in Figures 2.1–2.3. In interpreting these figures, we assume that mean values can be calculated using time averaging, detailed conditions for which are given in the next section. The first example, Figure 2.1, is typical of a turbulent quantity in a steady flow (statistical steadiness being one of the conditions for use of time averaging). The figure shows the time history, $U(t)$, the probability distribution of U, and the result of conversion of $U(t)$ to 0 and 1 values, with 1 when U lies in some range $U_- < U < U_+$, an operation that might, in practice, be performed by an electronic gate. As discussed above, the mean value of the resulting signal should give the probability that U lies in the given range. This mean, and hence the probability that $U_- < U < U_+$, can be obtained by time averaging the output of the gate. The distribution function,

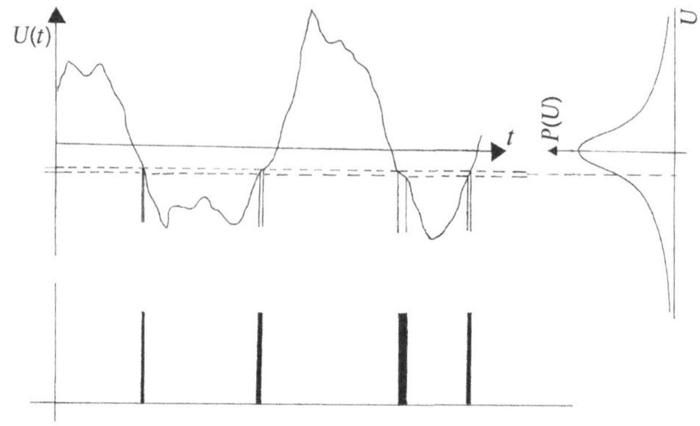

Figure 2.1. Sketch of the time history of a typical turbulent random quantity with its probability distribution function. Also shown is the result of passing the signal through a device that produces an output of 1 if the quantity lies in a given range and 0 if it lies outside the range. The average of the device output is the probability that the quantity lies in the range.

$P(U)$, perhaps determined in the above manner, is shown in the figure and has a single hump, typical of many turbulent flow quantities. The tails of the distribution, which represent rare values of U, are the most difficult to measure accurately, since such values only occur infrequently and so very long sampling times are needed to obtain converged statistics. Figure 2.2 illustrates a less common type with a double-humped, strongly asymmetric distribution. Figure 2.3 shows a sine wave, which might come from a nonrandom periodic flow if the amplitude and phase of the wave are repeatable from one experiment to another. In that case, U takes the same value in all realizations and is not truly random (nor statistically steady, so one cannot use time averaging to obtain mean values). It is then a discrete variable with only one possible value whose probability is 1, yielding a distribution function consisting of a single Dirac function, whose position is a sinusoidal function of time. If, on the other hand, and as implicit in the figure, the phase of the wave varies randomly from one realization to another with uniform probability over the range 0 to π, then the signal is statistically steady, with the probability distribution shown in the figure and mean value equal to zero. One could also allow the phase to vary between realizations over a different range of values or nonuniformly, in general yielding statistics that vary periodically with time. At first sight, such sine waves appear to have little to do with turbulence, but if one imagines adding a certain amount of the signal in Figure 2.1 to that in Figure 2.3, the result resembles a periodically modulated turbulent flow, such as that produced by blowing across the mouth of a bottle to produce a tone. A turbulent shear layer is produced over the mouth of the bottle, which, coupled to the Helmholtz resonance of the bottle, yields a turbulent flow with nearly periodic, self-sustained oscillations. If the phase of the oscillations varies uncontrollably from realization to realization of the flow, as is likely unless the oscillations are somehow phase locked, the oscillations themselves will appear as random fluctuations, whereas if the phase is the same in all realizations, they appear as a periodic mean flow superimposed on a periodically modulated random component due to turbulence in the shear layer. We will return to this example in the next section.

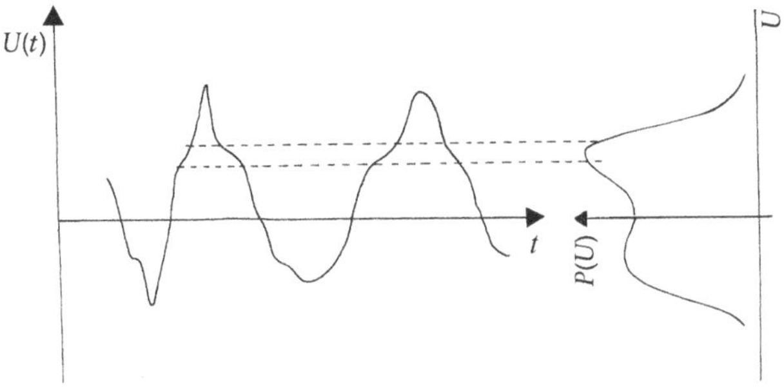

Figure 2.2. Time history and distribution function of a random quantity having a double-hump probability distribution.

Given several flow quantities, one can define a joint probability distribution function. For example, in the case of two variables, U_1 and U_2, the proportion of the

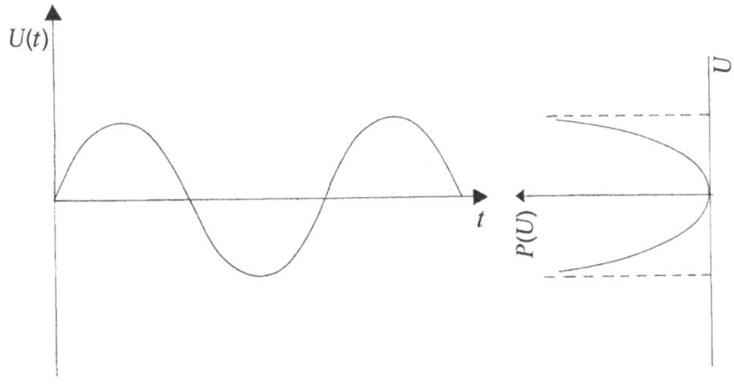

Figure 2.3. Time history of a single realization and probability distribution for an ensemble of randomly phased sinusoids.

ensemble in which U_1 takes values between U_1 and $U_1 + dU_1$, and U_2 takes values between U_2 and $U_2 + dU_2$ is given by $P(U_1, U_2)dU_1dU_2$. Thus,

$$\int_{U_{2-}}^{U_{2+}} \int_{U_{1-}}^{U_{1+}} P(U_1, U_2)\, dU_1 dU_2 \tag{2.11}$$

gives the probability that $U_{1-} < U_1 < U_{1+}$ and $U_{2-} < U_2 < U_{2+}$. Clearly

$$\int_{-\infty}^{+\infty} \int_{-\infty}^{+\infty} P(U_1, U_2)\, dU_1 dU_2 = 1 \tag{2.12}$$

and

$$P_1(U_1) = \int_{-\infty}^{+\infty} P(U_1, U_2)\, dU_2 \tag{2.13}$$

$$P_2(U_2) = \int_{-\infty}^{+\infty} P(U_1, U_2)\, dU_1 \tag{2.14}$$

give the distribution functions of U_1 and U_2 individually. The average of any function $f(U_1, U_2)$ can be obtained as

$$\overline{f(U_1, U_2)} = \int_{-\infty}^{+\infty} \int_{-\infty}^{+\infty} f(U_1, U_2) P(U_1, U_2)\, dU_1 dU_2 \tag{2.15}$$

The quantity $P(U_1, U_2)$ is often referred to as the joint distribution of U_1 and U_2, while the expression "joint statistics" is also sometimes used. Note that nothing stops us taking U_1 and U_2 as the values of a single flow quantity at different fixed times, t_1 and t_2, or at different spatial locations. It should be apparent how the above ideas can be extended to an arbitrary number of flow variables. The joint probability distribution of N variables, $U_1, \ldots, U_N$, is a function of those variables and can be thought of as a scalar field in the N-dimensional space with coordinates $U_1, \ldots, U_N$. For instance, the probability distribution, $P(\mathbf{U})$, of the vector velocity, $\mathbf{U} = (U_1, U_2, U_3)$, at a given point and time in a turbulent flow is a scalar function in the three-dimensional space spanned by the vector $\mathbf{U}$. Thus, the probability that the vector $\mathbf{U}$ falls within a small volume element, $d^3\mathbf{U}$, of that space is $P(\mathbf{U})d^3\mathbf{U}$.

Conditional probabilities and averages may be introduced as follows. The basic idea is to restrict attention to those realizations in which some flow quantity, U_2 say, takes on a particular value, that is, experiments in which U_2 has values other than the given one are ignored. Within this subensemble, the proportion of experiments in which U_1 has values between U_1 and $U_1 + dU_1$ is denoted by $P(U_1|U_2)dU_1$. Thus, $P(U_1|U_2)$ is the probability distribution of U_1, conditional on U_2 having the given value. When U_2 is a discrete value of the random variable on which the statistics are conditioned, this definition is satisfactory, but, as it stands, it does not work otherwise, because U_2 will never have the given value exactly. In that case, one allows U_2 to take on a small range of values between U_2 and $U_2 + dU_2$, and defines $P(U_1|U_2)$ as the proportion of such experiments that also give U_1 in the range U_1 and $U_1 + dU_1$. Now, the proportion of the total ensemble satisfying both conditions is $P(U_1, U_2)dU_1dU_2$, while the proportion of the total in which U_2 lies in the range U_2 to $U_2 + dU_2$ is $P_2(U_2)dU_2$. It follows, after a little thought, that

$$P(U_1|U_2) = \frac{P(U_1, U_2)}{P_2(U_2)} \tag{2.16}$$

which can be taken as the definition of the probability distribution of U_1, conditional on a nondiscrete value of U_2. Thus, given the joint distribution function, $P(U_1, U_2)$, of continuous variables, one can calculate the conditional distribution functions from (2.14) and (2.16). Notice that there is a difficulty when $P_2(U_2) = 0$, mathematically apparent through division by zero. Suppose, for instance, that $P_2(U_2) = 0$ over some range of values. It is evident that there is little sense in trying to define statistics conditional on values of U_2 in that range, since they never occur. Thus, $P(U_1|U_2)$ is undefined when U_2 is a continuous value of the conditioning variable and $P_2(U_2) = 0$. From (2.13) and (2.16), we obtain

$$P_1(U_1) = \int_{-\infty}^{\infty} P(U_1|U_2)P_2(U_2)\,dU_2 \tag{2.17}$$

showing that the single-variable distribution function can be calculated by combining the conditional distributions for all possible values of the conditioning variable, weighted by their probabilities. One may extend the above ideas to the joint distribution of any number of random variables with multiple conditioning variables.

Conditional averages can also be defined. Thus, the mean of U_1, taken only over realizations in which U_2 has a given value, is

$$\langle U_1|U_2\rangle = \int_{-\infty}^{+\infty} U_1 P(U_1|U_2)\,dU_1 \tag{2.18}$$

and can be used to construct the unconditional average via

$$\overline{U_1} = \int_{-\infty}^{\infty} \langle U_1 \mid U_2\rangle P_2(U_2)\,dU_2 \tag{2.19}$$

This result shows that one may calculate the average of U_1 in two stages. First determine the conditional averages with U_2 fixed, then average of all possible values of U_2. This may seem like a rather indirect way of proceeding, but such an approach sometimes proves the easiest way of determining average values. Once again, extension to multiple conditioning variables is straightforward.

Conditional averages are often useful in interpreting data from turbulent flows. Consider, for instance, the example of a boundary layer shown in Figure 2.4. The frontier of turbulence is sharp and mobile. Sometimes a given point finds itself inside the turbulence and sometimes it is outside. Data obtained inside and outside will be quite different in character and, if one simply takes the average of some flow quantity, the result does not generally reflect what is happening in either region, but instead gives some intermediate value. However, provided one can experimentally identify when a given sensor lies inside and outside the turbulent region, two conditional averages may be calculated: the first giving the average value inside the turbulence, the second outside. These conditional values will provide more detailed information than the unconditional average and the technique can be applied whenever one suspects statistics that differ significantly under identifiably different circumstances. In the case of the boundary layer, the variable, U_2, on which the averages are made conditional is discrete. Thus, although the frontier of turbulence is not really infinitely thin, a threshold value for some measure of turbulence intensity

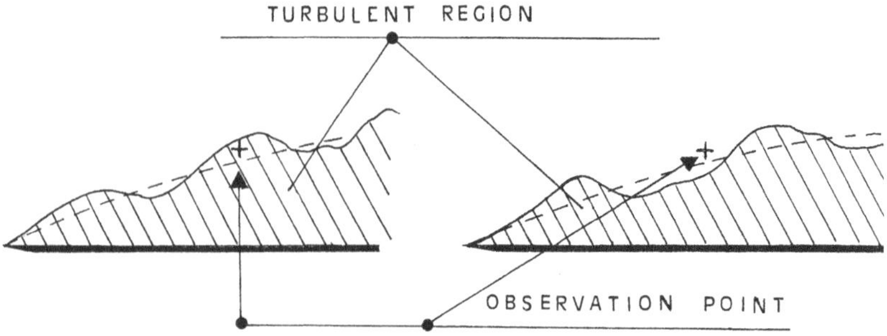

Figure 2.4. Sketch of a turbulent boundary layer at two different times, illustrating the use of conditional averaging. The hatched region represents the zone of turbulence. The dashed line is the mean location of the frontier between turbulent and laminar flow, while the continuous line is its instantaneous location.

is employed and one declares the measurement point to be inside the turbulence if it exceeds the threshold. The results should be insensitive to the choice of threshold, since the frontier of turbulence is thin.

Conditional averaging is often implemented by gating the signal, $U_1(t)$, whose conditional average one wishes to measure. Suppose, for instance, that we wanted to determine the average of some flow variable U_1 at a point in a boundary layer, conditional on being inside the turbulence. Let $U_2 = 0$ outside the turbulence and $U_2 = 1$ inside, denoting the probability of the latter by p. Thus p is the probability of turbulence at the given point. Gating simply takes U_1 and U_2 as inputs and outputs U_1 if $U_2 = 1$, and zero otherwise. The average of the gate output can be calculated as follows. In a large number, N, of realizations, the number in which the output is nonzero is pN. The average of the output over those nonzero realizations is $\langle U_1 \mid U_2 = 1 \rangle$, so the total sum of the different outputs over all N realizations is $pN\langle U_1 \mid U_2 = 1 \rangle$. Dividing by N, we obtain the average output

$$\overline{U_1 U_2} = p\langle U_1 \mid U_2 = 1 \rangle \tag{2.20}$$

while, as discussed earlier, $p = \overline{U_2}$. Thus, we can calculate the conditional average as

$$\langle U_1 \mid U_2 = 1 \rangle = \frac{\overline{U_1 U_2}}{\overline{U_2}} \tag{2.21}$$

This result indicates that we should divide the averaged gate output by the probability that the given measurement point lies inside the turbulence, which can be obtained by averaging U_2, to determine the required conditional average. The probability that a given point lies inside the turbulence is often called the turbulent intermittence and, in the case of the boundary layer, decreases with distance from the body surface owing to the decreasing frequency of turbulence, from a value very close to 1 in the region near the wall, to 0 outside the layer. The increasing rarity of turbulence at larger distances means that longer time samples are needed to obtain convergence of the averages in (2.21). Given that one can measure conditional mean values, conditional probabilities can be obtained by conditionally averaging functions which are 1 inside some range and 0 elsewhere, as for unconditional probability distributions. As the range used becomes narrower, one again needs to sample for longer to obtain converged statistics.

From (2.16), if U_1 and U_2 are statistically independent of one another, that is, specifying U_2 does not affect the distribution of U_1 and vice versa, we have

$$P_1(U_1) = P(U_1|U_2) = \frac{P(U_1, U_2)}{P_2(U_2)} \tag{2.22}$$

or

$$P(U_1, U_2) = P_1(U_1)P_2(U_2) \tag{2.23}$$

which expresses the important fact that the joint distribution function of independent variables is simply the product of their individual distribution functions. One consequence of (2.23) is that

$$\begin{aligned} \overline{U_1 U_2} &= \int_{-\infty}^{+\infty}\int_{-\infty}^{+\infty} U_1 U_2 P_1(U_1) P_2(U_2)\, dU_1 dU_2 \\ &= \int_{-\infty}^{+\infty} U_1 P_1(U_1)\, dU_1 \int_{-\infty}^{+\infty} U_2 P_2(U_2)\, dU_2 = \overline{U_1}\,\overline{U_2} \end{aligned} \tag{2.24}$$

showing that the mean of a product of statistically independent factors is the product of their means. In many turbulent flows, it is found that the velocity at widely separated times appears to approach statistical independence as the temporal separation increases. That is, if $U(t_1)$ and $U(t_2)$ are velocity components at the two times, they approach independence as $|t_1 - t_2| \to \infty$. Put another way, knowledge of the flow at the earlier time does not tell us much about its later behavior, which can be expressed by saying that the flow has only limited statistical memory. We will return to this topic in the next section when discussing correlation functions.

2.2 Statistical Moments and Correlations

Given a quantity U, the mean value of any power,

$$a_\nu = \overline{U^\nu} = \int_{-\infty}^{+\infty} U^\nu P(U)\, dU \tag{2.25}$$

is called the moment of order ν or νth moment of U. The moment of order 1 is, of course, the mean of U. The central moments are defined by

$$\mu_\nu = \overline{(U - \overline{U})^\nu} = \int_{-\infty}^{+\infty} (U - \overline{U})^\nu P(U)\, dU \tag{2.26}$$

and, apart from μ_1, which is zero, are generally more important than the a_ν in the theory of turbulence. Distribution functions exist for which the integrals in (2.25) and (2.26) fail to converge for some values of ν, which means that the corresponding moments do not exist. However, this is not usually the case for variables derived from turbulent flows, at least not for the positive orders of moments we have in mind here. The most important central moment is obtained for $\nu = 2$ and is called the variance, given by

$$\sigma^2 = \mathrm{Var}(U) = \mu_2 = \overline{(U - \overline{U})^2} \tag{2.27}$$

where $\sigma \geq 0$ is the standard deviation of U and measures how far about its mean $\overline{U}$ varies, that is, the magnitude of the random fluctuations in U. Furthermore, if U_1 and U_2 are statistically *independent*, it is not difficult, using (2.24), to derive

$$\mathrm{Var}(U_1 + U_2) = \mathrm{Var}(U_1) + \mathrm{Var}(U_2) \tag{2.28}$$

which can be extended to show that the variance of the sum of any number of independent variables is the sum of their variances.

The next two central moments can be nondimensionalized using σ to obtain the skewness

$$S = \frac{\mu_3}{\sigma^3} = \frac{\overline{(U - \overline{U})^3}}{\sigma^3} \tag{2.29}$$

and flatness factor (or kurtosis)

$$T = \frac{\mu_4}{\sigma^4} = \frac{\overline{(U - \overline{U})^4}}{\sigma^4} \tag{2.30}$$

of U. These higher-order moments are of considerably less importance than the variance, but the skewness is one possible, if coarse, measure of lack of symmetry of the distribution of U about its mean, whereas the flatness factor provides limited information about how extensive the tails of the distribution are.

Given two variables, U_1 and U_2, we can define their correlation

$$R = \overline{(U_1 - \overline{U_1})(U_2 - \overline{U_2})} \tag{2.31}$$

and, expressing the same quantity in nondimensional form, the correlation coefficient

$$\rho = \frac{R}{\sigma_1 \sigma_2} = \frac{\overline{(U_1 - \overline{U_1})(U_2 - \overline{U_2})}}{\sigma_1 \sigma_2} \tag{2.32}$$

which always lies between -1 and $+1$ and is zero for statistically independent variables. To derive the bounds on ρ, let $u_1 = U_1 - \overline{U_1}$, $u_2 = U_2 - \overline{U_2}$ be the fluctuations in U_1 and U_2, so that $R = \overline{u_1 u_2}$. Since

$$\overline{(u_1 + \lambda u_2)^2} \geq 0 \tag{2.33}$$

for any constant λ, expanding the square leads to

$$\sigma_2^2 \lambda^2 + 2R\lambda + \sigma_1^2 \geq 0 \tag{2.34}$$

for any λ, which implies that

$$|R| \leq \sigma_1 \sigma_2 \tag{2.35}$$

or, in other words, $|\rho| \leq 1$, as stated above. Observe that if ρ takes on one of its limiting values, $\rho = \pm 1$, then the above argument shows that $\sigma_2 u_1 = \pm \sigma_1 u_2$ and the random variables are deterministically related, since their fluctuations have the same ratio in all realizations. To show that $R = \rho = 0$ when the variables are statistically independent, we note that

$$R = \overline{u_1 u_2} = \overline{(U_1 - \overline{U_1})(U_2 - \overline{U_2})} = \overline{(U_1 - \overline{U_1})}\,\overline{(U_2 - \overline{U_2})} = 0 \tag{2.36}$$

where we have used (2.24) to write the mean of the product as a product of the means. It follows that R and ρ are measures of statistical dependence, although it should be noted that they might be zero even if the variables are not statistically independent.

As remarked at the end of the last section, many turbulent flows are thought to asymptotically approach statistical independence at wide temporal separations. This is reflected in correlation coefficients that go to zero at large temporal separation, though not necessarily monotonically. Thus, if $U(t)$ is some velocity component, $U_1 = U(t_1)$ and $U_2 = U(t_2)$ decorrelate as $|t_1 - t_2| \to \infty$, that is, the correlation coefficient $\rho(t_1, t_2) \to 0$. At zero time separation, $\rho(t_1, t_2) = 1$ takes its maximum value, from which it falls away at nonzero $|t_1 - t_2|$. The order of magnitude, Θ, of the temporal separation required for significant decorrelation is referred to as the correlation time. Decorrelation is also often observed between the velocities at two points in space at a single time as a function of spatial separation. The distance required for significant spatial decorrelation, or correlation length, is an important measure of the size of the large scales of turbulence, as discussed in Chapter 1. Chapter 3 examines the different time and space scales present in turbulent flows in some detail, but for the moment we want to consider the process of time averaging of flow quantities.

We noted earlier that time averaging is often used for experimental determination of the statistical properties of steady flows and we now want to pose the question as to when this leads to the same results as the ensemble definition. The flow is assumed statistically steady, for otherwise time averaging will mix together differing statistics from different times and there is no real hope that it will yield ensemble averages corresponding to a specific time. Consider some time-dependent flow quantity $U(t)$ and define the time average as

$$\hat{U}^{(T)} = \frac{1}{T}\int_0^T U(t)\,dt \tag{2.37}$$

whose ensemble average yields

$$\overline{\hat{U}^{(T)}} = \frac{1}{T}\int_0^T \overline{U}\,dt = \overline{U} \tag{2.38}$$

since the flow is supposed statistically steady. This shows that the ensemble average of $\hat{U}^{(T)}$ agrees with the ensemble average of U. Subtracting (2.38) from (2.37), squaring and ensemble averaging, we find

$$\mathrm{Var}\left(\hat{U}^{(T)}\right) = \overline{\left(\frac{1}{T}\int_0^T u\,dt\right)^2} = \frac{1}{T^2}\int_0^T\int_0^T \overline{u(t_1)u(t_2)}\,dt_1\,dt_2 \tag{2.39}$$

where $u(t) = U(t) - \overline{U}$ is the fluctuation in U. Introducing the correlation function

$$\overline{u(t_1)u(t_2)} = R(t_1 - t_2) \tag{2.40}$$

which is a function of the temporal separation $t_1 - t_2$ alone, thanks to statistical steadiness of $U(t)$. Changing integration variables to t_1 and $\tau = t_1 - t_2$, instead of t_1 and t_2, the integral over t_1 can be performed to give

$$\mathrm{Var}\left(\hat{U}^{(T)}\right) = \frac{\sigma^2}{T}\int_{-T}^{T}\left(1 - \frac{|\tau|}{T}\right)\rho(\tau)\,d\tau \tag{2.41}$$

where $\rho(\tau) = R(\tau)/\sigma^2$ is the correlation coefficient at time separation τ and σ is the standard deviation of U. Suppose that $\rho(\tau) \to 0$ sufficiently rapidly as $|\tau| \to \infty$ that the integral

$$\Theta = \int_{-\infty}^{\infty} |\rho(\tau)|\,d\tau \tag{2.42}$$

converges, yielding a correlation time.[2] One can then bound the integral in (2.41) to obtain

$$\sigma\left(\hat{U}^{(T)}\right) \leq \left(\frac{\Theta}{T}\right)^{1/2}\sigma \tag{2.43}$$

where $\sigma(\hat{U}^{(T)})$ is the standard deviation of $\hat{U}^{(T)}$. It follows that $\sigma(\hat{U}^{(T)}) \to 0$ as $T \to \infty$, that is, the random fluctuations of the finite time average, $\hat{U}^{(T)}$, about its mean value, $\overline{U}$, can be made as small as one likes by increasing the averaging time, T. That is, if T is taken large enough, the fluctuations in $\hat{U}^{(T)}$ are very small and the time average is a good estimate of the ensemble average. We conclude that time averaging over a sufficiently long period yields the same results as ensemble averaging for steady flows provided that there is rapid enough decorrelation that (2.42) converges. In that case, equation (2.43) can be used to estimate the error involved in the time average. The error decreases only slowly with increasing averaging time, proportional to $T^{-1/2}$. This type of calculation is often made when designing statistical measurements of turbulent flows. One asks how long an averaging time is needed for convergence of the average to within an acceptable margin of error. As we saw earlier, the determination of probability distributions can be reduced to the calculation of appropriate averages.

The time average (2.37) is inappropriate for unsteady flows; for instance, in the case of turbulent flow generated by an explosion, one must repeatedly carry out the experiment to produce the ensemble statistics. If a time average is employed it includes all stages of the explosion and neither converges nor yields results which meaningfully describe any given stage. However, consider the example sketched in Figure 2.5 of a cylinder inside a piston engine that is turning at constant speed and load. In such a flow, there are turbulent fluctuations from one cycle to the next, so a single realization of the flow is not periodic, but we might expect the statistical properties of the flow, for instance the mean velocity, to vary periodically with time. That is, the flow statistics vary throughout the cycle, but those at time t are the same as those at time $t + \tau$, where τ is the period of oscillation of the piston. In that case, a time average can be defined by

$$\hat{U}^{(N)}(t) = \frac{1}{N}\sum_{n=0}^{N-1} U(t + n\tau) \tag{2.44}$$

[2] Although precise expressions, such as (2.42), for correlation times arise in particular circumstances, in general it is better to think of them as order of magnitude scales. The same is true of correlation lengths.

rather than (2.37) and an argument very similar to that used above shows that, provided N is sufficiently large and there is rapid decorrelation of $U(t + n\tau)$ with increasing temporal separations, (2.44) will yield a good approximation to the ensemble average. As before, probabilities can be determined by appropriate averaging.

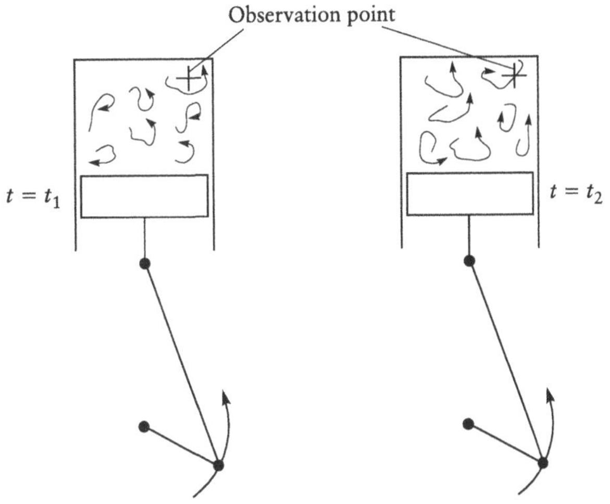

Figure 2.5. Illustration of a periodic turbulent flow. The two times shown, t_1 and t_2, are an integral number of periods apart so that the flow is nominally the same. It will, in fact, be different in detail because of turbulent fluctuations, as we have attempted to show schematically. The turbulent-velocity fluctuations sketched are superimposed on a periodic mean flow, which is the same at the two times.

Blowing across the mouth of a bottle at high Reynolds number to produce an audible tone provides an example in which, if one had never done the experiment, one might expect the flow to be non-oscillatory, whereas, in fact, it has important, nearly periodic oscillations, together with fluctuations from cycle to cycle of the oscillations due to turbulence in the periodically modulated jet/shear-layer over the mouth of the bottle. In this example, the flow itself generates self-sustained oscillations, in contrast with that of Figure 2.5, whose periodic variations are externally imposed by the piston. Repetition of the experiment to provide an ensemble yields oscillations whose phase varies from realization to realization, as in the earlier example of the randomly phased sine wave and reflecting the fact that the experimenter does not control the phase. As with the sine wave, the statistics may turn out to be independent of time, owing to random phasing, but this does not accurately reflect our intuition about the flow. The periodic oscillations, which are not really random in the intuitive sense, are lumped in with the turbulence as part of the random fluctuations and, in consequence, correlations extend to large temporal separations (in principle, to infinite separations if the oscillations were precisely periodic, although exact periodicity is unlikely in practice, given the possibility of random phase drifting over many cycles). This is an unsatisfactory situation because one would like to separate the physically distinct fluctuations due to the oscillations from those occurring from cycle to cycle, which one might identify with turbulence. One way of doing this is feasible if the phase of the oscillations can be experimentally identified in particular realizations, for the flow statistics may then be *conditioned by restricting attention to the subensemble of realizations in which the phase has a particular value* (or even by time-shifting the data so that the phase becomes the same in all realizations). The pressure fluctuations at some point within the bottle provide a good measure of the oscillations because they strongly focus attention on the Helmholtz resonance, rather than on the turbulent fluctuations in the jet. Thus, if the statistics are conditioned by the internal pressure fluctuation, we expect the oscillations to appear as a periodically varying mean flow, while periodically modulated fluctuations about the mean reflect turbulence within the jet and may now decorrelate rapidly with temporal separation. One can then use phase-locked sampling to calculate averages from (2.44). Conditioning the statistics in the above manner effectively makes the phase a controlled parameter and the flow becomes fundamentally similar to that of Figure 2.5, having statistics which vary periodically with time, more closely expressing our intuition about the nature of the flow. We should also remark that Fourier analysis in time allows one to identify periodic, or nearly periodic, components of a flow, which appear as sharp peaks in the Fourier transform, corresponding to the oscillation period and its harmonics.

The problem of uncontrolled parameters, like the phase in the above example, appears in spades when one considers naturally occurring flows, such as the atmospheric boundary layer in which we live. Furthermore, other than in the imagination, one cannot repeat the experiment to generate an ensemble of realizations, unless laboratory or numerical simulations of sufficient fidelity can be constructed, a possibility we ignore for the sake of argument. Thus, we are reduced to passive spectators, although the flow can be observed at different times and locations. In the case of the atmospheric boundary layer, Fourier analysis of the wind velocity with respect to time shows that its transform has well-separated peaks corresponding to time scales of the order of one minute and four days, which represent boundary-layer turbulence and the passage of meteorological systems. It is thus reasonable to employ time averages of the velocity with an averaging time large compared to the smaller of these scales and small compared with the larger, allowing statistical properties of boundary-layer turbulence to be measured. This procedure makes the longer time scales part of the average, while the shorter ones become fluctuations, hopefully decorrelating at temporal separations larger than a few minutes, if not to zero, then at least to small values. The layer statistics depend on the wind speed above the layer and its thermal stratification, to name but two time-varying parameters that the experimenter has no control over. In a theoretical model, one might consider an ensemble of realizations in which all such parameters are held fixed, presumably avoiding the difficulties associated with uncontrolled parameters and multiple time scales for variation of the real flow. Provided the parameters do not vary too rapidly in reality, one would expect such ensemble statistics to agree with those measured using appropriate conditioning or time averaging.

In summary, when ensembles of realizations are used to define the statistics of turbulence they should not be defined blindly, but with the physical properties of the flow in mind. In particular, all important parameters of the flow ought to be fixed, otherwise one may end up with fluctuations that include components of the flow other than turbulence and correlations that extend over large time or space separations.[3] This being said, most fundamental studies of turbulence concern more straightforward cases than those envisaged above, for instance, simple jets and boundary layers at high Reynolds numbers, or turbulence generated by passage of a uniform, steady flow through a grid. For these relatively simple flows, in which the experimenter is presumed to control all important parameters, it suffices to consider a full ensemble of realizations generated by repeating the experiment, or equivalently, since the flows are usually steady and decorrelate with temporal separation,

[3] To caricature, it has jokingly been said that, once one has eliminated all features of a flow that one understands, what remains is turbulence.

time-averaged statistics. One hopes and expects that experience gained with such flows will extend, at least in part, to turbulence in more complicated situations, such as atmospheric and oceanic flows.

We now turn to more mundane matters, namely the definition of characteristic functions and cumulants. The significance of cumulants lies in the fact that those of order higher than two are zero for Gaussian variables, an important class of statistics we discuss in the next section, but, for the moment, the definition of cumulants is purely formal.

The first characteristic function, $\varphi(s)$, of a random variable U is the complex-valued quantity given by

$$\varphi(s) = \overline{e^{isU}} = \overline{\cos sU} + i\overline{\sin sU} = \int_{-\infty}^{+\infty} e^{isU} P(U)\, dU \tag{2.45}$$

which will be recognized as the Fourier transform of the distribution function $P(U)$. Some mathematical properties of $\varphi(s)$ are as follows. Given $\varphi(s)$, we can determine $P(U)$ by Fourier inversion, so that $\varphi(s)$ contains the same information as the distribution of U. It is easily seen that $\varphi(0) = 1$ and $\varphi(-s) = \varphi^*(s)$, where "$*$" denotes complex conjugation, while if $P(U) = P(-U)$ then $\varphi(s)$ is real and $\varphi(-s) = \varphi(s)$. Writing the exponential in (2.45) as a power series, we obtain

$$\varphi(s) = \sum_{n=0}^{\infty} a_n \frac{(is)^n}{n!} \tag{2.46}$$

where a_n are the moments of U, while a similar procedure using $e^{isU} = e^{is\overline{U}} e^{is(U-\overline{U})}$ leads to

$$\varphi(s) = e^{is\overline{U}} \sum_{n=0}^{\infty} \mu_n \frac{(is)^n}{n!} \tag{2.47}$$

where μ_n are the central moments. The second characteristic function is defined by

$$\Psi(s) = \log \varphi(s) \tag{2.48}$$

where a principal value for the logarithm is implied. The power series expansion of $\Psi(s)$ is

$$\Psi(s) = \log\left\{ e^{is\overline{U}} \sum_{n=0}^{\infty} \mu_n \frac{(is)^n}{n!} \right\} = \sum_{n=1}^{\infty} \kappa_n \frac{(is)^n}{n!} \tag{2.49}$$

where κ_n are referred to as the cumulants of U. After some algebra, it can be shown that

$$\begin{aligned}
\kappa_1 &= \overline{U} \\
\kappa_2 &= \mu_2 = \overline{u^2} = \sigma^2 \\
\kappa_3 &= \mu_3 = \overline{u^3} \\
\kappa_4 &= \mu_4 - 3\mu_2^2 = \overline{u^4} - 3\overline{u^2}^2 \\
\kappa_5 &= \mu_5 - 10\mu_2\mu_3 = \overline{u^5} - 10\overline{u^2}\,\overline{u^3} \\
\kappa_6 &= \mu_6 - 10\mu_3^2 - 15\mu_2\mu_4 + 30\mu_2^3 = \overline{u^6} - 10\overline{u^3}^2 - 15\overline{u^2}\,\overline{u^4} + 30\overline{u^2}^3
\end{aligned} \tag{2.50}$$

The cumulants are mainly significant because they are zero, apart from κ_1 and κ_2 (i.e., $\Psi(s)$ is a quadratic function), in the case of a Gaussian probability distribution, an important class of statistics we will discuss in the next section. They therefore allow one to test whether a given variable is Gaussian or close to being Gaussian. It should be noted that there are many distributions which lead to divergence of the power series (2.49), but that, even in such cases, the cumulants are defined by the (now formal) process described above, leading to perfectly definite formulas, (2.50), for the cumulants, always presuming convergence of the moments.

Extensions can be made to the case of multiple variables. For instance, with two variables,

$$\varphi(s_1, s_2) = \overline{e^{i(s_1 U_1 + s_2 U_2)}} = \int_{-\infty}^{+\infty} \int_{-\infty}^{+\infty} e^{i(s_1 U_1 + s_2 U_2)} P(U_1, U_2)\, dU_1 dU_2 \tag{2.51}$$

is the first characteristic function, the Fourier transform of $P(U_1, U_2)$. The expansion of φ is

$$\varphi(s_1, s_2) = \sum_{n,m=0}^{\infty} a_{nm} \frac{(is_1)^n (is_2)^m}{n!m!} = e^{i(s_1 \overline{U_1} + s_2 \overline{U_2})} \sum_{n,m=0}^{\infty} \mu_{nm} \frac{(is_1)^n (is_2)^m}{n!m!} \tag{2.52}$$

where

$$a_{nm} = \overline{U_1^n U_2^m}, \qquad \mu_{nm} = \overline{u_1^n u_2^m} \tag{2.53}$$

are two-variable moments. For statistically independent variables

$$\varphi(s_1, s_2) = \overline{e^{is_1 U_1} e^{is_2 U_2}} = \overline{e^{is_1 U_1}}\;\overline{e^{is_2 U_2}} = \varphi_1(s_1)\varphi_2(s_2) \tag{2.54}$$

and, more generally, the characteristic function of any number of independent variables is the product of their characteristic functions. A second characteristic function is obtained by taking the logarithm of $\varphi(s_1, \ldots, s_N)$ and its (possibly formal) power series expansion yields the cumulants

$$\begin{aligned}
\kappa_1^{(i)} &= \overline{U_i} \\
\kappa_2^{(ij)} &= \overline{u_i u_j} \\
\kappa_3^{(ijk)} &= \overline{u_i u_j u_k} \\
\kappa_4^{(ijkl)} &= \overline{u_i u_j u_k u_l} - \overline{u_i u_j}\;\overline{u_k u_l} - \overline{u_i u_k}\;\overline{u_j u_l} - \overline{u_i u_l}\;\overline{u_j u_k}
\end{aligned} \tag{2.55}$$

as coefficients. For Gaussian variables, all cumulants above $\kappa_2^{(ij)}$ are zero, as for a single variable. Among other uses, this provides a basis for testing how near to joint Gaussian a given set of variables are.

2.3 Gaussian Statistics and the Central Limit Theorem

A single variable is said to be Gaussian (or normal) if $P(U)$ has the form

$$P(U) = \frac{1}{\sigma\sqrt{2\pi}} e^{-\frac{(U-\overline{U})^2}{2\sigma^2}} \tag{2.56}$$

which is shown in Figure 2.6a and has the well-known symmetric bell shape with a single hump centered on $\overline{U}$. The mean, $\overline{U}$, and standard deviation, σ, are the only parameters defining the Gaussian distribution and, if we multiply a Gaussian variable by a constant or add a constant to it, the result remains Gaussian, with different mean and standard deviation. From (2.56) it can be shown that the skewness and flatness factors of a Gaussian variable are $S = 0$, $T = 3$, departures from which can be used as measures of lack of normality of a given variable (although one can have $S = 0$, $T = 3$ for a non-Gaussian distribution). Figures 2.6b and 2.6c illustrate distributions with $S \neq 0$ and $T \neq 3$. Two random variables are jointly Gaussian if

$$P(U_1, U_2) = \frac{1}{2\pi\sigma_1\sigma_2\sqrt{1-\rho^2}} \exp\left\{-\frac{1}{2}\frac{1}{1-\rho^2}\left[\frac{u_1^2}{\sigma_1^2} + \frac{u_2^2}{\sigma_2^2} - 2\rho\frac{u_1 u_2}{\sigma_1\sigma_2}\right]\right\} \tag{2.57}$$

where $u_n = U_n - \overline{U_n}$, σ_n is the standard deviations of U_n, and ρ is the correlation coefficient of the two variables. Thus, in addition to the two means and standard deviations, the correlation coefficient enters as a parameter.

In general, an arbitrary number of random variables, $U_1, \ldots, U_N$, are said to have joint Gaussian (or normal) statistics if their joint probability distribution function has the form $\exp q(U_n)$, where $q(U_n)$ is a quadratic function of the U_i. The distribution function is then

$$[\det(2\pi R)]^{-1/2} \exp\left\{-\frac{1}{2}\mathbf{u}^T R^{-1}\mathbf{u}\right\} \tag{2.58}$$

where $\mathbf{u}$ represents the column vector formed from the N fluctuations $u_n = U_n - \overline{U_n}$ and $R_{nm} = \overline{u_n u_m}$ is the positive-definite, symmetric matrix of correlations. Thus, the means and correlations of jointly Gaussian variables suffice to fix their probability distribution function and hence the full statistics of the variables. The distribution (2.58) has a single maximum at the mean value, $U_n = \overline{U_n}$, and drops off rapidly as $|U_n - \overline{U_n}|/\sigma_n$ increases, as in Figure 2.6a. It can be shown that the sum of jointly Gaussian variables is Gaussian, while if $U_1, \ldots, U_N$ are *independent* variables that are individually Gaussian, they are also jointly Gaussian, since the joint distribution of independent variables is the product of the individual distributions and the product of exponentials is the exponential of the sum.

Suppose that the time-dependent variable $U(t)$ is statistically steady and Gaussian. By saying that the process $U(t)$ is Gaussian, we mean that, no matter what N is used, the values $U(t_1), \ldots, U(t_N)$ of $U(t)$ at any N times are jointly Gaussian. Steadiness implies that $\overline{U}$ is independent of time and that the correlation matrix $R_{nm} = R(t_n - t_m)$, where $\overline{u(t+\tau)u(t)} = R(\tau)$ is the correlation function of $U(t)$. Thus, for steady Gaussian processes, giving $\overline{U}$ and $R(\tau)$ suffices to determine the full (i.e., N-time, for any N) statistics of $U(t)$. From $\overline{u(t+\tau)u(t)} = R(\tau)$, it is easily shown that $R(-\tau) = R(\tau)$, but considerably harder to demonstrate Kinchin's theorem that $R(\tau)$ is the Fourier transform of a *positive* function, known as the frequency spectrum. Provided $R(\tau)$ has these properties, it can also be shown that a statistically steady, Gaussian process, $U(t)$, can be constructed that has the given $R(\tau)$. Similar results hold for statistically steady vector functions of space and time, such as the velocity, $U_i(\mathbf{x}, t)$, in steady flow.

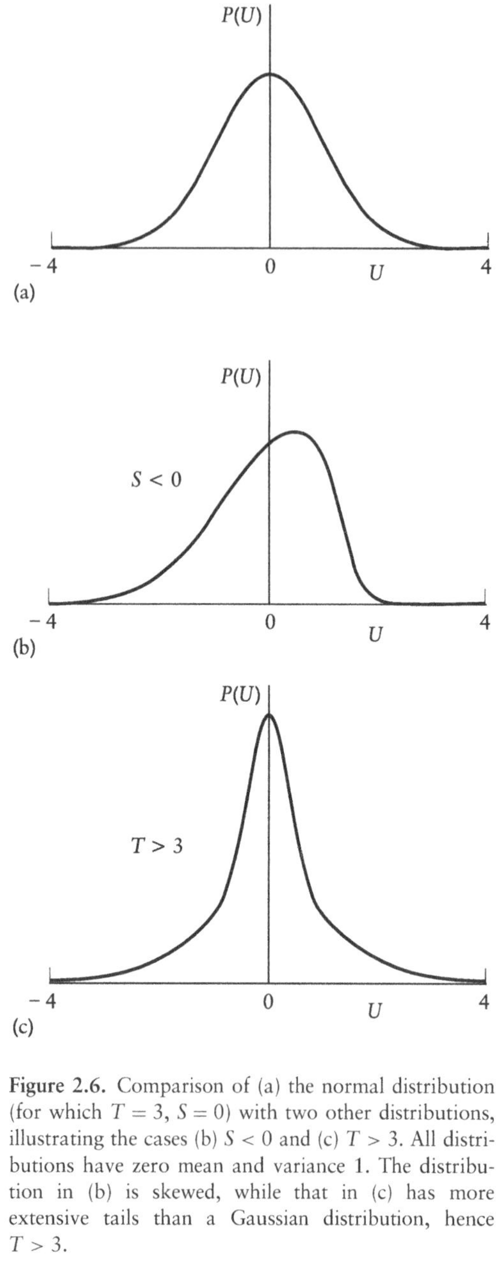

Figure 2.6. Comparison of (a) the normal distribution (for which $T = 3$, $S = 0$) with two other distributions, illustrating the cases (b) $S < 0$ and (c) $T > 3$. All distributions have zero mean and variance 1. The distribution in (b) is skewed, while that in (c) has more extensive tails than a Gaussian distribution, hence $T > 3$.

As noted in the previous section, the cumulants of a Gaussian distribution are zero at orders above two. This follows from taking the Fourier transform of (2.58) to obtain the first characteristic function. The result is the exponential of a quadratic function of the transform variables, so the second characteristic function yields that quadratic function and its series expansion terminates at order two. Setting the fourth line of (2.55) to zero yields

$$\overline{u_iu_ju_ku_l} = \overline{u_iu_j}\,\overline{u_ku_l} + \overline{u_iu_k}\,\overline{u_ju_l} + \overline{u_iu_l}\,\overline{u_ju_k} \qquad (2.59)$$

which allows us to express fourth-order central moments in terms of second-order correlations for jointly Gaussian variables. This result forms the basis of the so-called quasi-normal approximation for closing the statistical equations of turbulence, in which the turbulent velocity fluctuations are assumed to be a sufficiently good approximation to Gaussian variables that (2.59) can be used to express their fourth-order moments.

The importance of Gaussian statistics derives from a profound result of the theory of probability which concerns sums of independent variables: the celebrated *central limit theorem*. Let $U_1, \ldots, U_N$, be statistically independent variables with identical distributions functions, then $U = U_1 + \cdots + U_N$ approaches a Gaussian distribution as $N \to \infty$. The requirement that the variables be identically distributed, present in this, the basic version of the theorem, can, in fact, be relaxed considerably.[4] The more important condition is that of independence and what the theorem says is that the sum of a *large* number of *independent* variables will be close to Gaussian. When the variables have different distributions one can determine the limiting distribution from (2.56) with

$$\overline{U} = \sum_{i=1}^{N} \overline{U_i}, \quad \sigma^2 = \sum_{i=1}^{N} \sigma_i^2 \qquad (2.60)$$

[4] Proofs of the theorem require that the distribution functions of the random variables in the sum should satisfy certain mathematical conditions, the details of which we do not go into.

where a large number of terms must contribute significantly to the second sum, otherwise one can ignore the effects of all but the small number of variables which do, in which case there is no reason why U should be close to Gaussian.

There is a potentially important, and not widely appreciated, restriction on the central limit theorem, which is that the asymptotic description as a Gaussian distribution, although accurate over the most probable part of the distribution of $U = U_1 + \cdots + U_N$, is generally invalid in the tails of that distribution. This non-uniformity occurs when $U - \overline{U}$ is of order N, much larger than the standard deviation, $O(N^{1/2})$, of U and therefore representing rarely attained departures from the mean. Study of the asymptotic behavior when $U - \overline{U} = O(N)$ is known as large deviations theory (see, e.g., Varadhan (1984)), which has important applications to statistical mechanics and, more importantly for present purposes, to the statistical properties of the small scales of turbulence (see, e.g., Frisch (G 1995), section 8.6.4).

The reader may care for an example, demonstrating both the central limit theorem and its large deviation restriction. As assumed in the simplest version of the theorem, let U_i be identically distributed, independent random variables and $U^{(N)} = U_1 + \cdots + U_N$ be the sum of the first N, so that $U^{(N+1)} = U^{(N)} + U_{N+1}$ gives $U^{(N+1)}$ as the sum of two independent variables. The reader can show that, if V_1 and V_2 are independent, with distributions $P_1(V_1)$ and $P_2(V_2)$, then the distribution of $V = V_1 + V_2$ is given by

$$P_+(V) = \int_{-\infty}^{\infty} P_1(V_1)P_2(V - V_1)dV_1 \qquad (2.61)$$

which is the convolution integral of P_1 and P_2 (hint: owing to independence, the joint distribution of V_1, V_2 is $P(V_1, V_2) = P_1(V_1)P_2(V_2)$, while $P_+(V)dV$ is the integral of $P(V_1, V_2)$ over the infinitesimal strip $V < V_1 + V_2 < V + dV$ in the (V_1, V_2) plane). That is, the distribution of a sum of independent variables is the convolution of their distributions. Applying this result with $V_1 = U^{(N)}$ and $V_2 = U_{N+1}$, we have

$$P^{(N+1)}(U) = \int_{-\infty}^{\infty} P^{(N)}(V)P(U - V)dV \qquad (2.62)$$

where $P^{(N)}$ is the distribution of the sum $U^{(N)} = U_1 + \cdots + U_N$ and P denotes the distribution function of the U_i. In other words, each time an extra term is added to the sum, its distribution is convolved with P, beginning with $P^{(1)}(U) = P(U)$. In the example we have in mind, the terms, U_i, in the sum have the Poisson distribution $P(U) = e^{-U}$ for $U > 0$, $P(U) = 0$ when $U < 0$, and the reader may verify that

$$P^{(N)}(U) = \frac{U^{N-1}e^{-U}}{(N-1)!} \qquad (2.63)$$

for $U > 0$, $P^{(N)}(U) = 0$ when $U < 0$, reduces to $P(U)$ if $N = 1$ and satisfies the recurrence relation (2.62). The reader is encouraged to calculate and plot the probability distribution, $N^{1/2}P^{(N)}(N^{1/2}(V + N))$, of the normalized variable $V = (U - N)/N^{1/2}$ as a function of V for a series of increasing values of N, using a computer (the normalization is based on the mean, $\overline{U^{(N)}} = N$, and standard deviation, $N^{1/2}$, of $U^{(N)}$). This bears out the central limit theorem graphically, with quite small N sufficient for a roughly Gaussian-looking curve, although later convergence is rather slow.

At large N, one may rewrite (2.63) using $U^{N-1}e^{-U} = e^{(N-1)\log U - U}$, expand the argument of this exponential as a Taylor's series about its maximum at $U = N - 1$, and employ Stirling's formula to express $(N-1)!$, thus obtaining a Gaussian function for $P^{(N)}(U)$ in which the modified normalized variable $V' = (U - N + 1)/(N-1)^{1/2}$ appears, rather than V. Of course, in the limit $N \to \infty$, the variables V and V' coincide, confirming the central limit theorem. However, although the Taylor's series expansion gives a good approximation to the most probable part of the distribution of $U^{(N)}$ at large N, it fails to describe the behavior sufficiently far from the mean, breaking down when the departure from the mean is large, of $O(N)$. This is the region studied by large deviations theory and corresponds to events of low probability, but which, as noted above, can nonetheless be important in some applications, in particular the theory of small-scale turbulent intermittency.

Despite its large deviation restriction, the central limit theorem is the fundamental reason why Gaussian variables play an important role in the theory of probability and statistics. One can cite the classical example of experimental errors that are of multiple origins and can often be considered as many, independent, and additive, leading to a Gaussian distribution for the overall error in a measurement. Note, however, that, not only must the variables be independent, but it is their *sum* whose distribution is considered. Suppose, for instance, that some physical quantity is the *product* of positive, independent variables. Taking logarithms, the log of the result is the sum of the logs and hence should tend to a normal distribution as the number of terms, N, in the product increases. The standard deviation of the sum of logarithms increases proportional to $N^{1/2}$; thus, if N is large, widely different orders of magnitude can occur, once we take exponentials to reconstitute the product. Alternatively, one may say that the product, taken to the power $N^{-1/2}$ to suppress the rapidly growing fluctuations in its value, approaches a log-normal distribution (U is log-normal if $\log U$ is normal), rather than becoming Gaussian. While perhaps not as well known as the Gaussian distribution, the log-normal,[5] and other distributions are not without their uses. Indeed, log-normal behavior arises naturally in the theory of the small scales of turbulence at high Reynolds numbers when the energy cascade is modeled as a multiplicative random process (see, e.g., Frisch (G 1995), section 8.6.3). Departures from log-normality due to the large-deviation limitations of the central limit theorem also form an important aspect of such models, as alluded to above. Many other distribution functions have been investigated in the theory of probability, including the χ^2, Poisson, binomial, and gamma distributions, but are not used in this book. The interested reader may find details in any standard textbook on probability.

Consider the calculation of the time average, (2.37), of a statistically steady random quantity, $U(t)$, which decorrelates rapidly with temporal separation. As we saw in the last section, if the averaging time, T, is large enough, the time average is a good

[5] In fact, there is not a single log-normal distribution, but a family of log-normal distributions, corresponding to different variances for the Gaussian distribution of $\log U$. It is often forgotten that, strictly speaking, Gaussian distributions form a two-parameter family with different means and variances. While changes in these parameters have a trivial effect on a normally distributed random variable, corresponding merely to shifts and rescalings, different variances for $\log U$ lead to distinctly dissimilar distributions for U.

approximation to the ensemble average. We now make the stronger assumption that $U(t)$ approaches statistical independence at large time separations. Given large enough T, the integration range in (2.37) can be split into many equal intervals, each of sufficient length that it is plausible to neglect statistical dependence between different intervals. Thus, we have an approximation to the case envisaged in the central limit theorem, namely the sum of a large number of independent variables. It is therefore reasonable that the time average should tend towards a Gaussian distribution, whose mean we know to be $\overline{U}$ and whose standard deviation is given asymptotically by the right-hand side of (2.43). In other words, the time average should approach the ensemble average through a sequence of narrower and narrower Gaussian distributions, despite the fact that U itself need not be Gaussian.

2.4 Turbulent Mean Flow and Fluctuations

It is time to discuss the application of statistical methods to turbulent flow in a little more detail. We should first note that, whereas in earlier sections, U_i was used in a generic sense to represent any flow variables (velocities, pressures, etc.), here and throughout the remainder of this book, U_i will denote the fluid velocity. Since U_i is a function of position and time, it is a random vector field, varying from realization to realization of the flow within the supposed ensemble of different experiments. Its value at any particular position and time, $U_i(\mathbf{x}, t)$, yields a random vector, consisting of three components, which are single random variables. The full statistical properties of U_i are characterized by the *joint* probability distribution functions of its components at an arbitrary finite number of different points and times. That is, to specify the complete statistics of velocity requires that one give the joint distributions of the $3N$ components $U_i(\mathbf{x}_1, t_1), \ldots, U_i(\mathbf{x}_N, t_N)$ at all points $\mathbf{x}_1, \ldots, \mathbf{x}_N$ and times $t_1, \ldots, t_N$, and this for arbitrary values of N. Furthermore, one could also introduce the pressure field and consider its joint distributions with velocity. It is apparent that, in principle, a vast amount of statistical information is contained in a turbulent flow, but one is usually concerned with only a few of its simpler statistics, of which the mean-flow velocity is perhaps the simplest.

As its name implies, the mean velocity field at given $\mathbf{x}$ and t, denoted $\overline{U_i}(\mathbf{x}, t)$, is just the average of $U_i(\mathbf{x}, t)$ over the ensemble of flow realizations and is independent of time for a steady flow. The fluctuation of velocity is defined by

$$u_i = U_i - \overline{U_i} \tag{2.64}$$

and is usually interpreted as representing the turbulence, whose intensity in different directions can be measured by the standard deviations

$$u_i' = \left(\overline{u_i^2}\right)^{1/2} \tag{2.65}$$

or overall by the turbulent kinetic energy per unit mass

$$\frac{1}{2}\overline{u_i u_i} = \frac{1}{2}\sum_{i=1}^{3}\overline{u_i^2} = \frac{1}{2}\left(u_1'^2 + u_2'^2 + u_3'^2\right) \tag{2.66}$$

The related quantity u', defined by

$$u'^2 = \frac{1}{3}\overline{u_i u_i} = \frac{1}{3}\left(u_1'^2 + u_2'^2 + u_3'^2\right) \tag{2.67}$$

as an average over the three mean-squared components of velocity, is the most common quantitative measure of turbulent intensity. Observe that, since $\overline{U_i U_i} = \overline{U}_i \overline{U}_i + \overline{u_i u_i}$, the average of the total kinetic energy of the fluid consists of components arising from the mean and fluctuating flows. The latter component is usually identified with the energy of turbulence.

One can define velocity correlations, $R_{ij} = \overline{u_i u_j}$, at a single point and time, or introduce two-point or two-time correlations such as

$$R_{ij}(\mathbf{x}, t_1, t_2) = \overline{u_i(\mathbf{x}, t_1) u_j(\mathbf{x}, t_2)} \tag{2.68}$$

which, as discussed earlier, is generally taken to tend to zero as $|t_1 - t_2| \to \infty$, reflecting asymptotic statistical independence of the flow at widely separated times and leading to the idea of a correlation time. The same is true of velocity correlations between different points, giving rise to correlation lengths.

Higher moments of the u_i may also be introduced, such as their skewness and flatness factors

$$S_i = \frac{\overline{u_i^3}}{u_i'^3}, \quad T_i = \frac{\overline{u_i^4}}{u_i'^4} \tag{2.69}$$

which respectively measure asymmetry of the distribution of U_i about $\overline{U_i}$ and the prominence of the tails of the distribution. For a Gaussian distribution (Figure 2.6a), $S = 0$ and $T = 3$, while Figures 2.6b and 2.6c sketch possible distributions with $S <$ 0 and $T > 3$. Some turbulent fields have highly asymmetric distributions for one or other of the velocity components, giving $S_i \neq 0$ and, when one examines such velocity components as a function of time, there is a tendency towards shorter fluctuations of higher amplitude in one direction, as illustrated in Figure 2.7. The case $T_i > 3$ represents tails of the distribution that are more prominent than with Gaussian statistics, so atypically large velocity fluctuations are more frequent. It should be remarked that turbulent velocities are not usually Gaussian, although the large scales of grid turbulence are found to be approximately so. Moments of the u_i of any order involving many points and times may also be considered.

Statistical measures of turbulence, such as u', are usually taken to characterize orders of magnitude in *typical* realizations of a turbulent flow. In particular, we will often suppose that realizations have turbulent velocity fluctuations of order u', although the precise values of the fluctuating velocity components vary randomly from realization to realization. This requires that the turbulence in different realizations of the flow be sufficiently similar,

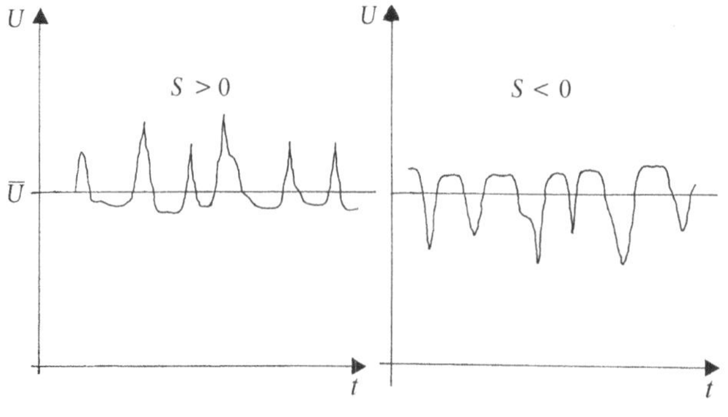

Figure 2.7. Sketch of the effects of nonzero S on a typical time history of a turbulent quantity.

so that the overall statistics of the ensemble describe the orders of magnitude in typical individual realizations. However, it is not difficult to construct examples in which the properties of typical realizations are not well represented by the statistics. For example, consider an infinite flow without mean velocity in which each realization consists of *widely separated*, well-defined, independent patches of turbulence distributed throughout the flow, whose positions vary randomly from realization to realization, so that the resulting flow is statistically homogeneous (one may imagine generating the patches using small, randomly placed and orientated, stirring devices, introduced into the fluid and then removed to form the initial conditions of the flow). Since, at a given point in space, a turbulent patch is only present for a small proportion of the ensemble of realizations, the average value, $\overline{u'^2}$, will greatly underestimate the intensity of turbulence within a patch, while overestimating it for the many realizations in which there is no patch present. Thus, $\overline{u'^2}$ is not a good measure of turbulence in typical realizations of such a flow. One result is that the time evolution of the turbulence, whose rate depends on the typical intensity within a patch, rather than the average $\overline{u'^2}$, is considerably more rapid than it would be if turbulence had the same value of $\overline{u'^2}$, but was spread more uniformly in individual realizations (for instance, by taking an initially Gaussian velocity field). This example may appear somewhat artificial,[6] and indeed this type of gross variability of turbulence from realization to realization is not what one usually envisages in a turbulent flow. Such grossly variable flows are not properly accounted for by much of the theory, which is based on a few low-order statistics. Thus, for the most part, we implicitly exclude them from consideration in this book, although the above example of homogeneous turbulence consisting of sparse patches will be used several times as an illustrative thought experiment. Another interesting aspect of the above flow is that it does not show the usual asymptotic statistical independence at temporal separations large compared with the correlation time. Briefly put, the reason for this is that, in particular realizations, turbulent patches remain near their initial locations, leading to long-term statistical memory, rather than relaxation of conditional statistics back towards those of the full ensemble.

As a concrete application of statistical ideas, let us consider an infinitesimal particle transported at the fluid velocity in a statistically steady, random flow that is assumed to approach statistical independence at large temporal separations. The particle position, $\mathbf{x}(t)$, satisfies

$$\frac{d\mathbf{x}}{dt} = \mathbf{U}(\mathbf{x}(t), t) \tag{2.70}$$

which can, in principle, be solved for $\mathbf{x}(t)$ if we know the velocity field $\mathbf{U}(\mathbf{x}, t)$ and the initial position $\mathbf{x}(0)$, assumed the same in all realizations. Formally one can write

$$\mathbf{x}(t) = \mathbf{x}(0) + \int_0^t \mathbf{U}(\mathbf{x}(t'), t')\, dt' \tag{2.71}$$

[6] This idealized example resembles the isolated, highly sporadic turbulent patches found to occur randomly in the oceanic thermocline, apparently due to occasional breaking of internal waves propagating in the density stratification, a process analogous to the familiar formation of intermittent whitecaps on the ocean surface in a wind just sufficient to produce breaking. However, the oceanic flow involves much else besides turbulent patches, whereas the example does not.

which is not an explicit solution for $\mathbf{x}(t)$ since it contains the unknown as an argument of U inside the integral. However, we now use an argument similar to that employed at the end of the last section when we showed that time averages are asymptotically Gaussian at large averaging times. Supposing t to be sufficiently large, we split the integral in (2.71) into a large number, N, of equal intervals of time, long enough that one can neglect statistical dependence of the flow between different intervals. Thus,

$$\mathbf{x}(t) = \mathbf{x}(0) + \sum_{n=1}^{N} \int_{(n-1)t/N}^{nt/N} \mathbf{U}(\mathbf{x}(t'), t')\, dt' \tag{2.72}$$

and, as we did earlier for the time average, it might be argued that this expresses $\mathbf{x}(t)$ as the sum of a large number of approximately independent random variables, and so should be asymptotically Gaussian. It is indeed observed that the positions of particles released from a source become approximately Gaussian in some cases; however, the above argument has a flaw, as becomes apparent if one considers turbulence within a closed volume, such as a vigorously stirred cup of coffee. A Gaussian distribution would imply that there was a finite probability of finding the particle outside the cup, which is clearly absurd. Furthermore, this probability grows with time, as the number of terms in the sum, and hence the variance of $\mathbf{x}(t)$, increases. This difficulty persists even if the physical velocity field in (2.70)–(2.72) is replaced by a mathematically ideal flow that is strictly statistically independent at different times (white noise), so the problem does not lie in the assumption of statistical independence of $\mathbf{U}(\mathbf{x}, t)$ between component intervals of time. The fallacy in the argument is that $\mathbf{x}(t')$ in (2.71) and (2.72) *depends on the entire history of the flow* up to time t'. Thus, even if $\mathbf{U}(\mathbf{x}, t_1)$ and $\mathbf{U}(\mathbf{x}, t_2)$ are statistically independent at any fixed $\mathbf{x}$, $\mathbf{U}(\mathbf{x}(t_1), t_1)$ and $\mathbf{U}(\mathbf{x}(t_2), t_2)$ need not be, since they share memory of the flow history via their dependence on the particle position. Different terms of the sum in (2.72) need not then be statistically independent, which vitiates application of the central limit theorem. However, if the flow is statistically homogeneous, dependence on $\mathbf{x}(t)$ is of no account, and in this case we expect the position of a particle released at fixed position into a steady, homogeneous flow[7] that is statistically independent at large time separations to asymptotically approach a Gaussian distribution. Neither the requirement of steadiness, nor that of asymptotic independence, is satisfied for the example of the flow with sparse patches of turbulence considered above, so one would not expect Gaussian behavior in that case.

2.5 Steadiness, Homogeneity, Isotropy, and Other Statistical Symmetries

Steady turbulent flows have already been defined as ones whose statistical properties do not change with time, as has homogeneous turbulence, in which the fluctuation statistics are the same at all spatial positions. These are examples of statistical symmetries, of which we now want to describe a number of important types.

[7] Homogeneous turbulent flows are generally statistically unsteady. Nonetheless, steady flows are often approximated as homogeneous when studying particle dispersion because it simplifies the analysis.

Homogeneity is first discussed in more detail, then we introduce a number of other classes of statistical symmetry.

Homogeneity implies more than uniformity of the one-point, single-time statistics of u_i, requiring, as it does, that, given any number of different spatial points and times, the joint probability distribution function of u_i will remain unchanged if all points are shifted by the same constant displacement. Implicit in this definition is that the flow domain should be infinite with no boundaries. It is also worth emphasizing that homogeneity refers to the fluctuations and not to the mean flow itself, which can be nonuniform. Although a general function $\overline{U_i}(\mathbf{x}, t)$ is incompatible with homogeneity, as we shall see in Chapter 4, it is quite possible to have homogeneous turbulence with a spatially linear mean velocity (i.e., with $\overline{U_i}(\mathbf{x}, t)$ varying as a linear function of $\mathbf{x}$).

The reason why the assumption of homogeneous turbulence is often made in theoretical studies is that it simplifies the equations and, more importantly, makes spectral analysis possible. Spectral analysis is introduced in Chapter 6 and is a powerful theoretical technique, allowing quantitative description of the continuum of different spatial scales of turbulence. However, no turbulent flow is really homogeneous and it has sometimes been said to be an academic case, irrelevant in practice. This is an extreme view and it is, in fact, possible to produce quite close approximations to homogeneity, for instance, in the turbulent flow behind a grid, many grid spacings downstream of the grid. The theory can be developed much further in the homogeneous case, while experience gained with such idealized turbulence, in particular the properties of the small scales, provides a basis for understanding some, but not all, of the features of more realistic, inhomogeneous flows. We shall often specialize to homogeneous turbulence in later chapters of this book.

Grid turbulence is approximately homogeneous in the sense that variations in its statistical properties take place over distances large compared with the length scales of the turbulence itself. Owing to the mean flow, which is nearly uniform, turbulence generated by the grid is convected downstream, decaying in intensity as it goes. It is in this frame of reference, moving with the mean velocity, that grid turbulence can be modeled as a homogeneous flow, decaying with time and of zero mean velocity. The change in reference frame causes the statistically steady flow seen in the laboratory frame to appear as unsteady, in agreement with the fact that homogeneous turbulence decays in the absence of a mean-flow gradient, as we will see in later chapters.

One may construct examples in which homogeneous turbulence has highly nonuniform properties in individual flow realizations, for instance, the flow with sparse, randomly located, turbulent patches described earlier. In that case, the flow *statistics* are uniform, despite gross nonuniformity in particular realizations. Although such flows fit the definition of statistical homogeneity, they are not what one generally has in mind when using the term homogeneous turbulence. Thus, as for grid turbulence, or the rather more academic case of a homogeneous velocity field whose initial statistics are Gaussian,[8] rough uniformity of turbulence properties in individual realizations is often implicitly assumed when considering homogeneous flows.

There are cases in which homogeneity occurs in just one or two spatial directions, rather than in all three dimensions, as assumed above. Thus, if all statistical proper-

[8] Although the statistics may be chosen Gaussian at the initial time, they will not remain Gaussian later.

ties of u_i remain the same when the flow is imagined displaced by an arbitrary distance in x_1, say, then the turbulence is homogeneous with respect to x_1. For example, sufficiently far from the entrance of a cylindrical pipe, the flow becomes essentially independent of position, x_1, along the pipe and is then referred to as developed (although the mean pressure is a linear function of x_1, of course, since it is this mean pressure gradient which drives the flow). Developed turbulent flow in a cylindrical pipe is thus homogeneous in x_1, but not with respect to position across the pipe. In other cases, there can be homogeneity with respect to two directions (developed plane channel flow between parallel walls being an obvious example).

Another important statistical symmetry is isotropy. In isotropic turbulence, all statistics of u_i are unchanged if we rotate the coordinate system by an arbitrary amount about an arbitrary line, or reflect the flow in any plane. Put another way, the turbulence has no preferred direction – it is rotationally (or spherically) symmetric. The definition also requires it to be statistically invariant under reflection, a condition that is really independent of the rotational symmetries, but forms part of the usual definition of isotropic turbulence. Isotropic turbulence is, in many ways, an even more idealized case than homogeneity, but again isotropy is often adopted because it simplifies the analysis.

Symmetry under reflection is part of the definition of isotropy, but can occur in the absence of rotational symmetry. Turbulence is symmetric under reflection in a plane if all statistics of u_i are unchanged when the flow field is imagined reflected in the plane. For instance, developed flow in a plane channel (a case we will consider in some detail in Chapter 4) is symmetric with respect to the center plane of the channel as well as being homogeneous with respect to in-plane translations. Obviously, one could consider other symmetries and combinations of symmetries, for instance, developed flow in a circular pipe possesses statistical symmetry with respect to rotation about the pipe axis, as well as being homogeneous in the direction of the pipe axis and, if there is no swirl, is also symmetric under reflection in any plane through the pipe axis. All such symmetries can potentially be used to infer properties of the turbulence without detailed calculation and hence simplify analysis, as we shall see in later chapters.

2.6 Conclusions

In this chapter we have tried to give the reader a basic understanding of statistical methods as applied to turbulence. Use of such methods will be made throughout the remainder of this book and they are amongst the most important elements in the repertoire of techniques for treating turbulent flows.

References

Lumley, J. L., 1970. *Stochastic Tools in Turbulence.* Academic Press, New York.
Varadhan, S. R. S., 1984. *Large Deviations and Applications.* SIAM, Philadelphia.

CHAPTER THREE

Space and Time Scales of Turbulence

Besides its apparent randomness, one of the important characteristics of turbulent flows at high Reynolds number (which we presume throughout this chapter) is the wide range of different space and time scales which they contain. As illustrated in Figure 3.1a, the velocity measured at a given point within a turbulent flow fluctuates randomly and does not yield a smooth function of time, but rather one with fine-scale "fur." If part of such a graph is examined at successively higher magnifications, instead of approaching a straight line, as would a smooth function, the fur is revealed as being itself "furry," and so on, through a succession of smaller and smaller time scales, until eventually the smallest of all is reached. At this scale, the velocity finally appears as a smooth function of time, with no further structure revealed by additional magnification. On the other hand, the largest time scales of turbulence are characterized by the time interval required for statistical decorrelation of the turbulent velocity fluctuations.[1] At high Reynolds number, the separation of scales between the largest and smallest is wide, and gets wider as the Reynolds number increases.

The continuum of time scales described above is also reflected in the spatial structure of the flow. The order of size of the largest spatial scales of turbulence is measured by the velocity correlation length L, which characterizes the spatial separation required for significant decorrelation of the velocity fluctuations. Successively higher magnification of the spatial structure of the velocity field shows smaller and smaller scales down to the smallest, known as the Kolmogorov scale, at which it appears smooth. Thus, turbulence contains a continuum of scales, ranging from the large ones, whose size is of order the correlation length, to those of order the Kolmogorov scale. Both the Kolmogorov scale, η, and correlation scale, L, are best considered as orders of magnitude: one talks of scales comparable with, much smaller than, or much larger than these quantities. The spatial and temporal "furriness" of the velocity field are related; indeed, one can think of the small time scales as resulting from convection of fine spatial structures past the measurement probe, as we will see later.

If one attempts to take the derivative of the velocity field, either with respect to time or one of the spatial coordinates, its furry nature becomes markedly more apparent. The spatial derivative of the velocity is the limit of the gradient,

[1] As elsewhere in this book, we presume decorrelation with temporal and spatial separation throughout the chapter.

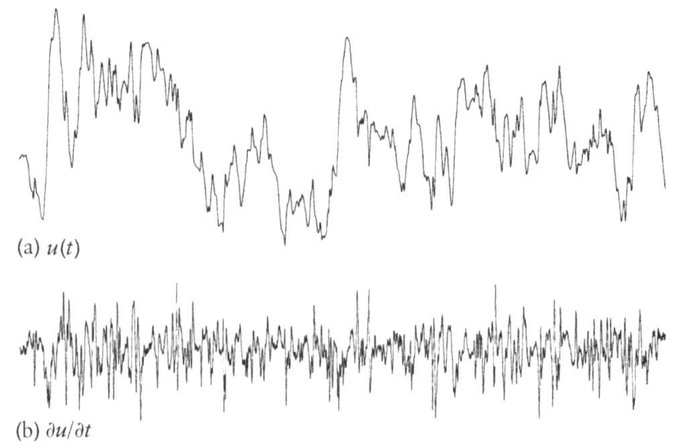

Figure 3.1. Measured time histories at a given point in space of (a) turbulent velocity, and (b) its time derivative in grid turbulence. Plots as a function of any of the spatial coordinates at fixed time are qualitatively similar. (Courtesy of F. Ladhari.)

$\Delta U/\Delta x$, but, because of the furry nature of the velocity field, rather than approaching a limit, this quantity fluctuates as Δx is decreased, until Δx becomes comparable with η, at which point a well-defined derivative begins to appear, finally tending to a limit for Δx small compared with η. As Δx is reduced, a typical velocity difference, ΔU, at separation Δx, shows decreasing fluctuations, but initially more slowly than Δx. Thus, before reaching the Kolmogorov scale, the gradient not only fluctuates with Δx, rather than approaching a limit, but has typical values which *grow* in magnitude as Δx decreases. When finally Δx is much smaller than η, the derivative obtained is considerably larger than would be estimated based on typical gradients at the correlation scale, and varies from point to point on the scale η (see Figure 3.1b). When examined on scales larger than η, the velocity seems to be continuous, but not differentiable (furry). Naturally, it is in fact differentiable, but the derivative is determined by the smallest scales of turbulence.

As discussed in Chapter 1, the kinetic energy of the flow is dissipated by viscous heating at the rate

$$\Delta = \frac{1}{2}\nu\left(\frac{\partial U_i}{\partial x_j} + \frac{\partial U_j}{\partial x_i}\right)\left(\frac{\partial U_i}{\partial x_j} + \frac{\partial U_j}{\partial x_i}\right) \tag{3.1}$$

per unit mass, where ν is the kinematic viscosity and **U** the velocity field. Since Δ is a function of the velocity derivatives, it is determined by the smallest scales of turbulence, which therefore dominate the dissipation of energy. On the other hand, the kinetic energy of the flow, $U_iU_i/2$ per unit mass, does not contain derivatives and so, although like U_i itself it has fine scale fur, it is mainly determined by the large scales of the flow. Thus, we have an essentially large-scale quantity, the kinetic energy, which is dissipated by viscosity at the smallest scales. The link between the two is provided by the turbulent energy cascade from large to small scales. The appearance of the viscosity in the expression, (3.1), for the dissipation rate reflects the importance of viscous effects at the smallest scales. It is the small, but nonzero viscosity (small in the nondimensional sense of high Reynolds number) which determines the Kolmogorov scale. Everything else being equal, the smaller the viscosity, the smaller the Kolmogorov scale required to make viscosity significant at scale η.

As was first done by Reynolds over a century ago, one may adopt statistical splitting of the velocity field as

$$U_i = \overline{U}_i + u_i \tag{3.2}$$

where $\overline{U}_i$ is the ensemble average velocity and u_i is the fluctuation, usually identified with the turbulence. In general, the mean velocity varies with time and spatial position, but, like all single-point, single-time averages, it does not show the furriness of particular realizations of the velocity, which disappears under averaging. The time and space scales over which $\overline{U}_i(\mathbf{x}, t)$ changes are characteristic of the flow as a whole, for instance the period of oscillation of the piston for turbulence subjected to periodic compression and rarefaction inside a cylinder, or the width of the wake of a body placed in a stream. Thus, the small-scale fur is contained in u_i rather than $\overline{U}_i$ when one carries out the splitting (3.2) and is therefore interpreted as a property of the turbulence. The large scales of turbulence are measured by statistical decorrelation of the fluctuating velocities, u_i, whereas variations of $\overline{U}_i$ may take place over larger space and/or longer time scales. For instance, a statistically steady flow will have a steady mean velocity field, $\overline{U}_i(\mathbf{x})$, which has an effectively infinite time scale for variations, while the turbulent velocity fluctuations nonetheless decorrelate over a finite time.

The average kinetic energy of the flow can be expressed as

$$\frac{1}{2}\overline{U_iU_i} = \underbrace{\frac{1}{2}\overline{U}_i\overline{U}_i}_{\substack{\text{Energy of}\\ \text{mean flow}}} + \underbrace{\frac{1}{2}\overline{u_iu_i}}_{\substack{\text{Turbulent}\\ \text{energy}}} \tag{3.3}$$

per unit mass. The second term in (3.3) is the average turbulent energy and is an important measure of the intensity of turbulence. Related measures of the intensity of the turbulent velocity fluctuations are given by the mean-squared value $q^2 = \overline{u_iu_i}$ or by u', defined by $u'^2 = q^2/3$ (the factor of one-third gives an average over the three components of velocity when we write $u'^2 = (\overline{u_1^2} + \overline{u_2^2} + \overline{u_3^2})/3$). These quantities are insensitive to the small-scale fur on the velocity graph in Figure 3.1a and are thus characteristic of the large scales of turbulence.[2] Combining u', L, and ν as a nondimensional parameter, we have the turbulent Reynolds number $\mathrm{Re}_L = u'L/\nu$, which is a measure of the significance of viscosity *for the large scales* of turbulence. The larger Re_L, the smaller the effects of viscosity on the large scales and the wider the range of scales present inside developed turbulence. In this chapter, we presume a large value of Re_L.

To see how the continuum of scales described above might arise, imagine setting up an initial high-Reynolds-number flow consisting of an ensemble of realizations containing *only* scales of order L, with velocity fluctuations of $O(u')$. Evidently, such an initial flow does not have the wide range of small scales described above, but, provided it gives rise to turbulence, it will develop them spontaneously. The large initial value of Re_L implies that the viscous term in the Navier–Stokes equation

[2] Here, as in most of this book, we implicitly assume that statistics such as u' are representative of typical realizations of the flow. This excludes "pathological" cases in which there is gross large-scale intermittency, such as the example with sparse turbulent patches discussed in Section 2.4.

$$\frac{\partial U_i}{\partial t} + \underbrace{U_j \frac{\partial U_i}{\partial x_j}}_{\text{Convective term}} = -\frac{1}{\rho}\frac{\partial P}{\partial x_i} + \nu \underbrace{\frac{\partial^2 U_i}{\partial x_j \partial x_j}}_{\text{Viscous term}} \tag{3.4}$$

is small compared with the convective term. At this stage, the flow evolves as if there were no viscosity. As discussed in Chapter 1, without viscosity, vortex lines move with the flow, and are stretched and folded by convection, while their vorticity may be amplified by stretching. In a high-Reynolds-number flow which is in the process of becoming fully turbulent, this complex process, which is far from being fully understood, leads to the appearance of small scales. The initial large vortical scales are believed to develop smaller-scale structures, conserving energy as they do so because viscosity is negligible at this stage. It should be understood that the small scales do not arise instantaneously. They do so *progressively*, taking a time of $O(L/u')$ to appear through evolution of the large scales. The small scales that are generated within the large scales are subject to the same mechanisms as their larger parents and they in turn give rise to yet smaller scales, and so on until viscosity is called into play at the smallest scales. Viscosity becomes more significant as the scale decreases because the viscous term in (3.4) grows relatively more important, thanks to its second spatial derivative, and the effective Reynolds number is no longer large at the smallest scales. Viscosity acts to damp the smallest scales, stopping the appearance of yet smaller scales. The above process generates smaller and smaller scales down to a minimum size determined by viscosity and eventually leads to developed turbulence having the continuum of scales described earlier. We will assume that the turbulence has been given the time to develop a full range of scales throughout the remainder of this chapter.

In developed turbulence, it is thought that a similar cascade process is continuously going on in which small scales are produced inside large ones, yet smaller ones form inside the small scales, and so on. At each stage, some of the energy of the parent eddy goes into creating its smaller-scale children, so there is an associated flux of energy from large to small eddies that cascades down to the smallest scales where viscosity is important and the energy flux is dissipated. A large-scale eddy evolves on a time scale[3] $O(L/u')$, acting as a continuous source of smaller eddies that evolve more quickly. Evolution of the large eddies is the slowest step in the cascade and controls the rate at which energy is fed through to be dissipated. The dissipation rate adjusts itself, via the size of the smallest eddies, and hence the velocity derivatives appearing in (3.1), according to the energy flux from the large scales. An internal equilibrium is set up in which the *statistical* properties of the small scales evolve at the rate determined by the large ones, even though a particular small-scale eddy evolves more rapidly. If turbulence is allowed to decay without any input of turbulent energy, the large scales decay progressively in intensity on a time scale $O(L/u')$, while they produce weaker and weaker small-scale offspring. That is, the whole turbulent flow decays at the rate dictated by evolution of the large scales. On the other hand, in a statistically steady flow, instabilities of the mean flow constantly

[3] It is possible for turbulence to evolve more rapidly if mean-flow velocity gradients are large compared with u'/L, leading to rapid distortion of turbulence by the mean flow. In this chapter we implicitly assume mean-flow gradients are $O(u'/L)$ or less.

replenish the large scales of turbulence, to which they supply energy, which is passed via the cascade to the dissipative scales. Note that large scales are always present in turbulent flow, representing most of the turbulent kinetic energy of the flow and acting as a continuous source of smaller, shorter-lived eddies.

The existence of the energy cascade from large to small turbulent scales was first conjectured by Richardson (1922) and, although the above description without doubt contains the essential physics of the phenomenon, it is evidently qualitative in nature. Such qualitative ideas, supported by quantitative theories based on them,[4] have proven very useful in understanding turbulence, but turbulent flows have proven tenaciously resistant to theoretical analysis. Little progress has been made on deriving rigorous, general theories of turbulence from the Navier–Stokes equations, rather than based on ad hoc hypotheses, for reasons which will appear in the course of this book. However, a considerable amount is known about particular turbulent flows, and some of the characteristics of general ones, thanks to a combination of measurements, analysis, inspired heuristics, and, more recently, numerical simulation.

For the moment, let us consider the definition of the velocity correlation functions, which allow us to develop a more quantitative description of the different spatial length scales present inside turbulence.

3.1 Velocity Correlations and Spatial Scales

One of the simplest classes of turbulent flows, and one we will return to repeatedly in this book, is *homogeneous*, that is, having velocity fluctuations whose statistical properties that do not depend on position. Strictly homogeneous flows are not usually met in practice, where flows are often *steady* (statistical properties independent of time), but inhomogeneous. However, a good approximation to homogeneity is provided by grid turbulence, which has consequently received considerable attention. The reason for studying homogeneous turbulence is its relative simplicity, in the hope that at least some of the features of more realistic flows can be understood from the simpler, homogeneous case. This hope will be borne out in later chapters in which the spectral analysis of homogeneous turbulence without mean flow is developed, providing a quantitative underpinning for the concept of different scales. Homogeneous turbulence can also coexist with special mean flows having a constant velocity gradient, as we shall see in the next chapter.

Consider then a field of homogeneous turbulence, $\mathbf{u}(\mathbf{x}, t)$, and define the velocity correlations as

$$R_{ij}(\mathbf{x}, \mathbf{x}', t) = \overline{u_i(\mathbf{x}, t)u_j(\mathbf{x}', t)} \tag{3.5}$$

By homogeneity, R_{ij} should not be changed if we shift both $\mathbf{x}$ and $\mathbf{x}'$ by the same constant vector, which implies that R_{ij} is solely a function of the separation, $\mathbf{r} = \mathbf{x} - \mathbf{x}'$:

$$\overline{u_i(\mathbf{x}, t)u_j(\mathbf{x}', t)} = R_{ij}(\mathbf{r}, t) \tag{3.6}$$

[4] Primarily Kolmogorov's theory, which we sometimes refer to in this chapter, but wait to discuss in detail in Chapter 7, after we have developed spectral theory.

The velocity correlations, R_{ij}, form a 3×3 matrix of correlations between different velocity components. It is a tensor having the elementary property that

$$R_{ij}(-\mathbf{r}, t) = \overline{u_i(\mathbf{x}', t)u_j(\mathbf{x}, t)} = R_{ji}(\mathbf{r}, t) \tag{3.7}$$

that is, the matrix at $-\mathbf{r}$ is the transpose of that at $\mathbf{r}$. In particular, the diagonal components of R_{ij} are even functions of $\mathbf{r}$, for instance, $R_{11}(-\mathbf{r}, t) = R_{11}(\mathbf{r}, t)$. The trace, $R = R_{ii} = R_{11} + R_{22} + R_{33}$, of R_{ij} is, of course, a scalar function of $\mathbf{r}$:

$$R_{ii}(\mathbf{r}, t) = \overline{\mathbf{u}(\mathbf{x}, t) \,.\, \mathbf{u}(\mathbf{x}', t)} \tag{3.8}$$

The turbulent velocity fluctuations are generally found to have the property that the values at two points, $\mathbf{u}(\mathbf{x}, t)$ and $\mathbf{u}(\mathbf{x}', t)$, decorrelate as the separation between the points, $|\mathbf{x} - \mathbf{x}'|$, increases. That is,

$$R_{ij}(\mathbf{r}, t) = \overline{u_i(\mathbf{x}, t)u_j(\mathbf{x}', t)} \to 0 \tag{3.9}$$

as $|\mathbf{r}| \to \infty$. As the separation, $r = |\mathbf{r}|$, increases, the correlation functions, R_{ij}, go from

$$R_{ij}(0, t) = \overline{u_i u_j} \tag{3.10}$$

the one-point correlation matrix, which applies at $\mathbf{r} = 0$ and is symmetric, positive definite, through some functional form that is characteristic of the spatial structure of turbulence, to tend to zero at infinite separation. The order of magnitude of $|\mathbf{r}|$ over which $R_{ij}(\mathbf{r}, t)$ falls to zero is known as the correlation scale and represents the largest length scales present in the turbulence. Figure 3.2 illustrates typical behavior of one of the diagonal components of R_{ij}, showing various order of magnitude scales, including the correlation length, L. Since it is difficult to show the behavior of all nine components of R_{ij} as a function of the three-dimensional quantity, $\mathbf{r}$, we have chosen to sketch just one diagonal component, $R_{11}(r_1, r_2 = r_3 = 0)$, say, as a function of r_1. The order of separation over which decorrelation occurs is L, the correlation length. The reader should bear in mind that this figure gives a particular line section through a three-dimensional field: the behavior is similar along other directions in $\mathbf{r}$-space. One can imagine surfaces of constant R_{11} (or other components) in that space.

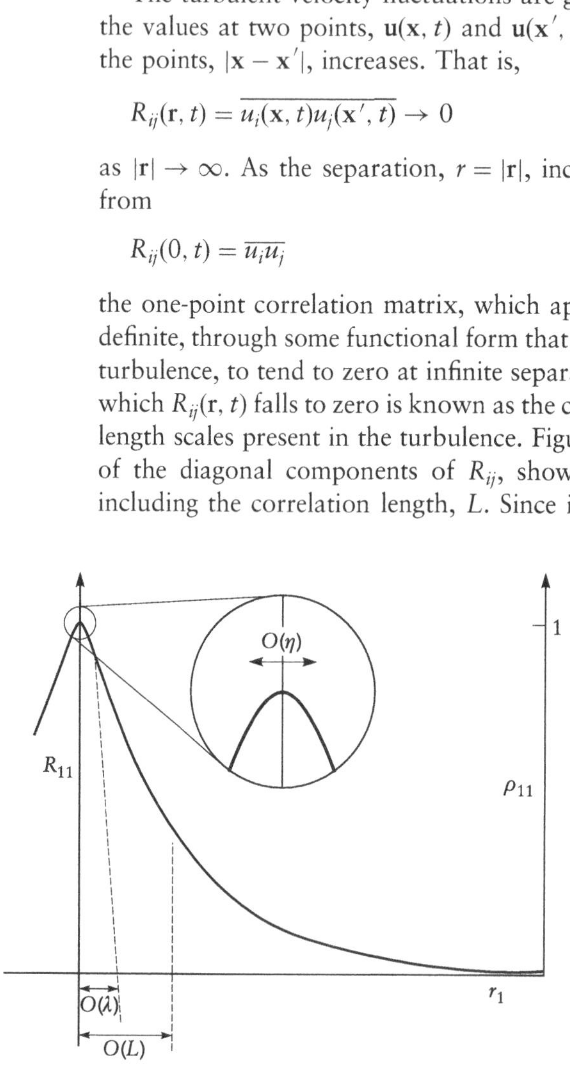

Figure 3.2. Sketch of the spatial velocity correlations as a function of $\mathbf{r} = (r_1, 0, 0)$, illustrating the correlation length, L, the microscale, λ, and the Kolmogorov scale, η. The dissipative range is shown magnified in the center of the figure and decreases in relative size as Re_L increases.

We can define a nondimensional version of the velocity correlations: the correlation coefficient of the random variables $u_i(\mathbf{x}, t)$ and $u_j(\mathbf{x}', t)$. The standard deviations, u_i' and u_j' (where $u_1' = \sqrt{\overline{u_1^2}}$, etc., do not depend on the spatial location at which the average is performed, by homogeneity) characterize the magnitude of turbulent velocity fluctuations in the x_i- and x_j-directions and all such components are typically of the same order of magnitude, $O(u')$. The correlation coefficients are

$$\rho_{ij}(\mathbf{r}, t) = \frac{R_{ij}(\mathbf{r}, t)}{u'_i u'_j} \tag{3.11}$$

not to be confused with fluid density, ρ. These coefficients always lie between -1 and $+1$, as we showed in Chapter 2. The diagonal elements of ρ_{ij} take on the limiting value $+1$ at $\mathbf{r} = 0$, representing complete correlation of $u_i(\mathbf{x}, t)$ with itself. The coefficients are simply a scaled version of the correlations, R_{ij}, and Figure 3.2 is equally illustrative of $\rho_{11}(r_1, r_2 = r_3 = 0)$. Evidently, the order of separation over which the correlation coefficients are significantly nonzero is L.

Since the ρ_{ij} are nondimensional, their integral with respect to any of the coordinates, r_1 say,

$$L_{ij}^{[1]} = \int_{-\infty}^{\infty} \rho_{ij}(r_1, r_2 = r_3 = 0) dr_1 \tag{3.12}$$

has the dimensions of length. Similar definitions can be given for $L_{ij}^{[2]}$ and $L_{ij}^{[3]}$ by integration over r_2 and r_3, respectively. The quantities $L_{ij}^{[k]}$ provide a quantitative measure of the correlation length scale and their definition in terms of integrals explains the terminology "integral scale" which is often used. The advantage of such scales is that they provide definite numbers, which can be measured and compared between different flows. However, the definiteness of these integral scales should not hide the fact that the correlation scale is not really a precise number, but an order of magnitude. Thus, for instance, which of the many $L_{ij}^{[k]}$ should one choose, why not integrate along lines in $\mathbf{r}$ other than the coordinate axes, and so on? When we use the terms correlation scale (or even integral scale) in this book, we will have an order of magnitude in mind, unless otherwise stated. All the $L_{ij}^{[k]}$ are supposed to be of the same order of magnitude, L, which characterizes the largest scales of turbulence.

More detailed examination of the ρ_{ij} as a function of $\mathbf{r}$ reveals that their behavior changes significantly in a region of $|\mathbf{r}|$ small compared with the overall scale for decorrelation, L. As apparent in the magnified zone of Figure 3.2, any of the diagonal elements of ρ_{ij}, ρ_{11} say, has zero derivative at the origin, $\mathbf{r} = 0$ (as follows from $\rho_{11}(\mathbf{r}) = \rho_{11}(-\mathbf{r})$). The graph curves downwards from $\rho_{11} = 1$ at $|\mathbf{r}| = 0$, as $|\mathbf{r}|$ increases (recall that $|\rho_{ij}| \leq 1$), but reaches a maximum gradient and then begins to curve in the other direction. This behavior occurs for $|\mathbf{r}|$ of order the Kolmogorov scale, η, defining the smallest scales of turbulence and is represented in the magnified part of the full correlation curve of Figure 3.2. The existence of this scale is due to (small) viscosity and the ratio η/L decreases[5] as the turbulent Reynolds number, $\mathrm{Re}_L = u'L/\nu$, increases. That is, the wide range of scales is asymptotic in large Re_L.

The mean-squared value of the difference of velocity at two points is the sum of components, such as

$$\overline{(u_1(\mathbf{x}, t) - u_1(\mathbf{x}', t))^2} = 2(u_1'^2 - R_{11}(\mathbf{r}, t)) = 2u_1'^2(1 - \rho_{11}(\mathbf{r}, t)) \tag{3.13}$$

[5] In fact, according to the Kolmogorov theory from Chapter 7, $\eta/L = O(\mathrm{Re}_L^{-3/4})$ at sufficiently large Re_L.

whose sum gives

$$(\Delta u)^2 = \overline{|\mathbf{u}(\mathbf{x}, t) - \mathbf{u}(\mathbf{x}', t)|^2} = 2\sum_{i=1}^{3} u_i'^2(1 - \rho_{ii}(\mathbf{r}, t)) \tag{3.14}$$

At $\mathbf{r} = 0$, $\rho_{11} = \rho_{22} = \rho_{33} = 1$ and $\Delta u = 0$, as of course it must be, since there is no difference of velocity at zero separation. When $|\mathbf{r}| = O(L)$, the correlations, ρ_{ii}, have decreased significantly from the value 1 (recall that L was defined via decorrelation) and so $\Delta u = O(u')$. Thus, the typical difference of velocity between two points separated by $O(L)$ is comparable with the overall turbulent velocity fluctuations. As $|\mathbf{r}| \to \infty$, $\rho_{ij} \to 0$ and $\Delta u \to 6^{\frac{1}{2}}u'$. Between $|\mathbf{r}| = 0$ and $|\mathbf{r}| = O(L)$, the ρ_{ii} decrease continuously from 1, as shown in Figure 3.2; thus Δu increases continuously from 0 up to $O(u')$ as $|\mathbf{r}|$ increases from 0.

For very small separations, $r = |\mathbf{r}| \ll \eta$, the correlation functions can be expanded as a Taylor's series about $\mathbf{r} = 0$, leading to the dashed parabola shown in Figure 3.2. For such very small r, $1 - \rho_{ii}$ is proportional to r^2 and hence Δu is proportional to r, according to (3.14). This linear dependence of the velocity difference on r is a consequence of the fact that the velocity field is smooth when looked at on scales of $O(\eta)$ or smaller. Thus, when $r \ll \eta$, the velocity field itself can be expanded as a Taylor's series in $\mathbf{r}$, giving leading-order proportionality of the velocity difference on separation. We see that the behavior of the correlation functions at $r = O(\eta)$, illustrated in the magnified part of Figure 3.2, reflects the change from an apparently furry velocity field at scales larger than η, to a smooth velocity field at η and below.

At scales intermediate between the smallest, η, and largest, L, the velocity field appears furry and, although the difference of velocities, Δu, decreases with separation, it does so more slowly than would be the case if it were a smooth function at these scales. In other words, $\Delta u/r$ increases with decreasing r. Looked at on these and larger scales, the velocity field appears to be a continuous, but nondifferentiable function of position (and, indeed, time). Of course, the velocity is really differentiable, but one needs to consider scales η and below for that to become apparent. The magnitude of the fur on a graph of velocity is measured by Δu.

According to Kolmogorov's theory of the cascade (described in Chapter 7), as Re_L is increased, an interval of r appears in $\eta \ll r \ll L$, within which $\Delta u \propto r^{1/3}$. This interval is known as the inertial range of separations; it appears at sufficiently large Re_L and grows in width as Re_L is further increased. Turbulence can exist at lower (but still large) Reynolds numbers without an inertial range, but the $\mathrm{Re}_L \to \infty$ asymptotic behavior only becomes evident with the appearance of the inertial range. The correlation functions and velocity differences then show asymptotic structure consisting of three ranges of separation: dissipative scales, $r = O(\eta)$, intermediate inertial-range scales, for which power laws apply, and large, energy-containing scales, $r = O(L)$.

We shall use the notation Δu_r to indicate a typical velocity difference at separation r. Since each of the terms in the sum of (3.14) has the same order of magnitude, we have $1 - \rho_{ii} = O((\Delta u_r/u')^2)$, linking the correlation coefficients and velocity difference at separation r. According to Kolmogorov's inertial range theory, $1 - \rho_{ii}$ is proportional to $r^{2/3}$ and Δu_r is proportional to $r^{1/3}$. This dependency of $1 - \rho_{ii}$ on

r leads to an infinite gradient at $r = 0$ for graphs such as Figure 3.2, singular behavior which is resolved at the Kolmogorov scale.

For future reference, we summarize the behavior of Δu_r and ρ_{ii} as a function of r at large Reynolds number and $\eta \ll r \ll L$ as follows:

Quantity	*General Behavior*	*Inertial Range*
$\dfrac{\Delta u_r}{u'}$	increases with r	$\sim \left(\dfrac{r}{L}\right)^{1/3}$
$\dfrac{\Delta u_r}{r}\dfrac{L}{u'}$	decreases with increasing r	$\sim \left(\dfrac{L}{r}\right)^{2/3}$
$1 - \rho_{ii}$	increases with r	$\sim \left(\dfrac{r}{L}\right)^{2/3}$

All the above quantities are $O(1)$ for the large scales ($r = O(L)$) and either small (for the first and third quantities) or large (for the second quantity) at smaller scales. We stress again that the existence of an inertial range within $\eta \ll r \ll L$, and hence the applicability of the third column in this table, requires a sufficiently large value of Re_L.

The second derivatives of R_{ij} at $\mathbf{r} = 0$ are directly related to the mean-squared velocity gradients. Thus, taking the derivative of

$$\overline{u_i(\mathbf{x}, t)u_j(\mathbf{x}', t)} = R_{ij}(\mathbf{x} - \mathbf{x}', t) \tag{3.15}$$

with respect to x_k and x_l' and evaluating at $\mathbf{x} = \mathbf{x}'$ yields

$$\overline{\frac{\partial u_i}{\partial x_k}\frac{\partial u_j}{\partial x_l}} = -\left.\frac{\partial^2 R_{ij}}{\partial r_k \partial r_l}\right|_{\mathbf{r}=0} \tag{3.16}$$

which gives, for instance,

$$\overline{\left(\frac{\partial u_1}{\partial x_1}\right)^2} = -u_1'^2 \left.\frac{\partial^2 \rho_{11}}{\partial r_1^2}\right|_{\mathbf{r}=0} \tag{3.17}$$

Referring to Figure 3.2, it is apparent that this second derivative is to be taken in the region of the Kolmogorov scale, reflecting the importance of the smallest scales in determining the mean-squared velocity derivatives, which increase as the viscosity, and hence η, decreases.

For $|\mathbf{r}| \ll \eta$, we can expand any of the diagonal elements of $\rho_{ij}(\mathbf{r}, t)$ as a Taylor's series in $\mathbf{r}$. For instance:

$$\rho_{11}(r_1, r_2 = r_3 = 0) = 1 - \frac{r_1^2}{\lambda^2} + \cdots \tag{3.18}$$

so that

$$\overline{\left(\frac{\partial u_1}{\partial x_1}\right)^2} = \frac{2u_1'^2}{\lambda^2} \tag{3.19}$$

according to (3.17). The quantity λ has the dimensions of length and is referred to as a Taylor microscale. From (3.19), it appears that the mean-squared velocity derivatives can be written using the overall turbulent fluctuating velocity and the microscale. Of course, since there are nine different velocity derivatives, there are generally a corresponding number of different possible microscales and, like L and η, it is usually better to think of microscales as orders of magnitude. From this point of view, the Taylor microscale is defined by the order of magnitude of the velocity derivatives:

$$\frac{\partial u_i}{\partial x_j} = O\left(\frac{u'}{\lambda}\right) \tag{3.20}$$

or from the energy dissipation as

$$\varepsilon = O\left(\frac{\nu u'^2}{\lambda^2}\right) \tag{3.21}$$

where the turbulent part of the energy dissipation rate, ε, is defined by (3.1) with the total velocity, $\mathbf{U}$, replaced by the fluctuating velocity, $\mathbf{u}$. In fact, since velocity derivatives are, in any case, dominated by the fluctuating component, ε equals Δ to a good approximation at the high Reynolds numbers we have in mind here. Thus, for present purposes it is unnecessary to make the distinction between the two and the reader can think of either one as representing viscous energy dissipation by the smallest scales of the flow. We will return to a detailed discussion of turbulence energetics in Chapter 4, from which it will appear that the mean value of ε is an important quantity, representing the average dissipation of turbulent kinetic energy.

As their name suggests, the microscales are typically small compared with the correlation length, L, defining the overall length scale of turbulence. However, it is important to recognize that they are *not* of the same order as the smallest scale, η. The velocity field is smooth at the Kolmogorov scale, η, and so we can estimate the velocity derivatives as

$$\frac{\partial u_i}{\partial x_j} = O\left(\frac{\Delta u_\eta}{\eta}\right) \tag{3.22}$$

which can be compared with (3.20), leading to

$$\frac{\lambda}{\eta} = O\left(\frac{u'}{\Delta u_\eta}\right) \gg 1 \tag{3.23}$$

which is large, since the velocity differences, Δu_η, at separation η, much less than L, are small compared with those, $\Delta u_L = O(u')$, at separation L. It follows that the microscale, λ, is large compared with η. One can also rewrite the estimate, (3.23), as

$$\frac{\lambda}{L} = O\left(\frac{\Delta u_L}{L} \Big/ \frac{\Delta u_\eta}{\eta}\right) \ll 1 \tag{3.24}$$

which is small since, as noted previously, $\Delta u_r / r$ is a decreasing function of r. Thus, $\eta \ll \lambda \ll L$, showing that the Taylor microscales are intermediate in order of magnitude between η and L and lie in the inertial range if Re_L is large enough. If the Kolmogorov inertial range power law

$$\Delta u_r = O\left(u'\left(\frac{r}{L}\right)^{1/3}\right) \tag{3.25}$$

is applied at the extreme limits, $r = O(\eta)$ and $r = O(L)$, we find

$$\lambda = O\left(\eta^{2/3}L^{1/3}\right) \tag{3.26}$$

according to (3.24).

Whereas there are definite physical processes associated with scales η and L, it is unclear what, if any, physical significance should be ascribed to λ. The definition, (3.20), in terms of velocity derivatives determined at the smallest scales, and u', which is an overall velocity scale associated with the large scales, makes λ an intermediate quantity, as is also apparent from the mix of η and L in (3.26). It is because λ contains quantities appropriate to both scales η and L that it is itself intermediate as $\mathrm{Re}_L \to \infty$. The microscales often appear in analyses of turbulence, usually to help estimate the energy dissipation via equations such as (3.21). A geometrical interpretation of the microscale is shown in Figure 3.2 and based on the Taylor's expansion, (3.18). A parabola is fitted to the correlation function at $r_1 = 0$ and intersects the r_1 axis at $r_1 = \lambda$.

As we noted earlier, the root-mean-squared fluctuating velocity, u', is mainly determined by the large scales of the velocity field, being insensitive to the small-scale fur apparent in Figure 3.1a. On the other hand, the root-mean-squared difference, Δu_r, is dominated by the fur at scales of order r and characterizes such scales of the flow. In other words, if one imagines the turbulence as a continuum of different spatial scales produced by the cascade, the quantity Δu_r gives the velocity scale associated with those of size r and is small compared with u', when $r \ll L$. It is believed that the *differences* of velocity, Δu_r, together with r itself, *characterize the dynamics of turbulence* at scales of order r. For instance, one can define a Reynolds number at scale r, based on Δu_r, as $r\Delta u_r/\nu$. This Reynolds number is a decreasing function of r, large of $O(\mathrm{Re}_L)$ at scale L and proportional to $r^{4/3}$ in the inertial range, according to the Kolmogorov theory. It falls to $O(1)$ at the Kolmogorov scale, $r = O(\eta)$, which is another way of saying that viscosity is important there.

In the course of describing the behavior of the correlation function, we have been obliged to illustrate three-dimensional behavior (as a function of $\mathbf{r}$) by plotting graphs as a function of just one spatial dimension. In the case of *isotropic* turbulence, statistical properties of u_i are independent of direction and reflection in any plane. As we shall see in Chapter 6, this allows us to write all the correlation components, $R_{ij}(\mathbf{r})$, in terms of just two scalar functions of $r = |\mathbf{r}|$. For the moment, however, we simply note that the *one-point* correlations have the form

$$\overline{u_i u_j} = u'^2 \delta_{ij} \tag{3.27}$$

when the turbulence is isotropic. This is a consequence of a general mathematical result that a tensor (such as $\overline{u_i u_j}$), whose components are unchanged under arbitrary rotation of axes, is a multiple of δ_{ij}. We shall use symmetry considerations similar to these in the next chapter, but we have no real need to assume isotropy here.

Although we have assumed homogeneity to make the presentation of correlation properties more definite, many of the concepts explored above do not depend on this assumption in any essential way. For instance, the correlation scale can be defined as the separation required for significant turbulent velocity decorrelation, and represents the large scales of inhomogeneous turbulence. Furthermore, it is observed that the properties of turbulence become more like the homogeneous case the smaller the scale considered. That is, even if it is inhomogeneous at the largest scales, smaller separations usually give approximately homogeneous results. It is as if the process of generation of small scales, described earlier, forgets the inhomogeneities present at the large scales, apart from large-scale nonuniformities of the rate of turbulent energy supply. This idea of a cascade to small scales, coupled with loss of memory of the specifics of the particular large-scale turbulent flow, forms one of the main elements of Kolmogorov's theory to be described in Chapter 7.

Notice that the correlation length, although obviously of the same order as, or smaller than, the total size of the region of turbulence, may in fact be much smaller. For instance, grid turbulence, which is a good approximation to homogeneity, has a correlation length which scales on the grid spacing, but occupies a much larger region overall. Strictly homogeneous turbulence is the extreme case: it occupies all space, but has a finite correlation length.

3.2 Temporal Correlations and Time Scales

Velocity correlations can be defined at a single fixed point and different times:

$$R_{ij}^{(t)}(\mathbf{x}, t, t') = \overline{u_i(\mathbf{x}, t)u_j(\mathbf{x}, t')} \tag{3.28}$$

Decorrelation takes place as the delay $|t - t'| \to \infty$, with the turbulence at widely differing times generally behaving as if it were taken from independent realizations of the flow, as for widely separated points in space.

Steady flows are the temporal analog of homogeneous turbulence. If the flow is steady, the correlations become functions of the delay, $\tau = t - t'$:

$$\overline{u_i(\mathbf{x}, t)u_j(\mathbf{x}, t')} = R_{ij}^{(t)}(\mathbf{x}, \tau) \tag{3.29}$$

and one finds that

$$R_{ij}^{(t)}(\mathbf{x}, -\tau) = \overline{u_i(\mathbf{x}, t')u_j(\mathbf{x}, t)} = R_{ji}^{(t)}(\mathbf{x}, \tau) \tag{3.30}$$

so that the diagonal elements of $R_{ij}^{(t)}$ are even functions of τ. The correlations at zero delay are, of course, just the one-point velocity moments

$$R_{ij}^{(t)}(\mathbf{x}, \tau = 0) = \overline{u_i u_j} \tag{3.31}$$

and coincide with the spatial correlations at zero separation. Correlation coefficients can again be defined by normalization as

$$\rho_{ij}^{(t)}(\mathbf{x}, \tau) = \frac{R_{ij}^{(t)}(\mathbf{x}, \tau)}{u_i' u_j'} \tag{3.32}$$

and lie between -1 and $+1$, with the diagonal components being 1 at $\tau = 0$.

A plot of any of the diagonal terms of $\rho_{ij}^{(t)}(\mathbf{x}, \tau)$ as a function of τ is very similar to Figure 3.2. Overall, decorrelation takes place on a correlation time scale, Θ, while the smallest time scales, at which the velocity field becomes a smooth function of time, appear as a distinct region of small time delays. One can define integral correlation times as, for instance,

$$\Theta_{ij} = \int_{-\infty}^{\infty} \rho_{ij}^{(t)}(\mathbf{x}, \tau)\, d\tau \tag{3.33}$$

which vary with position in the flow unless it is homogeneous as well as steady.

All in all, a discussion of temporal correlations and time scales for steady flows closely parallels that of spatial ones given earlier. Time scales and correlations such as these, which are defined at a *fixed* point in space are known as *Eulerian* and, in many cases, reflect the spatial structure of turbulence being convected past the given point by the flow, rather than time evolution of the turbulence itself. Consider, for instance, the case of turbulence generated by uniform flow passing through a grid of wires.

The turbulence produced downstream of the grid exists in an approximately uniform mean flow of velocity $\overline{U}$ and has a fluctuating velocity, u', small compared with $\overline{U}$. The uniform flow serves merely to convect the turbulence, whose large spatial scales are characterized by u' and a correlation length L. From these scales, one can construct a time scale, L/u', which describes the evolution of the large scales of turbulence. However, spatial scales of size L are convected past a fixed point in a time of order $L/\overline{U}$, and since $u' \ll \overline{U}$, this is much shorter than the time for evolution. Thus, to a first approximation, the turbulence can be thought of as simply convected at constant speed $\overline{U}$ *without evolution.* This approximation is often referred to as the Taylor hypothesis and the turbulence is said to be "frozen." A fixed, point sensor in the flow samples the velocity field along the line that is convected through the sensor at velocity $\overline{U}$. Time correlations at delay τ are then directly related to spatial correlations at separation $\overline{U}\tau$ in the direction of the mean flow. In particular, the correlation time is related to the spatial correlation scale by $\Theta = L/\overline{U}$. In this example, convection by the mean flow causes nearly frozen spatial structure of the turbulence to appear as time variations at a fixed point. This is typical of flows with significant mean velocity.

It should be made clear that Eulerian time scales depend on the frame of reference used. If, for instance, one examines the grid-generated turbulence in a frame of reference moving with the velocity $\overline{U}$, time variations no longer include the effect of convection at speed $\overline{U}$. Time scales are then considerably longer than in the frame of reference of the grid, being those for dynamics of the turbulence, $O(L/u')$. Of course, the flow is no longer steady when viewed in the new frame of reference: turbulence decays as it is convected downstream away from the grid, which is the source of the turbulence.

Even in the absence of significant mean flow, the small scales of turbulence (sizes small compared with L) are convected by the large scales, leading to Eulerian time scales short compared with their evolution times. The velocity differences at scale ℓ are $O(\Delta u_\ell)$ giving a time scale for evolution $O(\ell/\Delta u_\ell)$, since, as remarked earlier, Δu_ℓ is believed to characterize the turbulence dynamics at scale ℓ. That is, *turbulent scales of size ℓ are thought to evolve on the intrinsic time scale $\ell/\Delta u_\ell$*, which is often referred to as the eddy lifetime at scale ℓ. As we have seen, $\ell/\Delta u_\ell$ decreases with decreasing ℓ, leading to the conclusion that *smaller eddies have shorter lifetimes.* However, to see the time evolution of the small scales, rather than simply the effects of convection by the large ones or by the mean flow, one needs to move with the fluid. Choosing an infinitesimal fluid particle, one may follow it by moving at the fluid velocity **U** and examine a region of size ℓ centered on the particle. Time evolution in that region takes place on the time scale, $O(\ell/\Delta u_\ell)$, the intrinsic scale for turbulence dynamics at scale ℓ, but a Lagrangian description of the flow has been introduced by the requirement that one move with the chosen particle.

The large scales of turbulence produce convection velocities of $O(u')$. Thus, in the absence of significant mean flow, an eddy of size ℓ takes a time $O(\ell/u')$ to be swept past a fixed point. This time is short compared with the eddy lifetime, $O(\ell/\Delta u_\ell)$, by a factor of $O(u_\ell/u')$, which is small when $\ell \ll L$. That is, small scales do not have the time to evolve significantly in the time taken for convection past a fixed point. It follows that short-scale Eulerian time variations are mainly due to convection of nearly frozen small-scale spatial structures past a fixed point, as for the case with significant mean flow. However, it is the large eddies, rather than the mean flow, that now provide the convective sweeping.

In general, there is convection by both the mean flow and by the overall turbulent velocity fluctuations. If $\overline{U} \gg u'$, as is often the case, the mean flow wins and the time variations at a fixed point are mainly due to mean convection, even at the largest scales. On the other hand, when $\overline{U}$ is comparable to or small compared with u', short time-scale fluctuations result from convection of small spatial structures by larger scales, but the time scales for convection and evolution of the largest scales are both $O(L/u')$. Writing $u_c = \max(|\overline{U}|, u')$ as the dominant convection velocity, mean or turbulent, the Eulerian time scale associated with size ℓ, is $\tau = O(\ell/u_c)$. Thus, the largest and smallest Eulerian scales are $\Theta = O(L/u_c)$ and $O(\eta/u_c)$.

The smaller the spatial structure considered, the shorter the Eulerian time scale, with $\tau = O(\ell/u_c)$ giving rough proportionality between corresponding time and space scales. Within an inertial range, the velocity differences at scale ℓ are proportional to $\ell^{1/3}$ from Kolmogorov's theory and hence to $\tau^{1/3}$ according to the relationship $\tau = O(\ell/u_c)$. The analog of (3.14) for differing times, rather than points in space, indicates that $1 - \rho_{11}^{(t)}(\mathbf{x}, \tau) = O((\Delta u/u')^2)$, which is proportional to $\tau^{2/3}$ in Kolmogorov's theory. Thus, at time delays corresponding to $\eta \ll r \ll L$, the behavior of $\rho_{ij}^{(t)}(\mathbf{x}, \tau)$ as a function of τ is similar to that of $\rho_{ij}(\mathbf{r}, t)$ as a function of $|\mathbf{r}|$. Temporal behavior at a fixed point directly reflects spatial structure.

The intrinsic time scale, $O(\ell/\Delta u_\ell)$, characterizes evolution of eddies of size ℓ in particular realizations of the flow. In contrast, small-scale *averaged* properties of the flow, such as Δu_ℓ for $\ell \ll L$, do not evolve on this time scale in developed turbulence. Instead they may only change on the time scale, $O(L/u')$, for evolution of the large scales, while the case of a steady flow shows that such statistics may not evolve at all. The entire continuum of scales is in *statistical* equilibrium with the large scales once the turbulence has fully developed, while the small eddies in any particular realization evolve faster than this.

The mean kinetic energy contained in the largest scales is $O(u'^2)$ per unit mass and is transferred to the smaller scales in a time $O(L/u')$, thus giving an average energy flux $O(u'^3/L)$ per unit mass and time. The smallest scales rapidly adapt to dissipate this energy flux, requiring a mean dissipation rate $\overline{\varepsilon} = O(u'^3/L)$. We can combine this estimate with (3.21) to obtain

$$\frac{\lambda}{L} = O(\mathrm{Re}_L^{-1/2}) \tag{3.34}$$

again showing that $\lambda \ll L$ when Re_L is large. The result, $\bar{\varepsilon} = O(u'^3/L)$, which we obtained in the process of deriving (3.34) is often useful when one wants an estimate of the rate of energy dissipation of turbulence, as is the time scale, $O(L/u')$, for its decay in the absence of energy input.

In the literature on turbulence, one often sees a Reynolds number, $\mathrm{Re}_\lambda = u'\lambda/\nu$, constructed from λ and u', used instead of Re_L. According to (3.34), we have

$$\mathrm{Re}_\lambda = O(\mathrm{Re}_L^{1/2}) \tag{3.35}$$

showing that Re_λ is large, but not so large as Re_L. Use of Re_λ carries the same caveats as λ itself: it involves a mix of small- and large-scale quantities. On the other hand, it is relatively straightforward to translate between Re_λ and Re_L using expressions such as (3.35),[6] and Re_L *does* have a well-defined physical interpretation as a measure of the lack of importance of viscosity at the largest scales.

In the absence of an energy supply to the largest scales, it might be thought that the loss of energy of those scales to form smaller ones would cause the overall length scales of the turbulence to become shorter, thus making the correlation scale, L, decrease with time, but this is not the case. Amongst the largest-scale eddies, the larger ones live longer and it is these that determine the large-scale structure of later turbulence. Thus, in fact, the correlation length tends to increase as the turbulence decays. Of course, in a steady flow, the continuous supply of new eddies at the largest scale keeps the correlation scale constant, like all other statistical quantities.

As we saw earlier, temporal variations of turbulent velocity at a fixed point are often due to convection of nearly frozen spatial structures and not to the intrinsic time evolution of turbulent eddies. It is thus natural to look for a means by which the evolution itself might be observed. The effects of convection are due to flow and, if we imagine a point sensor that moves with the flow, such convective effects would be eliminated. This is one of the ideas behind *Lagrangian* descriptions of turbulent flow, which consider the time history following particles of fluid in their motion. Of course, it is experimentally well nigh impossible to take point measurements following the flow and so, whereas Eulerian fixed-point measurements are routinely made, Lagrangian time histories are inferred and often play the role of thought experiments, rather than real ones.

We do not want to enter into the detailed definition of Lagrangian velocity correlations here, since the technical points that raises are outside the scope of an introductory chapter. However, briefly put, the basic idea is that decorrelation of velocity following an infinitesimal fluid particle as it moves with the flow yields a Lagrangian correlation time scale that measures the evolution time of the large scales of turbulence and can be considerably longer than its Eulerian counterpart if the mean flow dominates. The Lagrangian correlation time may also be thought of as the order of magnitude of duration of statistical memory of the flow, which forgets its past behavior at longer temporal separations.

[6] Though note that there are $O(1)$ numerical factors hiding behind estimates such as (3.35).

3.3 Conclusions

In this chapter, we have attempted to introduce the reader to some important, fundamental properties of turbulence: the plethora of time and space scales that are characteristic of turbulent flows and the underlying physical mechanisms that produce them. The material given here forms a qualitative background for future chapters, which build and expand on the discussions given here.

From an experimental point of view, the existence of space and time scales down to the Kolmogorov scale means that one must use probes having a sufficiently high resolution and fast time response if one wants to fully capture a high-Reynolds-number turbulent flow. Depending upon the particular flow, this can place rather exacting requirements on the experimenter.

Reference

Richardson, L. F., 1922. *Weather Prediction by Numerical Process.* Cambridge University Press, Cambridge.

CHAPTER FOUR

Basic Theory and Illustrative Examples

Statistical averaging methods for turbulent flows were introduced by Reynolds (1894) over a century ago and remain crucial in the theory of turbulence. Such techniques may be used to study mean values involving any number of points in space and time; however, the simplest are the single-point, single-time averages. By considering one-point statistics, information about the flow is reduced to what might be presumed its essentials (e.g., mean velocities, root-mean-squared fluctuations) and the flow description is consequently simplified. However, one-point quantities do not encompass the full statistics of the flow, since multipoint, multitime averages cannot be deduced from one-point data. Statistical information is thus lost in going to one-point averages, information that may be crucial to an understanding of certain aspects of the flow. Even so, one-point averages include many of the more important physical measures of turbulent flow, for instance, the mean-flow velocity, $\overline{U_i}$, and turbulent kinetic energy per unit mass, $\overline{u_i u_i}/2$. The main aim of this chapter is to provide an introduction to one-point methods and to give the reader a feel for their use via simple, but representative examples. Flows cannot be fully described using such an approach (for instance, one-point methods do not capture the contributions of different scales to the overall energetics of turbulence, unlike the spectral analysis of later chapters) and we make no pretence at completeness here. Nonetheless, one-point techniques are fundamental to an understanding of the overall properties of turbulent flows, for example their energetics.

The simplest one-point averages are $\overline{U_i}$ and $\overline{P}$, where U_i and P are the velocity and pressure of the fluid. These quantities define a notional flow, known as the mean flow, the departures from which are called fluctuations and are generally identified with the turbulence. Thus, the statistical splitting consists of

$$U_i = \overline{U_i} + u_i \tag{4.1}$$

and

$$P = \overline{P} + p \tag{4.2}$$

where u_i and p are the fluctuating velocity and pressure respectively. Note that, since we are assuming an incompressible fluid throughout this book, there is no need to introduce splitting of other quantities, such as the density, as we would need to for a compressible fluid. The mean values $\overline{U_i}$ and $\overline{P}$ in (4.1) and (4.2) are ensemble averages, that is, a number of separate, independent experiments are imagined as being carried out with the same nominal flow conditions, and the results averaged.

For a nominally steady flow one can also usefully define time averages, which usually yield the same results as ensemble averages, as explained in Chapter 2.

By averaging the Navier–Stokes equations, one obtains those for the mean flow. As we shall see, the mean-flow equations contain what are called Reynolds stress terms involving second-order moments of the fluctuating velocity field (recall from Chapter 2 that moments are averages of products of random variables, their order being the number of terms in the product). Thus, the mean flow is coupled to the fluctuations by these terms, that is, it cannot, in general, be calculated independently of the turbulence.[1] The Reynolds stress terms arise from averaging the nonlinear convective term in the Navier–Stokes equation and are not truly part of the fluid stress, but rather represent the average momentum flux due to the turbulent velocity fluctuations. However, they appear formally like stress terms in the equations for the mean flow. The fact that the mean-flow equations are not closed in general, because they involve the fluctuations via the Reynolds stress terms, reflects incompleteness of the mean flow as a description of the one-point statistics, which, as noted above, do not encompass the full, multipoint, multitime statistics of the flow. Thus, the mean flow does not contain anything like full statistical information on the turbulent flow and the mean-flow evolution equations involve some lost one-point information, in the form of second-order moments of the fluctuating velocity.

In an attempt to complete the mean-flow equations, one may derive evolution equations for the second-order moments of fluctuating velocities. However, it turns out that, owing to the quadratic convective term in the Navier–Stokes equation, these equations contain third-order velocity moments, and so on for the evolution equations at all orders. The evolution equations for velocity moments of order n involves moments of order $n+1$, and no finite set of moment equations is closed. This closure problem is the main obstacle to the development of a rigorous theory of turbulence using averaging methods, and no fully satisfactory solution is known. The problem can be dealt with heuristically by completing the system of moment equations up to order n by assuming additional semiempirical relations for moments of order $n+1$ in terms of the lower-order ones. No rigorous justification has been found for any of the numerous closure assumptions which have been proposed and, although this type of approximate model may work quite well within a restricted class of flows for which it was intended and for which it has been parameterized, from a fundamental point of view, it is not a very satisfactory solution. However, until (and if) some better one is found, or computers grow powerful enough that reliable numerical simulation of the Navier–Stokes equations can be carried out for realistic Reynolds numbers and flow geometries, such closure models will no doubt be used when quantitative predictions of particular turbulent flows are required.

There is another problem associated with the use of the moment evolution equations, perhaps of less fundamental significance, but nonetheless creating technical difficulties. The evolution equations for second- and higher-order velocity moments at a single point involve two-point moments, those at two points introduce three-point moments, and so on. This is often referred to as nonlocality and means that

[1] An exception occurs for homogeneous turbulence, which leads to Reynolds stress terms in the mean-flow equations which are zero, as we will see later.

one cannot, for instance, restrict attention to single-point moments when formulating evolution equations for the velocity moment at orders above the first. Nonlocality is due to the fact that the Navier–Stokes equation contains pressure gradient and viscous terms, of which the former gives rise to pressure–velocity correlations when one formulates equations for velocity moments above the first order, while the latter produces a milder form of nonlocality that will be discussed later. The pressure can be expressed in terms of the velocity field, as we shall see in Section 4.3, but the pressure at a single point involves an integral over the entire flow field. Thus, when one derives evolution equations for the moments of velocities at a single point in space, they contain *spatial integrals* of two-point velocity moments via the velocity–pressure correlations. If one considers n-point velocity moments of order n, the evolution equations for the moments of all orders constitute an infinite hierarchy of integro-differential equations,[2] which appears to be thoroughly intractable unless closure hypotheses are introduced. The fact that this system is inextricably coupled leads to the closure problem, whereas the integrals appearing in the moment equations reflect nonlocality.

Of itself, nonlocality does not lead to any problems of principle and, once a closure approximation has been adopted, the resulting truncated set of moment evolution equations, involving n-point velocity moments up to some finite value of n, although perhaps technically demanding to solve, is at least complete. Nonetheless, when one restricts attention to one-point averages, as many routine turbulence prediction schemes do, quantities, such as pressure–velocity correlations, occur in the one-point equations that, in principle, require two-point velocity statistics for their evaluation. There is thus considerable practical interest in heuristic approximations of such quantities in terms of one-point ones, approximations that are often also referred to as closure hypotheses (or models), since they serve to close a set of one-point equations. The reader should nonetheless keep in mind the distinction between the fundamental closure problem, which is due to nonlinearity of the Navier–Stokes equations, leading to the appearance of moments of order $n + 1$ in the equations for those of order n, and incompleteness of the one-point equations as a result of nonlocality, which can be resolved, in principle, by considering multipoint moments. Fourier (or spectral) analysis, introduced in Chapter 6, allows nonlocality to be handled in a relatively straightforward way for homogeneous turbulence, but does not resolve the underlying closure problem due to nonlinearity of the Navier–Stokes equations.

Although they are fundamentally less satisfactory, because they require additional assumptions to make good the lack of two-point information, one-point turbulence prediction models are simpler, requiring less computational effort than multipoint ones. For this reason such methods are extensively used in industrial applications, where, due to the complex architecture of, for instance, piston engines, compressors, and turbines, practical multipoint models have yet to be developed. We shall describe

[2] There is another approach to the analysis of turbulence, describing the time evolution of all single-time statistics of the velocity field by a single linear equation, known as the Hopf equation, which is, however, a *functional* differential equation, i.e., it involves derivatives taken in a space of functions. Unfortunately, finding useful solutions of the Hopf equation is very difficult and little real progress seems to have been made using this approach. The interested reader is referred to Monin and Yaglom (G 1975), chapter 29, but should be prepared for some mathematical brain stretching.

one of the simplest and most commonly used one-point closure models, the k–ε model, in Chapter 8.

In this chapter, we shall develop the basic theory of turbulence using a physical space, rather than a spectral, formulation, mainly focusing on low-order, one-point moments. The aim is to provide physical interpretations and, via specific examples, to show the reader how one may extract useful information. Thus, the primary goal is not turbulence modeling, although closure and nonlocality problems are necessarily not far from the surface and, along the way, we will introduce some modeling ideas.

4.1 The Mean Flow

The motion in any realization of the turbulent flow obeys the incompressible Navier–Stokes equation

$$\frac{\partial U_i}{\partial t} + \underbrace{U_j \frac{\partial U_i}{\partial x_j}}_{\text{Convection}} = \underbrace{-\frac{1}{\rho}\frac{\partial P}{\partial x_i}}_{\text{Pressure gradient}} + \underbrace{\nu \frac{\partial^2 U_i}{\partial x_j \partial x_j}}_{\text{Viscosity}} \tag{4.3}$$

where ν is the kinematic viscosity, and the incompressibility condition

$$\frac{\partial U_i}{\partial x_i} = 0 \tag{4.4}$$

A turbulent flow is thus considered as an ensemble of different solutions of the Navier–Stokes equations, (4.3), (4.4).

Since (4.4) is linear, its average is closed and we have

$$\frac{\partial \overline{U_i}}{\partial x_i} = 0 \tag{4.5}$$

for the mean flow. Here and elsewhere, it should be recalled that ensemble averaging commutes with space and time derivatives (see Chapter 2). We will frequently change the order of partial differentiation and averaging without drawing special attention to the fact. Subtracting (4.5) from (4.4) yields the continuity equation for the fluctuations

$$\frac{\partial u_i}{\partial x_i} = 0 \tag{4.6}$$

Thus both the mean and fluctuating velocities satisfy the incompressible continuity equation as if they were independent flows.

This decoupling does not extend to the nonlinear Navier–Stokes equation, (4.3), which yields

$$\frac{\partial \overline{U_i}}{\partial t} + \overline{U_j}\frac{\partial \overline{U_i}}{\partial x_j} + \overline{u_j \frac{\partial u_i}{\partial x_j}} = -\frac{1}{\rho}\frac{\partial \overline{P}}{\partial x_i} + \nu \frac{\partial^2 \overline{U_i}}{\partial x_j \partial x_j} \tag{4.7}$$

where, in the convective term, we have used (4.1) and the fact that the average of the product of a fluctuating and an averaged quantity is zero. This equation contains both the mean and fluctuating velocities and is not closed. Because we wish to regard (4.7) as the equation for the mean flow, we prefer to shift the term

$$\overline{u_j \frac{\partial u_i}{\partial x_j}} = \frac{\partial}{\partial x_j}\left(\overline{u_i u_j}\right) \tag{4.8}$$

in which we have used (4.6), to the right-hand side of the equation which becomes the mean-flow equation

$$\begin{aligned}\frac{\partial \overline{U}_i}{\partial t} + \overline{U}_j \frac{\partial \overline{U}_i}{\partial x_j} &= -\frac{1}{\rho}\frac{\partial \overline{P}}{\partial x_i} + \nu \frac{\partial^2 \overline{U}_i}{\partial x_j \partial x_j} - \frac{\partial \overline{u_i u_j}}{\partial x_j} \\ &= \frac{1}{\rho}\frac{\partial}{\partial x_j}\left\{ \underbrace{-\overline{P}\delta_{ij}}_{\substack{\text{Mean}\\ \text{pressure}\\ \text{stress}}} + \underbrace{\mu\left(\frac{\partial \overline{U}_i}{\partial x_j} + \frac{\partial \overline{U}_j}{\partial x_i}\right)}_{\substack{\text{Mean viscous}\\ \text{stress tensor}}} \underbrace{-\rho \overline{u_i u_j}}_{\substack{\text{Reynolds}\\ \text{stress tensor}}} \right\}\end{aligned} \tag{4.9}$$

The convective term, arising from the fluctuating velocity, can be seen to appear here as an additional fictitious stress tensor, the Reynolds tensor, $-\rho\overline{u_i u_j}$, by which the fluctuating part of the flow interacts with and forces the mean flow, giving a force $-\partial\overline{u_i u_j}/\partial x_j$ per unit mass which the turbulence can be thought of as exerting on the mean flow. Although the Reynolds stress formally appears similar to the genuine stress terms of (4.9), it is not really part of the fluid stress, but instead represents the average momentum flux due to turbulent velocity fluctuations. The Reynolds stress tensor is one of the most important basic concepts in the theory of turbulence, whose divergence can be interpreted as forcing of the mean flow by turbulence. In all other respects, the equations for the mean flow, (4.5) and (4.9), are as if the mean flow was the whole flow and there were no turbulent fluctuations.

Thus, the nonlinear, convective term in the Navier–Stokes equation introduces coupling between the mean and fluctuating (or turbulent) parts of the velocity field through the Reynolds stress tensor, $-\rho\overline{u_i u_j}$, which is a second-order moment of the velocity components *at a single point in space*. In fully developed turbulence, the Reynolds stress tensor can easily be as much as 500 times larger than the mean viscous stress tensor, that is, we have

$$\rho\left\|\overline{u_i u_j}\right\| \gg \mu\left\|\frac{\partial \overline{U}_i}{\partial x_j} + \frac{\partial \overline{U}_j}{\partial x_i}\right\| \tag{4.10}$$

as first recognized by Boussinesq (1877). This is a consequence of the generally large-Reynolds-number turbulent flow, as we now briefly explain. In the last chapter, we saw that the velocity gradients of turbulence are associated with small dissipative scales of the flow, whose size decreases with the viscosity of the fluid (and hence with increasing Reynolds number). This has the effect of making the total velocity gradient $\partial U_i/\partial x_j$ grow large at high Reynolds numbers, but the mean flow gradient, $\partial\overline{U}_i/\partial x_j$, appearing in (4.10), does not, because, like all averages, $\overline{U}_i$ varies over length scales determined by the overall flow, not the small viscous scales of turbulence. Rather than $\overline{U}_i$, it is the other component of the total velocity, u_i, which develops small scales and large gradients, becoming a "furry" function of space and time. As the viscosity is progressively reduced, and the Reynolds number rises, the left-hand side of (4.10) does not drop to zero, being determined by the intensity of the turbulence, whose velocity fluctuations are typically some small, but significant fraction of the mean-flow velocity and do not tend to zero with the viscosity. The net result is that the term on the right of (4.10), which is multiplied by the viscosity of the fluid, goes to zero, but the left-hand side does not. Thus, at large Reynolds numbers, the viscous stress may generally be neglected compared with the Reynolds stress, as far as the mean-flow equations are concerned (although the viscous terms cannot be dropped in the equations for the fluctuating velocity, as we will see later).

There is, however, one important case (apart from the obvious one in which there is no turbulence) for which (4.10) ceases to hold. Towards boundaries of the flow, mean velocity gradients grow larger and there are very thin layers at solid surfaces in which viscous effects must be included in the mean-flow equations. That is, the right-hand side of (4.10) increases as the surface is approached and becomes comparable to the left-hand side in the viscous layer at the surface. Such viscous layers are obviously related to, but should not be confused with, turbulent boundary layers. As we shall see in the next chapter, there is a viscous layer *within* a turbulent boundary layer, which is considerably thinner than the overall boundary layer and will be referred to as the viscous sublayer.[3] Outside the viscous sublayer, but within the boundary layer, viscosity is negligible as regards the mean flow, and mean momentum transfer is dominated by the turbulent Reynolds stress terms. Turbulent boundary layers are discussed in detail in the next chapter.

The boundary condition at a solid surface is one of no-slip, that is, the total fluid velocity equals the velocity of the surface. Thus, if we assume that solid surfaces move in the same way in all realizations of the turbulent flow, or at least that the fluctuation in their positions and velocities are negligible, the fluid velocity at such a surface is independent of the realization considered and gives the mean velocity at the surface. It follows that the boundary conditions for the mean velocity are ones of no-slip, while the fluctuating velocity is zero. Imposition of the no-slip conditions requires viscosity, which is another way of seeing the need for a viscous layer at solid surfaces.

Note that if the turbulence is homogeneous, so that all one-point averages are independent of position, the Reynolds stress term in (4.9) is zero, and so *homogeneous turbulence has no effect on the mean flow*, which evolves as if there were no turbulent fluctuations. This illustrates the simplifications that can occur in homogeneous flows. However, not all mean flows will allow homogeneous turbulence: nonuniform generation and distortion of turbulence by a general mean flow will lead to inhomogeneous turbulence. This shows the need to examine the effects of the mean flow on the turbulence, as well as those of the turbulence on the mean flow.

[3] Note that we use the term viscous sublayer to mean the region in which viscosity is significant as regards the mean flow, thus encompassing the zone often called the buffer layer.

THE EDDY-VISCOSITY APPROXIMATION AND ONE-POINT MODELING

The Reynolds equation, (4.9), for the mean flow, shows that, in general, it is not possible to determine the mean field without knowing $\overline{u_i u_j}$, an ensemble average of the turbulent field. This reflects the closure problem, discussed in the introduction to this chapter. To close the problem, the crudest approach is to express the Reynolds stress tensor in terms of the mean velocity itself. The simplest such closure scheme is to draw an analogy between the Reynolds stress and the viscous stress in a Newtonian fluid. The idea of employing such an eddy-viscosity assumption to represent turbulence is one of the oldest in the subject (Boussinesq (1877)) and was developed further in the mixing-length theories of Taylor and Prandtl (discussed in, e.g., Hinze (G 1975), section 5.2). In those theories an analogy was sometimes drawn between the mean transfer of momentum by turbulent fluctuations, expressed by the Reynolds stress, and that by the chaotic motion of molecules of a gas according to the classical kinetic theory of viscosity. In kinetic theory, viscous stresses arise at a microscopic level from the mean transfer of momentum by random molecular motion, leading to Newtonian macroscopic behavior. Similarly, under the eddy-viscosity approximation one supposes that the turbulent Reynolds stress has a Newtonian expression in terms of the mean velocity gradients, a questionable assumption.

Supposing then that the Reynolds stress is analogous to that resulting from viscosity, one might at first sight be tempted to write

$$-\overline{u_i u_j} = \nu_T \left(\frac{\partial \overline{U_i}}{\partial x_j} + \frac{\partial \overline{U_j}}{\partial x_i} \right) \tag{4.11}$$

where ν_T is a coefficient, similar to the kinematic viscosity, called the turbulent eddy viscosity. The difficulty with (4.11) is that it implies

$$\overline{u_1^2} + \overline{u_2^2} + \overline{u_3^2} = \overline{u_i u_i} = -2\nu_T \frac{\partial \overline{U_i}}{\partial x_i} = 0 \tag{4.12}$$

from (4.5). Thus this rather naive version of the approximation implies that the turbulent velocity is zero, that is, that there is no turbulence! This objection can easily be catered for by changing (4.11) to read

$$-\overline{u_i u_j} = -\frac{1}{3}\overline{q^2}\delta_{ij} + \nu_T \left(\frac{\partial \overline{U_i}}{\partial x_j} + \frac{\partial \overline{U_j}}{\partial x_i} \right) \tag{4.13}$$

where $\overline{q^2} = 3u'^2 = \overline{u_i u_i}$ is the total, mean-squared turbulent velocity. Thus, there appears a term analogous to the pressure in the usual stress tensor for a viscous fluid, which can be absorbed into the real pressure term when (4.13) is used to close (4.9). Naturally, it is (4.13) rather than (4.11) that is used in practice, with an eddy viscosity, ν_T, which is positive, may vary with position and time and must be specified before (4.13) gives definite predictions for the Reynolds stress.

No matter what form for ν_T is used, there are several difficulties with the above approximation, one of which is that for a parallel shear flow (such as a channel flow, which we examine later), it implies that the three components of turbulent velocity have equal mean-squared values, which is observed not to be true. Thus, suppose that the mean flow is in the x_1-direction and varies only with the cross-stream coordinate, x_2, as is the case for fully developed channel flow. From (4.13), we have

$$-\overline{u_1^2} = -\frac{1}{3}\overline{q^2} + 2\nu_T \frac{\partial \overline{U_1}}{\partial x_1} = -\frac{1}{3}\overline{q^2} \tag{4.14}$$

$$-\overline{u_2^2} = -\frac{1}{3}\overline{q^2} + 2\nu_T \frac{\partial \overline{U_2}}{\partial x_2} = -\frac{1}{3}\overline{q^2} \tag{4.15}$$

$$-\overline{u_3^2} = -\frac{1}{3}\overline{q^2} + 2\nu_T \frac{\partial \overline{U_3}}{\partial x_3} = -\frac{1}{3}\overline{q^2} \tag{4.16}$$

so that all mean-squared values are predicted to be equal. This result is far from being in accord with experimental data on such flows.

A second failing of the approximation appears if we look at the components of (4.13) with $i \neq j$, again in a parallel shear flow, which are zero unless $i = 1, j = 2$, or vice versa. For the nonzero component, we have

$$-\overline{u_1 u_2} = \nu_T \frac{\partial \overline{U_1}}{\partial x_2} \tag{4.17}$$

giving the turbulent shear stress in terms of the mean-flow gradient. This result implies that the shear stress, $-\overline{u_1 u_2}$, should be zero where the mean-flow velocity has an extremum. In a wall jet, admittedly not a strictly parallel flow, but approximately so, the location of maximum velocity and zero Reynolds shear stress were found to be noncoincident more than forty years ago (Mathieu (1959), see Figure 4.1). The same was later discovered to be the case for flow in an asymmetric channel with one rough and one smooth wall (Hanjalic and Launder (1972)), which is a good approximation to parallel flow at distances from the rough wall large compared to the roughness elements. Such noncoincidence is not in agreement with the eddy-viscosity approximation.

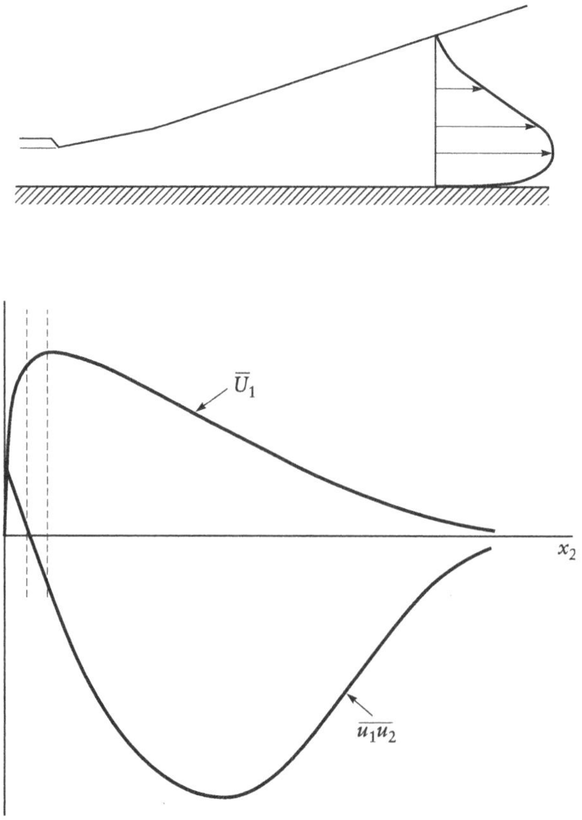

Figure 4.1. Sketch of a wall jet, with its mean velocity and shear stress, illustrating the noncoincidence of the locations of maximum mean velocity and zero shear stress, indicated by the dashed lines.

From the above discussion, it is apparent that the eddy-viscosity approximation has significant defects when compared with experiment. Furthermore, theoretical arguments by analogy with the kinetic theory of gases are not really justified, since, among other things, the clear separation of scales between microscopic and macroscopic which underlies kinetic theory is not present in turbulent flows, where the size of the large eddies is typically comparable with the distances for variation of the mean flow. The assumption (4.13) is thus one of many heuristic closure hypotheses which, although reflecting some aspects of the physics of turbulence, are quantitatively only approximate and may differ sufficiently from reality that important effects are missed for some flows. Attempts were made by Prandtl and Karman to go beyond the eddy-viscosity approximation, allowing the Reynolds stress to also depend on higher spatial derivatives of the mean velocity, in the manner of a Taylor's series. Introducing higher derivatives has the effect of making the description "less local," but the results remain unsatisfactory and such models are now of mainly historical interest.

Despite the significant shortcomings of the eddy-viscosity approximation, it forms one of the main ingredients of the classical semiempirical theory of jets, wakes, and mixing layers, described in Chapter 5, and is one of the building blocks of the most commonly used turbulence prediction model, namely the k–ε model. As noted above, to employ (4.13) as part of a quantitative model, one needs to know the eddy viscosity ν_T, which is expected to be determined by the properties of the turbulence, since it characterizes mean momentum transfer by turbulent fluctuations. The expression used for ν_T depends on the particular model employed. In the k–ε model, ν_T is supposed to be a certain function of the turbulent kinetic energy, $k = \overline{u_i u_i}/2$, and the mean dissipation rate of turbulent energy, $\overline{\varepsilon}$, defined precisely below. This, in turn, leads to the introduction of model equations for the time evolution of the turbulent quantities k and $\overline{\varepsilon}$, of which that for k is based on the turbulent energy equation derived later, together with a number of closure hypotheses, while the equation for $\overline{\varepsilon}$ is written down by analogy with that for k. Thus, in addition to supposing the eddy-viscosity closure (4.13), the k–ε model involves several other heuristic approximations, in the process introducing a variety of adjustable constants which can be tuned to fit the results of experimental observations in the best tradition of pragmatic engineering practice.

A class of more sophisticated one-point models, related to k–ε and known as $\overline{u_i u_j}$–ε, do not assume (4.13), but instead use the components of $\overline{u_i u_j}$ as basic variables of the turbulence model. That is, the mean-flow equations are not approximated, but, since the evolution equation for $\overline{u_i u_j}$, derived later in this chapter, contains a variety of terms that are not exactly expressible in terms of $\overline{u_i u_j}$ and the mean flow, closure hypotheses are needed. Some of these closure approximations will be touched on in the course of this chapter, but a more detailed description of the k–ε and $\overline{u_i u_j}$–ε models is deferred to Chapter 8 (in which the more sophisticated, but more computationally expensive direct and large-eddy simulation techniques are also discussed). We have nonetheless mentioned them at this stage to make the reader aware of the many heuristic assumptions implicit in such one-point modeling, assumptions aimed at overcoming the nonlinearity and nonlocality difficulties discussed in the introduction to this chapter, but which are questionable and should not be adopted uncritically.

ENERGETICS OF THE MEAN FLOW

We begin with the mean-flow equation, (4.9), which can be written

$$\frac{\partial \overline{U}_i}{\partial t} + \overline{U}_j \frac{\partial \overline{U}_i}{\partial x_j} = \frac{\partial T_{ij}}{\partial x_j} \tag{4.18}$$

where

$$T_{ij} = -\overline{u_i u_j} - \frac{\overline{P}}{\rho}\delta_{ij} + \nu\left(\frac{\partial \overline{U}_i}{\partial x_j} + \frac{\partial \overline{U}_j}{\partial x_i}\right) \tag{4.19}$$

is the effective stress tensor of the mean flow, divided by the fluid density.

We write

$$\frac{d}{dt} = \frac{\partial}{\partial t} + \overline{U}_j \frac{\partial}{\partial x_j} \tag{4.20}$$

to denote a material derivative, here with respect to the mean flow. We multiply (4.18) by $\overline{U}_i$ to find

$$\frac{d}{dt}\left(\frac{1}{2}\overline{U}_i\overline{U}_i\right) = \overline{U}_i \frac{\partial T_{ij}}{\partial x_j} \tag{4.21}$$

which can be integrated over a material volume of the mean flow, to give

$$\frac{d}{dt}\int \frac{1}{2}\overline{U}_i\overline{U}_i dv = \int \frac{\partial}{\partial x_j}(\overline{U}_i T_{ij})\, dv - \int T_{ij}\frac{\partial \overline{U}_i}{\partial x_j} dv \tag{4.22}$$

where d/dt is now a straightforward time derivative. Note that assuming a mean-flow material volume (i.e., one whose boundary moves at velocity $\overline{U}_i$) allows us to move the time derivative outside the volume integral to derive (4.22).

Assuming a closed flow volume with zero velocity at its boundary, the mean flow is also zero on the boundary, allowing us to use the divergence theorem to show that the first term on the right-hand side of (4.22) is zero. The second term on the right of (4.22) can also be rewritten using the definition of T_{ij} and the incompressibility condition, (4.5), to show that

$$\underbrace{\frac{d}{dt}\int \frac{1}{2}\overline{U}_i\,\overline{U}_i\, dv}_{\substack{\text{Rate of change}\\ \text{of energy of the}\\ \text{mean flow}}} = \underbrace{\int \overline{u_i u_j}\frac{\partial \overline{U}_i}{\partial x_j} dv}_{\substack{\text{Mean flow}\\ \text{coupling to}\\ \text{turbulence}}} \underbrace{- \frac{1}{2}\nu \int \left(\frac{\partial \overline{U}_i}{\partial x_j} + \frac{\partial \overline{U}_j}{\partial x_i}\right)\left(\frac{\partial \overline{U}_i}{\partial x_j} + \frac{\partial \overline{U}_j}{\partial x_i}\right) dv}_{\substack{\text{Viscous dissipation}\\ \text{rate of mean flow}}} \tag{4.23}$$

The form of the viscous dissipation here is the same as if the turbulence were absent, and is usually negligible, away from the thin viscous layers at solid surfaces discussed earlier, while the coupling term expresses the rate of working of the Reynolds stress, due to the fluctuating field, on the mean flow. This term represents energy taken from the mean flow to supply the turbulence.

Note that in the above derivation we assumed a closed volume with stationary boundaries. One can also allow for boundaries that move and can therefore do work to maintain the flow, or consider open flows, where energy can be supplied from outside. Taking these into account complicates the presentation needlessly for present purposes and we shall continue to assume zero velocity at the boundaries of the flow whenever we integrate the equations to perform global energy audits. The reader is invited to reconsider the various audits, allowing for moving boundaries. As noted earlier, the boundary conditions at a moving, solid boundary are no-slip for the mean velocity and fluctuating velocity zero. One finds that work done by motion of solid surfaces appears in the energy accounting for the mean flow, rather than for the turbulence. Open flows allow the further possibility of supply to both the mean flow and turbulence, coming from outside the flow domain considered.

ENERGETICS OF THE TOTAL FLOW

The energy budget of the total flow is derived in most textbooks on fluid mechanics. Thus, multiplying (4.3) by U_i and integrating over the flow volume, using (4.4) and the divergence theorem with the assumption that the flow velocity is zero on the boundary, we obtain

$$\frac{d}{dt}\int \frac{1}{2}U_iU_i\,dv = -\frac{1}{2}\nu\int\left(\frac{\partial U_i}{\partial x_j}+\frac{\partial U_j}{\partial x_i}\right)\left(\frac{\partial U_i}{\partial x_j}+\frac{\partial U_j}{\partial x_i}\right)dv = -\int \Delta\,dv \tag{4.24}$$

showing the viscous energy dissipation. Here, Δ is the rate of energy dissipation by viscosity per unit mass. Within the flow, kinetic energy is transferred from place to place by convection and work done by part of the fluid on neighboring fluid through the pressure and viscous stresses. This does not appear in (4.24), which, like (4.23) for the mean flow, gives the *overall* (integrated) rate of change of energy.

Using the decomposition, (4.1), into a mean and fluctuating velocity, one may write the *average* dissipation rate as

$$\overline{\Delta} = \frac{1}{2}\nu\left(\frac{\partial \overline{U}_i}{\partial x_j}+\frac{\partial \overline{U}_j}{\partial x_i}\right)\left(\frac{\partial \overline{U}_i}{\partial x_j}+\frac{\partial \overline{U}_j}{\partial x_i}\right)+\frac{1}{2}\nu\overline{\left(\frac{\partial u_i}{\partial x_j}+\frac{\partial u_j}{\partial x_i}\right)\left(\frac{\partial u_i}{\partial x_j}+\frac{\partial u_j}{\partial x_i}\right)} \tag{4.25}$$

showing the neat split of the overall average dissipation into mean and turbulent components, a decomposition that may be used to express the right-hand side of the averaged version of (4.24) as a sum of mean-flow and turbulent parts. The quantity appearing inside the average of the second term of (4.25) will be denoted ε, defined by

$$\varepsilon = \frac{1}{2}\nu\left(\frac{\partial u_i}{\partial x_j}+\frac{\partial u_j}{\partial x_i}\right)\left(\frac{\partial u_i}{\partial x_j}+\frac{\partial u_j}{\partial x_i}\right) \tag{4.26}$$

whose average value

$$\overline{\varepsilon} = \frac{1}{2}\nu\overline{\left(\frac{\partial u_i}{\partial x_j}+\frac{\partial u_j}{\partial x_i}\right)\left(\frac{\partial u_i}{\partial x_j}+\frac{\partial u_j}{\partial x_i}\right)} \tag{4.27}$$

appears in (4.25) and plays an important role in the theory of turbulence.

Splitting into mean-flow and turbulent components and averaging, (4.24) becomes

$$\frac{d}{dt}\int \frac{1}{2}\overline{U_i U_i}\, dv = -\frac{1}{2}\nu \int \underbrace{\left(\frac{\partial \overline{U_i}}{\partial x_j}+\frac{\partial \overline{U_j}}{\partial x_i}\right)\left(\frac{\partial \overline{U_i}}{\partial x_j}+\frac{\partial \overline{U_j}}{\partial x_i}\right)}_{\text{Mean flow viscous dissipation}} + \underbrace{\overline{\left(\frac{\partial u_i}{\partial x_j}+\frac{\partial u_j}{\partial x_i}\right)\left(\frac{\partial u_i}{\partial x_j}+\frac{\partial u_j}{\partial x_i}\right)}}_{\text{Turbulent viscous dissipation}}\, dv \tag{4.28}$$

showing how the total mean energy of the flow decays at a rate determined by the sum of the mean-flow and average turbulent dissipations. The dissipation of the total flow energy due to the mean-flow component is usually negligible compared to the turbulent contribution at the high Reynolds numbers associated with many turbulent flows, that is,

$$\frac{1}{2}\nu\left(\frac{\partial \overline{U_i}}{\partial x_j}+\frac{\partial \overline{U_j}}{\partial x_i}\right)\left(\frac{\partial \overline{U_i}}{\partial x_j}+\frac{\partial \overline{U_j}}{\partial x_i}\right) \ll \frac{1}{2}\nu\overline{\left(\frac{\partial u_i}{\partial x_j}+\frac{\partial u_j}{\partial x_i}\right)\left(\frac{\partial u_i}{\partial x_j}+\frac{\partial u_j}{\partial x_i}\right)} \tag{4.29}$$

because, as discussed earlier, the fluctuating velocity gradient, $\partial u_i/\partial x_j$, grows large as the Reynolds number increases, whereas the mean velocity gradient does not. That is, one can usually neglect the mean-flow contribution when calculating the overall dissipation rate because the mean-squared velocity derivatives are dominated by the small scales of turbulence, as explained in Chapter 3, rather than by the longer-scale variations of the mean flow. However, an exception occurs very close to a solid boundary of the flow, in the viscous layer or the sublayer inside a boundary layer, which we mentioned earlier. There, the derivative of the mean velocity in the direction perpendicular to the wall becomes comparable with the derivatives of the fluctuating velocity and (4.29) ceases to apply. However, such cases, although important, concern only a small part of any flow and (4.29) is true elsewhere.

The average kinetic energy itself also splits into a mean flow and turbulent contribution and, indeed, the same is true of each component of the tensor $\overline{U_i U_j}$. Thus,

$$\overline{U_i U_j} = \overline{U_i}\,\overline{U_j} + \overline{u_i u_j} \tag{4.30}$$

which, with $i = j$, can be used to write the left-hand side of (4.28) as a sum of mean and turbulent terms, like the viscous dissipation. It follows that the total energy equation contains a sum of mean-flow and turbulence contributions. The two parts of the flow are coupled together energetically by the first term on the right of (4.23), which does not appear in the total energy equation because it represents transfer of energy between the mean flow and turbulence, rather than a change in the total energy of the flow. As we shall see shortly, this coupling appears in the energetics of the turbulence as a production term.

4.2 Equations of the Second-Order Moments: Turbulence Energetics

We begin by considering the equation satisfied by the fluctuating field. This is obtained by subtracting the mean-flow equation, (4.7), from the equation for the total field, (4.3), using (4.6), (4.8) and the splits, (4.1) and (4.2). The result is the equation for the fluctuating part of the flow

$$\frac{\partial u_i}{\partial t} + \underbrace{\overline{U_k}\frac{\partial u_i}{\partial x_k} + u_k\frac{\partial \overline{U_i}}{\partial x_k}}_{\text{Coupling to mean flow}} + \underbrace{\frac{\partial}{\partial x_k}[u_i u_k - \overline{u_i u_k}]}_{\text{Nonlinear term}} = -\frac{1}{\rho}\frac{\partial p}{\partial x_i} + \nu\frac{\partial^2 u_i}{\partial x_k \partial x_k} \tag{4.31}$$

which, with (4.6) and the boundary condition that $u_i = 0$ on solid boundaries, governs the turbulent part of the field. Notice that the mean flow occurs in the equation for the fluctuations, and vice versa. The two are coupled together and calculation of the mean flow and turbulence cannot be performed separately, in general.

The nonlinear, convective term in the original equation, (4.3), has given rise to both linear and nonlinear terms in the equation for the fluctuating velocity. Here we follow the usual convention of referring to terms involving products of mean and fluctuating velocities as linear, because they are indeed linear in the fluctuating velocity. As we shall see later, when the fluctuating pressure, p, is expressed in terms of the velocity field, it contains two components, one of which is linear in the fluctuating velocity, the other nonlinear. Consequently, the pressure term in (4.31) implicitly involves both linear and nonlinear terms. The linear terms in (4.31) are thus: direct coupling of mean and fluctuating velocities, the gradient of the linear pressure component, both of which represent the influence of the mean flow on the turbulence, and lastly the viscous term, responsible for turbulent dissipation. The nonlinear terms consist of the one appearing explicitly in (4.31) and the gradient of the nonlinear pressure component, both of which can be thought of as representing the action of turbulence on itself.

The matrix of one-point velocity correlations, $\overline{u_i u_j}$, is a positive-definite, symmetric tensor, directly related to the Reynolds stress tensor, $-\rho\overline{u_i u_j}$. To obtain equations for the time evolution of $\overline{u_i u_j}$, we multiply (4.31) by u_j, add the result of interchanging the indices i and j, take the average, and use (4.6) to derive

$$\underbrace{\frac{\partial \overline{u_i u_j}}{\partial t}}_{\text{I}} + \underbrace{\overline{U_k}\frac{\partial \overline{u_i u_j}}{\partial x_k}}_{\text{II}} =$$

$$\underbrace{-\overline{u_i u_k}\frac{\partial \overline{U_j}}{\partial x_k} - \overline{u_j u_k}\frac{\partial \overline{U_i}}{\partial x_k}}_{\text{III}} - \underbrace{\frac{\partial \overline{u_i u_j u_k}}{\partial x_k}}_{\text{IV}} - \underbrace{\frac{1}{\rho}\left\{\overline{u_i\frac{\partial p}{\partial x_j}} + \overline{u_j\frac{\partial p}{\partial x_i}}\right\}}_{\text{V}} + \underbrace{\nu\left\{\overline{u_i\frac{\partial^2 u_j}{\partial x_k \partial x_k}} + \overline{u_j\frac{\partial^2 u_i}{\partial x_k \partial x_k}}\right\}}_{\text{VI}} \tag{4.32}$$

which is the evolution equation for the second moments of velocity at a single point, governing the time development of the Reynolds tensor. The role of each of the terms in this equation may be identified as follows:

(I) is the time rate of change of $\overline{u_i u_j}$ and allows for statistically unsteady flows.

(II) accounts for advection of the Reynolds stress by the *mean* flow, *not* the total flow.

(III) gives the interaction of the mean and turbulent parts of the flow and can be interpreted as responsible for the production and reorientation of the Reynolds stresses. This very important term will be considered in more detail later.

(IV) accounts for advection by the *fluctuating* part of the flow.
(V) describes the effects of the pressure. Pressure effects are complicated, because they are nonlocal and the consequence of both linear and nonlinear effects, as we shall describe later. The pressure is related to the velocity field nonlocally, by an integral over the flow domain, as we shall also see later. The domain of significant contributions to this integral is, in fact, limited in size because the velocity correlations between a pair of points tend to zero as the distance between the points increases, a property which is characteristic of turbulence.
(VI) represents dissipative and diffusive viscous effects and is dominated by the small scales of turbulence.

We saw in Section 4.1 that the equations for the mean flow are not closed because they contain second-order velocity moments. In an attempt to close the mean-flow equation (4.9), one may use (4.32) for the second-order moments. However, the term (IV) involves third-order velocity moments, as does (V) when the pressure is expressed in terms of the velocity, as we will see later. Thus, both (IV) and the nonlinear part of (V) introduce the next highest order of velocity moments, reflecting the fundamental closure problem discussed in the introduction to this chapter. These terms containing third-order moments arise from the nonlinear term in (4.31), which in turn comes from the convective term in (4.3), the ultimate source of the closure problem. Such terms in the moment evolution equations, arising from the nonlinear term of (4.31), are themselves referred to as nonlinear.

Observe that one cannot express the viscous term, (VI), which is linear in the above sense, in terms of single-point moments because it contains a second spatial derivative and is thus nonlocal. It can, however, be written using the second derivatives of the two-point moments, $\overline{u_i(\mathbf{x})u_j(\mathbf{x}')}$, evaluated at $\mathbf{x}' = \mathbf{x}$. Thus, a mild form of viscous nonlocality is present, which is severely compounded by the pressure term, (V), when it is expressed in terms of the velocity moments, introducing an integral over the entire flow field, as we shall see later.

None of the three terms (IV), (V), and (VI) is expressible in terms of the one-point moments $\overline{u_i u_j}$, thus (4.32) is not closed. However, if one is willing to work with two-point moments, deriving the equivalent of (4.32) for $\overline{u_i(\mathbf{x}, t)u_j(\mathbf{x}', t)}$, then all linear terms are closed, leaving the fundamental closure problem due to nonlinearity, expressed via third-order terms equivalent to (IV) and the nonlinear part of (V). On the other hand, if one sticks with one-point moments, as here, nonlocality of (VI) and the linear parts of (V) introduce an additional closure difficulty.

We can derive the equation for the turbulent kinetic energy by setting $j = i$ in equation (4.32) (with implicit summation over i) and dividing by 2. Thus we obtain the very important turbulent energy equation

$$\underbrace{\frac{\partial \frac{1}{2}\overline{q^2}}{\partial t}}_{\mathrm{I}} + \underbrace{U_k \frac{\partial \frac{1}{2}\overline{q^2}}{\partial x_k}}_{\mathrm{II}} = \underbrace{-\overline{u_i u_k}\frac{\partial U_i}{\partial x_k}}_{\mathrm{III}} - \underbrace{\frac{\partial \frac{1}{2}\overline{q^2 u_k}}{\partial x_k}}_{\mathrm{IV}} - \underbrace{\frac{1}{\rho}\overline{u_i \frac{\partial p}{\partial x_i}}}_{\mathrm{V}} + \underbrace{\nu \overline{u_i \frac{\partial^2 u_i}{\partial x_k \partial x_k}}}_{\mathrm{VI}} \tag{4.33}$$

Once again one can interpret the different terms:

(I) rate of increase of turbulent energy at a fixed point; zero for steady flow
(II) convection of turbulent energy by the mean flow
(III) production of turbulent energy by interaction between the mean flow and turbulence
(IV) advective transport of turbulent energy by the fluctuating motion (turbulent mixing)
(V) transfer of turbulent energy by pressure effects (work done by fluctuating pressure)
(VI) viscous effects (dissipation, diffusion)

The sum of terms (I) and (II) is the time rate of change of turbulent energy per unit mass, $\overline{q^2}/2$, following a point moving with the mean flow. Term (III), which is especially important, describes the production of turbulence. As we shall see from specific examples later, it can be either positive, leading to generation of turbulence, or (more rarely in practice) negative, representing extraction of turbulent energy. Notice that, if one were to use the eddy-viscosity assumption (4.13) in the production term of (4.33), the turbulent energy production is directly related to the mean-flow gradients and can be shown to be always positive. While often the case, this is not true of some flows, which is a further warning that the eddy-viscosity approximation should be treated with caution. By analogy with energy production, the corresponding term (III) in (4.32) is also interpreted as representing turbulence production, but contains more information about the nature of turbulence production owing to its tensorial nature. The final term (VI) can be rewritten as

$$\overline{\nu u_i \frac{\partial^2 u_i}{\partial x_k \partial x_k}} = \underbrace{\nu \frac{\partial}{\partial x_j}\left\{\overline{u_i\left(\frac{\partial u_i}{\partial x_j}+\frac{\partial u_j}{\partial x_i}\right)}\right\}}_{\text{Viscous transfer}} - \underbrace{\bar{\varepsilon}}_{\substack{\text{Dissipation of}\\ \text{turbulent energy}}} \tag{4.34}$$

where $\bar{\varepsilon}$ is defined by (4.27) and, as indicated by the labeling, is interpreted as the average turbulent energy dissipation rate. The viscous transfer term integrates to zero over the whole flow by the divergence theorem. Thus, it does not lead to any net change in the turbulent energy, hence its description as a transfer term. This term is sometimes also referred to as diffusive, because it is zero for homogeneous turbulence and can be thought of as roughly analogous to diffusion, which produces transfer due to inhomogeneities of concentration. In the analogy, turbulent energy replaces concentration. As we shall shortly see, the viscous transfer term is negligible at high Reynolds numbers, except within the thin viscous layers very near any solid surfaces that we referred to earlier. On the other hand, the dissipative term is of crucial importance to turbulence energetics everywhere and is definitely *not* negligible. Thus viscosity enters into the turbulence energetics in an important way, even if, as we saw earlier, it can be dropped in the mean-flow equations outside of viscous layers.

Equation (4.33) can be reexpressed using (4.34) and incompressibility of the mean and fluctuating flows, (4.5) and (4.6), to give the following useful form of the turbulent energy equation

$$\frac{\partial \frac{1}{2}\overline{q^2}}{\partial t} = \underbrace{-\overline{u_i u_j}\frac{\partial \overline{U_i}}{\partial x_j}}_{\text{Production}} - \underbrace{\overline{\varepsilon}}_{\substack{\text{Turbulent}\\ \text{dissipation}}} \underbrace{- \frac{\partial}{\partial x_j}\left\{ \underbrace{\frac{1}{2}\overline{q^2}\overline{U_j}}_{\substack{\text{Convection}\\ \text{by mean flow}}} + \underbrace{\overline{u_j\left(\frac{1}{2}q^2 + \frac{p}{\rho}\right)} - \nu\overline{u_i\left(\frac{\partial u_i}{\partial x_j} + \frac{\partial u_j}{\partial x_i}\right)}}_{\text{"Diffusion"}} \right\}}_{\text{Transfer of turbulent energy}} \tag{4.35}$$

In this form it is apparent that, when integrated over the flow, the term labeled as transfer yields zero by the divergence theorem (the fluctuating velocity being taken as zero at the boundaries). This term results in transfer of turbulent energy from place to place, but cannot change it overall. Thus, globally, the turbulent energy increases or decreases due to the difference between production, represented by the integral of the first term on the right of (4.35), and dissipation, given by the integral of the second. Equation (4.35) lends precise meaning to the production, dissipation, and transfer of turbulent energy. The sum of equations (4.23), for the mean-flow energy, and the integrated version of (4.35), for the turbulence energy, gives (4.28) for the total energy, after using (4.30) with $i = j$. This is, of course, as it should be: overall, energy supplied to the turbulence via the production term in (4.35) depletes the mean flow. For statistically steady flows, the right-hand side of (4.35) is zero. Thus, statistically steady turbulence is subject to balancing production, dissipation, mean-flow convection, and diffusion of energy due to inhomogeneity.

Multiplying (4.35) by the fluid density, integrating over an arbitrary *fixed* volume and using the divergence theorem on the transfer term, we may interpret the result as showing that transfer is represented by a vector

$$\underbrace{\frac{1}{2}\rho\overline{q^2}\overline{U_i}}_{\substack{\text{Mean-flow}\\ \text{convection}}} + \underbrace{\overline{u_i\left(\frac{1}{2}\rho q^2 + p\right)} - \mu\overline{u_j\left(\frac{\partial u_i}{\partial x_j} + \frac{\partial u_j}{\partial x_i}\right)}}_{\text{"Diffusion"}} \tag{4.36}$$

which gives the average flux of turbulent energy due to mean-flow convection and "diffusion" processes. In addition to this flux, there is turbulence production and viscous dissipation of turbulence over the volume considered. The transfer terms in (4.35), the divergence of the flux vector (4.36), are zero for homogeneous turbulence and, with the exception of the mean-flow convection term, are often thought of as diffusive in nature because they give a turbulent energy flux due to nonuniformities of the turbulence properties (inhomogeneities). The analogy with diffusion suggests that the term labeled "diffusion" in (4.36) should produce a flux of turbulent energy from regions in which it is high to those in which it is lower. This is usually the case, but is not universally true.

If the turbulence is homogeneous, (4.35) becomes

$$\underbrace{\frac{d\frac{1}{2}\overline{q^2}}{dt}}_{\text{Evolution}} = \underbrace{-\overline{u_i u_j}\frac{\partial \overline{U_i}}{\partial x_j}}_{\text{Production}} - \underbrace{\overline{\varepsilon}}_{\text{Dissipation}} \tag{4.37}$$

showing the roles of the remaining three terms very clearly. In particular, it is obvious that the term we have called production is the one responsible for generating (or destroying) turbulent energy, depending upon its sign (usually positive, corresponding to production of energy, rather than its destruction), and that the viscous dissipation term is always positive, producing a drain of turbulent energy via the fine-scale structures, which are responsible for most of the mean-squared velocity gradients, and hence the dissipation. Without mean velocity gradients, there is no production and the turbulent energy decreases continuously. The turbulent energy is $\overline{q^2}/2 = 3u'^2/2$, while its dissipation rate is $\overline{\varepsilon}$, leading to the time scale $u'^2/\overline{\varepsilon}$ for turbulent decay in the absence of production. As the turbulence decays due to dissipation, both u'^2 and $\overline{\varepsilon}$ change, as usually does the time scale $u'^2/\overline{\varepsilon}$ (this simply means that the decay of turbulent energy is not an exponential function of time). This time is characteristic of the decay of the large spatial scales of turbulence, a process which determines the energy supply for dissipation at the small scales, as discussed in Chapter 3. As described there, a second time scale, L/u', can be constructed from the large-scale quantities u' and L, L being the correlation length. Again in the absence of production, this second time scale also characterizes the decay of the large-scale turbulent structures. For the two time scales to be consistent, we must have $\overline{\varepsilon} = O\left(u'^3/L\right)$ as the rate of energy supply by the large scales, via the small-scale cascade described in Chapter 3, to be dissipation at the Kolmogorov scale. Thus energy dissipation is controlled by the large-scale energy supply, while the dissipation rate and viscosity in turn determine the velocity gradients and size of the dissipative scales.

By definition, the statistical properties of *isotropic* turbulence have no preferred direction. In particular, the quantity, $\overline{u_i u_j}$, should have the same components if the coordinate system is given an arbitrary rotation. By a well-known mathematical result of the theory of tensors, it is therefore a scalar multiple of δ_{ij}, that is, $\overline{u_i u_j} = u'^2 \delta_{ij}$, where the multiplier has been fixed using the definition, $u'^2 = \overline{u_i u_i}/3$, of u'. Thus, for isotropic turbulence, the Reynolds stress tensor is diagonal and all its diagonal elements are equal. This is often used as a criterion of isotropy, although the fact that $\overline{u_i u_j}/u'^2$ approaches δ_{ij} does not guarantee isotropy of other statistical quantities. As we shall see later, mean velocity gradients tend to induce anisotropy of $\overline{u_i u_j}$, whereas homogeneous turbulence left to itself (i.e., in a uniform mean flow) usually[4] shows decreasing anisotropy as it decays. Assuming homogeneity, (4.32) becomes

$$\frac{d\overline{u_i u_j}}{dt} = -\overline{u_i u_k}\frac{\partial U_j}{\partial x_k} - \overline{u_j u_k}\frac{\partial U_i}{\partial x_k} + \overline{\frac{1}{\rho}p\left\{\frac{\partial u_i}{\partial x_j} + \frac{\partial u_j}{\partial x_i}\right\}} - 2\nu\overline{\frac{\partial u_i}{\partial x_k}\frac{\partial u_j}{\partial x_k}} \tag{4.38}$$

which can be used to investigate the development of anisotropy in initially isotropic turbulence under the influence of gradients in the mean flow, as we will see in later examples. It may also be employed to study initially anisotropic, homogeneous turbulence in the absence of mean-flow gradients, for which the evolution of $\overline{u_i u_j}$ is determined by the pressure–rate-of-strain correlations and viscous terms in (4.38), both of which will shortly be discussed further.

[4] Cases are known in which, following rapid straining of homogeneous turbulence, $\overline{u_i u_j}$ shows transitory growth of anisotropy.

THE TURBULENT ENERGY DISSIPATION RATE

Here, we want to derive a number of expressions for the mean rate of turbulent energy dissipation that are essentially equivalent for large-Reynolds-number turbulence, away from viscous layers. We have already noted that, because velocity derivatives are dominated by the small scales of turbulence, the overall energy dissipation rate, Δ, is dominated by the turbulent part, ε, and so one does not usually need to distinguish between them. Similar reasoning is employed here. Using the relation

$$\overline{\frac{\partial u_i}{\partial x_j}\frac{\partial u_j}{\partial x_i}} = \frac{\partial^2 \overline{u_i u_j}}{\partial x_i \partial x_j} \tag{4.39}$$

which follows from (4.6), we can rewrite (4.27) as

$$\bar{\varepsilon} = \nu\overline{\frac{\partial u_i}{\partial x_j}\frac{\partial u_i}{\partial x_j}} + \nu\frac{\partial^2 \overline{u_i u_j}}{\partial x_i \partial x_j} \tag{4.40}$$

and a similar expression can be found in terms of the fluctuating vorticity, $\omega = \nabla \times \mathbf{u}$, using the definition of the curl operator and some straightforward algebraic manipulation (see the appendix to Chapter 6). Thus, we find

$$\bar{\varepsilon} = \nu\overline{\omega_i \omega_i} + 2\nu\frac{\partial^2 \overline{u_i u_j}}{\partial x_i \partial x_j} \tag{4.41}$$

For *homogeneous* turbulence, the spatial derivatives of $\overline{u_i u_j}$ are zero and (4.40), (4.41) give

$$\bar{\varepsilon} = \nu\overline{\frac{\partial u_i}{\partial x_j}\frac{\partial u_i}{\partial x_j}} = \nu\overline{\omega_i \omega_i} \tag{4.42}$$

which are the expressions, equivalent to (4.27) for homogeneous turbulence, that we wished to derive. In fact (4.42), which is strictly valid in homogeneous turbulence, is generally true to a good approximation away from viscous layers, provided the overall Reynolds number is large (as it is for a typical turbulent flow). This is because, as described in Chapter 3, the mean-squared velocity derivatives are dominated by the small scales of turbulence, whereas the derivatives of an averaged quantity, such as $\overline{u_i u_j}$, scale on the overall size of the flow, which is much bigger when the Reynolds number is large. Thus, for example, the first term on the right of (4.40) is much larger than the second. The vorticity, being defined in terms of derivatives of the velocity, also has the property that its mean-squared values are dominated by the fine structures of turbulence and one can neglect the second term on the right of (4.41). Similar reasoning shows that, if desired, one may replace the fluctuating velocity in (4.40) (or vorticity in (4.41)) by the total velocity (or vorticity), again assuming a large Reynolds number. Clearly then, a number of expressions for the dissipation can be given, differing only by terms which are negligibly small away from viscous layers. Much the same arguments show that the viscous part of the transfer term in (4.35) can be neglected outside of viscous layers.

A more quantitative formulation of such ideas can be given in terms of a length scale of turbulence known as the Taylor microscale, λ. As described in the previous chapter, this is defined by the magnitude of the squared derivatives of velocity

$$\overline{\frac{\partial u_i}{\partial x_j}\frac{\partial u_i}{\partial x_j}} \sim \frac{\overline{u_i u_i}}{\lambda^2} \tag{4.43}$$

The derivative of an averaged quantity has an associated length scale of order L or greater, where L is the turbulent correlation length. Thus, for instance

$$\frac{\partial^2 \overline{u_i u_j}}{\partial x_i \partial x_j} \sim \frac{\overline{u_i u_i}}{L^2} \tag{4.44}$$

or less, and the ratio of (4.44) to (4.43) is $O(\lambda^2/L^2)$ or smaller which, as we saw in the previous chapter, is small, of order Re_L^{-1}, where Re_L is the turbulent Reynolds number, based on the root-mean-squared fluctuating velocity, $u' = \sqrt{\overline{u_i u_i}/3}$, and correlation length scale, L, that is, $\mathrm{Re}_L = u'L/\nu$. Since Re_L is large, the derivative estimated by (4.44) is small compared with (4.43). In fact, (4.44) estimates the derivatives using the correlation length of the turbulence, whereas the length scale for variation of averaged quantities, although of the same order of size or larger than L, can be much bigger (becoming infinite in the extreme case of homogeneous turbulence). This can make the derivative, (4.44), even smaller.

In summary, the overall energy dissipation rate is dominated by the fine scales of turbulence when the Reynolds number is large and, to a good approximation, away from viscous layers, is unaffected by whether one takes into account the mean-flow contribution or not. The turbulent dissipation may be calculated from a number of essentially equivalent expressions, such as (4.27) and (4.42) (some more such expressions are given as an appendix to Chapter 6), as if the turbulence were homogeneous. However, within the viscous layers at a surface, one must carefully distinguish between these different expressions. Note that the turbulent energy equation, (4.33), only contains the mean flow via the convection and production terms, not in the viscous term. Within a viscous layer, the latter is not equivalent to $\bar{\varepsilon}$: there is also the viscous transfer term in (4.34) to take into account. In other words, viscous layers require careful analysis.

When the Reynolds number is large, turbulent energy dissipation can be thought of as due to a cascade of energy from larger to smaller scales, as discussed in the previous chapter. The large scales give up their energy progressively to form smaller ones, which, in turn, pass the energy through a cascade of still smaller ones to be dissipated at the smallest scales. Schematically:

Kinetic energy of mean flow
↓
Kinetic energy of large scales
↓
Energy flux through small scales
↓
Dissipation by viscosity

of which the first step represents turbulent energy production by mean-flow instabilities. However, we cannot distinguish between different scales of turbulence using the one-point methods of this chapter, still less give quantitative meaning to energy transfer between those scales. We will develop the necessary theory to do this in Chapter 6. Within the single-point theory of this chapter, the cascade appears simply via its net result, which is energy dissipation. As noted earlier, the dissipation rate can generally be estimated as $\overline{\varepsilon} = O(u'^3/L)$, which brings out the fact that it is the large scales that determine the rate of energy supply to the cascade, and hence the dissipation rate.

One final implication of the small-scale origin of dissipation at high Reynolds numbers should be mentioned. This concerns the equations for the Reynolds stress, (4.32) and (4.38). Tensorial viscous terms occur at the end of these equations. In calculating these terms, for reasons similar to those given above, it does not matter whether we consider homogeneous or inhomogeneous turbulence at large Reynolds number. Adopting the homogeneous form, (4.38), the viscous dissipation term is dominated by the small scales of turbulence. As we shall see in Chapter 7, at the small scales, turbulence is believed to be approximately isotropic when the Reynolds number is sufficiently large. This comes about through the process of production of smaller and smaller scales by the progressive drain of energy from large ones. During this process, the details of the large-scale properties of turbulence, which depend on the specific flow considered, are thought to be progressively forgotten and the small scales to become more and more closely isotropic. That is, even if it is anisotropic at the large scales, it may approach isotropy for the small ones. If we suppose such isotropy, the viscous term in (4.38) will be approximately given by the isotropic form

$$-2\nu\overline{\frac{\partial u_i}{\partial x_k}\frac{\partial u_j}{\partial x_k}} = -\frac{2}{3}\overline{\varepsilon}\delta_{ij} \tag{4.45}$$

an expression that is often used when modeling the Reynolds stress equation. Observe that this form is fixed by isotropy to within a constant and the constant determined from equation (4.42). This approximation is expected to hold at large enough Reynolds numbers, away from boundaries, and to improve the greater the Reynolds number.

4.3 The Effects of Pressure

Taking the divergence of the Navier–Stokes equation, (4.3), and using the incompressibility condition, (4.4), yields a Poisson equation for the pressure in terms of the velocity. In this way we can elucidate the close connection between the mean and fluctuating pressures and velocities. The Poisson equation obtained is

$$\nabla^2 P = -\rho\frac{\partial^2 U_i U_j}{\partial x_i \partial x_j} \tag{4.46}$$

which can be averaged, using (4.30), leading to

$$\nabla^2 \overline{P} = -\rho\frac{\partial^2}{\partial x_i \partial x_j}\{\overline{U}_i\,\overline{U}_j + \overline{u_i u_j}\} \tag{4.47}$$

for the mean pressure, which, when subtracted from (4.46), gives

$$\nabla^2 p = -\rho\frac{\partial^2}{\partial x_i \partial x_j}\{\overline{U}_i u_j + \overline{U}_j u_i + u_i u_j - \overline{u_i u_j}\} \tag{4.48}$$

for the fluctuating pressure. In general, in addition to (4.47) and (4.48), one requires boundary conditions on the pressure at any boundaries of the flow. For instance, the normal component of (4.3) at a boundary gives the normal derivative of the pressure in terms of the velocity field, thus leading to Neumann boundary conditions for (4.47) and (4.48) when split into mean and fluctuating parts.

The solution of the above Poisson equations and boundary conditions may be determined using Green's function techniques, with which the reader is assumed familiar. Thus, with a boundary condition of Neumann type, one may use the Green's function of the Laplace equation having zero normal derivative at the boundaries. The Green's function solution consists of two parts: a volume integral, representing the volumetric sources on the right of (4.47) or (4.48), and a surface integral over the boundaries, giving a solution of the Laplace equation satisfying the pressure boundary conditions. The simplest case, and the only one we consider in detail here, is that of unbounded, infinite flow. The solutions of (4.47) and (4.48) without boundaries can be expressed in terms of the standard Green's function of the Laplace equation as

$$\overline{P}(\mathbf{x}) = \frac{\rho}{4\pi}\int\!\!\int\!\!\int \frac{\partial^2}{\partial x_i'\partial x_j'}\{\overline{U}_i(\mathbf{x}')\overline{U}_j(\mathbf{x}') + \overline{u_i u_j}(\mathbf{x}')\}\frac{d^3\mathbf{x}'}{|\mathbf{x}-\mathbf{x}'|} \tag{4.49}$$

and

$$p(\mathbf{x}) = \frac{\rho}{4\pi}\int\!\!\int\!\!\int \frac{\partial^2}{\partial x_i'\partial x_j'}\{\overline{U}_i(\mathbf{x}')u_j(\mathbf{x}') + \overline{U}_j(\mathbf{x}')u_i(\mathbf{x}') + u_i(\mathbf{x}')u_j(\mathbf{x}') - \overline{u_i u_j}(\mathbf{x}')\}\frac{d^3\mathbf{x}'}{|\mathbf{x}-\mathbf{x}'|} \tag{4.50}$$

respectively. Here, the Cauchy kernel, $|\mathbf{x}-\mathbf{x}'|^{-1}$, accounts for the diminution with distance of the effects of a "pressure source," situated at $\mathbf{x}'$, whose strength is determined by the velocity field.

Using these expressions for the pressure, it is straightforward to calculate the one-point pressure–velocity correlations, which occur in equations such as the turbulent energy equation, (4.33). In fact, in (4.33) it is the correlation of the pressure derivative and velocity which is needed, but the derivative can be taken outside the average since $\mathbf{u}$ has zero divergence. Thus it is the divergence of the quantity

$$\begin{aligned}\overline{p(\mathbf{x})u_k(\mathbf{x})} = \frac{\rho}{4\pi}\int\!\!\int\!\!\int \frac{\partial^2}{\partial x_i'\partial x_j'}\{&\overline{U}_i(\mathbf{x}')\overline{u_j(\mathbf{x}')u_k(\mathbf{x})} + \overline{U}_j(\mathbf{x}')\overline{u_i(\mathbf{x}')u_k(\mathbf{x})}\\ &+\overline{u_i(\mathbf{x}')u_j(\mathbf{x}')u_k(\mathbf{x})}\}\frac{d^3\mathbf{x}'}{|\mathbf{x}-\mathbf{x}'|}\end{aligned} \tag{4.51}$$

which occurs in the turbulent energy equation. Equation (4.51) gives the one-point pressure–velocity correlation in terms of double and triple velocity correlations at *two* points. Similar behavior is found if one expresses the pressure–velocity term in (4.32) using (4.50). The appearance of third-order moments reflects the closure problem, while that of two-point quantities is due to nonlocality. This naturally leads to interest in methods that deal in multipoint moments, since the calculation of a one-point pressure–velocity correlation requires such moments. Integrals such as that occurring in (4.51) are limited in spatial extent because the velocity moments in the integrand drop to zero as the separation between the two points, $\mathbf{x}$ and $\mathbf{x}'$, increases beyond the correlation length of the turbulence. Boundary terms in the pressure–velocity correlation, if boundaries there were, would no doubt decay for similar reasons, and can thus be neglected many correlation lengths away from boundaries.

According to (4.48) and as reflected in (4.50) and (4.51), the fluctuating pressure is determined by two types of source terms. The first type consists of products of the mean and fluctuating velocities and is *linear* in the fluctuating velocity. The second type is quadratic (*nonlinear*) in the fluctuating velocity. The two types of source terms produce two components of the fluctuating pressure, which we denote by $p^{(1)}$ for the part due to the linear source term and $p^{(2)}$ for the nonlinear part. Thus

$$\nabla^2 p^{(1)} = -\rho \frac{\partial^2}{\partial x_i \partial x_j} \{\overline{U}_i u_j + \overline{U}_j u_i\} \tag{4.52}$$

and

$$\nabla^2 p^{(2)} = -\rho \frac{\partial^2}{\partial x_i \partial x_j} \{u_i u_j - \overline{u_i u_j}\} \tag{4.53}$$

The two pressure components, $p^{(1)}$ and $p^{(2)}$, can have quite different effects on the evolution of turbulence, as we shall see from examples later. The nonlinear pressure component, $p^{(2)}$, does not depend on the mean flow directly and will be present even in the absence of a mean flow (i.e., turbulence decaying alone). It is represented by the triple velocity correlation of equation (4.51).

Assuming homogeneous turbulence, there is no pressure term in the turbulent energy equation, (4.37), and thus the effects of pressure fluctuations only make themselves felt in the progressive redistribution of turbulent energy among different directions, either towards or away from isotropy. This is shown by the appearance of the pressure–rate-of-strain correlation term in the equation, (4.38), for the velocity correlation tensor, $\overline{u_i u_j}$, whose departure from the isotropic form $\overline{u_i u_j} = q^2 \delta_{ij}/3$ is an important measure of anisotropy. We can write the pressure–rate-of-strain term in (4.38) as

$$\overline{\frac{1}{\rho} p \left\{ \frac{\partial u_i}{\partial x_j} + \frac{\partial u_j}{\partial x_i} \right\}} = \overline{\frac{1}{\rho} p^{(1)} \left\{ \frac{\partial u_i}{\partial x_j} + \frac{\partial u_j}{\partial x_i} \right\}} + \overline{\frac{1}{\rho} p^{(2)} \left\{ \frac{\partial u_i}{\partial x_j} + \frac{\partial u_j}{\partial x_i} \right\}} \tag{4.54}$$

using the decomposition into linear and nonlinear parts of the fluctuating pressure.

If the turbulence happens to be isotropic at some instant of time and there are no boundaries, then the second term in (4.54), which arises from the nonlinear pressure, is zero. This comes about because $p^{(2)}$ does not depend on the mean flow, but only on the turbulence, and so the second term in (4.54) will be an isotropic tensor and hence has the form of a constant multiplying δ_{ij}. We also know that the trace of that tensor is zero because of incompressibility, (4.6), of the fluctuating velocity field. It follows that the second term in (4.54) is indeed zero if the turbulence is isotropic and it has often been supposed that, for anisotropic turbulence, there is a direct relationship between the tensor formed by this term and $\overline{u_i u_j}$. This is a closure hypothesis, since the term we are interested in is third order in the fluctuations and we are trying to relate it to a second-order quantity. The explicit form of this closure hypothesis (first proposed by Rotta (1951)) is

$$\frac{1}{\rho}\overline{p^{(2)}\left\{\frac{\partial u_i}{\partial x_j}+\frac{\partial u_j}{\partial x_i}\right\}} = -C\bar{\varepsilon}\, b_{ij} \tag{4.55}$$

where C is a numerical constant, and

$$b_{ij} = \frac{\overline{u_i u_j}}{\overline{q^2}} - \frac{1}{3}\delta_{ij} \tag{4.56}$$

is a nondimensional measure of anisotropy. The term in δ_{ij} is included in the definition, (4.56), so that the trace of b_{ij} is zero, like the pressure–rate of strain on the left of (4.55), which is thus satisfied identically when its trace is taken. The factor $\bar{\varepsilon}$ on the right of (4.55) is the mean energy dissipation, representing the rate of energy transfer by the turbulent cascade towards the small dissipative scales. The nonlinear pressure–strain term on the left of (4.55) has the same dimensions as the dissipation rate, $\bar{\varepsilon}$, and both are characteristic of the internal dynamics and intensity of turbulence. Furthermore, the two sides of (4.55) are zero in the case of isotropic turbulence in the absence of boundaries. They are each a measure of anisotropy, and it is thus at least plausible that they could be proportional for anisotropic turbulence. Nonetheless, (4.55) is merely a hypothesis and has not been justified rigorously, any more than other closure assumptions.

To see the consequences of the above closure of the nonlinear pressure–strain term for homogeneous turbulence, we suppose that there is no mean flow so that the turbulence is decaying alone. Without mean flow, $p^{(1)} = 0$ and so the fluctuating pressure consists of the nonlinear component, $p = p^{(2)}$, alone. We introduce (4.55) into equation (4.38) and use (4.45) to express the dissipative term. The result can be rewritten as an equation for the tensor b_{ij}, using the energy equation (4.37). Thus we find that

$$\frac{db_{ij}}{dt} = -(C-2)\frac{\bar{\varepsilon}}{\overline{q^2}} b_{ij} \tag{4.57}$$

which shows that each component of b_{ij} is decaying in magnitude if $C > 2$ and growing if $C < 2$. Since b_{ij} is a measure of anisotropy and it is thought that homogeneous turbulence, left to itself, will usually tend to move in the direction of isotropy, rather than away from it, we infer that $C > 2$. More specifically, integration of (4.57) using (4.37) yields the solution $b_{ij} \propto \overline{q^2}^{(C-2)/2}$ and, since $\overline{q^2} \to 0$ as $t \to \infty$, each nonzero component of b_{ij} grows without bound if $C < 2$, but decays to zero if $C > 2$. It can then be shown that, except in the isotropic case, $b_{ij} = 0$, $C < 2$ leads to a correlation matrix, $\overline{u_i u_j}$, which eventually ceases to be positive definite, thus violating a fundamental requirement and again showing that $C \geq 2$ for the closure to be at all believable. Observe that the pressure–strain term in (4.38) is essential to avoid increasing anisotropy in the absence of a mean flow, for without it $C = 0$ and we obtain unboundedly growing values of b_{ij} unless the initial $\overline{u_i u_j}$ is precisely isotropic. On the other hand, the viscous term tries to amplify anisotropy of $\overline{u_i u_j}$, because, according to (4.45), it produces decay of its diagonal components, but not of the off-diagonal ones. If $C = 2$, the closure approximation predicts that the two terms balance and that b_{ij}, and hence $\overline{u_i u_j}/\overline{q^2}$, do not change with time, but when $C > 2$, as it should be, b_{ij} tends towards zero and $\overline{u_i u_j}$ approaches an isotropic form. We should perhaps emphasize again the approximate nature of the closure hypotheses underlying (4.57) and caution the reader not to place too much confidence in the detailed results, which nonetheless give a qualitative feel for the behavior of homogeneous turbulence without mean flow and the general effects on its anisotropy of the viscous and pressure–strain terms in (4.38).

The linear pressure component, $p^{(1)}$, results from interaction of the turbulence with the mean flow and, in the presence of mean flow, the corresponding part of the pressure–strain correlation (the first term on the right of (4.54)) has no reason to be zero even for turbulence which happens to be isotropic at some instant of time. There is no real need to introduce a closure hypothesis for this term, since it is linear in the fluctuation velocity and so its correlation with the strain can be expressed in terms of second-order velocity moments, that is, moments of the same order as those whose evolution equation, (4.38), we are considering. Nonetheless, as was observed following equation (4.51), the pressure–strain correlation at a single point is an integral over the two-point velocity correlations, a fact that takes the detailed analysis beyond the one-point models considered in this chapter and into the realm of the spectral methods described in later chapters. For present purposes, it suffices to describe two limiting cases, for which one does not require any knowledge of later work.

Firstly, turbulence alone (i.e., with no mean flow) has no linear pressure component. This provides one limiting case. On the other hand, a mean flow will strain and distort the turbulence and, if mean-flow straining is sufficiently strong, there will be no time for interaction of the turbulence with itself during the rapid mean-flow straining of the turbulence. This is the opposite limit, resulting in rapid distortion theory of turbulence. It makes the nonlinear pressure component negligible and indeed, the evolution of the turbulence is then described by a linear theory, in which the term, $u_i u_k - \overline{u_i u_k}$, which is nonlinear in the fluctuations, is ignored in equation (4.31). That is, the nonlinear terms are dropped in rapid distortion theory, as also is viscosity, in keeping with the large Reynolds number of turbulence. We will consider rapid distortion theory in one of our examples later in this chapter and will find that the result is a reversible straining of the turbulence by the mean flow. It is the self-interaction of turbulence which leads to irreversible evolution of the turbulence, but such self-interaction is absent in the rapid distortion approximation, although it is the only mechanism for evolution of turbulence in the absence of mean flow and viscosity. The above remarks explain the nomenclature "rapid" and "slow" pressure terms, which are sometimes used to refer to the terms arising from the linear and nonlinear parts of the fluctuating pressure field.

It is also interesting to note that, given a turbulent field in a mean flow, if the mean flow is suddenly changed by, for instance, applying an impulsive mean pressure at the boundaries (see, e.g., Batchelor (G 1967) for a discussion of impulsive changes in incompressible flows, which represent limiting cases of rapid changes), the turbulence will remain the same. An impulsive mean pressure can instantaneously alter the mean velocity by any irrotational amount, but leaves the mean vorticity and fluctuating velocity the same. In principle, one could also envisage impulsive mean body forces, allowing arbitrary step changes in the mean velocity, although it is less obvious how such impulses might be achieved in practice. In these cases, the nonlinear fluctuating pressure field, which does not depend on the mean flow, will remain unchanged, while the linear part of the fluctuating pressure field undergoes a sudden change. This step change will result in a corresponding change in the Reynolds stress production, (4.54). The fact that part of the pressure term (the linear part) can be made to change suddenly, but the other cannot, indicates that the two play rather different roles in the development of the turbulence. We will return to questions of linear and nonlinear pressure effects in a later example. It should also be noted that terms (II) and (III) in the Reynolds stress equation, (4.32), are also linear (they arise from terms of (4.31) that are linear in the fluctuating velocity) and, like the linear pressure term, can undergo impulsive changes.

4.4 The Vorticity

Before deriving the equations governing the mean and fluctuating parts of the vorticity, we want to remind the reader of the means by which the vorticity in an inviscid fluid is amplified by stretching of vortex lines. This is thought to be one of the main physical mechanisms by which the small scales of turbulence are produced and maintained. The description we give here is not intended as a presentation to the reader who has never met the ideas before and provides an overview (see any basic textbook on fluid dynamics for more details). Although these concepts are not directly related to one-point averaging methods, they are of general importance for an understanding of the role of vorticity in turbulent flows.

In the basic equations of incompressible fluid flow, (4.3) and (4.4), both velocity and pressure terms occur and one can choose to focus on one quantity or the other, since both are interrelated. In Section 4.3, we took the divergence of (4.3) to derive the Poisson equation for the pressure in terms of the velocity. One can also take the curl of (4.3) to eliminate the pressure and obtain the Helmholtz equation, (4.63), which contains the velocity and vorticity fields. Since both the vorticity, $\mathbf{\Omega} = \nabla \times \mathbf{U}$, and the velocity, $\mathbf{U}$, are present, one needs to express one in terms of the other to close the problem. The velocity can be written as an integral over the vorticity field (Cauchy's formulation), from which a nonlinear integro-differential equation for the vorticity is obtained. The integral nature of this equation reflects the underlying nonlocality of the Navier–Stokes equations, which we saw in the previous section from consideration of the pressure. The equation is apparently more complicated than the original system, (4.3), (4.4). Nonetheless, the vorticity appears in several appealingly visual formulations of inviscid fluid dynamics.

In the absence of viscosity, Kelvin's circulation theorem tells us that the circulation of a closed curve that moves with the fluid is constant. For instance, if the curve shrinks in size due to flow convergence, and hence becomes shorter, the circulation stays the same and the velocity around the curve must increase to compensate. Some classical corollaries of this result can be described in terms of vortex lines, which are defined as everywhere tangential to the vorticity vector, and vortex tubes, which are thin tubelike surfaces made up of vortex lines. Vortex lines and tubes can be thought of as moving with the fluid flow as if they consisted of fluid particles. Furthermore, the strength of a tube can be defined as the circulation in a closed path on its surface going once around the tube and remains unchanged with time as the tube is convected. The circulation equals the strength of vorticity inside the tube times its cross-sectional area. In an incompressible fluid, if a tube is stretched by the flow, its diameter will decrease by conservation of volume, the velocity due to the tube will increase at the tube surface, and the vorticity it contains will be amplified. Both the reduction of diameter and the amplification of vorticity are precursors of the cascade process of turbulence.

As noted above, although these results are physically very appealing, the velocity field that convects vortex lines (or tubes) is itself, at least in part, generated by the vorticity they represent. Thus, one cannot, in general, specify the velocity field in advance and work out the effects on vorticity: motion of the vortex lines changes the velocity field by which they are, in turn, convected. This severe coupling is expressed by the integro-differential equation we mentioned above.

We want to briefly show the basic results that, in an inviscid fluid, vortex lines can be thought of as convected by the flow and that vorticity is amplified through stretching of vortex lines by the flow.

The vorticity is defined by $\mathbf{\Omega} = \nabla \times \mathbf{U}$, while a vortex line is a curve that is everywhere parallel to $\mathbf{\Omega}$. Consider the material points making up a vortex line at time $t = t_0$. These material points move at the flow velocity and we want to demonstrate that they continue to form a vortex line. This suffices to show that vortex lines can be thought of as being convected by the flow. Let two of the material points making up the initial line be separated by the infinitesimal vector displacement $\delta\mathbf{r}$. We need to prove that, as the two material points move with the flow, $\delta\mathbf{r}$ remains parallel to $\mathbf{\Omega}$, and, that $|\mathbf{\Omega}|/|\delta\mathbf{r}|$ does not vary with time. Thus, if $|\delta\mathbf{r}|$ increases, corresponding to stretching, then $|\mathbf{\Omega}|$ grows proportionately, representing amplification of vorticity. Of course, if a vortex line is shortened, rather than stretched, the same result shows that the vorticity is reduced.

The vorticity in an inviscid, incompressible fluid satisfies the equation

$$\frac{d\Omega_i}{dt} = \Omega_j \frac{\partial U_i}{\partial x_j} \tag{4.58}$$

where

$$\frac{d}{dt} = \frac{\partial}{\partial t} + U_j \frac{\partial}{\partial x_j}$$

is the material derivative, now with respect to the total flow, $\mathbf{U}$. The vector displacement, $\delta\mathbf{r}$, of material points evolves with time due to the difference in velocity between the two points considered. Thus we have

$$\frac{d\delta r_i}{dt} = \delta r_j \frac{\partial U_i}{\partial x_j} \tag{4.59}$$

where we have expressed the difference in velocity at the two points by a one-term Taylor's series. Equation (4.59) has exactly the same form as (4.58) (it is this fact that allows us to derive the required result). Suppose that $\delta\mathbf{r}$ is parallel to $\mathbf{\Omega}$ at time $t = t_0$ and define a vector, $\boldsymbol{\alpha}$, by $\alpha_i(t) = \Omega_i(t)|\delta\mathbf{r}|_0/|\mathbf{\Omega}|_0$, where the subscript 0 indicates $t = t_0$. Thus, we have $\boldsymbol{\alpha} = \delta\mathbf{r}$ at $t = t_0$ because the vectors $\delta\mathbf{r}$ and $\mathbf{\Omega}$ are parallel. From (4.58), we find that $\boldsymbol{\alpha}$ satisfies

$$\frac{d\alpha_i}{dt} = \alpha_j \frac{\partial U_i}{\partial x_j} \tag{4.60}$$

which is the same equation as governs $\delta\mathbf{r}$. Since $\boldsymbol{\alpha} = \delta\mathbf{r}$ at $t = t_0$ and the two vectors satisfy the same first-order equations, (4.59) and (4.60), they are equal at all times and the material line remains a vortex line with

$$\delta r_i(t) = \frac{|\delta\mathbf{r}|_0}{|\boldsymbol{\Omega}|_0} \Omega_i(t) \tag{4.61}$$

from which it follows that $|\delta\mathbf{r}|/|\boldsymbol{\Omega}| = |\delta\mathbf{r}|_0/|\boldsymbol{\Omega}|_0$ by taking the magnitudes of the two sides of the equation. Hence $|\boldsymbol{\Omega}|$ increases or decreases in proportion to $|\delta\mathbf{r}|$, which is the required result.

Amplification of vorticity by vortex line stretching can be important in laminar flows at high Reynolds number. For instance, the well-known bathtub vortex, which occurs when water runs out of the bottom of a container, is a result of this mechanism. Approximating the flow as axisymmetric about a vertical axis and imagining a horizontal circle, centered on the axis and moving with the fluid, this circular material curve is drawn towards the outlet, its radius decreases, and conservation of circulation implies that the circulating velocity increases. Alternatively, one can think of vortex lines as drawn out vertically and compressed in horizontal projection, which causes vortex lines starting far from the outlet to be reorientated toward the vertical by flow convection as they approach the outlet. At the same time, vorticity is amplified by vertical stretching of the vortex lines. It is, of course, vorticity preexisting in the container that is amplified by this mechanism, since vorticity cannot be created in an inviscid fluid. This example shows that even laminar flows can amplify existing vorticity by vortex line stretching, but the process is much more extreme and complicated in the development of turbulence.

Imagine a very high-Reynolds-number flow without solid boundaries, consisting of an intense, localized region of disorganized, large-scale vorticity, which has been produced by external forces and will become turbulent, but for which the velocity field as yet varies only over large scales. Owing to the assumed high Reynolds number at these scales, the effects of viscosity have not so far been called into play, and, thanks to the fluid motion, vortex line stretching may cause amplification of vorticity and thinning of vortex tubes. In such developing turbulence, convection stretches, distorts, and folds the thinning vortex tubes. Amplification of vorticity results in increasing velocity gradients (recall the definition of vorticity in terms of velocity derivatives) and thinning leads to smaller and smaller scales. This process, which is far from being fully understood, despite its central importance to turbulence dynamics, preserves the kinetic energy of the flow, because viscous dissipation has yet to act. The velocity itself cannot rise dramatically because the kinetic energy per unit mass is $U_iU_i/2$ and, when integrated over the flow, is constant. However, velocity gradients greatly increase, reflecting the appearance of smaller and smaller scales by vortex tube thinning and amplification. That is, the velocity field develops large derivatives and may become "furry," in the language of the previous chapter.

Through vortex stretching, the vorticity can be greatly amplified locally. If, however, the vorticity is integrated over some large-scale volume of the flow, the volume integral can be expressed in terms of a surface integral of the velocity field using the definition of the vorticity and vector calculus. Since, owing to energy conservation, the velocity field remains of the same order of magnitude during the production of the small vortical scales, it is clear that the volume-averaged vorticity cannot increase greatly, despite large amplification of the magnitude of local vorticity. The large-amplitude, small-scale vorticity variations average out when a large-scale volume integral is performed.

As the smallest length scales present in the flow decrease, viscosity eventually becomes important because it is represented in the equation of motion, (4.3), by a term that contains second derivatives, rather than first-order ones. In other words, the effective Reynolds number is then no longer large when determined at the smallest scales. Viscosity limits the amplification of vorticity and sets a lower bound for the smallest scale attainable by stretching. It introduces a diffusive term in the vorticity equation, which becomes (4.63) for a viscous fluid, rather than its inviscid version, (4.58): this diffusive term counterbalances the effects of vortex stretching at the finest scales, whose size is measured by the Kolmogorov scale, η. Viscosity also leads to kinetic energy dissipation at the smallest scales (see equation (4.42), which relates the dissipation rate and the turbulent vorticity fluctuations). Once such scales have come into existence, which takes a certain time, developed turbulence appears. The velocity field is furry and contains all sizes from the correlation scale down to the Kolmogorov scale, as described in the previous chapter. The vorticity is dominantly at the finest scales, like other quantities involving velocity derivatives, and there is a continuous flux of energy through the cascade from the decay of large-scale eddies, which is dissipated by viscosity at the smallest scales. Vortex line stretching is thought to play an important role in setting up, and no doubt in maintaining, the continuum of scales characteristic of high-Reynolds-number turbulence.

There is an important class of flows for which no vortex stretching occurs. Any strictly two-dimensional flow has vortex lines perpendicular to the plane of the flow and hence no line stretching is possible. In such a flow, vorticity is convected unchanged with the flow. The absence of vortex stretching and the associated vorticity amplification makes two-dimensional flows a very special case. Turbulence, as we understand the term, cannot take place in two dimensions. The case of a strictly axisymmetric, inviscid flow, although it may involve vortex stretching, is also rather special: vortex lines are circles centered on the axis and, as any one of these circles is convected by the flow, the associated vorticity increases or decreases proportional to its radius (as in the bathtub example). These flows are too special to produce the continuum of scales required for a turbulent flow. This is not to say that two-dimensional or axisymmetric *mean* flows cannot be turbulent; the intrinsically three-dimensional phenomenon of turbulence *can* occur in such mean flows.

Once turbulence has fully developed, the vorticity field has random variations on small length scales, comparable to the Kolmogorov scale, and fluctuates radically over such distances, both in magnitude and direction. Alongside the large-amplitude small-scale vorticity fluctuations there is small-scale dissipation of energy, supplied from the large scales of turbulence via the cascade. The two processes, vortex stretching and the cascade of energy, go hand-in-hand, but the precise relationship between them is far from clear in developed turbulence. It is much harder to identify the role of large-scale vortex stretching in developed turbulent flow, because there is already small-scale vorticity present and one cannot follow the process of vortex stretching of larger scales through to the production of smaller ones with higher vorticity, as the flow develops in time. Indeed, since vorticity is defined via derivatives of the velocity, which are dominated by the small scales, those scales are predominant when one considers the vorticity field. The vorticity is mainly associated with the small scales of turbulence and such intense, fine-scale vorticity variations mask the weaker, larger-scale vorticity field in developed flow. Naturally, one may speculate that vortex stretching should continue to play an important role in the generation and maintenance of the small scales.

The above discussion was aimed at turbulent flows in the absence of boundaries. Solid boundaries are often an important source of turbulence, and ultimately, all vorticity arises at boundaries, even if it is subsequently amplified greatly by vortex stretching in the body of the fluid. The dynamics of turbulence in the viscous layer at a wall are rather different from those away from the surface. The effective Reynolds number is relatively low, and both turbulent energy dissipation and production are high, the latter due to the large mean shear. However, it is still vortex stretching that is thought to amplify turbulent vorticity and produce smaller-scale vortical structures. The range of sizes present is comparatively limited, owing to the low Reynolds number. Outside the viscous layer, the effective Reynolds number grows with distance from the wall and the turbulence becomes qualitatively similar to that occurring well away from the wall, although it may still be significantly affected by the presence of the surface. In particular, it contains a growing range of scales, reflecting the increasing value of Re_L. If distance from the wall becomes large compared with the turbulent correlation length, L, the effects of the wall can be neglected.

Having now set the scene, we consider the equations for the mean, $\overline{\Omega_i}$, and fluctuating part, ω_i, of the vorticity in a turbulent flow. The fact that the vorticity is defined as a curl means that it has zero divergence

$$\frac{\partial \Omega_i}{\partial x_i} = 0 \tag{4.62}$$

which, being a linear equation, implies that both $\overline{\Omega_i}$ and ω_i have zero divergence considered separately. For the same reason, $\overline{\Omega_i}$ is the curl of $\overline{U_i}$ and ω_i is the curl of u_i. From earlier discussions, we expect that, at high Reynolds number, ω_i will be determined by the small turbulence scales and be large compared with $\overline{\Omega_i}$, because ω_i and $\overline{\Omega_i}$ are respectively defined in terms of velocity derivatives of u_i and $\overline{U_i}$, via the curl operator.

With viscosity, the Helmholtz equation for vorticity is

$$\frac{\partial \Omega_i}{\partial t} + \underbrace{U_j \frac{\partial \Omega_i}{\partial x_j}}_{\text{Convection}} = \underbrace{\Omega_j \frac{\partial U_i}{\partial x_j}}_{\text{Stretching}} + \underbrace{\nu \frac{\partial^2 \Omega_i}{\partial x_j \partial x_j}}_{\text{Viscous diffusion}} \tag{4.63}$$

where, as indicated by the annotation, the first term on the right is the vortex-stretching term, as in (4.58), and the second is due to viscous diffusion. Taking the average of (4.63), and using the split into mean and fluctuating velocities and vorticities gives

$$\underbrace{\frac{\partial \overline{\Omega_i}}{\partial t}}_{\text{Unsteady flows}} + \underbrace{\overline{U_j}\frac{\partial \overline{\Omega_i}}{\partial x_j}}_{\text{Mean flow advection}} = \underbrace{\overline{\Omega_j}\frac{\partial \overline{U_i}}{\partial x_j}}_{\text{Mean flow stretching}} - \frac{\partial}{\partial x_j}\left\{\underbrace{\overline{\omega_i u_j}}_{\text{Turbulent advection}} - \underbrace{\overline{u_i \omega_j}}_{\text{Turbulent stretching}}\right\} + \underbrace{\nu \frac{\partial^2 \overline{\Omega_i}}{\partial x_j \partial x_j}}_{\text{Viscous diffusion}} \tag{4.64}$$

which is the equation for the mean vorticity. The terms labeled turbulent advection and turbulent stretching represent the effects of the turbulence on the mean-flow vorticity (analogous to the Reynolds stress term in (4.9)). All others are as if there were no turbulence. An equation for the square of the mean vorticity can be easily obtained by multiplication of (4.64) by $\overline{\Omega_i}$. The result can be expressed as

$$\underbrace{\frac{\partial \frac{1}{2}\overline{\Omega_i}\,\overline{\Omega_i}}{\partial t}}_{\text{Unsteady flows}} + \underbrace{\overline{U_j}\frac{\partial \frac{1}{2}\overline{\Omega_i}\,\overline{\Omega_i}}{\partial x_j}}_{\text{Mean flow advection}} = \underbrace{\overline{\Omega_i}\,\overline{\Omega_j}\frac{\partial \overline{U_i}}{\partial x_j}}_{\text{Mean flow stretching}}$$

$$- \overline{\Omega_i}\frac{\partial}{\partial x_j}\left\{\underbrace{\overline{\omega_i u_j}}_{\text{Turbulent advection}} - \underbrace{\overline{u_i \omega_j}}_{\text{Turbulent stretching}}\right\} + \underbrace{\nu \frac{\partial^2 \frac{1}{2}\overline{\Omega_i}\,\overline{\Omega_i}}{\partial x_j \partial x_j}}_{\text{Viscous transfer}} - \underbrace{\nu \frac{\partial \overline{\Omega_i}}{\partial x_j}\frac{\partial \overline{\Omega_i}}{\partial x_j}}_{\text{Viscous dissipation}} \tag{4.65}$$

The equation for the fluctuating vorticity is

$$\frac{\partial \omega_i}{\partial t} + \overline{U_j}\frac{\partial \omega_i}{\partial x_j} =$$

$$\overline{\Omega_j}\frac{\partial u_i}{\partial x_j} + \omega_j \frac{\partial \overline{U_i}}{\partial x_j} - u_j \frac{\partial \overline{\Omega_i}}{\partial x_j} - \frac{\partial}{\partial x_j}\left\{\omega_i u_j - \omega_j u_i + \overline{\omega_j u_i} - \overline{\omega_i u_j}\right\} + \nu \frac{\partial^2 \omega_i}{\partial x_j \partial x_j} \tag{4.66}$$

which we can multiply by ω_i, average, and rearrange to obtain the equation for the mean-squared vorticity fluctuation. Thus, we obtain the somewhat lengthy equation

$$\frac{\partial \frac{1}{2}\overline{\omega_i\omega_i}}{\partial t} + \overline{U_j}\frac{\partial \frac{1}{2}\overline{\omega_i\omega_i}}{\partial x_j} = \underbrace{\overline{\omega_i\omega_j}\frac{\partial \overline{U_i}}{\partial x_j} + \overline{\Omega_j}\,\overline{\omega_i\frac{\partial u_i}{\partial x_j}} - \overline{\omega_j u_i}\frac{\partial \overline{\Omega_i}}{\partial x_j}}_{\text{Mean flow/turbulence coupling}}$$

$$+ \underbrace{\overline{\omega_i\omega_j\frac{\partial u_i}{\partial x_j}}}_{\substack{\text{Turbulent stretching}\\ \text{of fluctuating}\\ \text{vorticity}}} \underbrace{- \frac{\partial}{\partial x_j}\left(\frac{1}{2}\overline{u_j\omega_i\omega_i}\right) + \nu \frac{\partial^2 \frac{1}{2}\overline{\omega_i\omega_i}}{\partial x_j \partial x_j}}_{\substack{\text{Transfer terms}\\ \text{due to inhomogeneity}}} - \underbrace{\nu \overline{\frac{\partial \omega_i}{\partial x_j}\frac{\partial \omega_i}{\partial x_j}}}_{\substack{\text{Viscous}\\ \text{dissipation}}} \tag{4.67}$$

which is sometimes called the turbulent *enstrophy* equation – enstrophy being defined as $\overline{\Omega_i\Omega_i}/2 = \overline{\Omega_i}\,\overline{\Omega_i}/2 + \overline{\omega_i\omega_i}/2$, in which the component due to the mean vorticity is generally negligible in high Reynolds turbulence because, as noted above, the fluctuations contain the small scales, which dominate the velocity derivatives, and hence the vorticity. We leave it to the reader to determine which of the nonviscous terms in (4.67) arise from stretching and which from advection. Naturally, the term labeled dissipation does not represent *energy* dissipation, but rather the tendency of viscosity to damp out vorticity fluctuations (Taylor (1938)). This is related to energy dissipation, of course, since both processes occur at the finest scales of turbulence, where the vorticity fluctuations themselves take place. The fact that the spatial derivatives of fluctuating velocities in general, and ω_i in particular, are properties of the small scales of turbulence, whereas derivatives of mean-flow quantities are determined by the largest scales, means that different terms of (4.67) have widely disparate orders of magnitude, once the cascade has had time to create the small scales. This is thanks to the large Reynolds number which, as we have noted before, leads to large derivatives at the smallest scale, η. The fourth and final terms on the right-hand side of (4.67) can be shown to dominate the others in the limit $\mathrm{Re}_L \to \infty$. The fourth is dominant because it consists of a product of three velocity derivatives taken at the fine scales, whereas the other nonviscous terms contain at most two (note that the derivatives implicit in ω_i need to be accounted for here). Of the viscous terms, the second is a product of two second-order velocity derivatives at the small scales, thereby dominating the first of the viscous terms. If only the dominant terms are retained on the right of (4.67), the result is

$$\frac{\partial \frac{1}{2}\overline{\omega_i\omega_i}}{\partial t} + \overline{U}_j \frac{\partial \frac{1}{2}\overline{\omega_i\omega_i}}{\partial x_j} = \underbrace{\overline{\omega_i\omega_j \frac{\partial u_i}{\partial x_j}}}_{\text{Turbulent stretching}} - \underbrace{\nu \overline{\frac{\partial \omega_i}{\partial x_j}\frac{\partial \omega_i}{\partial x_j}}}_{\text{Viscous dissipation}} \tag{4.68}$$

of which the right-hand side is now precisely as obtained if we were to assume homogeneous turbulence without mean-flow gradients in (4.67).[5] This is because the terms on the right of (4.68) are dominated by the dissipative scales, whereas inhomogeneities or mean-flow gradients produce variations of the flow on much longer scales. Indeed, equation (4.68) tells us nothing about the dynamics of turbulence at larger scales: by considering the vorticity, one focuses on the small-scale properties of turbulence. For strictly homogeneous turbulence, the second term on the left of (4.68) is also zero. We stress that, with mean-flow gradients or in the absence of homogeneity, accuracy of the approximation leading from (4.67) to (4.68) requires a large Reynolds number.

One can go one step further in simplifying the enstrophy equation, once turbulence is fully developed and again assuming $\mathrm{Re}_L \to \infty$. The time evolution of the statistical properties of all turbulence scales are then controlled by that of the large scales, via the cascade. The left-hand side of (4.68), in which $\overline{\omega_i\omega_i}$ has only two small-scale velocity derivatives, is consequently negligible compared with the first term on the right, which contains three such derivatives. It follows that, to a good approximation, we may drop the left-hand side of (4.68), which becomes an expression of equilibrium between turbulent stretching and viscous destruction of vorticity fluctuations. Notice that, for these two terms to balance, the stretching term must be positive, that is, vortex stretching amplifies turbulent vorticity in the mean, as expected. Stretching is responsible for counteracting viscous diffusion of vorticity at the small scales that are described by (4.68). The stretching term presumably plays an important, if as yet poorly understood, role in the complex interactions of turbulence with itself. It is cubic in the velocity derivatives and, from a probability theory point of view, is a sum of velocity-derivative cubic moments. Thus, turbulent vortex-

[5] Bearing in mind the second equality of (4.42), multiplying the right-hand side of (4.68) by the twice the viscosity gives the time derivative of $\overline{\varepsilon}$. However, it seems to us that the interpretation in terms of vorticity is more physical.

stretching effects on vorticity fluctuations depend on rather subtle statistical properties of $\partial u_i/\partial x_j$.

While we are on the subject of small-scale vorticity in developed turbulence, we should mention the apparently ordered vortical structures first observed in numerical simulations (Figure 4.2a) and experimental visualizations of turbulent flows (Figure 4.2b). These structures take the form of persistent, long, thin filaments of concentrated vorticity, whose widths appear to be of order η and whose lengths are much greater than that. The existence of such structures is surprising because, as discussed above, one usually thinks of vorticity as randomly varying over the smallest scales, rather than showing structures extending over significant distances, albeit with small-scale widths. It is as yet far from clear what, if any, significance these filaments have for the overall dynamics of turbulence (see the discussion and references in Frisch (G 1995), section 8.9). They seem to carry little of the overall turbulent kinetic energy and, perhaps more surprisingly, little dissipation. Moreover, clearly identifiable, long filaments appear to become less frequent as the Reynolds number increases.

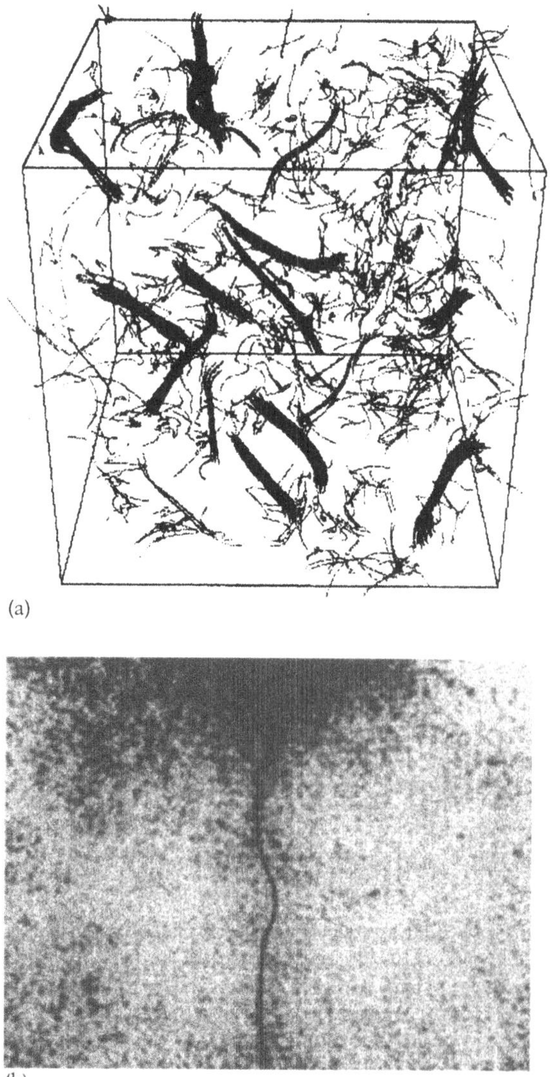

(a)

(b)

Figure 4.2. Ordered structures in the form of long vortex filaments, which are observed at small scales in (a) direct numerical simulation of turbulence, and (b) experimental visualization. (Reproduced with permission: (a) She, Jackson, and Orszag (1991); (b) Bonn et al. (1993).)

In the context of vorticity formulations, we should also mention the many studies of two-dimensional "turbulence" (for more details and references, see Lesieur (G 1990) and Frisch (G 1995), section 9.7). We have already made clear our view that strictly two-dimensional random flows are sufficiently different in nature from turbulence, as the term is usually understood, as to fall into a different category. In two dimensions and in the absence of viscosity, vorticity is convected unchanged by the flow (since there is no vortex stretching or reorientation) and the initial vortical structures are simply deformed by convection in the plane of the flow. High-Reynolds-number numerical simulations with random initial conditions show the development of thin, folded, sheet-like structures in which vorticity gradients are large, separating regions in which the vorticity gradients are lower. Overall, such two-dimensional flows appear to be quite highly organized. The sheets of high vorticity gradient tend to become thinner due to stretching in the plane of the flow and diffuse under viscosity, which limits the size of the vorticity gradients that can develop. At first sight, there would appear to be some similarity with the turbulent energy cascade in three dimensions, and indeed there is believed to be an associated cascade to smaller scales: not of energy, but of enstrophy, $\overline{\Omega_i \Omega_i}/2$ (Kraichnan (1967)). In the absence of viscosity, enstrophy, like energy, is conserved in two dimensions, and there appears to be an enstrophy cascade, formally similar to the energy cascade in three dimensions, with dissipation of enstrophy by viscosity at the smallest scales. However, although vorticity of smaller and smaller scales develops from an initially smooth vortical distribution, with the smallest scales limited by viscosity, it is found that the *energy* of the flow does the opposite: it goes into larger and larger scales of the motion, formed by agglomeration of the smaller scales of the velocity field. There are therefore two cascade processes at work in two dimensions, with energy (associated with the velocity field) going to larger scales, and enstrophy (associated with the vorticity field) to smaller ones. These and many other interesting results on two-dimensional "turbulence" are of considerable fundamental interest, but it should be recalled that strictly two-dimensional flow is generally unstable at high Reynolds numbers and breaks down to become three dimensional, at which point turbulence proper has different properties. Sufficiently strong constraints, such as stratification, can, however, produce flows that are at least approximately two dimensional at high Reynolds numbers and this remains an area of active research, particularly with atmospheric and oceanic applications in mind.

4.5 Some Examples of Simple Turbulent Flows

In this section we shall examine a number of relatively simple flows that are chosen because they illustrate behavior found in more general flows and, in some cases, lead to considerable simplifications in the averaged equations, thereby allowing a clearer view of the underlying physics.

TWO-DIMENSIONAL CHANNEL FLOW

The channel consists of two parallel, solid walls, occupying the planes $x_2 = 0$ and $x_2 = 2D$, at which the no-slip condition implies that the velocity is zero. The flow is assumed two dimensional, that is, $\overline{U}_3 = 0$ with all statistical properties independent of x_3 and unchanged under reflection in the plane $x_3 = 0$. This does not imply that single realizations of the flow are two dimensional, indeed the turbulence will be three dimensional, but simply that it is two dimensional in the mean. Overall, flow occurs in the x_1-direction, driven by a mean pressure gradient in that direction (see Figure 4.3). At the inlet of the channel, the flow profile may have any form independent of x_3, but at distances from the inlet of about $x_1 = 100D$ and greater, the flow takes on limiting asymptotic behavior independent of the inlet conditions. There, and further along the channel, the mean properties of the turbulent flow, with the exception of the mean pressure (which must have a streamwise gradient to drive the flow), depend only on x_2, there being no further evolution with streamwise distance, x_1, and everything being supposed independent of x_3. This flow, which is

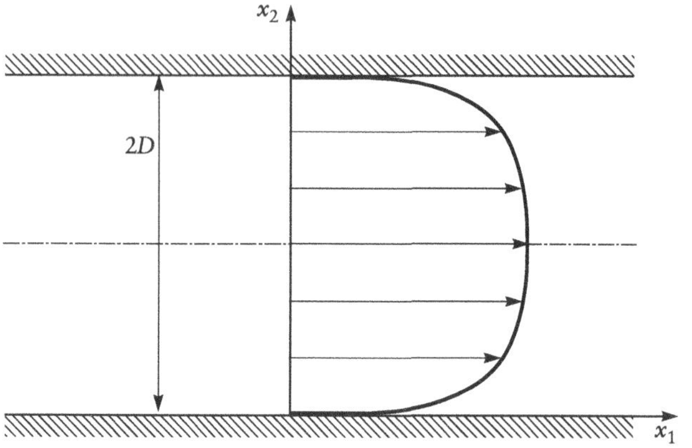

Figure 4.3. Coordinate system and sketch of the mean-flow profile for channel flow.

effectively infinite and homogeneous in x_1 and x_3, is approached in a long rectangular duct, subject to a constant pressure difference between its ends, and which is much wider in one transverse direction (x_3) than in the other (x_2). Well away from the entrance and side walls (those with constant x_3), the flow is in the direction of x_1, that is, along the duct, and is as if the duct were indeed infinite in x_1 and x_3.

The flow is symmetric under reflection in the plane $x_3 = 0$, from which it follows that $\overline{u_1 u_3} = 0$ and $\overline{u_2 u_3} = 0$ (by symmetry, the value of u_3 associated with any given u_1 and u_2 is as likely to be negative as positive). Thus, the only off-diagonal element of the Reynolds stress which is nonzero is $-\rho\overline{u_1 u_2}$, which like other average quantities, depends only on x_2. Furthermore, the mean flow is in the x_1-direction and so $\overline{U_2} = \overline{U_3} = 0$ and $\overline{U_1}$ depends only on x_2. The mean-flow equation, (4.9), therefore takes on the simpler forms

$$\mu \frac{d^2\overline{U_1}}{dx_2^2} = \frac{\partial \overline{P}}{\partial x_1} + \rho \frac{d\overline{u_1 u_2}}{dx_2} \tag{4.69}$$

$$\frac{\partial \overline{P}}{\partial x_2} + \rho \frac{d\overline{u_2^2}}{dx_2} = 0 \tag{4.70}$$

and

$$\frac{\partial \overline{P}}{\partial x_3} = 0 \tag{4.71}$$

Observe that the mean-flow continuity equation, (4.5), is automatically satisfied by the channel flow.

From (4.71), we infer that $\overline{P}(x_1, x_2, x_3)$ does not depend on x_3, while from (4.70) we see that

$$\overline{P} = P_w(x_1) - \rho\overline{u_2^2}(x_2) \tag{4.72}$$

where $P_w(x_1)$ is the mean pressure at the walls, where $u_2 = 0$ according to the no-slip condition, and depends only on the streamwise coordinate x_1. It is the gradient of P_w with respect to x_1 which drives the flow.

Substituting (4.72) into (4.69) we find that we can rewrite the equation as

$$\frac{dP_w}{dx_1} = \frac{d\tau}{dx_2} \tag{4.73}$$

where

$$\tau(x_2) = \mu \frac{d\overline{U_1}}{dx_2} - \rho\overline{u_1 u_2} \tag{4.74}$$

is total mean shear stress: a combination of the Reynolds shear stress and the viscous shear stress. In equation (4.73), the left-hand side depends only on x_1, whereas the right-hand side is solely a function of x_2. Thus both sides are constant: the mean gradient of wall pressure, dP_w/dx_1, which drives the flow, is constant (and negative since the flow is being driven in the direction of increasing x_1). The mean wall pressure is a linear function of x_1. Constancy of $d\tau/dx_2$ implies that

$$\tau = x_2 \frac{dP_w}{dx_1} + \tau_w \tag{4.75}$$

where

$$\tau_w = \mu \frac{d\overline{U_1}}{dx_2}\bigg|_{x_2=0}$$

is another (positive) constant, the value of the mean viscous shear stress at the wall, $x_2 = 0$ (recall that the velocity is zero at the walls).

The channel is symmetric about its center plane, $x_2 = D$, and so τ at the other wall, $x_2 = 2D$, must be $-\tau_w$. Thus, from (4.75), we find that

$$\tau_w = -D \frac{dP_w}{dx_1} \tag{4.76}$$

relating the constant mean wall stress to the constant streamwise wall pressure gradient. This condition reflects a balance of forces: the mean pressure gradient drives the flow and is resisted by the mean viscous stress (friction) at the walls. From (4.76), it follows that (4.75) reads

$$\frac{\tau}{\tau_w} = 1 - \frac{x_2}{D} \tag{4.77}$$

showing that $\tau = 0$ on the channel center plane, $x_2 = D$, and varies linearly across the channel. Observe that we have expressed both the total shear stress, τ, and the mean pressure gradient solely in terms of the single constant, τ_w, using equations (4.76) and (4.77).

Using (4.74) and (4.77), we obtain the equation

$$\tau_w\left(1 - \frac{x_2}{D}\right) + \rho\overline{u_1 u_2} = \mu \frac{d\overline{U_1}}{dx_2} \tag{4.78}$$

The large overall Reynolds number implies that the viscous term in (4.78) is small, except very close to the walls, in thin viscous layers for which the effective Reynolds number is lower. If we set the right-hand side of (4.78) to zero, we find a linear relation, $-\rho\overline{u_1 u_2} = \tau_w(1 - x_2/D)$, for the Reynolds shear stress as a function of x_2. Outside the viscous layers (i.e., over most of the channel), mean velocity gradients result in only small departures from this linear behavior, as borne out by measurements. Measurements further indicate that $\overline{U_1}(x_2)$ is approximately constant in the central part of the channel, so the derivative on the right of (4.78) is relatively small and the linear expression for the Reynolds stress is an even better approximation than one might at first think. Near the walls, there are steep gradients of $\overline{U_1}(x_2)$, similar to turbulent boundary layers, and the shear stress, although still approximately linear in x_2 until we enter the viscous layers, is not quite as close to linearity, becoming less and less so as the wall is approached. Within the viscous layers, the Reynolds shear stress is not even approximately linear and, like $\overline{U_1}(x_2)$, goes to zero at the wall to satisfy the no-slip condition. The mean velocity gradient at the wall is large and equal to τ_w/μ, according to (4.78) with $\overline{u_1 u_2} = 0$ and $x_2 = 0$.

We can also examine the equation, (4.32), for the Reynolds stress. The left-hand side is zero, because the flow is statistically steady, the mean flow has only the one component, $\overline{U_1}$, and $\overline{u_i u_j}$ depends solely on x_2. The diagonal components ($i = j$) of equation (4.32) are

$$0 = -2\overline{u_1 u_2}\frac{d\overline{U_1}}{dx_2} - \frac{d\overline{u_1^2 u_2}}{dx_2} - \frac{2}{\rho}\overline{u_1\frac{\partial p}{\partial x_1}} + 2\nu\overline{u_1\frac{\partial^2 u_1}{\partial x_k \partial x_k}} \tag{4.79}$$

$$0 = -\frac{d\overline{u_2^3}}{dx_2} - \frac{2}{\rho}\overline{u_2\frac{\partial p}{\partial x_2}} + 2\nu\overline{u_2\frac{\partial^2 u_2}{\partial x_k \partial x_k}} \tag{4.80}$$

and

$$0 = -\frac{d\overline{u_3^2 u_2}}{dx_2} - \frac{2}{\rho}\overline{u_3\frac{\partial p}{\partial x_3}} + 2\nu\overline{u_3\frac{\partial^2 u_3}{\partial x_k \partial x_k}} \tag{4.81}$$

which are the equations corresponding to the Reynolds stress components $-\rho\overline{u_1^2}$, $-\rho\overline{u_2^2}$, and $-\rho\overline{u_3^2}$, respectively. The first term in equation (4.79) describes the interaction of the mean and turbulent fields, and represents feeding of turbulence by the mean flow. All the other terms in (4.79)–(4.81) contain only fluctuating quantities and describe interaction of turbulence with itself and dissipation/diffusion by viscosity. They have a similar form in all three equations.

Since the mean flow occurs only in the first term of equation (4.79), which corresponds to the component $\overline{u_1^2}$, there is direct driving of u_1 by the mean flow, but not of the other turbulent components. The mean-flow driving term is observed to have a maximum within the viscous layer close to each wall, where it is considerably larger than towards the middle of the channel, owing to the high mean shear near the walls. Under these circumstances, it would be reasonable to expect that u_1' would also have a maximum within each viscous layer and that

$$u_1' > u_2' \approx u_3' \tag{4.82}$$

where u' denotes root-mean-squared fluctuation ($u_1'^2 = \overline{u_1^2}$, etc.). The maximum of u_1' is indeed found within the viscous layers and has an observed value

$$u_1' \approx 3u_* \tag{4.83}$$

while the other two components are lower than u_1', and ordered as $u_2' < u_3'$, rising from zero at a wall (due to no slip) to about

$$u_2' \approx u_* \tag{4.84}$$

and

$$u_3' \approx 1.5u_* \tag{4.85}$$

just outside the viscous layers. The quantity, u_*, occurring above is given by

$$u_* = \left(\frac{\tau_w}{\rho}\right)^{1/2} \tag{4.86}$$

and is known as the turbulent friction velocity. This quantity gives a velocity scale, directly related to the wall shear stress (friction), which describes the magnitude of the turbulent velocity fluctuations within a turbulent boundary layer, as we shall see in the next chapter, and also within the channel flow considered here. The above disparity between the three components of fluctuating velocity in the viscous layers clearly shows lack of isotropy, which is not surprising given the proximity of the wall and the high mean shear near the wall, which provides a strong orientating effect on the turbulence. The turbulent velocity components decrease further from the wall, but remain of order u_* throughout the channel. Their values are more nearly the same towards the middle of the channel, although still different. This indicates that the turbulence may be more nearly isotropic there, but remains anisotropic nonetheless.

Either directly from (4.35), or by adding equations (4.79)–(4.81) and dividing by 2, we obtain the turbulent energy equation, which contains the following terms that sum to zero:

$$\overline{u_1 u_2}\frac{d\overline{U_1}}{dx_2} \tag{4.87}$$

which gives (minus) the rate of turbulent energy production and is directly related to the mean-flow driving term in (4.79):

$$+\frac{d}{dx_2}\frac{1}{2}\overline{q^2 u_2} \tag{4.88}$$

represents transfer of turbulent kinetic energy by the x_2-component of the fluctuating velocity (this is nonzero due to inhomogeneity of the turbulence in the x_2-direction and, together with the next two terms, is diffusive);

$$+\frac{1}{\rho}\frac{d}{dx_2}\overline{pu_2} \tag{4.89}$$

expresses turbulent energy transfer across the channel due to pressure–velocity correlations and inhomogeneity in the x_2-direction (here, we have used the incompressibility condition for u_i, (4.6), to sum the pressure terms in (4.79)–(4.81));

$$-\nu\frac{d^2}{dx_2^2}\left(\frac{1}{2}\overline{q^2} + \overline{u_2^2}\right) \tag{4.90}$$

represents viscous diffusion effects (again due to inhomogeneity in x_2);

$$+\frac{1}{2}\nu\overline{\left(\frac{\partial u_i}{\partial x_k} + \frac{\partial u_k}{\partial x_i}\right)\left(\frac{\partial u_i}{\partial x_k} + \frac{\partial u_k}{\partial x_i}\right)} \tag{4.91}$$

is the turbulent energy dissipation. As described earlier, viscous diffusive terms such (4.90) can usually be neglected compared to dissipative terms, such as (4.91), because the Reynolds number is large. In fact, the viscous diffusive term is only needed very near the walls, in the viscous layers where the effective Reynolds number of the flow is lower.

If we integrate the turbulent energy equation with respect to x_2, across the duct, using the fact that the velocity at the walls is zero, we obtain

$$\int_0^{2D} -\overline{u_1 u_2}\frac{dU_1}{dx_2}\,dx_2 = \int_0^{2D} \frac{1}{2}\nu\overline{\left(\frac{\partial u_i}{\partial x_k}+\frac{\partial u_k}{\partial x_i}\right)\left(\frac{\partial u_i}{\partial x_k}+\frac{\partial u_k}{\partial x_i}\right)}\,dx_2 = \int_0^{2D} \bar{\varepsilon}\,dx_2 \qquad (4.92)$$

which expresses equality of the integrated rates of production and dissipation of turbulent energy. This is, of course, as it should be.

Among the most reliable and well-documented measurements of channel flow are those of Laufer (1951) and Comte-Bellot (1965). The detailed energy budget, that the sum of (4.87)–(4.91) is zero, should hold at each point in the channel. Figure 4.4 shows the measured values of each term in the energy budget, not for channel flow, but for the equivalent energy budget in the qualitatively similar case of flow in a circular pipe (Laufer 1954). One of the important features is that both the turbulence production and dissipation, (4.87), rise steeply as the wall is approached from the central part of the channel/pipe. The maximum production occurs close to the walls within the viscous layers (not apparent in the figure, since they are so thin), owing largely to the much higher values of mean shear there. This indicates that intense turbulence is generated near the walls, but is also primarily dissipated there, with the relatively small excess of production over dissipation serving to feed the middle part of the channel/pipe with turbulence via the diffusive terms, (4.88) and (4.89). On the other hand, dissipation exceeds production in the central part of the flow, where the mean shear is lower, absorbing the turbulent energy coming from near-wall region.

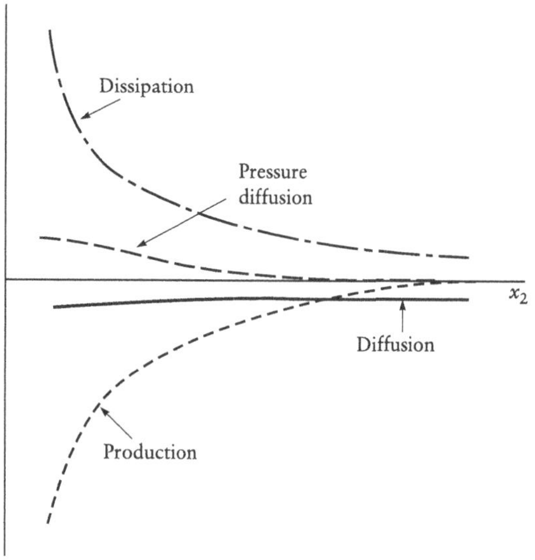

Figure 4.4. Measured terms in the turbulent-energy budget for circular pipe flow. Here, x_2 denotes distance from the pipe wall and the curves extend from the wall to the pipe axis. The budget for channel flow is qualitatively the same. The terms labeled "Diffusion" and "Pressure diffusion" are the pipe-flow equivalents of (4.88) and (4.89), respectively. (Laufer (1954), redrawn.)

Turbulence behaves somewhat differently in the wall region than in the middle part of the channel/pipe: near the boundaries it is dominated by the presence of the wall and is rather similar in character to other wall flows, such as the near-wall parts of turbulent boundary layers (see the next chapter for detailed treatment of turbulent boundary layers and a brief return to the channel problem considered here). In particular, it is strongly anisotropic and inhomogeneous in the direction normal to the wall. However, away from the walls the flow differs from a boundary layer.

As in all inhomogeneous turbulent flows, it is interesting to identify the regions of turbulence production, usually found near locations of maximum mean shear, which lie close to the walls for the channel flow. Measuring instruments in the middle part of the channel register turbulent activity coming from the two production centers, situated on either

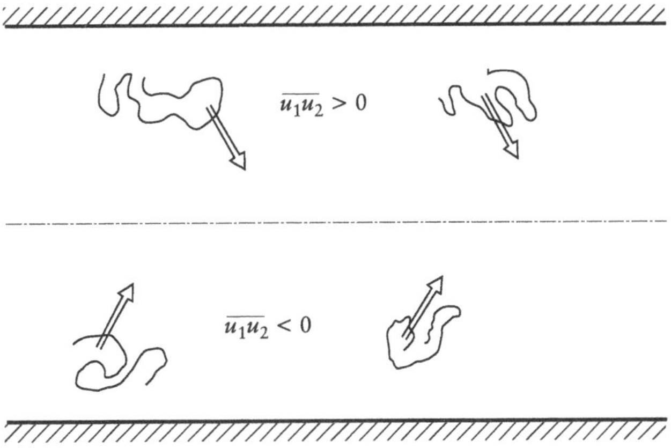

Figure 4.5. Turbulence production in a channel flow is sporadic and orginates at both walls with differently signed contributions to $\overline{u_1u_2}$.

side of the channel. These events, in which turbulence originating at the walls traverses a probe situated in the middle region, are clearly identifiable in the time records via rapid jittering of the measured velocities and are separated by periods of relative inactivity. Thus, turbulent activity in the middle part of the channel is found to be sporadic, but the main contributions to the Reynolds shear stress, $-\rho\overline{u_1u_2}$, evaluated as a time average since the flow is statistically steady, come from such intermittent turbulent events (see Sabot and Comte-Bellot (1976)). As we have seen, the Reynolds stress is highly significant because it determines the effects of turbulence on the mean flow. Turbulent events are observed to be associated either with positive or negative u_2. If one makes the reasonable assumption that turbulence coming from the wall at $x_2 = 0$ is associated with positive u_2 (since it has been ejected from the wall towards increasing x_2), while that arising from the other wall is associated with negative u_2, one may identify the source of turbulent events, and implement conditional sampling using the sign of the measured u_2 to trigger an electronic gate. As regards the Reynolds stress, either $u_1u_2 < 0$, which is found to be associated with events originating at the wall $x_2 = 0$, or $u_1u_2 > 0$, which is associated with the other wall (see Figure 4.5). Thus, turbulent eddies coming from the two walls tend to produce canceling contributions to the time average giving $\overline{u_1u_2}$, which is zero on the channel center plane, by symmetry (recall (4.78) and note that, by symmetry, $d\overline{U_1}/dx_2 = 0$ on the center plane, $x_2 = D$). Nearer the wall $x_2 = 0$, $\overline{u_1u_2} < 0$ because it is the eddies from that wall which tend to win, while, on the other side of the center plane, $\overline{u_1u_2} > 0$, owing to turbulence originating at the other wall, $x_2 = 2D$. The nearly linear distribution of the mean, $\overline{u_1u_2}$, which we found earlier, is therefore the result of rather subtle processes when examined in detail.

HOMOGENEOUS TURBULENCE SUBJECT TO UNIFORM SHEAR

Here we imagine a steady, infinite mean flow with uniform shear, and which, like the channel flow, only has a velocity component, $\overline{U_1}(x_2)$, in the x_1-direction ($\overline{U_2} = \overline{U_3} = 0$). The mean profile is taken to be linear in x_2, with

$$\overline{U_1} = sx_2 \tag{4.93}$$

where $s > 0$ is a constant, whose reciprocal has the dimensions of time. Thus, s^{-1} gives a time scale associated with the mean shear. We can see that this introduces the nondimensional time, st, which is a measure of the cumulative mean strain of the fluid.

The turbulence is supposed homogeneous, so that, in particular

$$\frac{\partial\overline{u_iu_j}}{\partial x_j} = 0 \tag{4.94}$$

and the turbulence does not enter into the mean-flow equation (4.9). In fact, the given mean flow is a solution of (4.5) and (4.9) with uniform mean pressure and therefore represents a possible mean velocity field. Moreover, initially homogeneous turbulence will remain so because a shift in the origin of coordinates, followed by a change to a new inertial frame based on the mean velocity at the new origin leaves the initial fluctuation statistics and the evolution equation, (4.31), for the fluctuations unchanged. Thus, the problem posed is self-consistent, giving a possible homogeneous turbulent flow. Similar reasoning applies to homogeneous turbulence in any mean flow for which the velocity derivatives are independent of position: such a mean flow will permit homogeneous turbulence to evolve as homogeneous turbulence, which, in turn, does not affect the mean flow. This fact forms the basis for the spectral analysis of homogeneous turbulence in mean flows with uniform velocity derivatives, of which the present flow is an example.

The turbulence will evolve with time, while remaining homogeneous, and it is this evolution which we wish to study: the effects of the mean-flow shear on the turbulence. We will suppose that the Reynolds number is sufficiently large that one can neglect the effects of viscosity on the large scales of turbulence, although one needs to take it into account at the small, dissipative scales, of course. The turbulent energy equation, (4.37), is

$$\frac{d\frac{1}{2}\overline{q^2}}{dt} = -s\overline{u_1u_2} - \overline{\varepsilon} \tag{4.95}$$

which gives an indication of the importance of the quantity $\overline{u_1u_2}$, which occurs in the term representing turbulence production by interaction with the mean flow. The other term corresponds to viscous dissipation and would lead to continuously decaying turbulence in the absence of the mean shear. In deriving (4.95), we have used the fact that only one component of the mean-flow gradient tensor is nonzero to simplify the equation.

The equation governing $\overline{u_iu_j}$, (4.38), becomes

$$\frac{d\overline{u_iu_j}}{dt} = -s\begin{bmatrix} 2\overline{u_1u_2} & \overline{u_2^2} & \overline{u_2u_3} \\ \overline{u_2^2} & 0 & 0 \\ \overline{u_2u_3} & 0 & 0 \end{bmatrix} + \frac{1}{\rho}\overline{p\left(\frac{\partial u_i}{\partial x_j} + \frac{\partial u_j}{\partial x_i}\right)} - 2\nu\overline{\frac{\partial u_i}{\partial x_k}\frac{\partial u_j}{\partial x_k}}. \tag{4.96}$$

We are mainly interested in the case in which the turbulence is initially isotropic. Of course, it will lose isotropy under the straining and orientating action of the mean shear. However, since, by initial isotropy, the problem is initially symmetric under reflection in the plane, $x_3 = 0$, it remains so at all times and hence $\overline{u_1u_3} = \overline{u_2u_3} = 0$ (symmetry implies that we are equally likely to find a positive as a negative value of u_3 associated with any given values of u_1 and u_2 and therefore that these moments are zero). Initially, we also have $\overline{u_1u_2} = 0$ (so that the turbulent energy at first decays, according to (4.95)) and $\overline{u_1^2} = \overline{u_2^2} = \overline{u_3^2}$, but these relations will cease to hold once anisotropy due to the mean shear begins to take effect. In particular, the quantity, $\overline{u_1u_2}$, which is zero for isotropic turbulence, is a measure of the anisotropy, although $\overline{u_1u_2} = 0$ *does not necessarily imply isotropy*. This component of $\overline{u_iu_j}$ corresponds to $i = 1, j = 2$ in (4.96), which can be written as

$$\frac{d\overline{u_1u_2}}{dt} = -s\overline{u_2^2} + \frac{1}{\rho}\overline{p\left(\frac{\partial u_1}{\partial x_2} + \frac{\partial u_2}{\partial x_1}\right)} - 2\nu\overline{\frac{\partial u_1}{\partial x_k}\frac{\partial u_2}{\partial x_k}} \tag{4.97}$$

of which the viscous term is usually found to be small because it is dominated by the small scales at high Reynolds numbers, which are approximately isotropic (recall the discussion leading to equation (4.45); according to (4.45), the off-diagonal elements of the viscous term, including that corresponding to $\overline{u_1u_2}$, are zero and we would therefore expect them to be small in practice). We will neglect the viscous term in (4.97) in what follows.

The evolution of an initially homogenous, isotropic turbulent field under the effects of a uniform mean shear, $s > 0$, has been widely studied experimentally and theoretically (see, e.g., Champagne, Harris, and Corrsin (1970) and other references in Tavoularis and Karnik (1989)). At sufficiently large Reynolds number, it is indeed observed that the viscous term in (4.97) is small and furthermore that $\overline{u_1u_2}$ becomes negative, due to the first term in (4.97), which represents the direct action of the mean shear on the Reynolds stress and is negative, since $s > 0$. Once $\overline{u_1u_2}$ has become negative, the production term in (4.95) begins to offset the dissipative term and the turbulence decays less rapidly. Eventually, $\overline{u_1u_2}$ is found to become sufficiently negative that production exceeds dissipation and the turbulence grows in intensity.

The second term in (4.97) tends to act in the opposite direction to the first, thus trying to reduce the positive $-\overline{u_1u_2}$ which is created by the first term. The second term contains the fluctuating pressure, which obeys the Poisson equation, (4.48), namely

$$\nabla^2 p = -2\rho s\frac{\partial u_2}{\partial x_1} - \rho\frac{\partial u_i}{\partial x_j}\frac{\partial u_j}{\partial x_i} \tag{4.98}$$

in the present case. Here, the first term on the right is linear in the fluctuating velocity and is responsible for the so-called rapid rate of strain–pressure correlation contribution to (4.97); the second term on the right is nonlinear in the fluctuating velocity. As described in Section 4.3, both terms produce contributions to the fluctuating pressure, but have fundamentally different effects on the evolution of the turbulence. We write the linear fluctuating pressure component as $p^{(1)} = s\Pi$, including the factor of s occurring in (4.98) in the definition of Π, and $p^{(2)}$ for the contribution due to nonlinearity. Corresponding to Π and $p^{(2)}$ there will be two components of the pressure–velocity gradient correlation in (4.97). Thus, (4.97) becomes

$$\frac{d\overline{u_1u_2}}{dt} = -s\overline{u_2^2} + \underbrace{\frac{s}{\rho}\overline{\Pi\left(\frac{\partial u_1}{\partial x_2} + \frac{\partial u_2}{\partial x_1}\right)}}_{\text{Linear}} + \underbrace{\frac{1}{\rho}\overline{p^{(2)}\left(\frac{\partial u_1}{\partial x_2} + \frac{\partial u_2}{\partial x_1}\right)}}_{\text{Nonlinear}} \tag{4.99}$$

where we have neglected the viscous term because, as noted above, it leads to only a small contribution in the equation for $\overline{u_1u_2}$. Equation (4.99) shows the linear and nonlinear parts of the pressure–rate-of-strain correlation term explicitly.

The effects of the two pressure terms in (4.99) can be partially understood as follows. As explained in Section 4.3, the nonlinear pressure term is zero for isotropic turbulence and is therefore initially zero here. This term is one result of the interaction of the turbulence with itself and tends to act to make the turbulence less anisotropic. If desired, an approximate quantitative description of the nonlinear pressure term can be obtained using the closure hypothesis (4.55) with $i = 1, j = 2$ (we stress that this is only a hypothesis and has no fundamental justification). Since C is positive, the closure shows explicitly that the nonlinear term in (4.99) resists the growth of $-\overline{u_1 u_2}$ implied by the term $-s\overline{u_2^2}$. The linear pressure term is nonzero when there is mean shear and, together with the term $-s\overline{u_2^2}$, represents the effects of the mean shear on the turbulence. The linear term is called into play immediately, whereas the nonlinear term increases as the anisotropy of the turbulence increases, but both pressure terms tend to resist the growth of $-\overline{u_1 u_2}$.

To better understand the differing roles of linear and nonlinear terms, we introduce the idea of rapid distortion of turbulence, which may apply when the mean-flow straining is so large that the characteristic time, s^{-1}, for straining is short compared with the typical time, $O(L/u')$, for the large scales of turbulence to act on themselves and to decay in the absence of production. If sL/u' is large, the straining is so rapid that large-scale turbulence does not have the time to act on itself, nonlinearity and viscosity become negligible, and the turbulence may be described by linearizing the basic equation (4.31) for the fluctuations and ignoring the viscous terms. Another way of saying the same thing is that the turbulence is so weak that the quadratic terms in (4.31) are negligible. Linear equations are mathematically more tractable and better understood and are consequently easier to analyze. They may also be applied to describe cases other than strongly sheared mean flows, such as flows with strong rotation (discussed later in this section) or density stratification.

The effects of mean-flow straining of turbulence are sometimes thought of as being similar to those produced by straining a solid material (Townsend G (1976); Lumley (1970)), for reasons to be made apparent. The cumulative strain of the "turbulent material" increases with time and is measured by the nondimensional time, st, in the case of the steady mean shear flow considered above. The idea behind the analogy with straining of a solid material is that the effects of *rapid* straining on turbulence are only apparent through the cumulative mean strain and will thus undo themselves if the straining is reversed and the "material" returned to its initial state of strain. If the turbulence were initially isotropic there would then be a return to isotropy at the instant when the material is brought back to its original state of strain. There is no doubt that initially isotropic turbulence tends to go back towards isotropy when an initial straining is later reversed, independently of any suppression of anisotropy by the nonlinear action of turbulence on itself, but, as we shall see, the effects of shear are never fully undone by reversing the shear and there is irreversibility of straining of the turbulent material, whose state is not simply a function of its cumulative mean strain. Below, we shall describe the effects of changing mean shear in more detail, but it should first be made clear that the temporal variations in mean shear that are envisaged are not easily achievable in practice, since they involve changes in the mean vorticity. This normally requires viscous diffusion from boundaries, which are not present in the infinite flow considered. One might imagine applying an appropriate mean rotational volume force to the fluid, but exactly how this might be achieved in practice is unclear. However, this is unimportant because what we have in mind here is more of a thought experiment, intended to illustrate a general principle. It is certainly the case that, from the point of view of a packet of turbulence being convected by the flow, changing mean shear is present in more realistic flows and the example given here is aimed at understanding the behavior of turbulence with such variable mean strain, rather than as a real-life flow.

In any case, we now allow s to vary with time and ask what effect this has on the turbulence. The easiest case to handle is the rapid-distortion limit in which the mean shear is so strong that the interaction of turbulence with itself and the effects of viscosity can be neglected. In the present case, linearization and neglect of viscosity in (4.31) yield

$$\frac{\partial u_1}{\partial t} + s\left\{x_2 \frac{\partial u_1}{\partial x_1} + u_2\right\} = -\frac{1}{\rho}\frac{\partial p}{\partial x_1} \tag{4.100}$$

$$\frac{\partial u_2}{\partial t} + s x_2 \frac{\partial u_2}{\partial x_1} = -\frac{1}{\rho}\frac{\partial p}{\partial x_2} \tag{4.101}$$

$$\frac{\partial u_3}{\partial t} + s x_2 \frac{\partial u_3}{\partial x_1} = -\frac{1}{\rho}\frac{\partial p}{\partial x_3} \tag{4.102}$$

which, together with the condition, (4.6), that $\mathbf{u}$ have zero divergence, form the governing equations for the turbulence. All viscous and nonlinear terms in the fluctuations have been thrown away in going to (4.100)–(4.102) and this means that it is an approximation whose validity needs to be carefully tested in any given case. However, regardless of the applicability of the theory in any particular flow, our objective here is to study the basic mechanism of mean shear effects on turbulence, a mechanism which is embodied in (4.100)–(4.102).

We set $\sigma = \int s\,dt$ as a new dimensionless time variable which measures the cumulative strain, just as the variable st does with constant shear. At the same time we introduce a new variable, $\pi = p/s$, to replace the pressure. The result is that (4.100)–(4.102) become

$$\frac{\partial u_1}{\partial \sigma} + x_2 \frac{\partial u_1}{\partial x_1} + u_2 = -\frac{1}{\rho}\frac{\partial \pi}{\partial x_1} \tag{4.103}$$

$$\frac{\partial u_2}{\partial \sigma} + x_2 \frac{\partial u_2}{\partial x_1} = -\frac{1}{\rho}\frac{\partial \pi}{\partial x_2} \tag{4.104}$$

$$\frac{\partial u_3}{\partial \sigma} + x_2 \frac{\partial u_3}{\partial x_1} = -\frac{1}{\rho}\frac{\partial \pi}{\partial x_3} \tag{4.105}$$

which have exactly the same form as (4.100)–(4.102), as if s had the constant value 1. The continuity equation, (4.6), is unchanged. It is not easy to solve these equations, even though they are closed and linear, but we do not need the detailed form of the solution for our purposes. All we need do is to remark that, according to (4.103)–(4.105), the fluctuating field, u_i, can be considered as a function of $\mathbf{x}$ and σ only. We now imagine beginning with a given initial turbulent field and allowing it to evolve with changing mean shear, in such a way that the net strain due to the shear is zero at some later time. In that case the variable σ has the same value at the start as at the later time and, since u_i is a function of σ, *the turbulence will have returned precisely to its initial state*. There is thus no hysteresis according to rapid distortion theory and the turbulence depends solely on the cumulative strain, just as it would for a hypothetical turbulent material subjected to straining.

This no longer holds true when nonlinear terms in the fluctuation equations are allowed for. Nonlinearity leads to memory of the past history of strain. Rapid distortion theory, and hence reversibility of turbulent straining, may apply when the straining effects (in our case the effects of shear) are sufficiently strong, so that the elapsed time is insufficient for turbulence to have significantly evolved in the absence of the imposed strain. According to the theory, reversing the shear should tend to undo the effects of the initial shear and this is generally true, even when rapid distortion theory itself is inapplicable. Furthermore, within the theory, it does not matter how one gets from one state of strain to another, it is only the value of σ, representing the cumulative strain, that counts. Thus the turbulence does indeed behave like a material under strain, within the approximations of rapid distortion theory. These results apply equally to inhomogeneous and homogeneous turbulence. Rapid distortion theory, that is, linearization of the equation of the fluctuating velocity and neglect of viscosity, can obviously be applied to more general flows than the simple one described here and similar conclusions reached.

If one assumes homogeneity and derives an equation for the Reynolds stress from rapid distortion theory, one can easily show that the result is equation (4.96) (which is exact) without the viscous term, which is in any case small. However, the rapid-distortion equivalent of the pressure equation does not have the exact form, (4.98), because it is missing the nonlinear term, which is not included in rapid distortion theory. As we have seen, this term has no effect on (4.96) when the turbulence is isotropic and so we might expect rapid distortion theory to describe the initial evolution of turbulence, which begins as isotropic, rather well, since it reproduces (4.96) precisely. Of course, (4.96) and (4.98) are not a closed set and we cannot really assert, when we recover these equations from some approximate model, that the model gives a good approximation to the exact solution of the turbulence problem we had in mind in other respects.

To go beyond rapid distortion theory, we need to include the effects of nonlinearity. We cannot then solve the equation (4.31) analytically for the fluctuations, but we can get an idea of how the solution might behave from a simple argument, as follows. Suppose that we begin with a constant mean shear, $s > 0$, applied to initially isotropic turbulence. We know, from our previous discussion, that the growth of $-\overline{u_1 u_2}$ is due to the first term on the right of (4.99), resisted by both the pressure–strain correlation terms of the same equation. Suppose now that the shear is suddenly reversed, so that, according to rapid distortion theory, $-\overline{u_1 u_2}$ would begin to decrease, retracing its steps in the opposite direction. Immediately after the sign of s is changed, the fluctuating velocities will be the same as they were just before and the first and second terms on the right of (4.99), which contain s, will change sign, but the third term will not, because it originates in the quadratic term on the right of (4.98), which does not contain the mean flow. Since the unchanged term in (4.99) is tending to reduce $-\overline{u_1 u_2}$, that quantity will return more quickly to zero than was implied by rapid distortion theory, as illustrated in Figure 4.6. The effect is one of hysteresis: the cumulative strain no longer completely describes the turbulence and there is no full return to isotropy, despite the fact that $-\overline{u_1 u_2}$ goes back towards zero. It is clear that, in general, there is a complicated interplay between the mechanisms of direct mean-flow–turbulence interaction, linear pressure effects, and nonlinear pressure effects.

Returning to the problem of constant shear that we set ourselves earlier, if the mean shear is sufficiently strong that rapid distortion theory applies, it can be shown, based on that theory, that the turbulence becomes increasingly anisotropic and the turbulent energy grows proportional to st asymptotically as $st \to \infty$. As the turbulence rises in intensity, nonlinear terms become more important and eventually the rapid distortion approximation breaks down. At this point, nonlinearity is significant and the evolution of turbulence changes character. At large times, it is observed to finally settle into a sort of equilibrium state in which the ratio of the different components of $\overline{u_i u_j}$ tend towards limiting values (see, e.g., Tavoularis and Karnik (1989)). In particular, the correlation coefficient, $|\overline{u_1 u_2}|/u_1' u_2'$, is found to have a limiting value around 0.5, typical of shear flows and apparent in Figure 4.6.

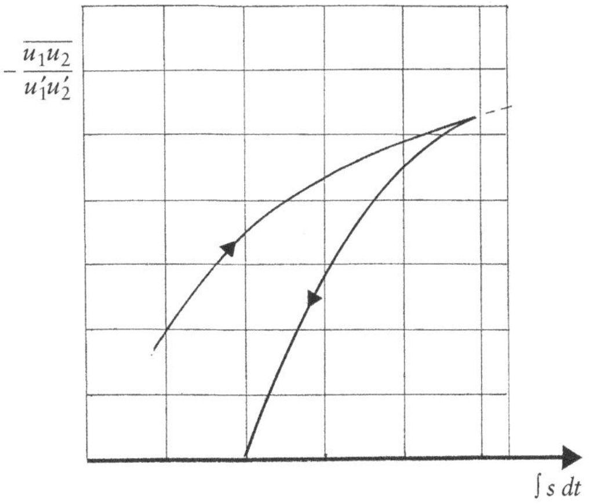

Figure 4.6. $|\overline{u_1 u_2}|/u_1' u_2'$ for a thought experiment in which uniform mean shear is suddenly reversed. The rising part of the curve represents experiments with constant shear and has an asymptote of about 0.5.

This asymptotic regime is also approached if the initial turbulence is insufficiently weak that one can apply rapid distortion theory in the initial phase and nonlinearity is important for turbulence evolution at all times. In the opposite limit to that of rapid distortion, for which the turbulence is so strong that one may initially neglect the mean shear, it decays to begin with, as if there were no mean flow, until the effects of the shear become significant, that is, $sL/u' = O(1)$, and the turbulence subsequently tends towards the above asymptotic regime. It therefore appears that, no matter what the initial conditions, high-Reynolds-number, homogeneous turbulence subject to a constant mean shear without boundaries will eventually approach the "equilibrium" state, thus forgetting its origins.

The asymptotic state of the flow is observed to have continuously growing correlation lengths and turbulent velocity fluctuations. Experimental results seem to be consistent with the idea that the large scales of turbulence become statistically self-similar, that is, that the large-scale statistical properties of

$$\mathbf{v}(\mathbf{y}, t) = \frac{\mathbf{u}(L(t)\mathbf{y}, t)}{u'(t)} \tag{4.106}$$

become independent of time, where $\mathbf{y} = \mathbf{x}/L(t)$ is the similarity variable. For instance, although u' continues to increase, the normalized components $\overline{u_i u_j}/q^2$ approach asymptotic limits at large times, as noted above. The quantities u' and L are found to be exponential functions of time[6] (see Tavoularis and Karnik (1989)).

[6] It is interesting to note that the growth of turbulent energy proportional to t predicted by rapid distortion (i.e., linear) theory appears to become exponential when nonlinear effects are important. One often finds exponential growth in the *linear* theory of hydrodynamic stability, which is usually assumed to be limited by nonlinearity.

As a consequence of increasing L, the large scales of turbulence grow in size without bound and, in practice, become comparable in size with the overall dimensions of the flow volume. Thus, the finite size of experimental flows sets an upper limit on the length of time for which a reasonable approximation to the ideal flow can be obtained. Afterwards, the flow can no longer be regarded as infinite, the turbulence in the middle of the flow volume senses the boundaries, and ceases to be approximately homogeneous. This takes us beyond the scope of the problem considered here and obviously limits the growth of quantities such as L.

The model discussed above supposes infinite, homogeneous turbulence in a uniform mean shear. Although the shear introduces a characteristic time scale, s^{-1}, there are no length scales, other than those defined by the initial turbulence itself. The existence of boundaries and associated length scales in more complex flows can significantly modify the evolution of turbulence. Furthermore, within the model, turbulence production is spread out uniformly over space (due to homogeneity) and the complication of having several spatially disparate turbulence production centers, which is another feature of more complex flows, is not present here. The model is simple and limited, but nonetheless representative of some facets of the evolution of more general turbulent shear flows. For instance, values similar to, though not identical with, $|\overline{u_1u_2}|/u_1'u_2' \approx 0.5$, obtained for the asymptotic state of the above model flow, are typical of general shear flows.

The concept of nonlinear memory effects, which was introduced above, can be used to better understand the noncoincidence of the positions of maximum mean flow and zero $\overline{u_1u_2}$ for a wall jet (Figure 4.1) which was briefly commented on when we described the eddy-viscosity approximation in Section 4.1. When following a packet of turbulence in the jet, which we think of as convected by the mean flow, the mean shear will change with time. Suppose that we can regard the evolution of the turbulence in this packet of inhomogeneous flow as similar to the homogeneous flow with variable mean shear described above. A packet of fluid that is convected by the flow will encounter the maximum of the mean flow, where the mean shear goes through zero, but will require further time, with reversed mean shear, to reduce the value of $\overline{u_1u_2}$ to zero. Thus we would expect the zero of $\overline{u_1u_2}$ to occur nearer the wall than the maximum of the mean flow, which is exactly as observed in practice. These ideas have been used as the basis of a quantitative model by Jeandel, Brison, and Mathieu (1978).

HOMOGENEOUS TURBULENCE SUBJECT TO SUCCESSIVE PLANE STRAINS

The flow considered here is, in some respects, similar to the sheared one of the previous subsection and we again consider issues of rapid distortion versus nonlinear reversibility. However, whereas a mean shear that varies with time requires rotational body forces, the mean flow considered here is irrotational and changes can be brought about in the straining field by pressure forces alone. This makes it possible to produce variations in the straining field in an experimentally straightforward manner and to carry out measurements designed to investigate the effects of such variable straining fields using flow in a duct.

The mean flow is the plane straining field

$$\overline{U_1} = \text{constant} \qquad \overline{U_\beta} = D_{\beta\gamma} x_\gamma \tag{4.107}$$

where the indices, β and γ, can take on the values 2 and 3, and the summation convention applies to repeated indices as usual (but with summation over the values 2 and 3 only when β or γ is repeated). Thus, the mean strain takes place in the x_2–x_3 plane. In general, the rate of strain components, $D_{\beta\gamma}(t)$, can depend on time and are symmetric, that is, $D_{23} = D_{32}$, so that the mean vorticity is zero. The trace, $D_{\beta\beta} = D_{22} + D_{33} = 0$ in order that the mean-flow continuity equation, (4.5), is satisfied. Observe that, since $D_{\beta\gamma}$ is symmetric, it can be diagonalized at any given instant of time by a suitable rotation of the coordinate system about the x_1-axis and, since it has zero trace, it then takes on the diagonal form $D_{22} = -D_{33} = D$ and $D_{23} = D_{32} = 0$. Thus, the straining field $D_{\beta\gamma}(t)$ may be specified by giving the time history of D and of the orientation of the principal axes of $D_{\beta\gamma}$ in the x_2–x_3 plane. The turbulence is homogeneous and therefore does not contribute to the mean-flow equation, (4.9), which is satisfied by the flow with $\overline{P}$ a quadratic function of x_2 and x_3. The given mean flow is such that turbulence remains homogeneous if it is initially so, for the same reasons as the shear flow in the previous subsection. It follows that the flow envisaged is a possible one (here, as noted above, the changes of the mean flow with time, being irrotational, can be brought about by the mean pressure field and need not involve rotational, external body forces). Our objective is to study how turbulence evolves under such plane mean straining.

Experimentally, this type of turbulent flow has been considered by Townsend (1954), Tucker and Reynolds (1968), and others for a *constant* straining field $D_{\beta\gamma}$, while Gence and Mathieu (1979, 1980) investigated the interesting generalization in which the principal axes of straining undergo a sudden rotation with $|D|$ maintained constant. In these experiments, grid-generated turbulence is carried by flow along a duct whose cross-section changes in shape along its length in such a way as to generate a close approximation to the mean flow (4.107), where x_1 measures distance along the duct axis and x_2, x_3 are transverse coordinates. For instance, Figure 4.7 shows a sketch of the Gence and Mathieu apparatus, in which turbulence produced by a grid enters an elliptical duct of varying eccentricity, subjecting the turbulence to transverse straining. Grid turbulence is locally nearly homogeneous, once near-grid effects have disappeared (i.e., many mesh spacings downstream of the grid, hence the initial section of uniform duct in the figure upstream of the nonuniform working section), but evolves continuously with downstream distance, and therefore cannot be exactly homogeneous. Provided the size of the large turbulent eddies, which scales on the mesh spacing of the grid, is small compared to the length scale for evolution of the turbulence (in particular $\overline{U_1}/|D|$, which is the distance for significant straining effects), one may imagine following a packet of turbulence moving along the duct: the elapsed time t is related to distance along the duct by $x_1 = \overline{U_1}t$. At any given duct position, the turbulence is locally homogeneous, and, to a good approximation, evolves with $t = x_1/\overline{U_1}$ as if it were exactly homogeneous. Thus, grid turbulence in an appropriately designed duct allows approximate realization of the idealized homogeneous turbulent flow described above, with $x_1/\overline{U_1}$ playing the role of time. In the absence of mean straining and once near-grid effects have disappeared, grid turbulence is found to be approximately isotropic, but, once it enters the mean straining

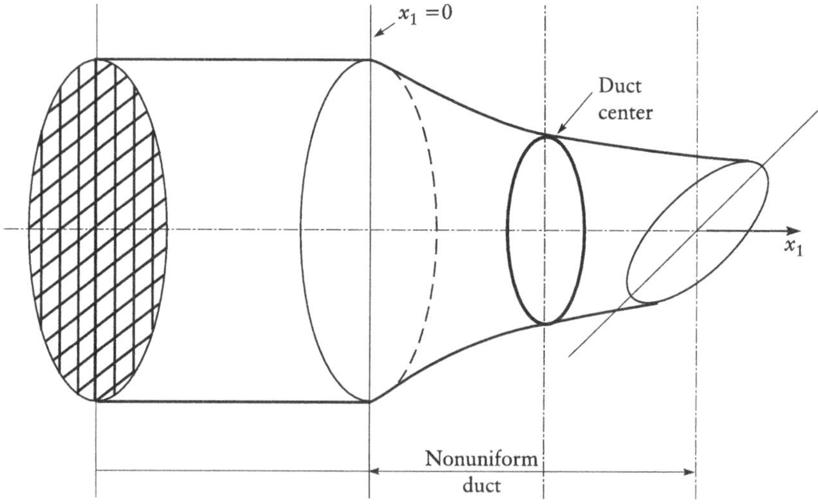

Figure 4.7. Sketch of the duct used by Gence and Mathieu (1979) to realize plane straining of turbulence.

field, it is distorted and loses isotropy. An appropriate theoretical model of the experiments is thus initially homogeneous, isotropic turbulence evolving in the mean flow described by (4.107).

To begin with, we consider the case of constant $D_{\beta\gamma}$, yielding a fixed orientation for the principal axes of straining. For simplicity sake, the coordinate system is chosen so that these principal axes lie in the x_2- and x_3-directions, corresponding to stretching at constant rate $D > 0$ in the x_2-direction and compression at rate D in the x_3-direction. The quantity, D^{-1}, has the dimensions of time, and the cumulative strain can be measured by the product Dt in the same way as st was used in the previous subsection. Given the symmetry of the initial turbulence and of the mean straining field under reflection in each of the coordinate planes, $\overline{u_i u_j} = 0,\ i \neq j$, and hence $\overline{u_1^2}$, $\overline{u_2^2}$, and $\overline{u_3^2}$ are the only nonzero components of $\overline{u_i u_j}$, whose principal axes are consequently x_1, x_2, x_3, those of straining. Initial isotropy implies that $\overline{u_1^2} = \overline{u_2^2} = \overline{u_3^2}$ and, in particular, that the initial turbulence is statistically axisymmetric about the x_1-axis. The nondimensional quantity

$$K = \frac{\overline{u_3^2} - \overline{u_2^2}}{\overline{u_3^2} + \overline{u_2^2}} \tag{4.108}$$

measures the extent to which axisymmetry of the turbulence has been removed by straining. Figure 4.8 shows experimental results for K as a function of e^{Dt}, initially close to zero owing to axisymmetry, becoming positive under the effects of straining and approaching an equilibrium value at large times. The small departures of K from zero apparent in some of the curves for $e^{Dt} = 1$, that is, $t = 0$, representing the start of the distorting duct, reflect lack of exact axisymmetry of the initial turbulence. A simple qualitative argument can be used to explain the observed sign of K at later times, based on the ideas of vortex stretching that were described earlier. The mean vorticity is zero and thus we have only to take account of turbulent vorticity. The mean flow produces stretching at rate D in the x_2-direction and compression at rate

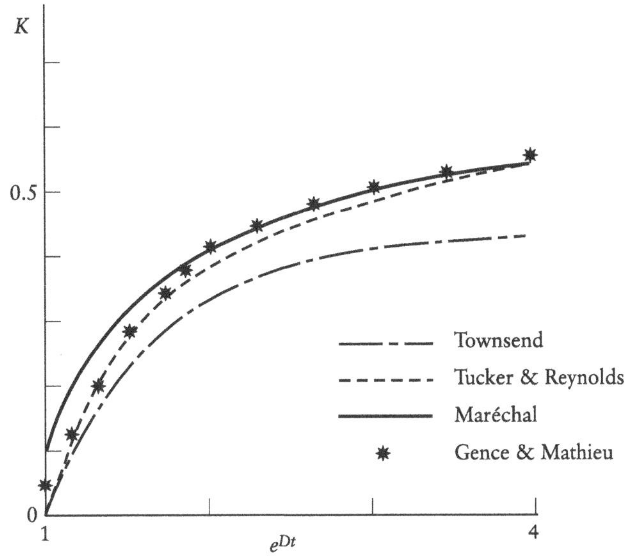

Figure 4.8. Plane strains applied to initially axisymmetric turbulence produce growing departures from axisymmetry, illustrated by the experimentally determined growth of the quantity K. In this and Figures 4.9–4.11, t is the time taken for convection at speed $\overline{U_1}$ to reach the measurement point from the beginning of the distorting duct. (Results of Townsend (1954); Tucker and Reynolds (1968); Maréchal (1967); Gence and Mathieu (1979).)

D in the x_3-direction, whereas there is no mean straining in the x_1-direction. Vortex stretching in the x_2-direction tends to cause the x_2-component of vorticity to increase (it behaves like e^{Dt} according to rapid distortion theory, i.e., using equation (4.31) without the nonlinear and viscous terms). Likewise, the x_3-component of vorticity tends to decrease (as e^{-Dt} under rapid distortion theory). The x_1-component of vorticity shows no such trends. Now, each component of vorticity is associated with the velocity components in the two other directions, for instance $\omega_2 = \partial u_1/\partial x_3 - \partial u_3/\partial x_1$, according to the definition of vorticity. Thus, increase of ω_2 due to stretching will tend to induce corresponding growth in the magnitudes of u_1 and u_3, while decrease of ω_3 tries to decrease the magnitudes of u_1 and u_2. It follows that $\overline{u_3^2}$ should increase, while $\overline{u_2^2}$ should decrease, and, since $K = 0$ initially, $K > 0$ at later times, as observed. The effects of straining on $\overline{u_1^2}$ are less clear cut, because there is both a tendency to increase, due to stretching in the x_2-direction, and to decrease, owing to compression in the x_3-direction. Viscous dissipation means that all turbulent velocity components tend to decay, a tendency which is superimposed on the increasing or decreasing trends due to mean straining that we have just described.

Coming back to the experiment of Gence and Mathieu (1979, 1980), only the nonuniform duct in Figure 4.7 concerns us here. Thanks to the circular cross-section at the duct center, the second half of the duct can be turned through an arbitrary angle (fixed in any given run of the experiment) about the duct axis. This corresponds to a discontinuous change in the principal axes of $D_{\beta\gamma}$, which suddenly rotate through an angle $0 \leqslant \alpha \leqslant \pi/2$, while the strain rate $|D|$ remains constant. When $\alpha = 0$, there is no change in the straining field at the center of the duct and one naturally recovers the results for constant $D_{\beta\gamma}$ described above. No matter what the value of α, straining in the first half of the duct is as for constant $D_{\beta\gamma}$ and consequently, prior to encountering the duct center, the turbulence is as before, taking on the initial principal axes of straining, x_1, x_2, x_3, with $\overline{u_3^2} > \overline{u_2^2}$. However, after undergoing the sudden change in straining field, the turbulence begins to readjust to the new axes of strain and, in general, the principal axes of $\overline{u_i u_j}$ rotate about the x_1-axis in an attempt to realign themselves with the new principal axes of straining. Note, however, that, if $\alpha = 0$ or $\alpha = \pi/2$, the principal axes do not change. The case $\alpha = \pi/2$ corresponds to switching the sign of D, while maintaining the same axes of mean strain.

The quantity K, defined by (4.108), is not a tensor invariant and, whereas for the earlier experiments x_2 and x_3 remain principal axes throughout the duct, so this does not matter greatly, tensor quantities are preferable in the general case. For this reason, we introduce the symmetric, traceless, nondimensional tensor

$$b_{ij} = \frac{\overline{u_i u_j}}{\overline{q^2}} - \frac{1}{3}\delta_{ij} \tag{4.109}$$

which is zero for isotropic turbulence and is a convenient measure of anisotropy. Since b_{ij} is symmetric, it can be diagonalized by rotation of coordinates to its principal axes, which are the same as those of $\overline{u_i u_j}$, namely the three coordinate axes prior to the sudden change in straining, making b_{ij} diagonal then. This is no longer the case after the rotation of principal axes of strain, when the off-diagonal element $\overline{u_2 u_3}$, and hence b_{23}, acquire a nonzero value in general, representing rotation of the principal axes of turbulence following those of straining. Observe that, since $K > 0$, we have $\overline{u_3^2} > \overline{u_2^2}$ and so

$$b_{33} - b_{22} > 0 \tag{4.110}$$

prior to the sudden change of straining field. Equation (4.37) for the turbulent energy becomes

$$\frac{d\frac{1}{2}\overline{q^2}}{dt} = D(b_{33} - b_{22})\overline{q^2} - \bar{\varepsilon} \tag{4.111}$$

in the present case, before the sudden rotation of the axes of straining. According to (4.110), the first term in (4.111), representing turbulence production, is positive and energy is fed to the turbulence, offset by viscous dissipation.

After the sudden rotation of axes, the behavior of the turbulence depends on the angle α. Let us first consider the simplest case, $\alpha = \pi/2$, so that one can think of the change of straining in terms of maintaining the same principal axes, but switching the sign of D. The straining in the x_2-direction becomes a compression instead of stretching and that in the x_3-direction, a stretching rather than a compression. Thus, the new straining motion attempts to undo the effects of the old one and, according to rapid distortion theory, would restore the turbulence to its initial state at the end of the second half of the duct. However, as described in the previous subsection, hysteresis (nonlinear and viscous) effects stop such complete reversibility in practice. Nonetheless, the trend should be towards reversal of the effects of the previous straining. In particular, the turbulence should move in the direction of isotropy. The energy equation, (4.111), still holds with the reversed straining, but, since D is now negative, the turbulent energy decreases rapidly in the second part of the duct, due both to the effects of straining and to the dissipation, which now act in consort, rather than opposing each other as before. Leaving aside dissipation, *energy is extracted from the turbulence* via the first term in (4.111) (so long as $b_{33} > b_{22}$). This type of behavior, in which energy flows away from turbulence, may, at first sight, appear paradoxical, since one is used to thinking of turbulence as a random phenomenon whose state of disorder would preclude (in some poorly defined Second Law of Thermodynamics sense) the extraction of energy. However, as we see here, such reverse transfers can and do occur.

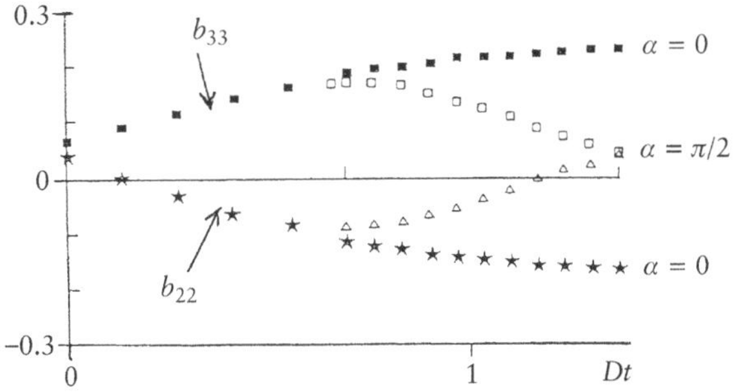

Figure 4.9. Experimental evolution of b_{22} and b_{33} in the distorting duct. (Figures 4.9–4.12 are based on results of Gence and Mathieu (1979, 1980).)

Figure 4.9 shows the measured evolution of b_{22} and b_{33} in the case $\alpha = 0$, that is, when there is no change in straining field, and $\alpha = \pi/2$, for which the straining field reverses at the duct center. When $\alpha = 0$, an asymptote is apparently approached at large time, confirming the behavior apparent in Figure 4.8. The results with reversal of straining are remarkably symmetric about the duct center, a symmetry predicted by rapid distortion theory and corresponding to reversibility of the effects of straining. Figure 4.10 illustrates the evolution of the turbulent energy for different values of α, showing a general decrease due to dissipation, upon which the effects of straining are superimposed. Thus, for the case $\alpha = 0$, in which the turbulence is continuously amplified by straining, the energy decreases initially, due to dissipation, but is subsequently so amplified that it returns to its original level by the end of the duct. The equality of the initial and final energies in this case is coincidental: the relative size of the effects of dissipation and straining depend on how rapidly the strain is applied. The net effect, along the whole duct, can be either an increase in turbulent energy, if straining wins, or a decrease if dissipation dominates. In the opposite extreme, $\alpha = \pi/2$, in which the strain is reversed at the center of the duct, the turbulent energy obviously follows the same curve as for $\alpha = 0$ until the duct center is reached, but subsequently declines rapidly as the combined effects of straining and dissipation both cause it to decrease. The other values of α show behavior intermediate between these two extreme cases. In the limit of strain applied infinitely rapidly, rapid distortion theory would predict continuous amplification with $\alpha = 0$ and exact symmetry about the duct center for $\alpha = \pi/2$. The fact that an irreversible process, such as dissipation by the turbulent energy cascade, is important in the experiment, as evident from Figure 4.10, suggests that rapid distortion theory ought to do rather poorly. However, as shown by Figure 4.9, the quantities b_{22} and b_{33} are remarkably symmetric about the duct center, a result in good accord with rapid distortion theory. As apparent from the definition, (4.109), of b_{ij}, it represents the *relative* magnitudes of the components of the Reynolds stress tensor. Why rapid distortion theory should yield reasonably satisfactory predictions of normalized Reynolds stresses when dissipation is significant (and hence the basis of the theory is suspect) is not fully understood.

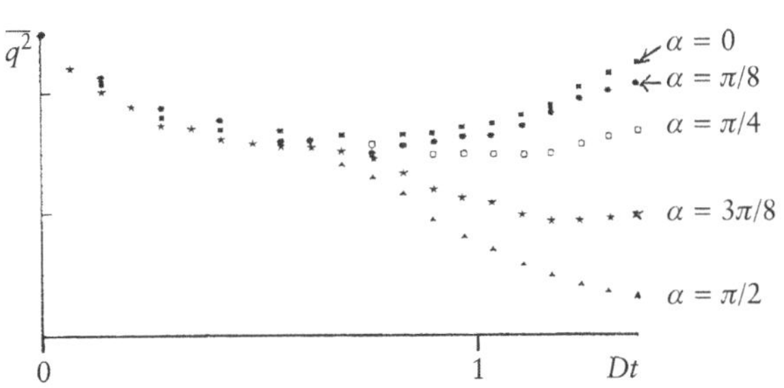

Figure 4.10. Experimental evolution of $\overline{q^2} = \overline{u_i u_i}$ in the distorting duct.

In the above flow, with $\alpha = \pi/2$, the turbulent production term is positive in the first half of the duct and anisotropy increases there, whereas in the second part, the turbulence production is negative and anisotropy decreases. There is, in fact, a close connection between the two for flows, such as this one, in which there are important effects of turbulent distortion by the mean flow. To see the relationship between turbulent energy production and increasing anisotropy more clearly, let us introduce the quantity $I = b_{ij}b_{ij}$, a scalar invariant under rotations of the coordinate system, which is the sum of squares of the components of b_{ij} and is therefore a measure of anisotropy. We wish to derive an equation for I, to approximate it, assuming the turbulence to be close to isotropy, and to compare the results with those for the turbulent energy. We consider a general, steady mean flow of the form

$$\overline{U}_i = \lambda_{ij} x_j \tag{4.112}$$

containing homogeneous turbulence. Here, in order that the mean flow satisfy the incompressibility condition, (4.5), we must have $\lambda_{ii} = 0$.

To derive an equation for I, we first use (4.112) to rewrite (4.38) as

$$\frac{d\overline{u_i u_j}}{dt} = -\lambda_{ik}\overline{u_j u_k} - \lambda_{jk}\overline{u_i u_k} + \frac{1}{\rho}\overline{p\left\{\frac{\partial u_i}{\partial x_j} + \frac{\partial u_j}{\partial x_i}\right\}} - 2\nu\overline{\frac{\partial u_i}{\partial x_k}\frac{\partial u_j}{\partial x_k}} \tag{4.113}$$

which describes the evolution of the turbulent velocity moments in the mean flow given by (4.112). We suppose that the small scales of turbulence, responsible for the dissipation at large Reynolds number, are isotropic so that (4.45) holds. We can also write the energy equation, (4.37), as

$$\frac{d\frac{1}{2}\overline{q^2}}{dt} = -\lambda_{ij}\overline{u_i u_j} - \overline{\varepsilon} \tag{4.114}$$

From the definition, (4.109), of b_{ij} and (4.45), (4.113), and (4.114), we obtain

$$\begin{aligned}\frac{db_{ij}}{dt} = &- \lambda_{ik}\left(b_{jk} + \frac{1}{3}\delta_{jk}\right) - \lambda_{jk}\left(b_{ik} + \frac{1}{3}\delta_{ik}\right) + 2\lambda_{kl}b_{kl}\left(b_{ij} + \frac{1}{3}\delta_{ij}\right)\\ &+ 2\frac{\overline{\varepsilon}}{\overline{q^2}}b_{ij} + \frac{1}{\rho\overline{q^2}}\overline{p\left\{\frac{\partial u_i}{\partial x_j} + \frac{\partial u_j}{\partial x_i}\right\}}\end{aligned} \tag{4.115}$$

Recalling that what we are after is an equation for I, we write

$$\frac{dI}{dt} = 2b_{ij}\frac{db_{ij}}{dt} \tag{4.116}$$

and use (4.115) to find that

$$\frac{dI}{dt} = -\frac{4}{3}\lambda_{ij}b_{ij} + \frac{2b_{ij}}{\rho\overline{q^2}}\overline{p\left\{\frac{\partial u_i}{\partial x_j} + \frac{\partial u_j}{\partial x_i}\right\}} - 4\lambda_{ik}b_{ij}b_{jk} + 4\left(\lambda_{ij}b_{ij} + \frac{\overline{\varepsilon}}{\overline{q^2}}\right)I \tag{4.117}$$

where we have exploited symmetry of b_{ij}.

The flows we have in mind here are those for which the turbulence is close to isotropy. Thus, b_{ij} is small and the last two terms on the right of (4.117), which are of quadratic order in b_{ij} (recall that I is a quadratic function of b_{ij}), may be neglected compared with the first term. Furthermore, for the second term, we adopt a closure model due to Lumley (1975) and Launder, Reece, and Rodi (1975), which implies that

$$\frac{1}{\rho}\overline{p\left\{\frac{\partial u_i}{\partial x_j}+\frac{\partial u_j}{\partial x_i}\right\}}=\frac{1}{3}C'\overline{q^2}\left(\frac{\partial \overline{U}_i}{\partial x_j}+\frac{\partial \overline{U}_j}{\partial x_i}\right)+\psi \tag{4.118}$$

where C' is a numerical constant of the closure and $\psi \to 0$ as $b_{ij} \to 0$ and may be neglected here because b_{ij} is assumed small. Thus, using (4.112), (4.117) becomes

$$\frac{dI}{dt}=-\frac{4}{3}(1-C')\lambda_{ij}b_{ij} \tag{4.119}$$

in which the bracketed expression is positive, since it is observed that $C' < 1$. This may be compared with the turbulent energy equation, (4.114), which can be written

$$\frac{d\frac{1}{2}\overline{q^2}}{dt}=-\overline{q^2}\lambda_{ij}b_{ij}-\overline{\varepsilon} \tag{4.120}$$

when the definition, (4.109), of b_{ij} is used. The first term on the right-hand side represents turbulence production and has the same sign as dI/dt, according to (4.119) with $C' < 1$. It follows that the turbulence moves towards isotropy, in the sense that I, and hence b_{ij}, decreases, when there is transfer of turbulent energy to the mean flow, but away from isotropy when there is positive turbulence production. Note that the use of a closure model (4.118) and the assumption that flow is already close to isotropy mean that this result should be regarded as qualitatively enlightening, rather than as providing an exact, quantitative description of the experiments.

In the case of the straining field considered in this subsection, Figure 4.11 shows the measured evolution of I for a number of different angles of rotation, α. Increasing I in the first half of the duct corresponds to positive production of turbulent energy there, whereas, depending on the value of α, I can either continue to increase, or begin to decrease in the second half, reflecting the sign of the energy production. Once again, we see the remarkable symmetry about the duct center when $\alpha = \pi/2$, while the dashed curves show the results of rapid distortion calculations for $\alpha = 0$ and $\alpha = \pi/2$, further illustrating the surprisingly good, if clearly approximate, description of the normalized Reynolds stresses obtained using a rapid-distortion approach.

Just after the first half of the duct, when the new principal axes of straining have come into effect, the principal axes of the Reynolds stress are still aligned with those of the first straining. When α differs from 0 and $\pi/2$, they rotate about the x_1-axis in the second half of the duct in an attempt to realign themselves with the new axes of straining. This is illustrated in Figure 4.12, which shows the evolution of the angle of the principal axes of $\overline{u_iu_j}$ in the second half of the duct. The angle plotted, φ, is that of the principal axes of the Reynolds stress, relative to the principal axes of the new strain (thus $\varphi = \alpha$ at the duct center). Complete readjustment of the

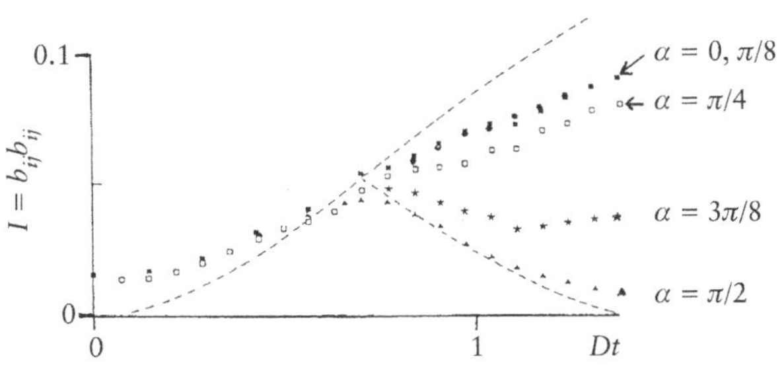

Figure 4.11. Experimental evolution of the quantity $I = b_{ij}b_{ij}$, measuring anisotropy. The dashed lines are the results of rapid distortion calculations for $\alpha = 0$ and $\alpha = \pi/2$.

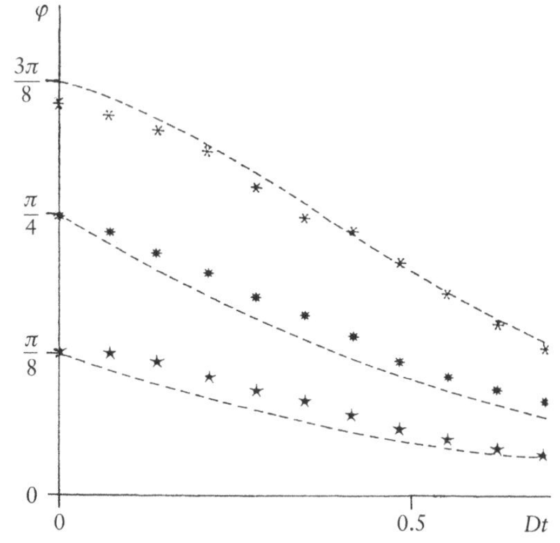

Figure 4.12. Experimentally determined evolution of the direction of the principal axes of turbulence following a sudden change of the direction of straining at the center of the duct. Here, unlike in previous figures, the time origin corresponds to the duct center. The dashed lines represent the results of a rapid distortion calculation.

principal axes to the new straining is not possible given the limited length of the duct (the distance from the center to the end of the duct represents a cumulative strain of only $Dt \approx 0.7$).

In summary, initially axisymmetric turbulence placed in a constant straining field tends to lose axisymmetry and to adopt the principal axes of that field. If the orientation of the field is suddenly changed, the turbulence slowly adjusts to the new principal axes. More importantly, energy can be extracted from the turbulence if the new straining field acts in the opposite sense to the original one. When this happens, the anisotropy developed during the original straining is partially removed. Straining of turbulence is thus somewhat reversible and is increasingly so the more rapidly the straining is applied.

ROTATING TURBULENCE

We consider turbulent flows in a steadily rotating fluid, such as found in rotating turbomachinery or in the earth's atmosphere and oceans, although without allowing for the effects of stratification which are important in geophysical applications. Sufficiently strong rotation is observed to considerably alter the turbulence dynamics, in particular leading to anisotropic structuring of turbulent eddies, which tend to be elongated in the direction of the axis of rotation. The importance of rotation for the large scales of turbulence is determined by the reciprocal of the turbulent Rossby number, $\mathrm{Ro} = u'/\Omega L$, where Ω is the angular velocity of rotation. The Rossby number is the ratio of a rotation time scale, Ω^{-1}, to a characteristic time, L/u', for evolution of the large scales of the turbulence in the absence of rotation. Thus, a large Rossby number corresponds to small effects of rotation and if it is sufficiently high one may neglect rotation altogether. On the other hand, a small enough Rossby number implies strong rotation and the large scales of turbulence do not have time to act on themselves significantly during a rotation period. One may then envisage applying rapid distortion theory, at least over intervals of time comparable with the rotation period. However, it turns out that rapid distortion theory predicts no development of anisotropy in initially homogeneous, isotropic, rotating turbulence without mean flow, a difficulty that it shares with a number of other turbulence models of varying degrees of sophistication. This and other characteristics make rotating turbulence somewhat of a challenge for turbulence modeling. We give an introduction to rotating turbulence here based on one-point methods, although the limitations of the description using one-point statistics will become apparent, thus illustrating the need for the more sophisticated multipoint techniques. Such techniques will be addressed in Chapter 6, albeit for the simplest case of turbulence in the absence of additional effects such as mean rotation and shear, whose detailed analysis goes beyond the scope of this volume.

A rotating fluid is most straightforwardly described using a frame of reference rotating with the fluid and we choose the coordinate system so that the rotation takes place about the x_3-axis. The angular velocity of rotation is constant and denoted by Ω, not to be confused with a vorticity component (we do not use the vorticity in this subsection). The effect of rotation is to introduce fictitious centrifugal and Coriolis forces into the equations of motion, as described in most textbooks on elementary mechanics. The Navier–Stokes equation in the rotating frame takes the form

$$\frac{\partial U_1}{\partial t} + U_j \frac{\partial U_1}{\partial x_j} = -\frac{1}{\rho}\frac{\partial \hat{P}}{\partial x_1} + \nu \frac{\partial^2 U_1}{\partial x_j \partial x_j} + 2\Omega U_2 \tag{4.121}$$

$$\frac{\partial U_2}{\partial t} + U_j \frac{\partial U_2}{\partial x_j} = -\frac{1}{\rho}\frac{\partial \hat{P}}{\partial x_2} + \nu \frac{\partial^2 U_2}{\partial x_j \partial x_j} - 2\Omega U_1 \tag{4.122}$$

$$\frac{\partial U_3}{\partial t} + U_j \frac{\partial U_3}{\partial x_j} = -\frac{1}{\rho}\frac{\partial \hat{P}}{\partial x_3} + \nu \frac{\partial^2 U_3}{\partial x_j \partial x_j} \tag{4.123}$$

where the pressure has been adjusted, $\hat{P} = P - \rho\Omega^2 R^2/2$, to take account of the fictitious centrifugal force and $R = (x_1^2 + x_2^2)^{1/2}$ is distance from the axis of rotation. All effects of the centrifugal force are allowed for simply by this redefinition of the pressure. Equations (4.121)–(4.123) should be compared with (4.3): the additional terms on the right-hand side represent the Coriolis force. It is straightforward to allow for the new (Coriolis) terms when averaging, because they are linear. The continuity equation, (4.4), is unaffected by rotation.

Taking the average of (4.121)–(4.123) gives the mean-flow equations

$$\frac{\partial \overline{U_1}}{\partial t} + \overline{U_j}\frac{\partial \overline{U_1}}{\partial x_j} = -\frac{1}{\rho}\frac{\partial \overline{\hat{P}}}{\partial x_1} + \nu \frac{\partial^2 \overline{U_1}}{\partial x_j \partial x_j} - \frac{\partial \overline{u_1 u_j}}{\partial x_j} + 2\Omega \overline{U_2} \tag{4.124}$$

$$\frac{\partial \overline{U_2}}{\partial t} + \overline{U_j}\frac{\partial \overline{U_2}}{\partial x_j} = -\frac{1}{\rho}\frac{\partial \overline{\hat{P}}}{\partial x_2} + \nu \frac{\partial^2 \overline{U_2}}{\partial x_j \partial x_j} - \frac{\partial \overline{u_2 u_j}}{\partial x_j} - 2\Omega \overline{U_1} \tag{4.125}$$

$$\frac{\partial \overline{U_3}}{\partial t} + \overline{U_j}\frac{\partial \overline{U_3}}{\partial x_j} = -\frac{1}{\rho}\frac{\partial \overline{\hat{P}}}{\partial x_3} + \nu \frac{\partial^2 \overline{U_3}}{\partial x_j \partial x_j} - \frac{\partial \overline{u_3 u_j}}{\partial x_j} \tag{4.126}$$

which are the equivalent of (4.9) in a rotating fluid. Observe that, as in the absence of rotation, the only effects of turbulence on the mean flow appear via the divergence of the Reynolds stress tensor.

We can obtain equations for the fluctuating field in the same way as we derived the equivalent, (4.31), without rotation. Thus,

$$\frac{\partial u_1}{\partial t} + \overline{U_k}\frac{\partial u_1}{\partial x_k} + u_k \frac{\partial \overline{U_1}}{\partial x_k} + \frac{\partial}{\partial x_k}[u_1 u_k - \overline{u_1 u_k}] = -\frac{1}{\rho}\frac{\partial p}{\partial x_1} + \nu \frac{\partial^2 u_1}{\partial x_k \partial x_k} + 2\Omega u_2 \tag{4.127}$$

$$\frac{\partial u_2}{\partial t}+\overline{U_k}\frac{\partial u_2}{\partial x_k}+u_k\frac{\partial \overline{U_2}}{\partial x_k}+\frac{\partial}{\partial x_k}[u_2u_k-\overline{u_2u_k}]=-\frac{1}{\rho}\frac{\partial p}{\partial x_2}+\nu\frac{\partial^2 u_2}{\partial x_k\partial x_k}-2\Omega u_1 \tag{4.128}$$

$$\frac{\partial u_3}{\partial t}+\overline{U_k}\frac{\partial u_3}{\partial x_k}+u_k\frac{\partial \overline{U_3}}{\partial x_k}+\frac{\partial}{\partial x_k}[u_3u_k-\overline{u_3u_k}]=-\frac{1}{\rho}\frac{\partial p}{\partial x_3}+\nu\frac{\partial^2 u_3}{\partial x_k\partial x_k} \tag{4.129}$$

The equations for the Reynolds stress components can now be obtained by multiplying (4.127)–(4.129) by different components of u_i and averaging. We find that

$$\begin{aligned}\frac{\partial \overline{u_iu_j}}{\partial t}+\overline{U_k}\frac{\partial \overline{u_iu_j}}{\partial x_k}&=2\Omega\beta_{ij}-\overline{u_iu_k}\frac{\partial \overline{U_j}}{\partial x_k}-\overline{u_ju_k}\frac{\partial \overline{U_i}}{\partial x_k}-\frac{\partial \overline{u_iu_ju_k}}{\partial x_k}\\&-\frac{1}{\rho}\left\{\overline{u_i\frac{\partial p}{\partial x_j}}+\overline{u_j\frac{\partial p}{\partial x_i}}\right\}+\nu\left\{\overline{u_i\frac{\partial^2 u_j}{\partial x_k\partial x_k}}+\overline{u_j\frac{\partial^2 u_i}{\partial x_k\partial x_k}}\right\}\end{aligned} \tag{4.130}$$

where the rotational term contains the matrix

$$\beta_{ij}=\begin{bmatrix}2\overline{u_1u_2} & \overline{u_2^2}-\overline{u_1^2} & \overline{u_2u_3}\\ \overline{u_2^2}-\overline{u_1^2} & -2\overline{u_1u_2} & -\overline{u_1u_3}\\ \overline{u_2u_3} & -\overline{u_1u_3} & 0\end{bmatrix} \tag{4.131}$$

Equation (4.130) is the equivalent of (4.32) with the inclusion of fluid rotation. In many practical flows the Rossby number is large enough for the rotational term in (4.130) to be neglected, rotation having a significant effect on the mean flow only, if at all. However, in compressors, to name one example, the rotational terms are often of the same order of size as the production terms, that is, the first two terms on the right-hand side of (4.130). From here on, we suppose that the rotation is sufficiently strong that it affects the turbulence significantly.

The turbulent energy equation can be derived by setting $j=i$ in (4.130), with an implied summation. The result is equation (4.33), as before, and so rotation *does not change the turbulent energy equation*. This is a reflection of the fact that the Coriolis force is perpendicular to the fluid velocity and hence does no work. Rotation only enters into the energetics of turbulence because it changes the various correlation terms which appear in equation (4.33). For homogeneous turbulence we obtain equation (4.37), that is,

$$\frac{d\frac{1}{2}\overline{q^2}}{dt}=-\overline{u_iu_j}\frac{\partial \overline{U_i}}{\partial x_j}-\overline{\varepsilon} \tag{4.132}$$

as if there were no rotation. Rotation can, however, modify both the production and dissipation terms in this equation by acting on the mean and fluctuating motions.

When the turbulence is homogeneous, as assumed from here on, equation (4.130) can be rewritten as

$$\frac{d\overline{u_iu_j}}{dt}=2\Omega\beta_{ij}-\overline{u_iu_k}\frac{\partial \overline{U_j}}{\partial x_k}-\overline{u_ju_k}\frac{\partial \overline{U_i}}{\partial x_k}+\frac{1}{\rho}\overline{p\left\{\frac{\partial u_i}{\partial x_j}+\frac{\partial u_j}{\partial x_i}\right\}}-2\nu\overline{\frac{\partial u_i}{\partial x_k}\frac{\partial u_j}{\partial x_k}} \tag{4.133}$$

which is the equivalent of (4.38) with rotation. To bring out the effects of rotation more clearly, *we take the mean flow to be zero* from now on, so that there is no longer any mean-flow–turbulence interaction, such as the distorting effects which dominated the previous two examples (since the turbulence is homogeneous, there is no Reynolds stress forcing of the mean flow, which consequently remains zero if it is so initially). Note, however, that zero mean velocity in the rotating frame is equivalent to a nonzero mean velocity in the nonrotating frame, representing solid-body rotation at angular velocity Ω. Without mean flow, equation (4.133) becomes

$$\frac{d\overline{u_iu_j}}{dt} = 2\Omega\beta_{ij} + \frac{1}{\rho}\overline{p\left\{\frac{\partial u_i}{\partial x_j} + \frac{\partial u_j}{\partial x_i}\right\}} - 2\nu\overline{\frac{\partial u_i}{\partial x_k}\frac{\partial u_j}{\partial x_k}} \tag{4.134}$$

which now contains only the rotational, pressure–rate-of-strain correlation, and viscous terms. When, as here, the mean flow is zero, so there is no turbulence production, the turbulent energy equation, (4.132), simply tells us that the turbulent energy decays under the effects of dissipation, as also follows from (4.134) with $j = i$ and $\beta_{ii} = 0$.

If the turbulence is initially isotropic,

$$\overline{u_iu_j} = \frac{1}{3}\overline{q^2}\delta_{ij}, \qquad \nu\overline{\frac{\partial u_i}{\partial x_k}\frac{\partial u_j}{\partial x_k}} = \frac{1}{3}\bar{\varepsilon}\delta_{ij}, \qquad \text{and} \qquad \beta_{ij} = 0$$

according to (4.131). Thus, on examining (4.134), we see that the turbulent velocity correlations, $\overline{u_iu_j}$, can only become anisotropic via the pressure–velocity term. Symmetry implies that initially isotropic turbulence with no mean flow will maintain $\overline{u_1^2} = \overline{u_2^2}$ and $\overline{u_1u_2} = \overline{u_1u_3} = \overline{u_2u_3} = 0$ at all times. Thus, the only effect of developing anisotropy on $\overline{u_iu_j}$ is to make the values of the mean-squared velocities, $\overline{u_3^2}$ and $\overline{u_1^2} = \overline{u_2^2}$, respectively parallel and perpendicular to the axis of rotation, different. In fact, it is found experimentally (Jacquin et al. (1990)) that initially isotropic turbulence leads to only relatively minor departures of $\overline{u_iu_j}$ from an isotropic form at later times. That is, although the internal structure of turbulence is rendered very significantly anisotropic by rotation, this is not apparent if one restricts attention to $\overline{u_iu_j}$, but other quantities, for instance correlation lengths parallel and perpendicular to the axis of rotation, are found to reflect the development of gross anisotropy. In particular, larger axial correlation lengths indicate elongation of turbulent structures parallel to the rotation vector. This illustrates the limited scope of the one-point moments $\overline{u_iu_j}$ as a measure of turbulence and should caution one against assuming that, because the Reynolds stress tensor is close to $\overline{u_iu_j} = \overline{q^2}\delta_{ij}/3$, or equivalently b_{ij} is small, the turbulence is nearly isotropic. Two-point, spectral formulations contain considerably more statistical information and allow a much more complete description of turbulent anisotropy, including its distribution over differently sized and orientated turbulent eddies.

Another important experimental observation is that rotation tends to reduce the rate of decay of turbulence by inhibiting the cascade of energy to small scales and hence the dissipation. As discussed earlier, turbulence left to itself decays on a time scale of $O(L/u')$, but in the presence of strong rotation (i.e., small Rossby number) decay takes considerably longer than this. The detailed explanation of this phenomenon requires spectral analysis, but can be roughly understood as follows. A rotating fluid supports waves (Greenspan (1968)), known as inertial waves, which oscillate in time with frequencies of order Ω. At high rotation rates, turbulence can be usefully thought of as a superposition of plane inertial waves (Fourier analysis) whose nonlinear interactions produce transfer of energy among different wave orientations and wavelengths, leading to the development of anisotropy, noted above, and to the energy cascade to smaller scales. However, the nonlinear terms inherit the inertial-wave oscillations and the cumulative effects of these oscillations tend to be nearly self-canceling except for a restricted class of wave pairs, whose nonlinear interaction happens to drive a third wave, producing significant cumulative energy transfer, a process known as wave resonance. Resonance (or near resonance) yields efficient exchange of wave energy, but, being confined to only certain wave interactions, the net effect of rotation is to inhibit energy transfer. One result of this is that the energy cascade, and hence the dissipation, are reduced. It should be clear from this brief discussion that the main observed consequences of rotation applied to initially isotropic turbulence, namely the development of anisotropic structuring and the reduction in the dissipation rate, involve rather subtle effects that are beyond the scope of a one-point description, thus pointing the way toward the need for spectral formulations.

To further understand the characteristics and limitations of one-point descriptions of rotating turbulence, we may apply one-point closures to express the pressure and viscous terms in (4.134) in terms of $\overline{u_i u_j}$. The pressure fluctuation, p, satisfies a Poisson equation which is derived by taking the divergence of (4.127)–(4.129). Thus we find that, in place of (4.48), we have

$$\frac{1}{\rho}\nabla^2 p = 2\Omega\left(\frac{\partial u_2}{\partial x_1} - \frac{\partial u_1}{\partial x_2}\right) - \frac{\partial u_i}{\partial x_j}\frac{\partial u_j}{\partial x_i} \tag{4.135}$$

where we have again assumed homogeneous turbulence and no mean flow. The rotational term in (4.135) will be recognized as proportional to the x_3-component of the vorticity, while the second term on the right is nonlinear in the fluctuation. As before, we write the corresponding two components of the fluctuating pressure as $p^{(1)}$, for the linear part due to rotation, and $p^{(2)}$ for the nonlinear component. The pressure–rate-of-strain term in (4.134) can be decomposed into linear and nonlinear components and the traditional one-point closure hypothesis for the nonlinear part is given by (4.55). The usual one-point closure for the linear pressure term (Launder et al. (1975)) assumes that it is proportional to the production term[7] in (4.134), i.e.,

$$\overline{\frac{1}{\rho}p^{(1)}\left\{\frac{\partial u_i}{\partial x_j} + \frac{\partial u_j}{\partial x_i}\right\}} = -2C'\Omega\beta_{ij} \tag{4.136}$$

where C' is a numerical constant of the model. Employing the closures (4.45), (4.55), and (4.136) to express the viscous and pressure terms in (4.134) gives a one-point model for rotating homogeneous turbulence. It is convenient to express the resulting system of equations for $\overline{u_i u_j}$ using the variables

[7] Actually the closure involves the traceless part of the Reynolds-stress production, but, in the case of solid-body rotation, the production term is already traceless, since $\beta_{ii} = 0$.

$$\Gamma_1 = \overline{u_3^2} - \frac{1}{2}\left(\overline{u_1^2} + \overline{u_2^2}\right) \tag{4.137}$$

which is the difference between the mean-squared velocities parallel and perpendicular to the axis of rotation, and the complex quantities

$$\Gamma_2 = \frac{1}{2}\left(\overline{u_1^2} - \overline{u_2^2}\right) + i\overline{u_1 u_2} = \frac{1}{2}\overline{(u_1 + iu_2)^2} \tag{4.138}$$

and

$$\Gamma_3 = \overline{u_1 u_3} + i\overline{u_2 u_3} = \overline{u_3(u_1 + iu_2)} \tag{4.139}$$

where, as usual, $i = \sqrt{-1}$. In terms of these variables, the model takes the form

$$\frac{d\Gamma_1}{dt} = -C\frac{\overline{\varepsilon}}{\overline{q^2}}\Gamma_1 \tag{4.140}$$

$$\frac{d\Gamma_2}{dt} = -C\frac{\overline{\varepsilon}}{\overline{q^2}}\Gamma_2 - 4i(1 - C')\Omega\Gamma_2 \tag{4.141}$$

$$\frac{d\Gamma_3}{dt} = -C\frac{\overline{\varepsilon}}{\overline{q^2}}\Gamma_3 - 2i(1 - C')\Omega\Gamma_3 \tag{4.142}$$

with the exact energy equation

$$\frac{d}{dt}\left(\overline{u_1^2} + \overline{u_2^2} + \overline{u_3^2}\right) = \frac{d\overline{q^2}}{dt} = -2\overline{\varepsilon} \tag{4.143}$$

In the case of initially isotropic turbulence, $\Gamma_1 = \Gamma_2 = \Gamma_3 = 0$ remains so at all times, according to (4.140)–(4.142). Thus, the model predicts that $\overline{u_i u_j}$ retains the isotropic form $\overline{u_i u_j} = \overline{q^2}\delta_{ij}/3$, a prediction which is not far from the truth, as discussed above. Equation (4.143) is then the only nontrivial member of the system (4.140)–(4.143), describing the decay of turbulent energy under dissipation. In addition to the closure hypotheses introduced above for the terms in the equation for $\overline{u_i u_j}$ which are not exactly expressible in terms $\overline{u_i u_j}$ itself, traditional one-point models include an evolution equation for $\overline{\varepsilon}$ (Launder et al. (1975)), based, like the closures, on heuristic assumptions. In the present case, this takes the form

$$\frac{d\overline{\varepsilon}}{dt} = -C''\frac{\overline{\varepsilon}^2}{\overline{q^2}} \tag{4.144}$$

where C'' is a further numerical constant of the model, supposed, like C and C', to be universal, but adjustable to fit experimental data. To make (4.144) reproduce the reduction in the dissipation rate in the presence of rotation noted above, the supposedly universal constant C'' may be increased. However, it should be recalled that, although we have here applied the model to a specific flow, namely rotating homogeneous turbulence, it is supposed to apply to general flows with the same constants.[8] Ad hoc adjustment of the constants of one-point models depending on the particular flow considered is a rough and ready way of allowing for physical

[8] Although we have formulated the problem in the rotating frame, it should perhaps be reiterated that, in the nonrotating frame of reference, fluid rotation merely appears as a special type of mean flow.

effects which are not properly accounted for by the models and illustrates the need for more refined approaches. Of course, from a practical engineering point of view, adjustment of the modeling constants for particular classes of flows using empirical data fitting may be the best way of producing reasonably accurate flow predictions, but it is fundamentally unsatisfactory and does not necessarily help to explain the physics, unlike, for instance, the description in terms of inertial waves, outlined above.

Anisotropic initial values for $\overline{u_i u_j}$ bring into play equations (4.140)–(4.142) of the model in a nontrivial way. There are two types of terms, namely those containing $\bar{\varepsilon}$, which arise from the nonlinear pressure and dissipative terms in (4.134) and are assumed to be controlled by $\bar{\varepsilon}$ under the one-point closures (4.45) and (4.55), and those involving Ω, which come from the production and linear pressure terms of (4.134) and represent the effects of rotation according to the model. The former terms may be neglected in the limit of rapid rotation (small Rossby number) over intervals of time comparable to the rotation period, leading to

$$\Gamma_2 = \Gamma_2(0)\exp\{-4i\Omega(1 - C')t\} \tag{4.145}$$

$$\Gamma_3 = \Gamma_3(0)\exp\{-2i\Omega(1 - C')t\} \tag{4.146}$$

together with a constant value of Γ_1 and, neglecting dissipation in (4.143), of $\overline{u_1^2} + \overline{u_2^2} + \overline{u_3^2}$ as well. Thus, the mean-squared velocities, $\overline{u_3^2}$ and $\overline{u_1^2} + \overline{u_2^2}$, parallel and perpendicular to the axis of rotation are predicted by the model to be constant in this, the rapid distortion limit. The exponentials in (4.145) and (4.146) can be written in terms of sines and cosines, indicating oscillations of $\overline{u_1^2} - \overline{u_2^2}$ and the off-diagonal components of $\overline{u_i u_j}$ via (4.138) and (4.139). Such oscillations of the Reynolds stresses are indeed to be expected, owing to the presence of inertial waves, but the model predicts that they are undamped in the small Rossby number limit, whereas rapid distortion calculations show that the oscillations are in fact damped in this limit. Furthermore, although $\overline{u_1^2} + \overline{u_2^2} + \overline{u_3^2}$ is obviously constant in the above limit, since dissipation can be neglected over times comparable to the rotation period, individual constancy of $\overline{u_3^2}$ and $\overline{u_1^2} + \overline{u_2^2}$ is not borne out by rapid distortion theory, thus allowing rapid exchange of energy between axial and transverse turbulent motions, again with damped oscillations. In conclusion, the usual one-point models do not describe the evolution of anisotropic $\overline{u_i u_j}$ for rapidly rotating turbulence at all well. Cambon, Jacquin, and Lubrano (1992) have examined the problem of one-point modeling of rotating turbulence using rapid distortion theory and more sophisticated spectral closure techniques, suggesting some modifications to improve the one-point description.

As implied above, rapid distortion theory, based on linearizing equations (4.127)–(4.129) for the fluctuations and neglecting viscosity, may be employed in the limit of small Rossby number. Rapid distortion analysis of homogenous flows, including rotating turbulence, can be carried out based on spectral theory, which describes two-point moments and hence provides a more complete description of the turbulence. In the case of a rotating fluid, rapid distortion theory expresses the turbulent fluctuations as a superposition of noninteracting inertial waves and, as noted above, leads to damped oscillations[9] if one calculates $\overline{u_i u_j}$ for initially anisotropic turbulence. However, it predicts no generation of anisotropy from initially isotropic turbulence, not even anisotropy of more subtle statistics than $\overline{u_i u_j}$, such as correlation lengths. In consequence, rapid distortion theory alone, although more sophisticated than one-point descriptions, is also insufficient to explain the observed appearance of anisotropy. Anisotropic, nonlinear spectral formulations, allowing interaction of inertial waves and leading to the appearance of anisotropy and the cascade of energy to smaller wavelengths, are needed before a fully satisfactory description of rotating turbulence is obtained. As was briefly explained earlier, nonlinear exchanges of energy mainly involve certain resonant families of waves in the limit of small Rossby number. Thus, anisotropy and the energy cascade in rotating turbulence are brought about by a subtle combination of linear and nonlinear effects acting together.

This example has indicated the difficulties associated with the construction of one-point models of turbulence and, we hope, has shown the need to go to multipoint models and hence spectral methods. This need is not confined to rotating fluids, for the pressure is always a nonlocal function of the velocity field and one-point pressure–velocity moments are determined by integrals over the two-point moments of velocity, both second order, for the linear component of pressure, and third order for the nonlinear part. Of course, the appearance of third-order moments due to nonlinearity implies closure. However, the need to go to multipoint formulations is a quite separate issue from nonlinearity, increasing the mathematical complexity, but, unlike the closure problem, does not present a fundamental restriction on our ability to describe turbulence using averaging methods. For instance, linear theory allows spectral calculations to be carried through in detail without explicit closure assumptions or, more accurately, employs the simplest closure of dropping the nonlinear terms entirely.

4.6 Conclusions

Despite suffering from significant limitations, single-point statistical formulations are among the most important techniques in the study of turbulence. As we have seen, they provide useful and physically appealing methods for analyzing such quantities as the mean-flow, turbulent energy, and Reynolds stresses. An appreciation of these methods, their application and their limitations is essential for anyone who wants to undertake work in the field of turbulence, from fundamental research through to industrial applications. This being said, multipoint statistics contain more information about the flow and, although multipoint theory, usually based on spectral analysis for homogeneous flows, is technically more demanding, it is, in many cases, essential to a fuller understanding of turbulent flows. In particular, a description of the contributions of differently sized and orientated turbulent structures to overall, one-point statistics such as the average kinetic energy $\overline{u_i u_i}/2$ requires multi-

[9] Damping of the oscillations of $\overline{u_i u_j}$ originates, not from dissipation of wave energy, which is neglected in rapid distortion theory, but from the fact that these one-point moments are integrals over all possible Fourier components, representing waves of all frequencies whose destructive interference produces decaying oscillations.

point methods, as does the theory of the energy cascade via which turbulent dissipation takes place.

The separation into a mean and fluctuating field, of which the latter is identified with turbulence, is the basis of statistical methods, be they single or multipoint. The mean flow sees the turbulence as like an externally applied stress field, the Reynolds stress, which dominates the viscous stress outside of thin viscous layers near boundaries. The Reynolds stresses are proportional to the one-point velocity correlations $\overline{u_i u_j}$, for which an evolution equation, (4.32), is derived using the equations of motion. This equation reduces to (4.38) in the homogeneous case and contains terms representing Reynolds stress production due to the mean-flow gradients, dissipation by viscosity, and a pressure–rate-of-strain contribution whose nonlocality when the pressure is expressed in terms of the velocity severely complicates the problem and underscores the need for multipoint formulations. Inhomogeneity leads to a variety of additional terms which are usually lumped together and loosely regarded as "turbulent diffusion," an expression of the fact that they are less well understood than those present in homogenous flows.

The turbulent energy equation can be obtained by taking the trace of the equation for $\overline{u_i u_j}$. It contains a production term involving the mean flow which is usually positive, representing turbulent energy production, but can be negative, expressing destruction of turbulent energy. Viscous dissipation represents a drain on the turbulent energy and, although not apparent from a one-point description, takes place via the cascade from large scales, where production occurs, to the smallest scales, where viscosity acts and the energy is dissipated. Additional terms appear in inhomogeneous flows, which can be expressed as the divergence of a turbulent energy flux vector and therefore represents transfer of energy from place to place, rather than its creation or destruction. As for the Reynolds stress equation, this term is normally thought of as diffusive in character, expressing the average flow of turbulent energy due to nonuniformity of statistical properties.

The trace of the tensor $\overline{u_i u_j}$ yields the turbulent energy, but it also contains information on anisotropy of turbulence. Anisotropy is present in most flows and develops from initially isotropic turbulence under the action of mean velocity gradients, as we have seen in a number of examples. In the isotropic case, $\overline{u_i u_j} = u'^2 \delta_{ij}$ and the nondimensional, traceless tensor $b_{ij} = \overline{u_i u_j}/q^2 - \delta_{ij}/3$, which along with the turbulent energy suffices to determine $\overline{u_i u_j}$, is often used as a measure of anisotropy. However, as we saw in the case of rotating turbulence, it is quite possible for the form of $\overline{u_i u_j}$ to be approximately isotropic, or equivalently for b_{ij} to be small, while other statistics show gross anisotropy. This illustrates one of the limitations of $\overline{u_i u_j}$ as a description of turbulence, pointing again to the need for more complete descriptions, such as two-point ones. Indeed, rotating turbulence introduces wave dynamics into the flow physics, whose natural mathematical expression is via Fourier decomposition.

Nonlinearity leads to lack of closure of the velocity moment equations even in multipoint formulations. The simplest approach to this closure problem is to linearize the equations for the fluctuating field, which, combined with neglect of viscosity, yields rapid distortion theory. The rapid distortion approximation describes the distorting effects of the mean flow on the turbulence, but not the interactions of the latter with itself. Thus, it is only strictly applicable in the limit in which the mean velocity gradients (or other linear effects, such as gravity combined with density stratification) are sufficiently strong that one can neglect such self-interactions. This not only limits the allowable intensity of turbulence, but the length of time for which rapid distortion theory holds, since nonlinear effects, for instance the energy cascade or the appearance of anisotropy in rotating turbulence, can be significant given sufficient time, even if they are negligible over shorter time scales. However, as we saw from the example of straining by mean stretching in one direction and compression in another, it is found that rapid distortion results, in particular reversibility of the effects of straining, can hold surprisingly well for certain quantities, such as the normalized Reynolds stresses, outside the range of formal applicability of the theory. When the effects of nonlinearity are important, the closure problem must be tackled head on, generally by introducing heuristic hypotheses to close the moment equations.

Nonlocality means that the one-point moment equations are incomplete, a different type of closure difficulty which is present even if one neglects nonlinearity, as in rapid distortion theory. The production term is the only one in the evolution equation, (4.32), for $\overline{u_i u_j}$ that is closed and one-point turbulence modeling based on $\overline{u_i u_j}$ (i.e., at the level of sophistication above k–ε type models) consists of supposing heuristic expressions for the other terms, with the pressure–rate-of-strain being first decomposed into linear and nonlinear parts. In so doing, incompleteness of the one-point moment equations owing to nonlocality and the more fundamental closure problem due to nonlinearity are tackled together. The closure hypotheses involve the turbulent energy dissipation rate $\overline{\varepsilon}$, for which a model evolution equation is also assumed, and introduce a variety of adjustable constants, providing flexibility to match experimental data. Such models are of great practical utility, since they allow routine calculations of flows in complicated geometries, but are fundamentally questionable owing to the many hypotheses they entail.

As we have previously implied and will see in detail in later chapters, beginning with Chapter 6, spectral analysis describes the way in which the energy and anisotropy of turbulence are distributed over the differently sized and orientated scales of turbulence. It allows the relatively straightforward treatment of two-point equations for homogeneous flows via Fourier analysis. The three-dimensional spatial Fourier transforms of the two-point velocity moments yield a spectral tensor, which is a function of the wavenumber vector **k**, whose direction and inverse magnitude, $|\mathbf{k}|^{-1}$, may be thought of as representing the orientation and size of the corresponding structures of turbulence. The one-point correlations $\overline{u_i u_j}$ can be calculated from the spectral description by integrating the spectral tensor over all wavenumber directions and magnitudes, showing that $\overline{u_i u_j}$ lumps together scales of all sizes and orientations, whose dynamics may differ significantly. Furthermore, two flows can have the same $\overline{u_i u_j}$, yet differ significantly in the form of their spectral tensors, again showing the limitations of $\overline{u_i u_j}$ alone as a measure of turbulence. Whereas the spectral tensor has infinite degrees of freedom corresponding to the three-dimensional spectral space of different wavenumbers, $\overline{u_i u_j}$ has but six. Turbulence has still more, since the spectral tensor only contains information about second-order, one-time velocity moments, not higher-order moments and those involving two or more times.

Finally, as discussed in Section 4.4, the vorticity is an appealing quantity because it allows one to think in terms of amplification by stretching of vortex lines and the reduction of scale of turbulence that occurs in the cascade. However, because vorticity is defined in terms of velocity derivatives, once the turbulence has developed, vorticity is dominantly a small-scale quantity, making it quantitatively difficult to identify large-scale vortices, whose stretching might give rise to the smaller scales. Furthermore, the equations for the mean and fluctuating vorticity, given in Section 4.4, are relatively poorly understood.

References

Bonn, D., Couder, Y., van Dam, P. H. J., Douady, S., 1993. From small scales to large scales in three-dimensional turbulence: the effect of diluted polymers. *Phys. Rev. E*, **47**, R28–R31.

Boussinesq, J., 1877. Essai sur la théorie des eaux courantes. *Mém. prés. par div. savants à l'Acad. Sci. Paris*, **23**, 1–680.

Cambon, C., Jacquin, L., Lubrano, J. L., 1992. Toward a new Reynolds stress model for rotating turbulent flows. *Phys. Fluids A*, **4**, 812–24.

Champagne, F. H., Harris, V. G., Corrsin, S., 1970. Experiments on nearly homogeneous turbulent shear flow. *J. Fluid Mech.*, **41**, 81–139.

Comte-Bellot, G., 1965. Ecoulement turbulent entre deux parois parallèles. Publication number PST 419, "*Ministère de l'Air*", France.

Gence, J. N., Mathieu, J., 1979. On the application of successive plane strains to grid-generated turbulence. *J. Fluid Mech.*, **93**, 501–13.

Gence, J. N., Mathieu, J., 1980. The return of isotropy of an homogeneous turbulence having been submitted to successive plane strains. *J. Fluid Mech.*, **101**, 555–66.

Greenspan, H. P., 1968. *The Theory of Rotating Fluids*. Cambridge University Press, Cambridge.

Hanjalic, K., Launder, B. E., 1972. Fully developed asymmetric flow in a plane channel. *J. Fluid Mech.*, **51**, 301–35.

Jacquin, L., Leuchter, O., Cambon, C., Mathieu, J., 1990. Homogeneous turbulence in the presence of rotation. *J. Fluid Mech.*, **220**, 1–52.

Jeandel, D., Brison, J. F., Mathieu, J., 1978. Modeling methods in physical and spectral space. *Phys. Fluids*, **21**, 169–82.

Kraichnan, R. H., 1967. Inertial ranges in two-dimensional turbulence. *Phys. Fluids*, **11**, 265–77.

Laufer, J., 1951. Investigation of turbulent flow in a two-dimensional channel. *NACA*, report no. 1033.

Laufer, J., 1954. The structure of turbulence in fully developed pipe flow. *NACA*, report no. 1174.

Launder, B. E., Reece, G. J., Rodi, W., 1975. Progress in the development of a Reynolds-stress turbulence closure. *J. Fluid Mech.*, **68**, 537–66.

Lumley, J. L., 1970. Toward a turbulent constitutive relation. *J. Fluid Mech.*, **41**, 413–34.

Lumley, J. L., 1975. Prediction methods for turbulent flows. *Von Karman Institute*, Lecture series 76.

Maréchal, J., 1967. Anisotropie d'une turbulence de grille déformée par un champ de vitesse moyenne homogène. *C. R. Acad. Sci. Paris*, **265**, 478–81.

Mathieu, J., 1959. Etude d'un jet plan frappant sous une incidence de 7° une plaque plane lisse. *C. R. Acad. Sci. Paris*, **248**, 2713–15.

Reynolds, O., 1894. On the dynamical theory of turbulent incompressible viscous fluids and the determination of the criterion. *Phil. Trans. R. Soc. Lond.*, **A186**, 123–61.

Rotta, J., 1951. Statistische Theorie nichthomogener Turbulenz. *Z. Phys.*, **129**, 547–72.
Sabot, J., Comte-Bellot, G., 1976. Intermittency of coherent structures in the core region of fully developed turbulent pipe flow. *J. Fluid Mech.*, **74**, 767–96.
She, Z. S., Jackson, E., Orszag, S. A., 1991. Structure and dynamics of homogeneous turbulence: models and simulations. *Proc. Roy. Soc. Lond.*, **A434**, 101–24.
Tavoularis, S., Karnik, U., 1989. Further experiments on the evolution of turbulent stresses and scales in uniformly sheared turbulence. *J. Fluid Mech.*, **204**, 457–78.
Taylor, G. I., 1938. Production and dissipation of vorticity in a turbulent fluid. *Proc. Roy. Soc. Lond.*, **A164**, 15–23.
Townsend, A. A., 1954. The uniform distortion of homogeneous turbulence. *Quart. J. Mech. Appl. Math.*, **7**, 104–27.
Tucker, H. J., Reynolds, A. J., 1968. The distortion of turbulence by irrotational plane strain. *J. Fluid Mech.*, **32**, 657–73.

CHAPTER FIVE

Classical Models of Jets, Wakes, and Boundary Layers

Three important classes of statistically steady turbulent flows will be examined in some detail in this chapter, namely jets, wakes, and boundary layers. These flows are of considerable practical interest and are illustrated in Figure 5.1, together with a shear layer, which is not specifically considered here, but whose treatment is not essentially different from that of jets and wakes. Much is known about these flows from numerous experimental and theoretical studies, stretching back over many decades, and the resulting theoretical models are semiempirical, being based on the experimentally observed properties of the flows, such as the small angles of divergence of jets and wakes and the existence of self-similar behavior far downstream of the nozzle, in the case of jets, or body, in the case of wakes. We attempt to give a unified presentation of these models, emphasizing their common origins in the boundary-layer approximation and various types of mean-flow self-similarity.

The basis of the theories developed in this chapter is the mean-flow equations, (4.9), to which simplifying assumptions, similar to those of laminar boundary-layer theory, are applied, leading to neglect of some of the terms in the equations. This turbulent boundary-layer approximation, which applies to jets and wakes as well, is justified by slowness of mean-flow development with streamwise distance, as witnessed by the small angles of divergence of turbulent jets and wakes. Since the approximations used are similar to Prandtl's theory of laminar boundary layers, with which we assume the reader is familiar, and since, despite their quite different mechanisms of growth, the behavior of turbulent boundary layers in many ways resembles that of laminar ones, we devote Section 5.1 to reminding the reader of the principal characteristics and theory of laminar boundary layers, before developing the turbulent analog of Prandtl's theory in Section 5.2.

Having applied the boundary-layer approximation to simplify the mean-flow equations, one is still faced with their lack of closure, apparent from the Reynolds stress terms. Here, empirical information is introduced, whose details depend on the class of flows considered. In the case of jets and wakes, we use the observed fact that the mean flow becomes self-similar sufficiently far downstream, together with an eddy-viscosity approximation and an assumed eddy viscosity[1] that does not vary across the flow, but only with streamwise distance. This allows us to solve for the mean flow in far jets and wakes, as we shall do in Sections 5.3 and 5.4.

[1] This is the simplest assumption and the one we adopt. However, more complicated eddy-viscosity models exist, such as those based on mixing-length ideas (see, e.g., Hinze (G 1975)).

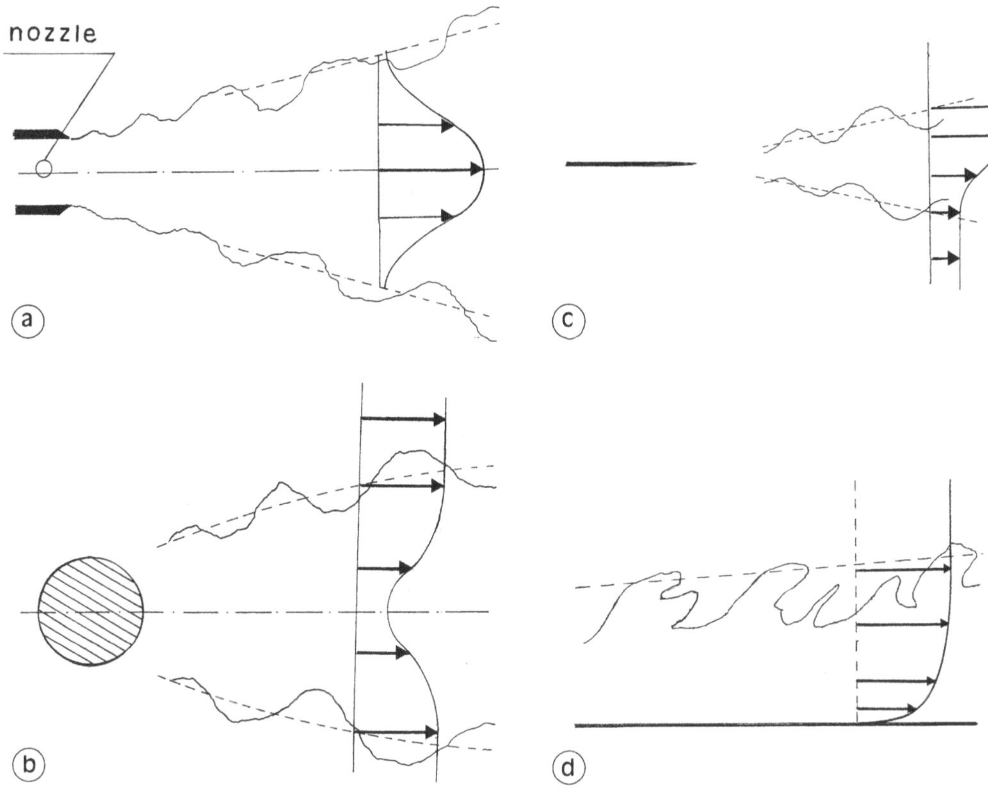

Figure 5.1. Illustrations of a turbulent (a) jet, (b) wake, (c) shear layer, and (d) boundary layer. The streamwise mean-flow profiles are also shown. The rate of spreading of the boundary layer has been exaggerated for clarity; in reality, the angle of spreading is only of the order of 1°.

Turbulent boundary layers are treated in Section 5.5 and are more complicated because they possess internal structure, with an outer layer, in which one may neglect the viscous terms in the mean-flow equations, as for jets and wakes, and a very thin viscous sublayer at the surface of the body,[2] whose existence was noted in Chapter 4. For a flat plate at zero incidence in a uniform stream, the mean flow is found to be self-similar in both the outer layer and viscous sublayer, having universal forms for the velocity profiles, although, needless to say, the similarity variables and velocity profiles are different in the two regions. The two expressions for the mean flow, one for the outer region, one for the viscous sublayer, agree with one another at intermediate distances from the wall, small compared with the overall boundary-layer thickness and large compared with the viscous sublayer thickness, a zone known as the inertial layer. As we shall see, if one wants to think in those terms, one can regard the two expressions as matched asymptotic expansions, applicable in distinct asymptotic regions – the outer region and viscous sublayer – whose asymptotic separation of scale results from a large boundary-layer Reynolds number. In any case, the net

[2] As remarked in Chapter 4, we choose to include the conventionally named buffer layer in the definition of the viscous sublayer, for reasons of parsimony. Thus, the viscous sublayer is the zone in which viscosity is important in the mean-flow equations.

result is that, for the flat plate, one has the form of the mean flow in the boundary layer expressed in terms of two universal, empirically determined functions, with just one unknown parameter (e.g., the boundary-layer thickness) remaining to determine. The evolution of this parameter with streamwise distance is governed by a momentum balance equation, known as the Von Karman equation, which is obtained by integration of the streamwise mean-flow equation across the boundary layer and is the only input provided by the equations of motion to the model. Once one has solved the Von Karman equation, one has the solution for the mean flow throughout the boundary layer.

As we will see in Section 5.5, in which the details of boundary-layer modeling are described, the case of turbulent boundary layers over surfaces with curvature, or having external pressure gradients or vorticity, is more complicated and, indeed, it will involve us in first deriving a more sophisticated version of the turbulent boundary-layer approximation. It turns out that the outer part of the boundary layer is considerably more sensitive to such complicating influences than the viscous sublayer, for which the universal flat-plate form of the mean velocity may continue to apply, allowing us to describe the viscous sublayer even when the outer layer has been significantly affected. The main difficulty in such cases is that the outer part of the layer is no longer self-similar and one has no equivalent of the universal functional form that describes the outer region of the flat-plate boundary layer.

Figure 5.1 illustrates the three classes of statistically steady, turbulent flows considered in this chapter, as well as a shear layer, which will not be discussed in detail. One can identify a streamwise direction, such that the streamwise component of mean velocity is dominant, and a cross-stream (or transverse) direction, in which turbulent statistical properties vary most rapidly. For instance, in a boundary layer, the transverse and streamwise directions are respectively normal and tangential to the body surface, and therefore vary with position if the surface is curved. We will mostly restrict attention to two-dimensional[3] plane flows in this chapter, that is, flows whose statistical properties depend only on the coordinates x_1 and x_2, with $\overline{U}_3 = 0$.

The flows considered are turbulent inside a well-defined interface which separates laminar fluid outside from turbulent fluid within. This boundary moves randomly as bulges of turbulent fluid, whose size is comparable to the large scales of the turbulence, are convected downstream, as we have tried to indicate in Figure 5.1. Laminar fluid is engulfed by these turbulent bulges and itself becomes turbulent, a process known as entrainment. The entrained fluid is accelerated in the case of jets, or decelerated in wakes and boundary layers. There is transfer of mean momentum across the flow, due to the turbulence, which causes the mean velocity profile (also shown in Figure 5.1) to develop with streamwise distance, spreading out downstream. This cross-stream transfer of mean momentum by turbulent mixing appears in the mean-flow equations, (4.9), via the Reynolds stress term. It attempts to make the mean velocity profile uniform as it develops downstream, producing lateral spreading and, in the case of jets and wakes, a decreasing difference between the mean velocity outside and at the center of the jet or wake.

[3] Needless to say, the turbulence in any one realization is three-dimensional. It is the statistical properties of the flow, such as the mean velocity, which are two-dimensional.

At the same time as the turbulence affects the mean flow through transfer of streamwise momentum across the flow, mean shear acts as a source of turbulence, which would otherwise decay rather quickly. The turbulence is symptomatic of mean-flow instabilities, occurring in regions of high mean shear, which are consequently centers of turbulent production. For boundary layers, turbulence production is concentrated near the wall (where the mean shear is high), and in a less localized manner around the points of maximum mean-velocity gradient in jets and wakes. Because the maximum of mean shear is much less intense in the absence of a wall, turbulence production is more evenly distributed across the flow, but still tends to peak where the mean shear is high.

Thus, there is a two-way interaction between mean flow and turbulence: the turbulence is produced by local mean-flow instabilities near positions of high mean shear, while the resulting turbulence tends to reduce the mean shear which created it, through mean momentum transfer.[4] The distance required for streamwise development of the flow due to mean momentum transfer by turbulence is considerably greater than the transverse length scale (width) of the flow. This is a reflection of the relative slowness of momentum transfer across the flow, compared with streamwise convection by the mean flow. That is, the mean flow convects the turbulence downstream considerably faster than the mean-velocity profile can develop sideways. The net result of rapid streamwise convection and slower cross-stream spreading is a statistically steady flow in which streamwise distances are long compared with transverse ones. Transverse transfer of streamwise mean momentum is slow because turbulent velocities are small compared with streamwise mean velocities. It is the small ratio of the turbulent root-mean-squared velocity,[5] u', to an appropriate mean-velocity scale (to be made more precise later) which makes transverse gradients larger than streamwise ones.

The larger gradients in the transverse direction(s) are reminiscent of laminar boundary layers, with which we shall suppose that the reader is acquainted (a short review is given in Section 5.1; for more information, see chapter 5 of Batchelor (G 1967), or the more extensive treatment in Schlichting (1987)). If a body is placed in a stream at high Reynolds number, it develops a thin boundary layer at its surface in which the fluid moves more slowly than outside the layer. The layer is laminar provided that the Reynolds number is not too high. Viscosity is important within the layer, despite the large Reynolds number, owing to large velocity gradients. The higher the Reynolds number, the thinner the layer compared with the body dimensions, and hence the larger the velocity gradients across the layer. The length scale for variations along the surface is determined by the overall body size, whereas, normal to the surface, it is dictated by the small layer thickness. Thus, transverse variations are more rapid, a property which is exploited in the Prandtl theory of laminar boundary layers to simplify the Navier–Stokes equation by neglecting terms which are small. The result is the Prandtl equations for laminar boundary layers, which we review in Section 5.1.

[4] The result of the instability tries to alleviate its cause, as is often the case.

[5] Although u' varies across the flow, the notation u' is used as a characteristic velocity scale for turbulence in this chapter. In so doing, we have in mind some cross-stream averaged value of u'.

The laminar layer comes about because of the small, but nonzero, viscosity of the fluid. At the body surface, a viscous fluid must satisfy the no-slip condition. The boundary layer allows adjustment of the essentially inviscid flow outside the layer, which would not satisfy this condition, so that no-slip is respected at the surface. Viscosity is thus crucial to the existence of the layer and the manner in which it evolves with streamwise distance along the surface. It provides a mechanism for diffusion of streamwise momentum across the layer and tends to cause the boundary layer to thicken with downstream distance. Viscosity produces friction between neighboring sheets of fluid due to their differential motion, which tries to equalize the velocities of the fluid within the boundary layer as it travels downstream. This process is analogous to that of cross-stream momentum transfer by turbulence, described above, although its physical origins are quite different.

The other physical mechanism controlling the development of a laminar boundary layer is the streamwise pressure gradient, which is imposed from outside the layer and tends to accelerate or decelerate the fluid, as dictated by the external flow. Denote the velocity and pressure just outside the layer by $U_\infty(x)$ and $P_\infty(x)$, where x is streamwise distance along the body surface. Since the essentially inviscid flow external to the boundary layer has the surface as a streamline, Bernoulli's theorem implies that $P_\infty + \rho U_\infty^2/2$ is constant to a good approximation (more precisely, the external flow approaches an inviscid limit having the surface as streamline when the Reynolds number goes to infinity). Thus, the pressure gradient is directly related to the derivative of $U_\infty(x)$ by $dP_\infty/dx = -\rho U_\infty \, dU_\infty/dx$, which shows that a pressure gradient is synonymous with nonuniformity of U_∞. That is, according to its sign, a pressure gradient causes the fluid to accelerate or decelerate outside the layer as it flows downstream. It also attempts to accelerate/decelerate the fluid within the boundary layer, although viscous friction between neighboring sheets of fluid is important there also. Such pressure gradient effects are also present in the turbulent flows of this chapter when there is significant streamwise variation of the pressure external to the flow. However, for jets and wakes, we will be concerned with cases in which there is no external pressure gradient, so that streamwise development of the flow is due solely to cross-stream transfer of mean momentum.

In the turbulent flows of this chapter, transverse transfer of streamwise mean momentum results primarily from turbulent mixing, rather than viscosity. In fact, for the high-Reynolds-number flows considered here, the viscous terms in the mean-flow equations are negligibly small, except in the viscous sublayer of a turbulent boundary layer. Elsewhere, turbulent momentum transfer, represented in the mean-flow equations by Reynolds stress terms, dominates over viscous transfer, and the rate of transfer of momentum is dictated by the turbulent intensity, measured by u'. Whereas, for a laminar boundary layer, smallness of the viscosity (or equivalently a large Reynolds number) is responsible for weak momentum transfer, and hence a thin layer, it is the small ratio of u' to a mean-flow velocity scale that leads to thin turbulent boundary layers, as well as to jets and wakes whose streamwise development takes place slowly on a transverse length scale (or width).

Based on the more rapid variation of the statistical properties of the flow in the transverse direction, a turbulent analog of the Prandtl equations can be developed, as we will see in Sections 5.2 and 5.5. In carrying over ideas from laminar boundary-layer theory to turbulent flows, they must be applied to the equations for statistical quantities, not to the raw Navier–Stokes equation, for it is the *statistics* of turbulence which show slow variations in the streamwise direction, *not* the turbulence in any one realization. Although the turbulence is somewhat anisotropic in the flows considered here, even the largest eddies have roughly similar properties in different directions and are by no means as strongly elongated in the streamwise direction as would be required for direct application of the boundary-layer approximations to the Navier–Stokes equation.

The turbulent analog of Prandtl's theory, applied to the statistical equations, will be referred to as the "turbulent 'boundary-layer' approximation." We will mainly be concerned with applying the approximation to the mean-flow equations, for which, as noted above, one can neglect the viscous terms compared with the Reynolds stresses, except within viscous sublayers. However, even outside viscous sublayers, viscous terms cannot be neglected when we come to consider the equations describing the statistical properties of the turbulence itself (e.g., energy and Reynolds stresses). This is as one might expect, because the loss of turbulent energy via the cascade and its dissipation by viscosity at the smallest scales is crucial to the dynamics of turbulence. There is nothing specific to the boundary-layer approximation in this; indeed, we saw in Chapter 4 that viscosity could generally be neglected for the mean flow at high Reynolds number outside viscous wall layers, but not for the turbulent energy and Reynolds stress equations.

The basic turbulent "boundary-layer" approximation is developed in Section 5.2 for jets and wakes. An order-of-magnitude analysis is carried out for the mean-flow equations, based on slow streamwise versus transverse variations of the turbulence statistics, and the result is a simplified set of equations, similar to the Prandtl's equations, but with a Reynolds stress term in place of the viscous one. These equations are intended to describe turbulent jets, wakes, and shear layers, although we do not discuss the latter class of flows. It might be remarked that, although Prandtl's equations are usually associated with boundary layers, they can also be employed for laminar jets, wakes, and shear layers at high Reynolds number (not to mention pipe and channel flows).

The reason why "boundary layer" appears within quotes in this context is that the basic approximation to the mean-flow equations, as developed in Section 5.2, turns out to be inadequate to describe general turbulent boundary layers, having significant surface curvature or mean vorticity of the flow external to the boundary layer. This is in contrast with laminar boundary layers, for which the corresponding basic approximation, namely Prandtl's equations, suffices. One of the main differences between turbulent and laminar boundary layers is that the former have an internal structure consisting of a thin viscous sublayer at the surface, an outer layer which occupies most of the total boundary-layer thickness, and an intermediate region known as the inertial layer. The variations of mean velocity in the outer layer are found to be small, of the same order of magnitude as the turbulent velocities. The small parameter on which the turbulent boundary-layer approximation is based is u'/U_0, where U_0 is a typical value of the mean-flow velocity. Thus, the variations of the mean flow in the outer layer, $O(u')$, are formally of higher order, that is, the mean flow is uniform at leading order. To describe the outer-layer variations in the mean flow, which are crucial to the theory, one therefore needs either to introduce higher-order boundary-layer approximations of the mean-flow equations, or, more simply, to work with the equations governing the velocity defect. The velocity defect is defined as the difference between the actual mean flow and a notional, inviscid velocity field obtained by suitably extrapolating the velocity field from outside the boundary layer to the surface. The variations in mean velocity, although small, then show up at leading order even in the outer layer and, as in deriving Prandtl's equations for the laminar case, one examines the governing equations for the velocity defect, neglecting small terms, to obtain a refined turbulent boundary-layer approximation. In the absence of significant curvature and external mean vorticity, this refined approximation turns out to give the same equations as the basic approximation. However, to correctly describe the outer region of general turbulent boundary layers, the refined formulation, developed in Section 5.5, is required.

The mean-flow equations of the boundary-layer approximation are considerably simpler than the original ones, but they still contain a Reynolds stress term, representing cross-stream transfer of streamwise mean momentum by turbulence. The presence of this Reynolds-stress term means that the mean-flow equations are incomplete, reflecting the closure problem discussed in the previous chapter, which appears whether or not one adopts the boundary-layer approximation: the equations describing the moments of order n contain those of order $n+1$. Thus, the mean-flow equations depend on the Reynolds stresses, while if one formulates the equations for the Reynolds stresses, in an attempt to obtain a complete system, third-order moments and nonlocal terms show up and the combined mean-flow and Reynolds stress equations are still not closed.

For the flows considered here, the usual approach to overcoming the difficulty of lack of closure, and the one we adopt, is to introduce experimental information. For instance, jets exhausting into infinite, quiescent fluid and wakes in an infinite, uniform external stream are found to become self-similar sufficiently far downstream of the nozzle, in the case of a jet, or the body, in the case of a wake. In the self-similar region of a wake, the cross-stream profile of the velocity defect (i.e, the difference between the external velocity and the streamwise mean velocity in the wake) has the same form at different streamwise locations, but its width and velocity scaling vary with the streamwise coordinate according to simple power laws. Likewise, the mean-velocity profile of a jet exhausting into stationary fluid becomes self-similar far from the nozzle and has power law behavior there. Thus, the streamwise development of these flows can be supposed known and the problem becomes that of the determination of the cross-stream profile. A variety of approaches exist and we adopt the simplest, consisting of a turbulent eddy-viscosity closure (see "The Eddy-Viscosity Approximation and One-Point Modeling," Section 4.1) for the Reynolds-stress term, in which the eddy viscosity is assumed only to depend on the streamwise coordinate. When the supposed similarity behavior and eddy-viscosity approximation are used in the mean-flow equations resulting from the boundary-layer approximation, the mean-flow similarity profile can be solved for analytically, as we shall see in Sections 5.3 and 5.4. The results agree quite well with experiments on jets and wakes, but, from a fundamental point of view, the lack of theoretical explanations for self-similarity and (approximate) transverse uniformity of the eddy viscosity is somewhat less than fully satisfactory.

For turbulent boundary layers, the use of experimental information to overcome the lack of closure is somewhat different. As noted earlier, a turbulent boundary layer has an internal structure with a thin viscous sublayer at the surface and an outer region taking up most of the overall thickness of the layer. Within the viscous sublayer it is found that the mean-velocity profile has a universal form when suitably scaled and expressed in terms of an appropriately nondimensionalized distance from the body surface. Specifically, the scaling used for the streamwise velocity, $\overline{U}_x$, is the wall friction velocity, u_*, which is defined in terms of τ_w, the mean tangential stress at the surface (or skin friction), by $u_* = \sqrt{|\tau_w|/\rho}$, where ρ is the fluid density, as usual. The distance, y, from the wall is nondimensionalized on the laminar sublayer thickness scale, ν/u_*. Thus, when the scaled mean velocity, $\overline{U}_x/u_*$, is plotted as a function of $y_+ = u_* y/\nu$, known as inner scalings, the experimental data for viscous sublayers is found to collapse to a universal form, $\overline{U}_x/u_* = f(y_+)$ (except close to a boundary-layer separation point, as we shall see later). The behavior in the viscous sublayer is therefore determined by u_* and ν alone, at least as far as the mean flow is concerned.

The outer region of a turbulent boundary layer without significant pressure gradient, surface curvature, or external vorticity (for instance, the classical problem of a flat plate at zero incidence in a uniform stream) is observed to have universal behavior when the *velocity defect*, nondimensionalized using u_*, is expressed in terms of $\eta = y/\delta$, where y, is distance from the surface and δ is the overall thickness of the boundary layer. Thus, for the flat-plate boundary layer, if U_∞ is the streamwise mean velocity outside the boundary layer, $(\overline{U}_x - U_\infty)/u_* = F(\eta)$, where $F(\eta)$ is a universal function (which is negative, since fluid moves more slowly in the layer). Turbulent velocities are found to be of order u_* in both the viscous sublayer and outer layer. That is, they have the same order of magnitude throughout the boundary layer, although they are somewhat larger near the surface.

The ratio of thicknesses of the outer layer and viscous sublayer is the Reynolds number $\mathrm{Re}_* = u_*\delta/\nu$, a large parameter, responsible for the separation of scales between the two regions. Thus, the viscous sublayer is a rather small fraction, Re_*^{-1}, of the overall layer thickness. In fact, the entire theory of turbulent boundary layers, described in Section 5.5, is based on a large Reynolds number, that is, the theory is asymptotic in large Re_*. In particular, flat-plate universality of the scaled velocity defect in the outer region, noted above, requires that the Reynolds number be sufficiently large.

For the experimentally observed universal forms in the outer layer and viscous sublayer to agree on the form of the mean-velocity profile at distances from the surface intermediate between ν/u_* and δ, it will be shown in Section 5.5 that the mean velocity there must have a logarithmic form as a function of distance from the wall. This is the celebrated log law. Such intermediate distances are collectively known as the inertial layer and, indeed, a log law is found to describe the mean velocity for $\nu/u_* \ll y \ll \delta$. The argument leading to the log law is very similar to the application of the matching principle in the method of matched asymptotic expansions and those readers acquainted with that technique may like to regard the theory as based on two matched asymptotic regions: an inner viscous sublayer, described by the variable $y_+ = \nu y/u_*$, and an outer layer, characterized by $\eta = y/\delta$, with the small parameter Re_*^{-1} controlling the asymptotic separation of scales.[6] The limit of the

[6] We do not mean to suggest that an analytical theory of turbulent boundary layers, matched expansions or otherwise, exists. However, it may be helpful to some readers to think informally in these terms.

mean velocity in the outer region at small $\eta = y/\delta$ and that of the inner region at large $y_+ = \nu y/u_*$ coincide as the inertial-range log law. Thus, the inertial layer plays the role of a region of overlap between the inner and outer asymptotic regions.

For a boundary layer without pressure gradient, surface curvature, or external vorticity, the velocity defect law, $(\overline{U_x} - U_\infty)/u_* = F(\eta)$, implies that the departures of $\overline{U_x}$ from the uniform value U_∞ are of order u_* in the outer layer and consequently of the same order of magnitude as the turbulence velocities. The ratio, u_*/U_∞, measures the relative magnitudes of the turbulent and mean velocities and is small. Indeed, as well as the log law, matching of the viscous sublayer and outer layers turns out to imply that U_∞/u_* is a logarithmically increasing function of Re_*, and since the latter is large, so is the former (though less so, since the dependence is logarithmic). Smallness of u_*/U_∞, and hence of the turbulent velocities compared to the mean ones, is thus a consequence of the large Reynolds number. As noted earlier, slow development of the boundary layer, which is the basis of the boundary-layer approximation, comes about because the turbulent velocities are small, that is, it is due to small u_*/U_∞. At the same time, this implies that the departures of $\overline{U_x}$ from U_∞ are small within the outer layer, which is the motivation for using a formulation of the boundary-layer approximation in terms of the velocity defect $U_\infty - \overline{U_x}$ rather than $\overline{U_x}$ itself. One consequence is that, as one traverses the whole boundary layer, most of the changes in streamwise mean velocity, from zero at the surface (due to the no-slip condition), to U_∞ at the edge of the boundary layer, occur near the surface, rather than in the outer part of the layer.

So far, the above description of turbulent boundary layers consists of experimental observations and a matching-style argument. No use has been made of the equations describing the flow. Those equations are now employed in an integrated, momentum balance form, known as the Von Karman equation, to determine the streamwise development of the boundary layer, as described in detail in Section 5.5. It turns out that, given the observed universal forms in the viscous sublayer and outer region, together with the matching condition relating U_∞/u_* to Re_*, the only remaining unknown quantity is the boundary-layer thickness, δ. The Von Karman equation can be used to derive a differential equation governing the streamwise evolution of δ, and hence the boundary-layer flow is completely determined in the absence of pressure gradients, surface curvature, or external vorticity. The Von Karman equation also holds in the presence of such complicating effects, but the mean velocity defect in the outer layer is no longer described by flat-plate universality, and there is insufficient information to determine the streamwise evolution of a general turbulent boundary layer. However, as we shall see, some progress can be made for a special class of layers with pressure gradients for which the outer-layer velocity defect profile is self-similar. Such self-preserving turbulent boundary layers are somewhat analogous to the Falkner–Skan similarity solutions of the Prandtl equations for laminar layers with pressure gradients.

The first effect of significant pressure gradients, surface curvature, or external vorticity is to modify the outer region of the boundary layer. That is, the viscous sublayer is comparatively insensitive to such complicating factors, except via the resulting modifications of u_*, and hence the sublayer scalings. In developing the general theory of turbulent boundary layers in Section 5.5, we will implicitly suppose that the magnitude of the pressure gradient, curvature, and external vorticity are such that their effects on the outer layer are not dominant, implying that they only perturb the viscous sublayer a little. As we shall see, nondimensional parameters can be constructed which measure the importance of these complicating effects in the outer region. These parameters should not be too large, for otherwise the outer-layer scalings and viscous sublayer mean velocity profile may be completely different. This happens, for instance, near separation, where the properties of the boundary layer are quite different from those of a flat plate, more closely resembling a shear layer. For this reason, apart from a qualitative discussion, we will implicitly exclude from consideration layers near separation.

The Prandtl equations of laminar boundary layers do not contain surface-curvature terms, which are of higher order. At first sight, the same is true of the leading-order turbulent boundary-layer equations for the mean velocity defect. However, as implied earlier, curvature *can* have an effect at leading order, even when the radius of curvature is large compared with the layer thickness. As we shall see, it does so by modifying the turbulence, and hence the mean-flow equations, via the Reynolds-stress term. The same is true of external vorticity and pressure gradients, although, since the pressure gradient occurs explicitly in the leading-order mean-flow equations, it has more far-reaching effects than either curvature or external vorticity, which enter only indirectly through the Reynolds stresses.

We shall mainly consider two-dimensional, plane flows, that is, ones for which the statistical properties are independent of x_3 and $\overline{U_3} = 0$. It is not, of course, implied that the turbulence is two-dimensional in particular realizations of such a flow, but rather that its statistical properties are independent of x_3. This requires that the body, in the case of wakes and boundary layers, or the nozzle, for jets, be effectively infinite in the x_3-direction. The analysis of two-dimensional flows is simpler than their three-dimensional counterparts, especially for boundary layers, but illustrates most of the features of the corresponding three-dimensional flows. The theory of three-dimensional boundary layers is beyond the scope of this book, but we will discuss nonplanar jets and wakes (resulting from finite-width nozzles and bodies), and in particular the important case of axisymmetric ones.

5.1 Laminar Boundary Layers

In studying the turbulent flows of this chapter, especially turbulent boundary layers, it is often helpful to have the theory of laminar boundary layers in mind. Consequently, we include a brief review here. The reader will find further material on this subject in chapter 5 of Batchelor (G 1967) and a much more extensive treatment in Schlichting (1987).

If a streamlined body is placed in a steady flow at high Reynolds number, the small viscosity means that the flow behaves as inviscid away from the body surface. However, the inviscid solution of the problem does not satisfy the no-slip condition at the surface, and a thin boundary layer, in which viscosity is called into play, develops, starting from the upstream stagnation point, downstream along the surface (see Figure 5.2). Outside the boundary layer, the flow can be determined by inviscid

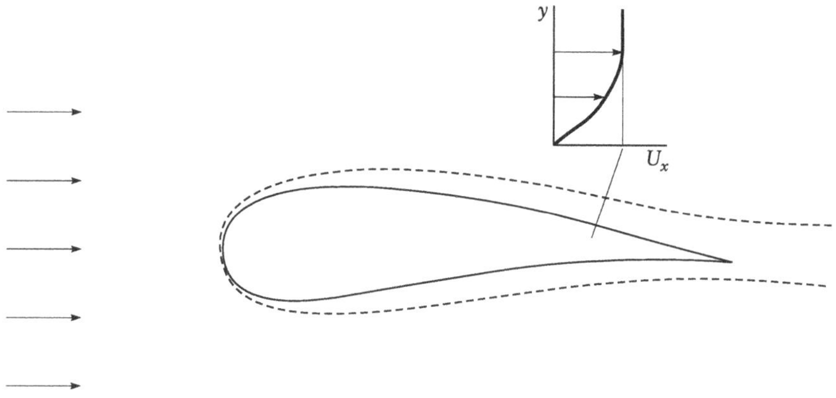

Figure 5.2. Laminar boundary layer on a streamlined body at high Reynolds number.

theory, while the boundary layer contains large velocity gradients so that viscosity is important, allowing satisfaction of the no-slip condition at the surface.

In terms of vorticity, which always originates at a boundary for incompressible flow without body forces, the boundary layer represents a balance between two competing mechanisms. On the one hand, vorticity is convected by the flow and, on the other, it diffuses away from the surface due to viscosity. The two effects can be estimated as follows. Let U_0 be a typical velocity scale in the (inviscid) flow away from the surface, for instance the velocity far from the body if, as is often supposed, it has been placed in a uniform stream. The velocity tangential to the surface within the boundary layer is also of $O(U_0)$ (whereas the normal velocity is smaller, owing to the proximity of the surface). The time taken for a fluid particle in the boundary layer to move a distance $O(D)$, comparable to the body scale D, is $t = O(D/U_0)$. Recall (see equation (4.63)) that viscosity causes diffusion of vorticity. In the time t, vorticity, like any quantity diffusing outwards from the surface with diffusivity ν, will have diffused a distance $O((\nu t)^{1/2})$ from the surface.[7] Thus, a boundary layer of thickness $\delta = O((\nu D/U_0)^{1/2})$ forms. The ratio of layer thickness to body size is $\delta/D = O(\mathrm{Re}^{-1/2})$ where $\mathrm{Re} = U_0 D/\nu$ is the body-scale Reynolds number, assumed large so that an identifiable boundary layer exists. The Reynolds number based on the boundary-layer thickness is $\mathrm{Re}_\delta = U_0\delta/\nu = O(\mathrm{Re}^{1/2})$ and increases proportional to $\mathrm{Re}^{1/2}$. A typical boundary-layer Reynolds number, Re_δ, is therefore large, but nonetheless small compared with the body-scale Reynolds number.

For simplicity sake, we consider two-dimensional flow, independent of x_3, with $U_3 = 0$. Three-dimensional boundary layers are more complicated and beyond the scope of this book. For any point in the boundary layer, let y denote the distance of the nearest point of the surface, and x the distance of that surface point from some arbitrary origin near the front of the body,[8] measured along the surface in the streamwise direction (see Figure 5.3). The coordinates x and y are generally curvi-

[7] This discussion does not allow for the effect of vorticity convection by the normal velocity, which is present in the Prandtl equations. It nonetheless contains the essential physics and gives a correct order of magnitude estimate.

[8] Usually taken as the upstream stagnation point.

linear, and the corresponding components of velocity, parallel and normal to the surface, are written U_x and U_y. The continuity equation takes the approximate form

$$\frac{\partial U_x}{\partial x} + \frac{\partial U_y}{\partial y} = 0 \tag{5.1}$$

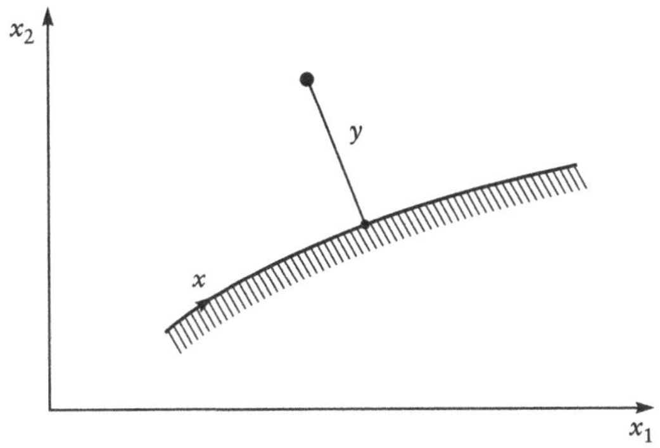

Figure 5.3. Boundary-layer coordinates: x along the surface, y normal to the surface.

since, for a thin laminar boundary layer, any curvature of the surface can be neglected to a first approximation. Equation (5.1) is the first of the Prandtl boundary-layer equations, asymptotically valid even for curved surfaces when the Reynolds number is large and, consequently, the layer is thin. The remaining Prandtl equations arise from the two components of the steady momentum equation, parallel and normal to the surface. To leading order in the large Reynolds number, they are

$$U_x \frac{\partial U_x}{\partial x} + U_y \frac{\partial U_x}{\partial y} = -\frac{1}{\rho}\frac{\partial P}{\partial x} + \nu \frac{\partial^2 U_x}{\partial y^2} \tag{5.2}$$

and

$$\frac{\partial P}{\partial y} = 0 \tag{5.3}$$

We do not give detailed derivations of these approximate equations here. The most systematic technique which has been found for the analysis of laminar boundary layers is the method of matched asymptotic expansions, which we now summarize (see Van Dyke (1975) for details of the method).

The objective of the method is to obtain asymptotic expansions for the velocity components and pressure in the limit $\mathrm{Re} \to \infty$. Two expansions in power of $\mathrm{Re}^{-1/2}$ are developed, the first (outer expansion) is valid outside the boundary layer, at body-scale distances from the surface, the second (inner expansion) applies inside the layer. The inner expansion employs the coordinates x and a scaled version of y, which is chosen according to the boundary-layer thickness, $O(\mathrm{Re}^{-1/2}D)$. More precisely, as before, let D be a body-size scale and U_0 be a velocity scale in the outer region, for instance the velocity far from the body if it has been placed in a uniform stream. The body-scale Reynolds number is then $\mathrm{Re} = U_0 D/\nu$, while the velocity and pressure are nondimensionalized on U_0 and ρU_0^2. In the inner region, nondimensional velocity and pressure are considered as functions of the coordinates $x^* = x/D$ and $y^* = \mathrm{Re}^{1/2} y/D$, whereas, in the outer region, they are treated as functions of $\hat{x}_1 = x_1/D$, $\hat{x}_2 = x_2/D$. Expansions in powers of $\mathrm{Re}^{-1/2}$ for both regions are introduced into the governing equations and coefficients of powers of $\mathrm{Re}^{-1/2}$ equated. At leading order, the outer region yields the inviscid (Euler) equations, while the inner region gives the Prandtl equations (5.1)–(5.3), once it is recognized that the inner expansion for U_y begins with a term in $\mathrm{Re}^{-1/2}$, unlike the other expansions, which all start with a Re^0 term. This procedure allows the systematic determination of equations for the two regions at any order in $\mathrm{Re}^{-1/2}$, but the two expansions remain to be matched.

In the inner (boundary layer) region, the no-slip condition implies that the velocity is zero at $y^* = 0$, while another boundary condition applies as $y^* \to \infty$ and is obtained by matching to the outer (inviscid at leading order) flow. We do not want to go into the details of asymptotic matching here (see Van Dyke 1975), but the idea is the following. The inner expansion applies in the boundary layer and the outer expansion at distances $O(D)$ from the surface. They are both assumed to hold in a region of overlap, at distances from the surface small compared with the body scale, D, but large compared with the boundary-layer thickness, $\mathrm{Re}^{-1/2}D$. The requirement that the two expansions give the same result in this overlap region yields the matching conditions and hence completes the problem.

The matching conditions can be applied at any order, but become increasingly cumbersome and difficult to interpret physically as the order is increased. At leading order, matching yields the usual inviscid boundary condition, $U_y = 0$ at $y = 0$, for the outer problem. It also gives following boundary conditions for the inner problem:

$$\begin{aligned} U_x &\to U_\infty \\ P &\to P_\infty \end{aligned} \tag{5.4}$$

as $y^* \to \infty$, where, $U_\infty(x)$ and $P_\infty(x)$ are the velocity and pressure *at the surface*, obtained from an inviscid calculation (i.e., from the leading-order outer flow). The matching conditions, (5.4), supplement (5.1)–(5.3) and the no-slip conditions

$$\begin{aligned} U_x(y = 0) &= 0 \\ U_y(y = 0) &= 0 \end{aligned} \tag{5.5}$$

thus completing the leading-order boundary-layer problem. The meaning of (5.4) is, of course, that the boundary-layer flow should approach the inviscid solution as one leaves the boundary layer and emerges into the outer region. This is the sense in which one should interpret the limit $y^* \to \infty$. It might be thought that there should also be a condition on U_y, but the corresponding matching condition is, in fact, automatically satisfied. A second thought might be that it is not the inviscid solution precisely at the surface, but at some small distance away, which should appear in (5.4). However, such small corrections are regarded as higher-order effects in the asymptotic theory and thus do not appear at leading order.

From (5.3), we see that the pressure is a function of streamwise position, x, only. Applying (5.4), we have

$$P = P_\infty(x) \tag{5.6}$$

and the pressure in the layer is thus *imposed from outside* and can be considered as known. Furthermore, since the leading-order flow in the outer region is supposed inviscid and steady, Bernoulli's theorem applied to the surface streamline of that flow implies that $P_\infty + \rho U_\infty^2/2$ is constant. Hence, (5.2) can be rewritten as

$$U_x \frac{\partial U_x}{\partial x} + U_y \frac{\partial U_x}{\partial y} = U_\infty \frac{dU_\infty}{dx} + \nu \frac{\partial^2 U_x}{\partial y^2} \tag{5.7}$$

which, together with (5.1), (5.5), and

$$U_x \to U_\infty(x) \tag{5.8}$$

as $y^* \to \infty$, give the final version of the laminar boundary-layer problem. Of the boundary-layer equations, (5.1) and (5.7), the former can be thought of as playing the subsidiary role of determining[9] U_y in terms of U_x, whereas the latter describes the evolution of the velocity profile, $U_x(y)$, with streamwise distance, x.

Before carrying out a boundary-layer calculation, one must first determine the steady inviscid flow away from the surface. This provides the input data, $U_\infty(x)$, for the boundary-layer problem, (5.1), (5.5), (5.7), and (5.8). Unlike the full Navier–Stokes equations, which are elliptic in nature, the boundary-layer equations are parabolic and, although a few similiarity solutions (e.g., Blasius, Falkner–Skan; see Batchelor (G 1967), sections 5.8 and 5.9) are known, solutions are generally obtained numerically by marching downstream in x. The profile, $U_x(y)$, is specified at some initial value of x (for instance, derived from one of the above self-similar profiles), allowing the marching process to begin. The boundary layer then develops with x according to the prescribed external velocity, $U_\infty(x)$. Note that, apart from the initial profile, only $U_\infty(x)$ is needed as input data, differentiating one laminar boundary layer from another.

It can be objected that the steady, boundary-layer flow considered here becomes unstable and eventually turbulent as the boundary-layer Reynolds number, $\mathrm{Re}_\delta = O(\mathrm{Re}^{1/2})$, grows larger with increasing body-scale Reynolds number, Re. That is, the large value of Re on which boundary-layer theory is based leads to large Re_δ, and hence to possible instability of the steady boundary-layer flow. This is certainly true for sufficiently large Re_δ, but the critical value of Re_δ for instability is found to be large, and the corresponding body-scale Reynolds number, $\mathrm{Re} = O(\mathrm{Re}_\delta^2)$, is still larger. Thus, there is, after all, a range of large Re in which the asymptotic theory describes the boundary layer without instability. Furthermore, the boundary layer thickens as we go downstream from the forward stagnation point. The Reynolds number, Re_δ, based on the layer thickness is thus smaller near the stagnation point and grows as the layer develops downstream. In consequence, the boundary-layer flow is usually stable at the nose of the body, even if it becomes unstable downstream. If the body-scale Reynolds number is sufficiently large that the critical value of Re_δ is exceeded at some point, the layer initially develops with distance from the nose according to Prandtl's theory until the instability arises and the flow becomes unsteady, generally followed by transition to turbulence further downstream. The theory of laminar boundary-layer stability and transition is beyond the scope of this book (see, e.g., Schlichting (1987) for further information, although somewhat dated), but we might note that linear stability calculations are routinely used to predict transition even though the fundamental justification for such calculations is far from clear because the transition process is essentially nonlinear. It should also be remarked that the critical value of Re_δ for instability is not a fixed universal number, but depends on the details of the particular flow considered, hence the need for stability calculations. The theory of boundary-layer stability and transition remains an area of active research and a fully satisfactory transition model

[9] In practice, this is most easily expressed by introducing a stream function, rather than working with the velocity components themselves.

has yet to be developed. Of course, steady laminar boundary-layer theory does not apply beyond the point of instability.

Boundary-layer separation should not be confused with transition. Separation occurs when the boundary layer leaves the body surface at some location, in the manner sketched in Figure 5.4. There is then a stagnation point on the surface, characterized by $\partial U_x/\partial y = 0$, and called the separation point. Reverse flow occurs near the surface, downstream of the separation point, and a dividing streamline branches out from that point, separating fluid coming towards the separation point from upstream and downstream. The vorticity of the boundary layer is swept off the surface and enters the outer flow as a thin shear layer. Separation is always present in high-Reynolds-number flow past a bluff body, such as a sphere (see Figure 1.8a). There is a thin boundary layer over the front part of the body surface, which separates to form the wake behind the body. Streamlined bodies, such as the wing illustrated in Figure 5.2, are designed so that there is no separation at small enough angles of attack, but the boundary layer can separate if the angle of attack becomes too large (leading to stalling in the case of the wing).

If separation occurs, the boundary layer upstream of, but not too close to, the separation point is still governed by the Prandtl equations given above, but the determination of the appropriate $U_\infty(x)$ is far from straightforward. Away from the body surface, a simple inviscid flow calculation that does not take into account separation (e.g., solution of Laplace's equation for the velocity potential in the irrotational case) will not allow for the wake, whose existence affects the whole flow. Thus, one can no longer perform a calculation of the outer flow first, followed by a boundary-layer computation. The two flows are coupled together in a complicated way: boundary-layer development is conditioned by the flow away from the surface, which, in turn, depends on boundary-layer vorticity shed by separation to form the wake. However, one can, for instance, imagine using measured values of $U_\infty(x)$ upstream of the separation point. Whatever method is used to obtain $U_\infty(x)$, it is found that the solution of the Prandtl equations generates a singularity at or near separation. This is symptomatic of a breakdown in the approximations used to derive the Prandtl equations, which consequently cannot be used to describe the immediate vicinity of the separation point.[10] A vari-

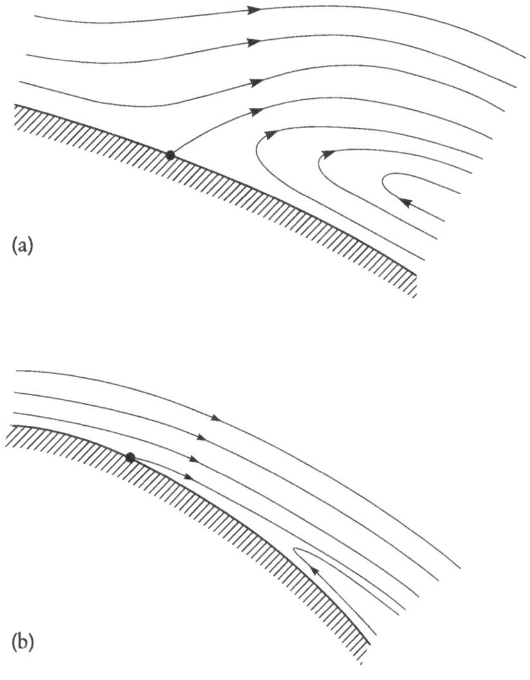

(a)

(b)

Figure 5.4. Sketches of streamlines near a separation point of a boundary layer: (a) at boundary-layer thickness distances, (b) at body-scale distances.

[10] In fact, the theory also needs refinement near the upstream stagnation point, but this does not affect the calculation elsewhere (see Batchelor (G 1967), section 5.5b).

ety of attempts have been made to develop a more sophisticated asymptotic theory of boundary layers for cases, such as separation, in which the usual theory breaks down. The resulting methodology is known as "triple deck" theory and the interested reader is referred to Smith (1982) for a general review, and to Smith (1979) for an analysis of separation in particular. Such techniques are well beyond the scope of this book and we continue to use the Prandtl equations, thus restricting attention to boundary layers upstream and away from the separation point (if there is separation at all).

The fact that separation is often unsteady, owing to global instability of the steady flow, further complicates the issue. A striking example is provided by the periodic shedding of vortices forming the Von Karman vortex street behind a cylinder (see Figure 1.1). This is an instability of the steady flow, which sets in once the Reynolds number exceeds a certain critical value (about $\mathrm{Re}_D = 45$), and in which wake oscillations couple to periodic fluctuations of the separation point, leading to unsteady shedding of boundary-layer vorticity into the wake. This instability is *not* localized in the boundary layer, as is the one described above which gives rise to boundary-layer transition, but involves the whole flow. Furthermore, the critical Reynolds number is much lower (transition of the boundary layer on a cylinder happens at around $\mathrm{Re}_D = 10^5$). Although this instability does not lead to boundary-layer transition, the laminar boundary layer upstream of separation becomes unsteady, a case which is not described by the steady Prandtl equations given above, but may be allowed for by including the time-derivative term from the Navier–Stokes equation in (5.7). Calculation of the flow field away from the surface, to determine $U_\infty(x, t)$, is even harder than for the steady case.

The two terms on the right of (5.7) represent the pressure gradient and viscosity, while the second term on the left accounts for convection by the normal velocity. Taken together, these three terms determine $\partial U_x/\partial x$, while (5.1) plays the secondary role of allowing U_y to be calculated in terms of U_x. The sign of dU_∞/dx, and hence the pressure gradient term in (5.7) (or (5.2)), is particularly important in determining the qualitative behavior of the boundary layer. Starting at the upstream stagnation point, where the pressure, P_∞, is a maximum according to Bernoulli's theorem the fluid is accelerated by the pressure gradient term in (5.7). So long as this acceleration continues, the layer remains relatively thin and does not separate.

Let $U_y^{(inv)}(x, y)$ denote the normal component of the leading-order *outer* flow (here the superscript (*inv*) stands for inviscid), so that the outer region continuity equation at the surface can be written

$$\left.\frac{\partial U_y^{(inv)}}{\partial y}\right|_{y=0} = -\frac{dU_\infty}{dx} \tag{5.9}$$

At the surface we have $U_y^{(inv)} = 0$ and a one-term Taylor's series reads

$$U_y^{(inv)} = -\frac{dU_\infty}{dx}\, y \tag{5.10}$$

describing the behavior of the normal component of velocity as the boundary layer is approached from outside.

When the sign of the pressure gradient is such that the fluid just outside the layer is accelerated (a favorable pressure gradient), $dU_\infty/dx > 0$. In that case, fluid is drawn in from outside the layer, because $U_y^{(inv)} < 0$ near the surface according to (5.10), and transverse convection *towards* the surface (represented by the second term in (5.7)) tends to keep the layer from thickening too quickly under the competing effect of viscous diffusion and may actually result in thinning of the layer if the pressure gradient is sufficiently strong. Thus, for a favorable pressure gradient, thinning due to the gradient and viscous thickening are working in opposition. If, on the other hand, the sign of the pressure gradient changes (becoming "adverse" in the jargon), the fluid just outside the layer is decelerated, $dU_\infty/dx < 0$, $U_y^{(inv)} > 0$, and transverse convection *away* from the surface acts *with* viscosity to cause the layer to thicken rather quickly. At the same time, the adverse gradient tends to decelerate the fluid within the layer. If an adverse pressure gradient is sufficiently intense and applied for sufficiently long, the slower moving fluid near the surface in the layer can be decelerated to the point where it comes to rest, so that reverse flow and separation occur. An adverse pressure gradient therefore favors separation by decelerating the fluid near the surface, but must act against viscous friction with the external flow, which is still moving forwards outside the boundary layer and trying to drag the fluid nearer the surface along with it. In summary, a favorable gradient does not induce separation and can produce thinning of the boundary layer, whereas an adverse gradient leads to rapid thickening of the layer and possible separation. In practice, laminar boundary layers cannot resist much in the way of an adverse gradient before they separate. As described earlier, separation, when it occurs, means that the boundary-layer approximation breaks down in the neighborhood of the separation point and a more sophisticated model is needed there. The outer flow also needs to be reassessed, because of the shear layer of shed vorticity generated by separation.

The classical example of a flat plate in a uniform stream at zero incidence (the Blasius problem) is a rather special case, since it has no pressure gradient. The solution of Prandtl's equation has a similarity form in that case, with similarity variable $y/x^{1/2}$ (i.e., the layer thickens proportional to $x^{1/2}$). No separation occurs on the plate, but transition may come about if the Reynolds number based on the layer thickness grows large enough. The layer is still attached when it reaches the trailing edge of the plate, where it continues to form the wake. This is not usually called separation, but is similar in nature: the vorticity from the boundary layer leaves the body and enters the outer flow. Such shedding of vorticity at a sharp edge is an important feature of wings and turbomachinery. Needless to say, the boundary-layer approximation breaks down near such a sharp edge, where a more complicated asymptotic representation is necessary, if indeed the layer is still laminar. In the wake of a flat plate, away from the trailing edge, a Prandtl-type description can again be applied, although there is now no solid surface. This is similar to the approach for turbulent jets and wakes that we use later.

The point of the above review of laminar boundary layer is to provide the reader with general background when we come to consider the turbulent flows that are the main subject of this chapter. However, some major dissimilarities between the turbulent and laminar cases should be noted. Firstly, the mechanism for transverse diffusion of momentum in a laminar layer is viscosity, whereas turbulent mixing plays that role in the turbulent flows of this chapter. Physically, the two mechanisms are quite different, even though many simple models make them appear similar in mathematical structure using the concept of an eddy viscosity.

Mathematically, the mean-flow equations of turbulent flow are incomplete because they contain Reynolds stress terms representing turbulent momentum transfer, rather than the viscous terms of laminar flows. If one is to solve the equations, they must first be closed, using one of the many heuristic closure approximations which have been proposed. However, rather than relying entirely on such closures, in this chapter we will introduce experimental information, as discussed in the introduction. In the case of jets and wakes, a simple eddy-viscosity closure is used to determine the transverse structure of the mean flow, with its streamwise development constrained to be self-similar, as indicated by measurements. For boundary layers, no closure is used. The transverse structure of the layer is dictated by experiments, while its streamwise development is determined by applying an integrated momentum balance derived directly from the mean-flow equations. This is not to say that closure models cannot be used to completely calculate the flows in turbulent jets, wakes, and boundary layers, but, given the approximate and semiempirical nature of existing closures for inhomogeneous flows, such as these, it is probably better to rely directly on experimental information where available. Fortunately the flows considered have been much studied experimentally, so their general properties are known.

A second difference between laminar boundary layers and the turbulent flows of this chapter is that Prandtl's theory can be derived as the first term of an inner expansion in the small parameter $\mathrm{Re}^{-1/2}$. The larger the Reynolds number, the better the approximation becomes. This is in contrast with the corresponding approximations for turbulent jets and wakes, to be described shortly, which rely upon the smallness of u'/U_0 (where U_0 is a suitable mean-velocity scale), a parameter which describes the relative rates of turbulent mixing and streamwise convection. The flow itself selects the intensity of turbulence, u', and u'/U_0 just happens to be small. The errors committed in going to the boundary-layer approximation are fixed by the size of u'/U_0 and cannot be decreased asymptotically to zero:[11] we just have to live with them or go to a more accurate model.

In summary, two levels of approximation are involved in this chapter. The first is a consequence of the general closure problem for turbulence and appears unavoidable unless numerical simulation of the Navier–Stokes equations is invoked. To get around this difficulty, one can either use a closure approximation alone, or, as here, combine closure with experimental input. The second approximation is specific to dominantly unidirectional, slowly developing flows of the type considered here. The turbulent boundary-layer approximation is analogous to that used for laminar boundary layers and simplifies the equations, but unlike the closure model, is a convenience rather than the result of a fundamental difficulty of the theory. The turbulent-boundary-layer approximation for jets and wakes, unlike its laminar counterpart, is not a truly asymptotic theory, but is nonetheless a rational approximation, based on smallness of u'/U_0. Closure approximations, on the other hand, are heur-

[11] As noted in the introduction, the value of the corresponding parameter for boundary layers, u_*/U_∞, decreases slowly (logarithmically) with increasing Reynolds number, so the boundary-layer approximation *is* formally asymptotic in that case.

istic assumptions, whose worth is judged by results when compared with experiment or more exact theories.

5.2 The Turbulent-"Boundary-Layer" Approximation for Jets and Wakes

For the sake of simplicity, in this section we restrict attention to *two-dimensional* jets and wakes, that is, ones whose statistics are independent of x_3 and have $\overline{U_3} = 0$. Furthermore, we suppose that the streamwise direction is the same everywhere and choose x_1 as the streamwise coordinate. The remaining coordinate, x_2, represents the transverse direction, in which the flow properties vary more rapidly. We will examine the order of magnitude of terms in the mean-flow equations to eliminate those which are of lower order. To this effect, streamwise and transverse length scales, d and Δ, are introduced, where $\Delta \ll d$ expresses the more rapid variations across the flow, width $O(\Delta)$, than those produced by streamwise development over distances of $O(d)$. Thus, derivatives with respect to x_1 will be estimated as $O(d^{-1})$, whereas $\partial/\partial x_2 = O(\Delta^{-1})$ is larger in order of magnitude. We also introduce a mean-flow velocity scale, U_0, which characterizes the order of magnitude of the x_1-component of the mean velocity, $\overline{U_1}$. In the course of the analysis, it turns out that the small quantity Δ/d is related to u'/U_0 and smallness of Δ/d is in fact a consequence of the weakness of turbulence compared with the mean flow, as discussed earlier.

It should first be made clear that the name turbulent-"boundary-layer" approximation does not mean that the approximation developed in this section is necessarily adequate for turbulent boundary layers! The name "boundary-layer" approximation is used because it is similar to Prandtl's approach for laminar boundary layers. The basic approximation, derived in this section, describes jets and wakes, but is *not* sufficiently accurate for turbulent boundary layers with significant surface curvature or external vorticity and needs to be refined to handle such general turbulent boundary layers. Such a refined approximation is described in the later section devoted to boundary layers.

Our starting point is the mean-flow equations

$$\underbrace{\overline{U_j}\frac{\partial \overline{U_i}}{\partial x_j}}_{\text{Convection}} = -\frac{1}{\rho}\frac{\partial \overline{P}}{\partial x_i} - \underbrace{\frac{\partial \overline{u_i u_j}}{\partial x_j}}_{\text{Reynolds stress}} + \underbrace{\nu\frac{\partial^2 \overline{U_i}}{\partial x_j \partial x_j}}_{\text{Viscosity}} \tag{5.11}$$

and

$$\frac{\partial \overline{U_i}}{\partial x_i} = 0 \tag{5.12}$$

where the Reynolds stress term on the right of (5.11) represents the effect of turbulence on the mean flow.

Since $\overline{U_3} = 0$ and $\partial/\partial x_3 = 0$, equation (5.12) gives

$$\frac{\partial \overline{U_1}}{\partial x_1} + \frac{\partial \overline{U_2}}{\partial x_2} = 0 \tag{5.13}$$

in which we can estimate $\partial/\partial x_1 = O(d^{-1})$ and $\overline{U_1} = O(U_0)$, so that

$$\frac{\partial \overline{U_2}}{\partial x_2} = O\left(\frac{U_0}{d}\right) \tag{5.14}$$

and $\partial/\partial x_2 = O(\Delta^{-1})$, leading to the estimate

$$\overline{U_2} = O\left(\frac{\Delta}{d} U_0\right) \tag{5.15}$$

for the transverse mean velocity, which is of smaller order than $\overline{U_1} = O(U_0)$ since $\Delta \ll d$. Equation (5.13), which is exact, is the first of the turbulent-"boundary-layer" equations. It is identical in form with its laminar equivalent, (5.1), except that we are now dealing with the *mean* velocity of a turbulent flow.

The streamwise ($i = 1$) component of (5.11) is

$$\overline{U_1}\frac{\partial \overline{U_1}}{\partial x_1} + \overline{U_2}\frac{\partial \overline{U_1}}{\partial x_2} = -\frac{1}{\rho}\frac{\partial \overline{P}}{\partial x_1} - \frac{\partial \overline{u_1 u_j}}{\partial x_j} + \nu\frac{\partial^2 \overline{U_1}}{\partial x_j \partial x_j} \tag{5.16}$$

and we immediately neglect the viscous term by virtue of the assumed large Reynolds number. Viscosity can become important for the thin viscous sublayer at the surface in a boundary layer, but is otherwise negligible for the mean-flow equations at high Reynolds number (see the discussion in Section 4.1). Neglect of viscosity in the mean-flow equations is independent of the "boundary-layer" approximation and is generally considerably more accurate.

Both terms on the left of (5.16) can be estimated as $O(U_0^2/d)$, using (5.15) for the second term. The Reynolds stress contribution to (5.16) is the sum of two terms:

$$\frac{\partial \overline{u_1 u_j}}{\partial x_j} = \frac{\partial \overline{u_1^2}}{\partial x_1} + \frac{\partial \overline{u_1 u_2}}{\partial x_2} \tag{5.17}$$

The diagonal components of $\overline{u_i u_j}$ are found to be of the same order of magnitude $O(u'^2)$, while off-diagonal components, such as $\overline{u_1 u_2}$, are somewhat smaller, particularly for jets. In any case, the order of the first term in (5.17) is u'^2/d, which is $O(u'^2/U_0^2)$ smaller than the terms on the left of (5.16). Thus, since u'/U_0 is small, we can neglect this term, leading to

$$\overline{U_1}\frac{\partial \overline{U_1}}{\partial x_1} + \overline{U_2}\frac{\partial \overline{U_1}}{\partial x_2} = -\frac{1}{\rho}\frac{\partial \overline{P}}{\partial x_1} - \frac{\partial \overline{u_1 u_2}}{\partial x_2} \tag{5.18}$$

which is the second of the turbulent-"boundary-layer" equations. The quantity $\rho\overline{u_1 u_2}$ is the average turbulent flux of the x_1-component of momentum in the x_2-direction. Variations of this flux across the flow transfer mean-flow momentum, as appears from (5.18). Comparing equation (5.18) with its laminar equivalent, (5.2), we see that the turbulent Reynolds stress has replaced the viscous term. Turbulent transport of streamwise momentum across the flow takes over from transverse viscous diffusion, as one might intuitively expect.

Since turbulent momentum transport is expected to be important in the flows considered here, the Reynolds stress term on the right of (5.18) should be comparable to those on the left. Equating the orders of magnitude, U_0^2/d of the left-hand side and $\overline{u_1 u_2}/\Delta$ of the Reynolds stress term, gives

$$\frac{\Delta}{d} = O\left(\frac{\overline{u_1 u_2}}{U_0^2}\right) \le O\left(\frac{u'^2}{U_0^2}\right) \tag{5.19}$$

showing that smallness of Δ/d, which we have assumed in deriving the "boundary-layer" approximation, is a consequence of small u'/U_0. Smallness of u'/U_0 is the fundamental requirement of the "boundary-layer" approximation developed here. As discussed earlier, because turbulent velocities are small compared with the mean velocity, the flow spreads slowly in the transverse direction. Observe that together (5.15) and (5.19) give

$$\frac{\overline{U_2}}{u'} \le O\left(\frac{u'}{U_0}\right) \ll 1 \tag{5.20}$$

and so the ordering is $\overline{U_2} \ll u' \ll \overline{U_1}$, that is, the turbulent velocities are intermediate in magnitude between the transverse and streamwise mean-flow velocities. The transverse mean flow is therefore rather small, smaller even than the turbulent velocities. If one follows a packet of fluid as it is convected downstream at speed $O(U_0)$, the velocity of transverse spreading of the turbulence is $O(U_0\Delta/d)$, which is the same order as $\overline{U_2}$, according to (5.15). Thus, the mean-velocity profile and turbulent statistical properties spread out at a velocity *small* compared with u'. This is perhaps surprising because one might expect turbulence to spread out at speed $O(u')$ due to entrainment. However, turbulence cannot continuously spread faster than the mean-flow profile because it gets its sustenance from mean shear and decays rather rapidly in its absence.

As a final consequence of the streamwise momentum equation, assuming that the pressure gradient term in (5.18) is of the same order of magnitude, $O(U_0^2/d)$, as all the other terms, we find

$$\overline{P} = O(\rho U_0^2) \tag{5.21}$$

as an estimate of the mean pressure variations in the flow.

We now turn to the transverse ($i = 2$) component of (5.11), which becomes

$$\overline{U_1}\frac{\partial \overline{U_2}}{\partial x_1} + \overline{U_2}\frac{\partial \overline{U_2}}{\partial x_2} = -\frac{1}{\rho}\frac{\partial \overline{P}}{\partial x_2} - \frac{\partial \overline{u_2 u_j}}{\partial x_j} \tag{5.22}$$

when we neglect the viscous term. One can again estimate the order of different terms, using (5.15) and (5.21). Those on the left are $O(\Delta U_0^2/d^2)$, while the first term on the right is $O(U_0^2/\Delta)$. The left-hand side of (5.22) is thus $O(\Delta^2/d^2)$ smaller than the pressure gradient term and can therefore be neglected compared with the latter.

The Reynolds stress term in (5.22) can be written as

$$\frac{\partial \overline{u_2 u_j}}{\partial x_j} = \frac{\partial \overline{u_1 u_2}}{\partial x_1} + \frac{\partial \overline{u_2^2}}{\partial x_2} \tag{5.23}$$

of which the first term is $O(u'^2/d)$ or smaller, while the second is $O(u'^2/\Delta)$ and dominates. This Reynolds stress term is smaller than the pressure gradient term in (5.22) by a factor $O(u'^2/U_0^2)$ and, to leading order, (5.22) gives

$$\frac{\partial \overline{P}}{\partial x_2} = 0 \tag{5.24}$$

which is the third turbulent-"boundary-layer" equation, to be compared with its laminar counterpart, (5.3).

Equations (5.13), (5.18), and (5.24) form the turbulent-"boundary-layer" equations and, as for the laminar case, require boundary conditions determined by (now informal) matching to an outer flow. For the jets and wakes we consider, these take the form

$$\overline{U_1} \to U_\infty(x_1) \tag{5.25}$$

$$\overline{P} \to P_\infty(x_1) \tag{5.26}$$

as $x_2 \to \pm\infty$, where U_∞ and P_∞ are the streamwise velocity and pressure outside the jet or wake.[12] A boundary condition should also be imposed on $\overline{U_2}$ at some value of x_2 as the analog of the surface boundary condition $U_y(y=0)=0$ in a laminar boundary-layer calculation. Again, by analogy with the laminar boundary layer, one should also specify the velocity profile, $\overline{U_1}(x_2)$, at some value of x_1. For instance, the profile could be imposed at the nozzle of a jet or from the measured wake profile behind a body. This is not usually critical, however, because the flow appears to "forget" the details of such upstream conditions as it spreads and develops downstream.

From (5.24) and (5.26), we deduce that

$$\overline{P} = P_\infty(x_1) \tag{5.27}$$

that is, the mean pressure is imposed from outside, as for the pressure in laminar boundary layers. This can be used in (5.18) to obtain

$$\overline{U}_1 \frac{\partial \overline{U}_1}{\partial x_1} + \overline{U}_2 \frac{\partial \overline{U}_1}{\partial x_2} = -\frac{1}{\rho}\frac{dP_\infty}{dx_1} - \frac{\partial \overline{u_1 u_2}}{\partial x_2} \tag{5.28}$$

which, combined with (5.13), give the turbulent-"boundary-layer" approximation. Equation (5.28) shows that any streamwise gradient of external pressure would tend to make the flow accelerate or decelerate. However, we will mainly be concerned with wakes in an infinite, uniform stream and jets exhausting into infinite fluid at rest, for which the external pressure is uniform and the pressure gradient term in (5.28) is zero. In that case it is the Reynolds stress term alone, representing turbulent momentum transfer across the flow, which causes streamwise development of the mean flow. Equations (5.13) and (5.28), without pressure gradient, form the basis for the analysis of plane jets and wakes in Sections 5.3 and 5.4.

It is enlightening to consider higher-order corrections to the turbulent-"boundary-layer" approximation, partly for their intrinsic interest, but mainly to determine the limitations of the approximation. First consider the transverse momentum equation, (5.22), which, as we have seen, leads to (5.24) at leading order. Terms other than the pressure gradient can be estimated and compared. Of the two terms making up the Reynolds stress contribution, (5.23), the second dominates and is larger than the left-hand side of (5.22) by a factor $O(u'^2 d^2/\Delta^2 U_0^2)$. This factor can in turn be estimated, using (5.19), to be at least $O(U_0^2/u'^2)$, and therefore large. We conclude that, among

[12] More generally one could have different limiting values for $\overline{U}_1$ as $x_2 \to -\infty$ and $x_2 \to +\infty$, representing a turbulent shear layer, for example.

the lower-order terms of (5.22), the second term in (5.23) is the largest, and so the next higher order version of (5.24) is

$$\frac{\partial}{\partial x_2}\left(\overline{P} + \rho\overline{u_2^2}\right) = 0 \tag{5.29}$$

Thus, $\overline{P} + \rho\overline{u_2^2}$ is uniform across the flow and

$$\overline{P} = P_\infty(x_1) - \rho\overline{u_2^2} \tag{5.30}$$

where P_∞ is the externally imposed pressure, since turbulence is supposed absent (or at least of much smaller intensity) outside the jet or wake. According to (5.30), the mean pressure should drop slightly as one enters the turbulent flow from outside. The term $\rho\overline{u_2^2}$ gives the next order correction to (5.27).

For the streamwise component of momentum, the next order approximation to (5.16) above (5.18) keeps both Reynolds stress terms in (5.17), but one continues to neglect the viscous term. Thus, the improved version of (5.18) is

$$\overline{U_1}\frac{\partial\overline{U_1}}{\partial x_1} + \overline{U_2}\frac{\partial\overline{U_1}}{\partial x_2} = -\frac{1}{\rho}\frac{\partial\overline{P}}{\partial x_1} - \frac{\partial\overline{u_1u_2}}{\partial x_2} - \frac{\partial\overline{u_1^2}}{\partial x_1} \tag{5.31}$$

which we can combine with (5.30) to obtain the next order equivalent of (5.28):

$$\overline{U_1}\frac{\partial\overline{U_1}}{\partial x_1} + \overline{U_2}\frac{\partial\overline{U_1}}{\partial x_2} = -\frac{1}{\rho}\frac{dP_\infty}{dx_1} - \frac{\partial\overline{u_1u_2}}{\partial x_2} + \frac{\partial}{\partial x_1}\left(\overline{u_2^2} - \overline{u_1^2}\right) \tag{5.32}$$

Equation (5.32) and the exact equation (5.13), give the improved turbulent-"boundary-layer" equations. Compared with the leading-order equation (5.28), an additional term has appeared, whose smallness, relative to the other terms, should indicate the accuracy of the leading-order form of the equations. It is because this additional term is small compared with the other ones that the approximation (5.28) is useful. The new term depends on $\overline{u_2^2} - \overline{u_1^2}$, which is usually noticeably smaller than either $\overline{u_1^2}$ or $\overline{u_2^2}$ taken separately, so the leading-order approximation, (5.28), is somewhat better than one might expect.

The main difficulty in applying the "boundary-layer" equations, (5.13) and (5.28), is that the Reynolds stress, $\overline{u_1u_2}$, occurring in (5.28) is unknown. If we want to use the equations to predict flows, this term must be evaluated in some way, for instance, by a closure hypothesis. In the context of jets and wakes, the most commonly used closure is the eddy-viscosity approximation

$$\overline{u_iu_j} = \frac{1}{3}\overline{q^2}\delta_{ij} - \nu_T\left(\frac{\partial\overline{U_i}}{\partial x_j} + \frac{\partial\overline{U_j}}{\partial x_i}\right) \tag{5.33}$$

discussed in the previous chapter, where ν_T is the turbulent eddy viscosity, which generally varies with position in the flow. We have already given a number of health warnings concerning this closure in Chapter 4. In the present case, we use (5.33) with $i = 1, j = 2$, neglecting the term $\partial\overline{U_2}/\partial x_1$, since it is of lower order than $\partial\overline{U_1}/\partial x_2$. Thus, we find

$$\overline{u_1u_2} = -\nu_T\frac{\partial\overline{U_1}}{\partial x_2} \tag{5.34}$$

and so (5.28) becomes

$$\overline{U_1}\,\frac{\partial \overline{U_1}}{\partial x_1} + \overline{U_2}\,\frac{\partial \overline{U_1}}{\partial x_2} = -\frac{1}{\rho}\frac{dP_\infty}{dx_1} + \frac{\partial}{\partial x_2}\left(\nu_T\,\frac{\partial \overline{U_1}}{\partial x_2}\right) \tag{5.35}$$

according to the eddy-viscosity closure. Comparison with the laminar boundary-layer equation, (5.2), shows a strong formal similarity. The difference between the Prandtl equations and the turbulent-mean-flow "boundary-layer" equations with eddy-viscosity closure is that, unlike the real viscosity, ν_T is not a physical property of the fluid and can vary with position in the flow in a way which is not known a priori. Observe that, since only one component, namely $\overline{u_1 u_2}$, of the Reynolds stress enters into the mean-flow equations in the "boundary-layer" approximation considered here, one may regard (5.34) as an *exact* definition of ν_T, rather than a closure approximation. However, in many flows, such as the wall jet considered towards the end of this chapter, the resulting eddy viscosity has pathological behavior, with regions of negative values and infinite singularities where $\partial \overline{U_1}/\partial x_2 = 0$. Such behavior happens not to occur for the self-similar free jets and wakes which are the subject of Sections 5.3 and 5.4 because symmetry implies that the quantities $\overline{u_1 u_2}$ and $\partial \overline{U_1}/\partial x_2$ are both zero at $x_2 = 0$.

If the eddy-viscosity model is accepted, there remains the problem of determining ν_T. Analogies with the calculation of viscosity in the kinetic theory of gases were often employed in the early work on turbulence, leading to the concept of a turbulent mixing length, analogous to the molecular mean free path, and expressions for ν_T in terms of the mean flow (see, e.g., the cases of jets and wakes described in Hinze (G 1975)). Such mixing-length theories had some success, but were never very satisfactory from a fundamental point of view. Qualitatively, the idea that mean momentum transfer by turbulent mixing has similarities with that due to the random motion of molecules may be conceptually helpful, but the analogy is not upheld in detail. Stripped of their kinetic-theory ideas, mixing-length theories are eddy-viscosity closures, with particular expressions for ν_T in terms of the mean flow.

Rather than using a mixing-length, or still more complicated model, we adopt a simple form of eddy viscosity to describe jets and wakes, independent of x_2 and consistent with self-similarity. As noted above, experimental measurements may be used to determine $\overline{u_1 u_2}$ and $\partial \overline{U_1}/\partial x_2$ in (5.34), allowing ν_T to be calculated. The resulting ν_T is indeed found to vary little with respect to x_2 across the central regions of free jets and wakes and can thus be approximated as a function of the streamwise location, x_1, alone. The dependence of the eddy viscosity on streamwise distance is determined using self-similarity, another observed feature of these flows.

For the wakes of bodies placed in an infinite, uniform stream, it is found that the velocity defect takes on a self-similar form sufficiently far downstream, in which the width of the wake increases according to a power law as a function of streamwise distance. Similarity also holds for jets exhausting into infinite, quiescent fluid.[13] Thus, near the body in the case of a wake, or the nozzle for a jet, self-similarity

[13] The case of a jet exhausting into an infinite, uniform flow parallel to the jet leads to different self-similarity far downstream. This flow is rather similar to a wake, although there is a velocity *excess*, rather than a deficit. It is the velocity excess profile which becomes self-similar and the theory of the self-similar jet in a coflowing stream is much the same as for wakes, but with a change of sign.

does not apply, but in the far wake or jet, the streamwise development of the mean flow is determined by similarity. As we shall see in Sections 5.3 and 5.4, if one substitutes the self-similar mean flow into the turbulent-"boundary-layer" equations without a pressure gradient, adopting the closure (5.34) with the form of $\nu_T(x_1)$ chosen to be consistent with similarity, the self-similar mean-velocity profile can be calculated analytically. For the moment, however, we leave the question of closure to one side.

Consider the turbulent wake of a body placed in an infinite, uniform stream, U_∞, in the x_1-direction. The fluid in the wake is slower moving than the external stream, U_∞, but turbulent momentum transport across the wake causes the difference to decrease with increasing streamwise distance from the body. The velocity defect introduced by the wake falls and the uniform flow "heals" itself after the passage of the body, that is, $\overline{U_1} \to U_\infty$ as $x_1 \to \infty$. Thus, in the far wake, not only is the mean velocity defect self-similar, but $\overline{U_1}$ is close to U_∞, a feature which can be used to obtain a simplified form of (5.28) for far-wake calculations, as we now show.

We briefly reexamine the order of magnitude estimates with the case of the far wake of a body in an infinite, uniform stream in mind. Let $V = \overline{U_1} - U_\infty$, so that $-V$ gives the velocity deficit in the wake and tends to zero as $x_1 \to \infty$. In the far wake, $|V| \ll U_\infty$ and we adopt a velocity scale, V_0, for V, small compared with U_∞. From (5.13), written in terms of V, rather than $\overline{U_1}$, we obtain the estimate

$$\overline{U_2} = O\left(\frac{\Delta}{d} V_0\right) \tag{5.36}$$

instead of (5.15).

Dropping the viscous term in (5.16), as usual, we may estimate the orders of magnitude of the remaining terms. The first term on the left-hand side is $O(U_\infty V_0/d)$ and dominates the second one, which is $O(V_0^2/d)$ according to (5.36). One may also approximate the first term by $U_\infty(\partial\overline{U_1}/\partial x_1)$, since $\overline{U_1}$ is close to U_∞. Of the Reynolds stress terms in (5.17), the second is larger, for the same reasons as before. Thus, (5.16) becomes

$$U_\infty \frac{\partial \overline{U_1}}{\partial x_1} = -\frac{1}{\rho}\frac{\partial \overline{P}}{\partial x_1} - \frac{\partial \overline{u_1 u_2}}{\partial x_2} \tag{5.37}$$

to leading order. This is a simplified far-wake version of (5.18), in which transverse mean-flow convection has disappeared, and the streamwise convection velocity has been approximated by the uniform value U_∞.

To make the Reynolds stress term of (5.37), which is $O(\overline{u_1 u_2}/\Delta)$, of the same order, $O(U_\infty V_0/d)$, as the left-hand side,

$$\frac{\Delta}{d} = O\left(\frac{\overline{u_1 u_2}}{U_\infty V_0}\right) \le O\left(\frac{u'^2}{U_\infty V_0}\right) \tag{5.38}$$

showing that Δ/d is small, as assumed in the "boundary-layer" approximation used here, provided $u' \ll (U_\infty V_0)^{1/2}$, which is a more stringent requirement than that of small u'/U_∞ for jets exhausting into fluid at rest. However, the basic reason for the requirement, that the turbulence be sufficiently weak that transport of mean momentum across the flow be slow, remains the same. The condition, $u' \ll (U_\infty V_0)^{1/2}$, is found to be well verified in far wakes.

For the pressure gradient term in (5.37) to be significant, we find that

$$\overline{P} = O(\rho U_\infty V_0) \tag{5.39}$$

Order of magnitude analysis of (5.22) in the far wake, using (5.39), shows that (5.24) still holds to leading order. Thus, $\overline{P} = P_\infty(x_1)$ and we can replace the pressure derivative in (5.37) by dP_∞/dx_1. Since the pressure, P_∞, in the external flow is constant for wakes in a uniform stream, (5.37) becomes

$$U_\infty \frac{\partial \overline{U_1}}{\partial x_1} = -\frac{\partial \overline{u_1 u_2}}{\partial x_2} \tag{5.40}$$

which is the simplified form of (5.28) for the far wake of a body placed in an infinite, uniform stream. Observe that, if one uses the eddy-viscosity closure, (5.34), in (5.40), a linear differential equation for $\overline{U_1}$ results, containing the unknown diffusivity ν_T. Equation (5.40) forms the basis of the treatment of self-similar wakes in Section 5.4, where it is assumed that ν_T is a function of x_1 alone.

Note that, aside from allowing us to use the simplified "boundary-layer" equation (5.40), far wakes have scalings different from those of jets or near wakes. For instance, the ratio of the width to downstream spreading distance becomes (5.38) for the far wake, instead of (5.19). The transverse velocity scale is also changed from (5.15) to (5.36) for the far wake. These different scalings reflect the differing physics of a jet exhausting into external fluid at rest, which must do its own streamwise convection, and a far wake which rides on the back of the external stream.[14]

Until now we have been concerned with the mean-flow equations, (5.11) and (5.12). The boundary-layer approximation also has something to say about the turbulent energy equation, which we can write as

$$\underbrace{\overline{U_j}\frac{\partial \frac{1}{2}\overline{q^2}}{\partial x_j}}_{\text{Convection}} = \underbrace{-\overline{u_i u_j}\frac{\partial \overline{U_i}}{\partial x_j}}_{\text{Production}} - \underbrace{\varepsilon}_{\text{Dissipation}} \underbrace{-\frac{\partial}{\partial x_j}\left\{\overline{u_j\left(\frac{1}{2}q^2 + \frac{p}{\rho}\right)} - \overline{\nu u_i\left(\frac{\partial u_i}{\partial x_j} + \frac{\partial u_j}{\partial x_i}\right)}\right\}}_{\text{"Diffusive" transfer}} \tag{5.41}$$

for a steady flow (equation (4.35), bearing in mind (4.5)), where

$$\varepsilon = \frac{1}{2}\nu\overline{\left(\frac{\partial u_i}{\partial x_j} + \frac{\partial u_j}{\partial x_i}\right)\left(\frac{\partial u_i}{\partial x_j} + \frac{\partial u_j}{\partial x_i}\right)} \tag{5.42}$$

is the usual turbulent energy dissipation.

According to the boundary-layer approximation, the production term can be approximated by noting that $\partial\overline{U_1}/\partial x_2$ is of higher order than the other mean velocity derivatives, so we drop all but the $i = 1, j = 2$ contributions in the implied sum. The diffusive transfer term is likewise simplified by neglecting all but $j = 2$ in the implied sum, on the grounds that $\partial/\partial x_2$ is greater than $\partial/\partial x_1$. Finally, since we are here

[14] At first sight, it might be thought that the wake can be turned into a jet by changing to a frame of reference moving with velocity U_∞. However, it should be recalled that the flows considered here are steady and the change of reference frame turns a steady flow into an unsteady one.

concerned with jets and wakes, there is no viscous sublayer and we can ignore the viscous transfer term to obtain

$$\underbrace{\overline{U}_1 \frac{\partial \frac{1}{2}\overline{q^2}}{\partial x_1} + \overline{U}_2 \frac{\partial \frac{1}{2}\overline{q^2}}{\partial x_2}}_{\text{Convection}} = \underbrace{-\overline{u_1 u_2} \frac{\partial \overline{U}_1}{\partial x_2}}_{\text{Production}} - \underbrace{\overline{\varepsilon}}_{\text{Dissipation}} - \underbrace{\frac{\partial}{\partial x_2}\left\{\overline{u_2\left(\frac{1}{2}q^2 + \frac{p}{\rho}\right)}\right\}}_{\text{Transverse "diffusive" transfer}} \tag{5.43}$$

This equation shows that, as a small packet of turbulence is convected by the mean flow, it gains energy by production, loses it by dissipation and that there is diffusive transfer of turbulent energy across the flow due to turbulent mixing and work done by the fluctuating pressure.

Note the important general point that, even outside a viscous sublayer, where we can neglect viscosity as far as the mean-flow momentum equations are concerned, viscous energy dissipation associated with the smallest scales of turbulence remains important. This energy dissipation is apparent in equations, such as (5.41) and (5.43), for the turbulence itself.

Further order-of-magnitude analysis of (5.43) would tend to indicate that the production term is larger than both the left-hand side and diffusive transfer term, leading to approximate local equilibrium between production and dissipation. This is not true for jets and wakes. The reasons for this vary, depending upon the flow considered: for instance, in the case of a jet exhausting into quiescent fluid, $\overline{u_1 u_2}$ is considerably smaller than u'^2, so production is actually less than one would naively estimate. Thus, although order-of-magnitude estimation correctly gives the form, (5.43), the differences of magnitude between the terms of (5.43) are too subtle to be captured by such estimation and all terms of (5.43) can be, and often are, needed.

As we shall see in Section 5.5, the boundary-layer approximation, (5.43), for the turbulent energy equation needs to be refined in the outer part of a turbulent boundary layer if there is a significant pressure gradient, wall curvature, or external vorticity. In addition, the viscous transfer term

$$\nu \frac{\partial}{\partial x_2}\left\{\overline{u_i\left(\frac{\partial u_i}{\partial x_2} + \frac{\partial u_2}{\partial x_i}\right)}\right\} \tag{5.44}$$

should be added to (5.43) within the viscous sublayer of a turbulent boundary layer.

5.3 Jets

In this section, we consider the case of a jet exhausting into infinite, quiescent fluid, beginning with the case of a plane jet. We first want to show that the jet momentum flux is constant. To this effect, we rewrite (5.28), using (5.13) and constancy of P_∞, as

$$\frac{\partial}{\partial x_1}(\overline{U}_1^2) = -\frac{\partial}{\partial x_2}(\overline{U}_1\overline{U}_2 + \overline{u_1 u_2}) \tag{5.45}$$

which can be integrated with respect to x_2, that is, across the jet, using the fact that $\overline{U}_1$ and the turbulent velocities tend to zero outside the jet, to show that

$$\frac{dQ_m}{dx_1} = 0 \tag{5.46}$$

where Q_m is the jet momentum flux (divided by the fluid density)

$$Q_m = \int_{-\infty}^{\infty} \overline{U}_1^2 dx_2 \tag{5.47}$$

It follows that the momentum flux, Q_m, is independent of streamwise position, according to the "boundary-layer" approximation, and is thus fixed by its value at the nozzle, supposed given. The reader is encouraged to rederive (5.46) via a momentum audit (starting from the x_1-component of the exact, volume-integrated mean momentum equation) using a control volume consisting of a slice of fluid bounded by two planes of constant x_1 and noting the terms that must be neglected in the process.

Constancy of the jet momentum flux can be contrasted with streamwise variation of the volume flux. Integration of (5.13) with respect to x_2 gives

$$\frac{dQ_v}{dx_1} = \overline{U}_2(x_2 = -\infty) - \overline{U}_2(x_2 = +\infty) \tag{5.48}$$

where

$$Q_v = \int_{-\infty}^{\infty} \overline{U}_1 dx_2 \tag{5.49}$$

is the volume flux (equal to the mass flux divided by the density). In general, the transverse velocity, $\overline{U_2}$, is nonzero outside the jet, representing entrainment of fluid that must be swept in to increase the mass flux of the jet as it develops downstream. Such entrainment is a characteristic feature of jets and creates a large-scale, relatively slowly moving flow in the region external to the jet (see Figure 5.5 and bear in mind that the transverse velocity, $\overline{U_2}$, is small compared with the turbulent fluctuations, which are themselves small compared with the streamwise mean velocity of the jet). Entrainment is also a feature of laminar jets and it is illuminating to read the discussion given in Batchelor (G 1967, section 4.6) of an exact solution of the Navier–Stokes equations. That solution describes an axisymmetric jet rather than the plane

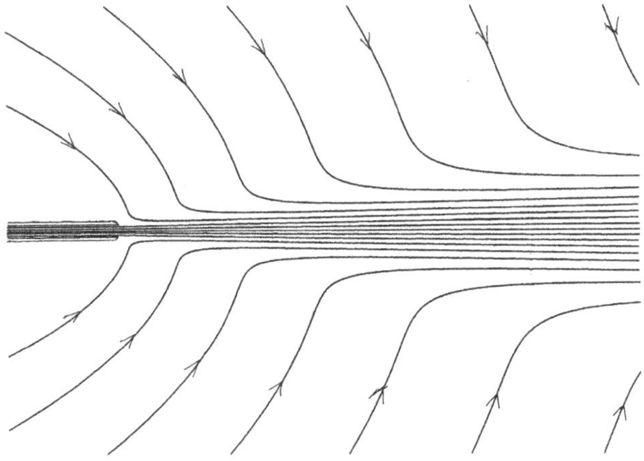

Figure 5.5. Sketch of mean-flow streamlines illustrating entrainment flow caused by a jet. Note that the incoming flow outside the jet is comparatively weak.

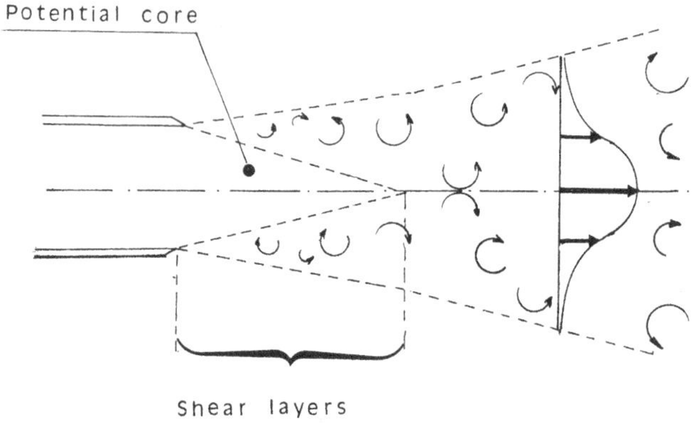

Figure 5.6. Sketch of a jet flow near the nozzle. For clarity, the horizontal scale has been compressed: the end of the potential core is located at about five nozzle widths, as stated in the text.

one considered here and has many features in common with the axisymmetric turbulent jet we discuss later.

To proceed further, we introduce some experimental observations of the plane turbulent jet. Such observations indicate that the flow becomes self-similar sufficiently far downstream of the nozzle, with similarity variable $\xi = x_2/(x_1 - x_0)$, where x_0 is a constant that scales on the nozzle width and corresponds to an arbitrariness in the origin of x_1 for the self-similar jet. The value of x_0 is not fixed by imposing conditions at the nozzle, because similarity only applies many nozzle widths downstream. More precisely, it is found that of the order of twenty nozzle widths or more are required for similarity of the mean flow, considerably further for turbulent quantities such as $\overline{u_i u_j}$. Caution must therefore be exercised in assuming full statistical similarity of the flow just because the mean flow is self-similar.

The flow at the nozzle has a comparatively low level of turbulence and is often supposed approximately uniform across the nozzle exit. It emerges into the external fluid, creating shear layers originating at the lips of the nozzle. These layers separate the fast-moving jet from the quiescent external fluid and, presuming the jet Reynolds number is sufficiently high, they break down rapidly to turbulence.[15] The shear layers spread out by entrainment until they meet at the center plane of the jet (see Figure 5.6) about five nozzle widths downstream (the approximately laminar zone between the mixing layers is often referred to as the "potential core"). Subsequently, the jet develops as a whole, with a bell-shaped velocity profile, $\overline{U_1}(x_2)$, which spreads and evolves with downstream distance until it eventually becomes self-similar. Incidentally, it should be no surprise that the streamwise evolution of the jet takes many nozzle widths: this is the basis of the approximation developed in the previous section and, as we saw there, is due to smallness of $u'/\overline{U}$.

Assuming self-similarity, the jet streamwise velocity profile has the form

$$\overline{U}_1 = (x_1 - x_0)^{-1/2} f(\xi) \tag{5.50}$$

where the factor $(x_1 - x_0)^{-1/2}$ is required to make the jet momentum flux, Q_m, independent of x_1. If the form, (5.50), is introduced into (5.13), we may integrate with respect to x_2 to obtain

[15] Figure 1.9 illustrates a lower Reynolds number for which transition occurs further downstream.

$$\overline{U_2} = (x_1 - x_0)^{-1/2} g(\xi) \tag{5.51}$$

The self-similar plane jet considered here is supposed symmetric about $x_2 = 0$ and so $f(\xi)$ is an even function of ξ, while $g(\xi)$ is an odd function of ξ, given by

$$g(\xi) = \xi f(\xi) - \frac{1}{2}\int_0^{\xi} f(\xi')d\xi' \tag{5.52}$$

Note that the above use of symmetry imposes the boundary condition $\overline{U_2} = 0$ at $x_2 = 0$. Applying (5.50) and (5.51) in (5.28) with zero pressure gradient, we can integrate with respect to x_2 and again use the symmetry of the flow (which makes $\overline{u_1 u_2}$ an odd function of x_2) to obtain

$$\overline{u_1 u_2} = (x_1 - x_0)^{-1} h(\xi) \tag{5.53}$$

where

$$h(\xi) = \frac{1}{2} f(\xi) \int_0^{\xi} f(\xi')d\xi' \tag{5.54}$$

We choose to express $\overline{u_1 u_2}$ according to the eddy-viscosity approximation, that is, using (5.34), so that

$$\nu_T = (x_1 - x_0)^{1/2} \chi(\xi) \tag{5.55}$$

where

$$\chi \frac{df}{d\xi} + h = 0 \tag{5.56}$$

Equation (5.55) gives the form of the turbulent eddy viscosity required for self-similarity. As noted earlier, it is remarkable that measurements are rather well described by assuming a value of ν_T which does not depend on x_2, that is, constant χ.

Let us therefore suppose that χ is a constant. Elimination of h between (5.54) and (5.56) gives

$$2\chi \frac{df}{d\xi} + f(\xi) \int_0^{\xi} f(\xi')d\xi' = 0 \tag{5.57}$$

or, if we set

$$F(\xi) = \int_0^{\xi} f(\xi')d\xi' \tag{5.58}$$

and integrate,

$$4\chi \frac{dF}{d\xi} = 16\alpha^2\chi^2 - F^2 \tag{5.59}$$

where α is a nondimensional constant of integration. The integral of this equation yields

$$F(\xi) = 4\alpha\chi \tanh \alpha\xi \tag{5.60}$$

so that

$$f(\xi) = \frac{dF}{d\xi} = 4\alpha^2 \chi \operatorname{sech}^2 \alpha\xi \tag{5.61}$$

gives the streamwise velocity profile via (5.50),

$$g(\xi) = \xi f - \frac{1}{2} F = 4\alpha\chi\left(\alpha\xi \operatorname{sech}^2 \alpha\xi - \frac{1}{2}\tanh \alpha\xi\right) \tag{5.62}$$

provides the transverse velocity from (5.51), and, finally,

$$h(\xi) = \frac{1}{2} fF = 8\alpha^3 \chi^2 \operatorname{sech}^2 \alpha\xi \tanh \alpha\xi \tag{5.63}$$

gives $\overline{u_1 u_2}$ from (5.53).

The jet momentum flux is

$$Q_m = \int_{-\infty}^{\infty} f^2(\xi) d\xi = \frac{64}{3}\alpha^3 \chi^2 \tag{5.64}$$

which can be used to calculate χ if Q_m, measuring the intensity of the jet, and α, which determines its rate of spreading, are known. The solution then depends only on Q_m, which is fixed by the jet momentum flux at the nozzle, α, which is a purely numerical parameter, and x_0 giving the apparent origin of the self-similar jet. Experimentally, it is found that $\alpha = 8$ fits the data quite well.

The jet described by the above similarity solution spreads out to form a wedge with straight sides, its width being proportional to $x_1 - x_0$. The wedge semiangle at which the streamwise velocity is one half its maximum at the same streamwise location (i.e., on the center plane) is about 6° (assuming $\alpha = 8$), which gives an idea of the extent of the jet. Mean velocities decrease with streamwise distance proportional to $(x_1 - x_0)^{-1/2}$, as do turbulent velocities, such as u' (although the measurements of turbulent velocity moments are less conclusive than those for mean velocities). For this reason, quantities, such as $\overline{u_1 u_2}$, which are quadratic in the turbulent velocities, are proportional to $(x_1 - x_0)^{-1}$. Figure 5.7 shows a comparison of the streamwise velocity profile derived from (5.61) with some experimentally determined points. The reasonable agreement should not be surprising, since the underlying assumption that ν_T is independent of x_2 was originally ascertained from such data. Note that the measured profile drops off considerably faster than (5.61) toward the outskirts of the jet, where the ν_T required to fit the data decreases. In fact, it could be argued that the agreement is spurious: given that the parameter α is available to produce a fit, different curves with the overall bell shape of (5.61) can produce results that are as good or better, for instance a Gaussian distribution. On the other hand, the assumption of ν_T independent of x_2 is relatively simple and appears to apply to other wake and jet flows.

Regardless of the precise form of $f(\xi)$, the fact that the jet approaches self-similarity is perhaps surprising in itself. One interpretation is that the jet eventually forgets about the details of its production and that the only parameter remaining far downstream is the constant momentum flux, Q_m. If Q_m, x_2, and $x_1 - x_0$ are taken as the only dimensional parameters, dimensional analysis indicates that the flow velocity must be of the form (5.50), (5.51), with f and g proportional to $Q_m^{1/2}$.

The observed self-similarity of the mean properties of the far jet suggests that the large scales of turbulence are self-similar. Thus, for instance, the correlation lengths

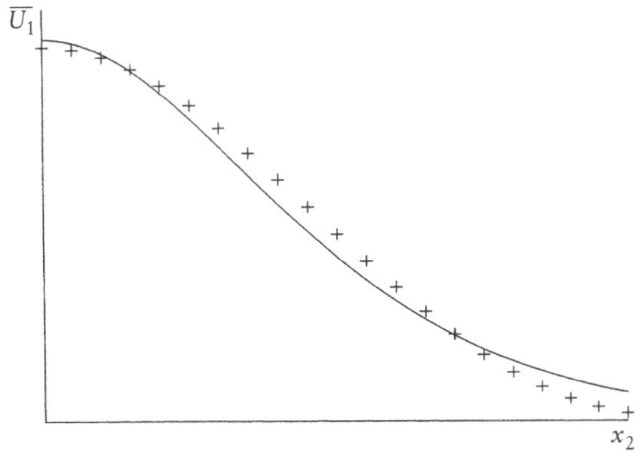

Figure 5.7. Comparison of theoretical (continuous line, $\alpha = 8$) and measured (points) streamwise velocity profiles in the self-similar region of a plane turbulent jet (the points are based on experimental results of Bradbury (1965)).

of turbulence ought to be proportional to $x_1 - x_0$. However, this similarity of turbulence properties cannot extend to the smallest scales of turbulence in the plane jet because the turbulent Reynolds number, Re_L, of the jet, based on a correlation length scale increasing as $x_1 - x_0$ and a turbulent velocity decreasing proportional to $(x_1 - x_0)^{-1/2}$, grows like $(x_1 - x_0)^{1/2}$. This increase of Reynolds number with downstream distance must be reflected in a decrease of η/L, where η and L are the Kolmogorov and correlation length scales (see Chapter 3), so that the smallest (dissipative) scales do not share the self-similarity of the large ones. Presumably, this lack of similarity of the dissipative scales has little effect on the large ones responsible for mixing, for otherwise the observed overall similarity of the jet would not occur. This suggests that the large-scale structure of the turbulent jet is insensitive to Reynolds number, provided it is large enough. That is, viscosity determines the size and dynamics of the smallest scales of turbulence, but little else. Incidentally, the increase of Reynolds number with x_1 indicates that the jet should remain turbulent indefinitely far downstream.

Two features of the self-similar plane jet are observed to apply to self-similar jets and wakes in general. Both mean and turbulent velocities are proportional to the *same* power of $x_1 - x_0$, for instance $(x_1 - x_0)^{-1/2}$ in the case of the plane jet. This is found to be true in general, although the exponent of the power law depends on the particular flow considered, and for wakes, the appropriate mean velocity, whose variation with $x_1 - x_0$ mirrors that of the turbulence, is the difference, $V = \overline{U}_1 - U_\infty$, rather than $\overline{U}_1$ itself. This is natural, since it is the variations of mean velocity across the flow (mean shear) that generate and maintain turbulence. The jet considered above, which exhausts into fluid at rest, has $U_\infty = 0$, so that $V = \overline{U}_1$ in that case. In all cases, the turbulent velocities and mean-flow variations across the flow are found to have the same power-law dependence on $x_1 - x_0$ in the self-similar regime. A second general property is the observed approximate independence of ν_T on x_2. Based on these two properties, which are purely empirical, and the turbulent-"boundary-layer" equations, one can calculate the mean-velocity profile in the self-similar range for all the jets and wake flows considered here.

The case of a circular jet is of considerable practical interest, but for this flow we need the axisymmetric equivalents of (5.13) and (5.28). One can use cylindrical polar coordinates, r, ϕ, x, based on the jet axis and express the averaged Navier–Stokes equations, (5.11) and (5.12), in terms of these coordinates (see Batchelor (G 1967), appendix 2). Axisymmetry implies that the flow statistics are independent of ϕ and we further assume no swirl, that is, $\overline{U}_\phi = 0$. The mean-flow continuity equation, (5.12), gives

$$\frac{1}{r}\frac{\partial r\overline{U}_r}{\partial r}+\frac{\partial \overline{U}_x}{\partial x}=0 \tag{5.65}$$

After order-of-magnitude estimation of the steady, axisymmetric momentum equations, using the "boundary-layer" approximation as before, one finds that, to leading order,

$$\overline{P}=P_\infty(x) \tag{5.66}$$

is imposed from outside, as in the plane case, and that

$$\overline{U}_x\frac{\partial \overline{U}_x}{\partial x}+\overline{U}_r\frac{\partial \overline{U}_x}{\partial r}=-\frac{1}{\rho}\frac{dP_\infty}{dx}-\frac{1}{r}\frac{\partial r\overline{u_r u_x}}{\partial r} \tag{5.67}$$

is the axisymmetric equivalent of (5.28). Equations (5.65)–(5.67) are the inviscid, axisymmetric turbulent-"boundary-layer" equations. The pressure gradient term is zero for the circular jet exhausting into infinite, quiescent fluid considered here.

From (5.65) and (5.67) with zero pressure gradient, one can show that the jet momentum flux (divided by the density):

$$Q_m=\int_0^\infty 2\pi r\overline{U}_x^2\,dr \tag{5.68}$$

is a constant, fixed by its value at the nozzle.

The similarity variable for the far, circular jet is $\xi=r/(x-x_0)$ and, to obtain constancy of Q_m,

$$\overline{U}_x=(x-x_0)^{-1}f(\xi) \tag{5.69}$$

while (5.65) and (5.67) imply

$$\overline{U}_r=(x-x_0)^{-1}g(\xi) \tag{5.70}$$

and

$$\overline{u_r u_x}=(x-x_0)^{-2}h(\xi) \tag{5.71}$$

of which (5.71) is consistent with proportionality of turbulence velocities to $(x-x_0)^{-1}$, like the mean flow. As noted above, this appears to be a general property of such self-similar flows. In fact, if one assumes self-similarity with the general similarity variable $\xi=r/W(x)$, so that

$$\overline{U}_x=W^{-1}f(\xi) \tag{5.72}$$

to make Q_m constant, the requirement that

$$\overline{u_r u_x}=W^{-2}h(\xi) \tag{5.73}$$

(which follows from imposing the same dependency of turbulent and mean velocities with respect to x), together with (5.65) and (5.67) without pressure gradient, imply that $W(x)$ is a linear function of x. Thus, requiring the same dependency of mean and turbulent velocities with respect to x yields the correct similarity variable. Alternatively, one can assume that the jet forgets its origins and depends only on Q_m, r, and $x-x_0$ in the far jet. Dimensional analysis then leads to the forms (5.69)–(5.71), with f and g proportional to $Q_m^{1/2}$ and h to Q_m.

Using (5.69)–(5.71) in (5.65) and (5.67), we can show that

$$\xi g(\xi) = \xi^2 f(\xi) - F(\xi) \tag{5.74}$$

and

$$\xi h(\xi) = f(\xi)F(\xi) \tag{5.75}$$

where

$$F(\xi) = \int_0^{\xi} \xi' f(\xi')d\xi' \tag{5.76}$$

The eddy-viscosity approximation, (5.33), in cylindrical coordinates leads to

$$\overline{u_r u_x} = -\nu_T \frac{\partial \overline{U}_x}{\partial r} \tag{5.77}$$

to leading order and so

$$\nu_T = -h(\xi)\left(\frac{df}{d\xi}\right)^{-1} \tag{5.78}$$

is automatically independent of x. If we further suppose it independent of ξ, ν_T is a constant. We can substitute for $h(\xi)$ in (5.78), using (5.75), and integrate to obtain

$$\xi \frac{dF}{d\xi} = 2F - \frac{F^2}{2\nu_T} \tag{5.79}$$

Integrating again leads to

$$F(\xi) = \frac{4\nu_T(\alpha\xi)^2}{1 + (\alpha\xi)^2} \tag{5.80}$$

from which we obtain

$$f(\xi) = \frac{8\nu_T\alpha^2}{(1 + (\alpha\xi)^2)^2} \tag{5.81}$$

$$g(\xi) = \frac{4\nu_T\alpha^2\xi(1 - (\alpha\xi)^2)}{(1 + (\alpha\xi)^2)^2} \tag{5.82}$$

$$h(\xi) = \frac{32\nu_T^2\alpha^4\xi}{(1 + (\alpha\xi)^2)^3} \tag{5.83}$$

and the jet momentum flux,

$$Q_m = \frac{64}{3}\pi\nu_T^2\alpha^2 \tag{5.84}$$

which can be used to calculate ν_T, given Q_m (determined at the nozzle) and α, a nondimensional constant characterizing the rate of spread of the jet.

The circular jet begins at the nozzle as a cylindrical shear layer that spreads out to fill the jet, in much the same way as for the two shear layers of the plane jet (see Figure 5.6). The potential core disappears about six diameters from the nozzle and the jet develops downstream, attaining mean-flow similarity at distances of about twenty diameters and greater. In the self-similar regime, the jet spreads in a conical fashion and the velocity profile is well represented by (5.81) with $\alpha = 8$, apart from the outskirts of the jet where the approximation by a uniform ν_T no longer fits the data well. The cone on which the streamwise velocity is one-half its maximum at the same streamwise location has a semiangle of about 5°.

It will be noted that the circular jet has much in common with the plane one, in particular, they share the same linear dependence of jet width on downstream distance in the self-similar zone and nearly the same angle of spreading. However, the velocities now decrease like $(x - x_0)^{-1}$, which has the effect of maintaining a constant (large) Reynolds number. This means that the smallest scales of turbulence may form part of the similarity solution for the circular jet. The fact that the Reynolds number remains large suggests that the jet will remain turbulent indefinitely far downstream.

Whereas the eddy-viscosity model for the plane jet had ν_T proportional to $(x_1 - x_0)^{1/2}$, the circular jet has *constant* ν_T. This makes the turbulent mean flow appear like a laminar one in a fictitious fluid having a much greater viscosity than the real one. The solution represented by (5.81), (5.82) is, in fact, equivalent to that for a laminar jet (as described in Batchelor (G 1967), section 4.6) at high Reynolds number, when the jet angle becomes small, so that the "boundary-layer" approximation can be applied (as assumed in deriving (5.81), (5.82)). The Reynolds number of the equivalent laminar jet, based on the maximum streamwise velocity, the distance from the axis of the half-velocity point, and the viscosity ν_T is a numerical constant, proportional to α, of about 40. The true Reynolds number of the jet is much higher than this, of course (by a factor that equals the ratio of ν_T to the real viscosity of the fluid). A similar calculation for the plane jet gives a constant value of the "Reynolds number" based on ν_T of about 30. Self-similar turbulent jets thus appear to select an eddy viscosity that makes the "apparent Reynolds number" constant and not too large, even though the eddy viscosity must change with jet momentum flux and in the case of the plane jet, with streamwise location, to achieve this.

Entrainment of fluid from outside plane and circular self-similar jets can be investigated using (5.62) and (5.82) in the limit $\xi \to \infty$. For the plane jet, fluid is swept towards the jet at velocity

$$2\alpha\chi(x_1 - x_0)^{-1/2} \tag{5.85}$$

which is a factor $1/(2\alpha) \approx 6\%$ of the maximum streamwise velocity. To the fluid outside the jet, it looks like a surface distribution of volume sinks of strength twice (5.85) per unit area, resulting in a flow like that sketched in Figure 5.5. For the circular jet, we find that

$$\overline{U}_r \approx -\frac{4\nu_T}{r} \tag{5.86}$$

outside the jet. This corresponds to a semi-infinite uniform line sink of volume flux $8\pi\nu_T$ per unit length of jet. The flow will again be qualitatively as shown in Figure 5.5.

Turbulent jets that are neither axisymmetric nor plane have an interesting feature that neither of the two more widely studied special cases possess. For such flows, one can no longer use the two-dimensional "boundary-layer" equations derived previously, but must allow for general variation of averaged quantities with respect to the transverse coordinates, x_2 and x_3. We saw earlier that an improved turbulent-"boundary-layer" approximation for the transverse-momentum equation leads to (5.29) for a plane flow, and a similar equation can be derived for the axisymmetric case. These equations express a balance between the transverse mean pressure gradient and the *transverse* part of the Reynolds stress force, $\partial(\overline{u_i u_j})/\partial x_j$. The latter is in the x_2-direction for plane flow and the r-direction for axisymmetric flow. In either of these special cases, the Reynolds stress can be balanced by a pressure gradient in the corresponding direction. However, in general, the transverse part of $\partial(\overline{u_i u_j})/\partial x_j$ will *not* be expressible as the transverse part of a scalar gradient because there is no good reason why it should be irrotational.

In mathematical terms, the nonplanar "boundary-layer" approximation for the transverse momentum equations, taken to the next order above the basic approximation (which gives $\overline{P}$ independent of x_2 and x_3) leads to

$$\frac{1}{\rho}\frac{\partial \overline{P}}{\partial x_2} = -\frac{\partial \overline{u_2^2}}{\partial x_2} - \frac{\partial \overline{u_2 u_3}}{\partial x_3} \tag{5.87}$$

$$\frac{1}{\rho}\frac{\partial \overline{P}}{\partial x_3} = -\frac{\partial \overline{u_2 u_3}}{\partial x_2} - \frac{\partial \overline{u_3^2}}{\partial x_3} \tag{5.88}$$

Differentiating (5.87) with respect to x_3, (5.88) with respect to x_2 and subtracting (which is the same as taking the x_1-component of the curl), we find

$$\left(\frac{\partial^2}{\partial x_3^2} - \frac{\partial^2}{\partial x_2^2}\right)\overline{u_2 u_3} = \frac{\partial^2}{\partial x_2 \partial x_3}\left(\overline{u_3^2} - \overline{u_2^2}\right) \tag{5.89}$$

and there is no pressing reason why the turbulence should arrange itself so that the velocity moments satisfy this equation (they "just happen" to do so for plane and axisymmetric flows). Without (5.89), equations (5.87) and (5.88) are inconsistent and something must go wrong with the reasoning leading to these equations. Clearly we should make terms in the transverse momentum equations, other than those present in (5.87), (5.88), play a role.

To resolve this difficulty, we reassess the order-of-magnitude arguments leading to the standard form of the turbulent-"boundary-layer" approximation. The mean-flow continuity equation is

$$\frac{\partial \overline{U}_1}{\partial x_1} + \frac{\partial \overline{U}_2}{\partial x_2} + \frac{\partial \overline{U}_3}{\partial x_3} = 0 \tag{5.90}$$

and, as before (equation (5.15)), we take the transverse velocity scale

$$\overline{U}_2, \overline{U}_3 = O\left(\frac{\Delta}{d}\, U_0\right) \tag{5.91}$$

so that all three terms of (5.90) are of the same order. The x_2-component of the transverse momentum equation is

$$\overline{U}_1\frac{\partial \overline{U}_2}{\partial x_1} + \overline{U}_2\frac{\partial \overline{U}_2}{\partial x_2} + \overline{U}_3\frac{\partial \overline{U}_2}{\partial x_3} = -\frac{1}{\rho}\frac{\partial \overline{P}}{\partial x_2} - \frac{\partial \overline{u_1 u_2}}{\partial x_1} - \frac{\partial \overline{u_2^2}}{\partial x_2} - \frac{\partial \overline{u_2 u_3}}{\partial x_3} \tag{5.92}$$

after neglecting the viscous term. Of the Reynolds stress terms, the first is of lower order because it involves a streamwise rather than a transverse derivative, and we therefore drop this term. The right-hand side of (5.92) then gives (5.87) if it is equated to zero, while (5.88) arises from the x_3-component in a similar fashion.

To avoid the problem described above, the left-hand side of (5.92) must be of the same order as the terms on the right, which are dominant under the standard "boundary-layer" approximation. One can estimate the left-hand side as $O(\Delta U_0^2/d^2)$, using (5.91), while the second of the Reynolds stress terms is $O(u'^2/\Delta)$. Equating these order of magnitudes, we have

$$\frac{\Delta}{d} = O\left(\frac{u'}{U_0}\right) \tag{5.93}$$

to be compared with (5.19) for the standard approximation. Adopting (5.93), the only term we drop in (5.92) is $\partial(\overline{u_1 u_2})/\partial x_1$, and likewise for its x_3-component analog.

The x_1-component of the inviscid momentum equation is

$$\overline{U}_1 \frac{\partial \overline{U}_1}{\partial x_1} + \overline{U}_2 \frac{\partial \overline{U}_1}{\partial x_2} + \overline{U}_3 \frac{\partial \overline{U}_1}{\partial x_3} = -\frac{1}{\rho}\frac{\partial \overline{P}}{\partial x_1} - \frac{\partial \overline{u_1^2}}{\partial x_1} - \frac{\partial \overline{u_1 u_2}}{\partial x_2} - \frac{\partial u_1 u_3}{\partial x_3} \tag{5.94}$$

of which the Reynolds stress terms are at most $O(u'^2/\Delta)$, while the left-hand side is $O(U_0^2/d)$. From (5.93), their ratio is $O(u'/U_0)$, which makes the Reynolds stress terms negligible in (5.94) at leading order, whereas they are present in the standard approximation. Indeed, earlier we based our order-of-magnitude estimates on the assumption that the Reynolds stresses should appear in the streamwise momentum equation at leading order, whereas now we find that they are negligible unless the flow is close to being planar or axisymmetric. As before, equating the order of magnitude of the pressure gradient term to that, $O(U_0^2/d)$, of the left-hand side of (5.94) gives

$$\overline{P} = O(\rho U_0^2) \tag{5.95}$$

as an estimate for the mean pressure.

Returning to the transverse momentum equation, (5.92), we find that (5.95) makes the pressure gradient term dominant, as in the standard approximation. Thus, to leading order

$$\overline{P} = P_\infty(x_1) \tag{5.96}$$

is still imposed from outside the flow. The next order approximation to (5.92) involves a corrected form

$$\overline{P} = P_\infty(x_1) + \tilde{P}(x_1, x_2, x_3) \tag{5.97}$$

of (5.96), where $\tilde{P}$ is of lower order than P_∞. We can use this expression in (5.92), dropping the term $\partial(\overline{u_1 u_2})/\partial x_1$ compared to the other Reynolds stresses, to obtain

$$\overline{U}_1 \frac{\partial \overline{U}_2}{\partial x_1} + \overline{U}_2 \frac{\partial \overline{U}_2}{\partial x_2} + \overline{U}_3 \frac{\partial \overline{U}_2}{\partial x_3} = -\frac{1}{\rho}\frac{\partial \tilde{P}}{\partial x_2} - \frac{\partial \overline{u_2^2}}{\partial x_2} - \frac{\partial \overline{u_2 u_3}}{\partial x_3} \tag{5.98}$$

with a similar equation for the x_3-component of momentum, while neglecting all Reynolds stress terms in (5.94) gives

$$\overline{U}_1 \frac{\partial \overline{U}_1}{\partial x_1} + \overline{U}_2 \frac{\partial \overline{U}_1}{\partial x_2} + \overline{U}_3 \frac{\partial \overline{U}_1}{\partial x_3} = -\frac{1}{\rho}\frac{dP_\infty}{dx_1} \tag{5.99}$$

Equations (5.90), (5.98) (with its x_3 analog), and (5.99) are the final set of modified turbulent-"boundary-layer" equations describing jets that are not plane or axisymmetric. Observe that $\tilde{P} = O(\rho u'^2)$ to make the pressure gradient term in (5.98) comparable with the Reynolds stress contribution. From (5.90) and (5.99), one can show that, as before, the streamwise momentum flux

$$Q_m = \int\int \overline{U}_1^2 dx_2 dx_3 \tag{5.100}$$

is constant in a jet having no external pressure gradient.

Let us interpret the new equations. There is no direct effect of the turbulence on the streamwise mean velocity of the jet at leading order, as is apparent from (5.99); however, the Reynolds stress forcing of the transverse mean flow via (5.98) will set up transverse mean motions within the jet. These mean motions change the streamwise velocity profile through the convective terms in (5.99). Observe that the mean transverse velocities created are $O(u')$, according to (5.91) and (5.93), and are thus considerably stronger than those present in plane and axisymmetric jets (recall that the transverse mean velocity was small compared with the turbulent velocities for those cases). The distance over which the jet now develops is $O(U_0/u')$ times its width, according to (5.93), compared with the rather greater $O((U_0/u')^2)$ (or more) of standard type jets (from (5.19)). Thus, the asymmetric jet develops more rapidly downstream (though still slowly on a jet width scale, so the "boundary-layer" approach remains valid), under the effects of transverse mean flow, driven by the transverse Reynolds stress gradients of the turbulence. Such transverse mean circulations are well known to exist in turbulent flow through ducts which are not circular or plane (see Schlichting 1987).

We might have guessed that the result of transverse Reynolds stress forcing which cannot be balanced by a pressure gradient would be significant transverse mean flow. Experiments on elliptic jets exhausting into infinite, quiescent fluid show that such jets quite rapidly approach axisymmetry (see, e.g., Hussain and Husain 1989). The induced transverse mean flow thus appears to bring the jet back towards the circular case, but in the process it may also lead to more rapid mixing of the fluid in the jet, which can have beneficial effects in some applications. Once the jet is close to axisymmetry, the transverse forcing mechanism described above becomes weaker and the jet develops more slowly. It should then approach the regime of applicability of the standard "boundary-layer" aproximation, no doubt passing through an intermediate range of small eccentricities in which the Reynolds stress terms, $-\partial(\overline{u_1 u_2})/\partial x_2 - \partial(\overline{u_1 u_3})/\partial x_3$, should be included in (5.99), and the jet is nearly, but not quite, circular. It should eventually approach the self-similar circular jet considered earlier. Incidentally, we might expect that elliptic jets with very large eccentricities would also develop towards axisymmetry, but more slowly than less eccentric ones because they locally approximate a plane jet as the eccentricity becomes large.

A wide rectangular jet would usually be considered as plane, well away from its edges, but near the edges of the jet there will be transverse mean motions due to Reynolds stress gradients, which cause the jet to develop more rapidly there. Such a jet may therefore need a considerably larger aspect ratio than might at first be thought to avoid nonplanar effects at its center. In practice, side walls are often used to reduce end effects in measurements on nominally plane jets.

Until now, we have been concerned with the mean flow of the jet, and have only considered the turbulence in as much as the Reynolds stress enters into the mean-flow equations. We now want to briefly consider the behavior of the turbulence itself. Downstream of the nozzle, at a Reynolds number high enough that transition occurs, turbulence is present in the shear layer(s) and the turbulent region expands with the layer(s) to fill the jet. Sufficiently far downstream, the turbulence approaches similarity, but is observed to do so rather more slowly than the mean flow. In the self-similar regime, the turbulent intensity, as measured by $\overline{q^2}$ for instance, has a bell-shaped profile with a single maximum at the jet axis for a circular jet, and a symmetric double-humped profile whose maxima lie off the central plane of symmetry for a plane jet. Of course, in both cases, $\overline{q^2}$ goes to zero outside the jet.

A more detailed picture of the turbulence in the self-similar regime emerges from consideration of the various terms in the turbulent energy equation, (5.41) or, after simplification using the "boundary-layer" approximation (5.43), or its axisymmetric equivalent. Figure 5.8 shows the various measured contributions to (5.41) for a circular jet. The sign of the production term has been switched in the figure to make the sum of all contributions shown zero. Turbulent energy production is positive everywhere and has a peak near the location of maximum shear, that is, maximum $|\partial \overline{U}_x/\partial r|$. This can be interpreted qualitatively in two ways. Firstly, high mean shear is usually associated with turbulence production and, indeed, we can see from (5.43) (or better, its axisymmetric analog), that the production term contains the mean shear as a factor. Secondly, the velocity profile has an inflection point when the shear is a maximum and such inflections are destabilizing according to laminar-flow stability theory. Thus, localized instability may be thought of as leading to turbulence production. This explanation, although often employed, has not been given a rigorous justification for turbulent flows. The production term for a plane jet shows similar behavior. In that case, the peak in production feeds through into an off-center maxima in turbulent intensity, noted above.

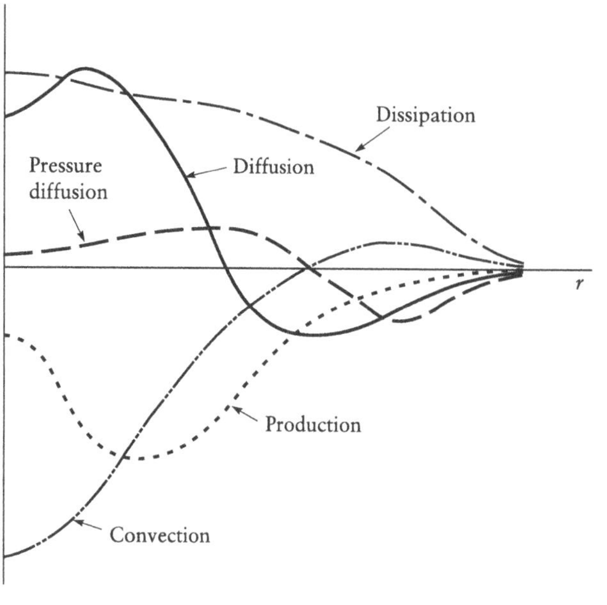

Figure 5.8. Different terms in the turbulent energy balance of a circular jet in the self-similar regime. Note that the sign of the production term has been switched here to make the contributions shown sum to zero. The viscous contribution to the diffusive term in (5.41) is negligible outside the viscous sublayer, as here, while the remainder is the sum of a pressure–velocity term ("Pressure diffusion" in the figure) and a cubic velocity term ("Diffusion" in the figure). (Wygnanski and Fiedler (1969), redrawn.)

Viscous dissipation occurs fairly uniformly across the jet, but, of course, decreases to zero at the outskirts. Despite being able to neglect viscosity in the mean-flow equations it *is* crucial for the turbulence energetics.

Turbulent "diffusive" transfer occurs mainly in the cross-stream, rather than in the streamwise direction, due to the greater inhomogeneity of the turbulence across the jet. This is apparent in (5.43), where only the transverse diffusion term is retained. There are two contributions in (5.43). One, due to work done by the turbulent pressure and velocity, is comparatively small over most of the jet, according to Figure 5.8. The other is more significant and is the result of turbulent advection (it is referred to simply as "diffusion" in the figure). Taken together, the sum of these two terms is positive in the central part of the jet and negative in its outer part. Given the sign convention of the figure, this represents transfer of turbulent energy from the center to the outside, which is reasonable for a diffusive process, since the turbulent intensity is larger towards the jet axis. Via diffusive transfer, turbulent energy is lost from the jet center and transferred to its outer part.

Finally, the convection term, which equals minus the sum of the three considered above, can be interpreted in two ways. In the frame of reference of the nozzle and external fluid, the jet is statistically steady and the convective term represents the net rate at which a small fixed element of volume gains turbulent energy by mean-flow convection through its boundaries. This point of view is expressed mathematically by rewriting the convective term in (5.41), using (5.12), as

$$\frac{\partial}{\partial x_j}\left(\frac{1}{2}\overline{U}_j\overline{q^2}\right) \tag{5.101}$$

corresponding to a flux equal to $\overline{U}_j\overline{q^2}/2$ (cf. equation (4.35)). For instance, if a small, cylindrical volume lies on the axis of the jet, on the average more turbulent energy arrives through its upstream face than leaves via its downstream face or through its curved sides (recall that the intensity of turbulence diminishes downstream and away from the jet axis). This net influx of turbulent energy due to mean-flow convection is augmented by some production inside the volume, and the two are balanced by dissipation in the volume and diffusion away from the jet axis through the sides. The result is that the mean turbulent energy in the small cylinder remains constant, as it must, since the flow is statistically steady. Similar descriptions of the energy balance can be given off-axis, of course.

A second interpretation of the convection term is that it represents the rate of change of the average turbulent intensity following a packet of turbulence undergoing mean-flow convection. The fact that the convective term is negative on the axis is not then surprising because turbulent intensity decreases with streamwise distance. Following a packet of turbulence convected downstream along the axis by the mean flow, turbulence is produced on the average, but this is more than offset by dissipation and transverse diffusion away from the axis. The result is that the mean intensity of the packet decreases with streamwise distance. Needless to say, the two interpretations of the convective term given above are actually different ways of expressing the same thing, but the reader may intuitively prefer one or the other.

When considering results such as those described above for turbulence energetics, it is easy to forget that averaging can create a somewhat misleading impression of the behavior of a turbulent flow. As noted at the beginning of the chapter, at any instant of time, a wake or jet has a quite well-defined boundary separating turbulent fluid inside from laminar fluid outside (Figure 5.1a). This boundary has a randomly corrugated form, with large-scale bulges that are convected downstream, causing the position of the boundary to fluctuate with time. The bulges develop and engulf laminar fluid as they are convected, leading to entrainment. When looked at more closely, the boundary is seen to have corrugations at finer and finer scales, down to a small limiting scale at which one can no longer distinguish a sharp boundary. For flows like the jets considered above, which take place in irrotational external fluid, the turbulence can, in principle, be distinguished from the laminar flow by its non-zero *vorticity*. One can imagine an experimental probe capable of measuring the vorticity[16] at a point. As the turbulence boundary moves past the probe, it will register a rapid change in the level. A threshold value of vorticity could be defined to distinguish between turbulent and laminar flow at the probe. Of course, the boundary is not infinitely sharp (no doubt owing to viscosity) and both the fine-scale fluctuations of vorticity inside the turbulent zone and fine-scale corrugations of the boundary as it passes the probe would cause rapid flickering of the measured vorticity, making it hard to determine exactly when the turbulence boundary passes the probe. Nonetheless, one can obtain quite precise experimental results showing intervals of time in which there is turbulence at the probe, separated by intervals in which the flow is laminar there.

Such turbulence is said to be *intermittent* and the proportion of the time for which it is turbulent is known as the *intermittence* (often denoted Ω and not to be confused with the notation Ω_i, used in this book for vorticity, or Ω, which is often used to denote the rotation rate when studying rotating fluids). For jets, the intermittence is indistinguishable from 1 at the jet axis, indicating that the flow is always turbulent there, and drops off continuously to zero well outside the jet. It remains close to 1 in the central part of the jet and only falls off in the outer part of the jet where the turbulence boundary spends most of its time.

Average turbulent quantities, such as $\overline{q^2}$, include contributions from both the turbulent and laminar parts of the flow as the turbulence boundary crosses and recrosses a given point. The contribution from the laminar phases is *not* zero (which is perhaps a weakness of considering all fluctuations in velocity as "turbulence"), but appears to be considerably smaller than that of the turbulent phases. It has been proposed that a significant proportion of the decrease in $\overline{q^2}$ with distance from the axis can be explained by the decreasing proportion of the time for which the flow is turbulent. This suggests that the turbulence may be rather more homogeneous than appears from averaged quantities like $\overline{q^2}$.

5.4 Wakes

The wake of a bluff body in an infinite uniform stream at high Reynolds number originates by separation of thin boundary layers[17] that develop on the front part

[16] Although the vorticity is probably the best choice here, in practice it is hard to measure, since it involves spatial derivatives of all components of velocity (see, e.g., Sunyach and Mathieu (1969)). Thus, other quantities involving velocity derivatives are often used instead.

[17] The boundary layer itself can be either laminar or, if the Reynolds number is large enough (above about 10^5 for a circular cylinder or sphere) turbulent over at least part of the body surface prior to separation.

of the body. As the Reynolds number is increased, the steady laminar flow becomes unstable and the position of separation begins to fluctuate, leading to an oscillating wake. For cylindrical bodies, the (plane) wake takes the form of a vortex street if the Reynolds number is not too high. As the Reynolds number is increased, the vortex street becomes turbulent, and the transition zone approaches the body with growing Reynolds number until the entire wake is turbulent (at about $\mathrm{Re}_D \approx 2{,}500$ for a circular cylinder). In the case of an axisymmetric bluff body, such as a sphere, there appears to be no identifiable vortex street and separation leads to a region of slowly recirculating fluid behind the body if the Reynolds number is not too high, which begins to oscillate as the Reynolds number is increased, eventually giving way to a turbulent wake. However, as we shall see, the Reynolds number of the turbulent wake of a finite-sized body, such as a sphere, falls slowly with downstream distance and such a wake eventually becomes laminar very far downstream.

In the turbulent regime, although oscillations may still be discernible in the wake, the flow is steady in a statistical sense. Separation produces shear layers that divide the fast-moving fluid outside the wake from slowly moving fluid behind the body. As for jets, the shear layers entrain fluid and spread both outwards into the external flow and inwards towards the axis of the body. In so doing, the slower-moving fluid near the axis is accelerated by turbulent momentum transfer (i.e., via the Reynolds stress term in the mean-flow equations). Eventually, it is found that the wake becomes self-similar, but this takes a much longer distance than for a jet (of the order of 100 body widths, or more). The profile of streamwise mean velocity has a bell shape, as for jets, but now it approaches the free stream velocity, U_∞, outside the wake rather than zero as it does for jets exhausting into fluid at rest. Furthermore, the fluid in a wake obviously moves more slowly than that outside the wake, whereas for a jet the opposite is true. Needless to say, when we later consider the self-similar turbulent wake of a finite body, we assume a sufficiently large streamwise distance that the wake is self-similar, but not so far that the Reynolds number of the wake has become too low to maintain turbulence.

We begin by considering a plane wake, that is, one produced by a body that is infinite in the x_3-direction and whose cross-section does not vary with x_3. Thus, the flow is supposed to have statistical properties that do not vary with respect to x_3 and $\overline{U_3} = 0$. In the far wake, we adopt $V = \overline{U}_1 - U_\infty$ to represent the streamwise velocity variations across the wake (note that V is negative). Equation (5.40) applies sufficiently far downstream that $|V| \ll U_\infty$, that is,

$$U_\infty \frac{\partial V}{\partial x_1} = -\frac{\partial \overline{u_1 u_2}}{\partial x_2} \qquad (5.102)$$

whose integral with respect to x_2 gives

$$\frac{dQ_w}{dx_1} = 0 \qquad (5.103)$$

where

$$Q_w = -\int_{-\infty}^{\infty} V dx_2 = \int_{-\infty}^{\infty} (U_\infty - \overline{U}_1) dx_2 \tag{5.104}$$

is the (constant) volume flux deficit of the wake. Writing (5.13) in terms of V we have

$$\frac{\partial V}{\partial x_1} + \frac{\partial \overline{U}_2}{\partial x_2} = 0 \tag{5.105}$$

whose integral yields

$$\overline{U}_2(x_2 = +\infty) - \overline{U}_2(x_2 = -\infty) = \frac{dQ_w}{dx_1} = 0 \tag{5.106}$$

according to (5.103). Thus, to leading order, there is no entrainment in the far wake and one can take $\overline{U}_2 \to 0$ as $x_2 \to \pm\infty$ if there is no external cross-flow velocity (which would be represented by the common value of $\overline{U}_2(x_2 = \pm\infty)$).

It should be clear that constancy of Q_w in the far wake is a consequence of momentum, not mass, conservation: after all, it was derived from (5.102), which is the simplified form of the momentum equation in the x_1-direction. A momentum audit (using the x_1-component of momentum) on a rectangular control region whose boundaries are lines of constant x_1 and x_2 far from the body, and which includes the body, yields the result that the average drag force on the body (per unit length) is

$$F_D = \rho U_\infty Q_w \tag{5.107}$$

This value is the difference between the momentum deficit, $2\rho U_\infty Q_w$, of the far wake, and an efflux of momentum, $\rho U_\infty Q_w$, distributed around the remainder of the boundary *external* to the wake. The former corresponds to the integral of the momentum flux deficit, $\rho(U_\infty^2 - \overline{U}_1^2) \approx 2\rho U_\infty(U_\infty - \overline{U}_1)$, across the far wake. The latter is the result of a mass efflux, ρQ_w, from the vicinity of the body, which is needed to balance the mass flux deficit of the wake. That is, far from the body and external to the wake, the body appears like a line source of volume flux Q_w. The velocity due to this line source is much smaller than the velocity deficit of the wake, since their volume fluxes are equal, whereas line-source effects are spread over all directions. Thus, the line source is negligible when we restrict attention to the wake region, but nonetheless enters into the overall mass and momentum audits. The reader is encouraged to carry out these mass and momentum audits in detail. Formally, the quickest way of obtaining the total momentum flux is to use the (unaveraged) mass audit

$$\oint U_i n_i ds = 0 \tag{5.108}$$

where n_i is a unit vector normal to the boundary of the control region. Thus, the total instantaneous outflow of the x_1-component of momentum can be written

$$\rho \oint U_1 U_i n_i ds = \rho \oint (U_1 - U_\infty) U_i n_i ds \tag{5.109}$$

which can be evaluated by restricting attention to the wake region and the approximation $U_i n_i \approx U_\infty$. This leads to (5.107) for the mean drag force, via the momentum audit, after averaging and neglecting small terms. The formulation, (5.109), by which a multiple of (5.108) is subtracted from the momentum flux, takes account of the body line source, placing all significant contributions to the overall momentum flux in the wake.

In the frame of reference of the body, which we are using here, and which makes the flow steady, the far-wake deficit and line-source velocity fields appear as small corrections to a uniform flow, U_∞. The uniform flow disappears if we instead use a frame of reference in which the fluid at infinity is at rest and the body moves. The flow is unsteady in this second frame, but what were perturbations to uniform flow now show up as the flow itself. As the body moves, it lays down a wake in which the fluid is in motion in the direction of body movement. The fluid moves outwards from the body in all other directions owing to the body line source (see Figure 5.9). The reader may find the description of a laminar wake (which is similar in many respects) given in Batchelor (G 1967, section 5.12) helpful. Observe that if the body had a lift force (i.e., a component of force perpendicular to the line of motion), this implies circulation about the body.[18] In that case, outside the wake the body appears as a combination of a line source *and* a line vortex; the velocity field of the latter causes the wake to be deflected sideways, owing to cross-flow. We exclude lift forces and cross-flow in the remainder of this section and revert to using a frame of reference in which the body is fixed and the fluid moving (forming a steady flow).

Turning now to the self-similar behavior of the wake sufficiently far from the body, we adopt the similarity variable $\xi = x_2/W(x_1)$, so the wake velocity excess (which is negative) is given by

$$V = W^{-1} f(\xi) \qquad (5.110)$$

where the factor, W^{-1}, is required to make Q_w independent of x_1. Integrating the continuity equation, (5.105), we find

$$\overline{U}_2 = W^{-1} \frac{dW}{dx_1} \xi f(\xi) \qquad (5.111)$$

assuming no cross-flow (i.e., $\overline{U}_2 \to 0$ as $x_2 \to \pm\infty$). The simplified momentum equation, (5.102), can likewise be used to show that

$$\overline{u_1 u_2} = U_\infty W^{-1} \frac{dW}{dx_1} \xi f(\xi) \qquad (5.112)$$

gives the Reynolds shear stress.

Suppose that the turbulent velocities decrease with x_1 in the same way as the velocity defect

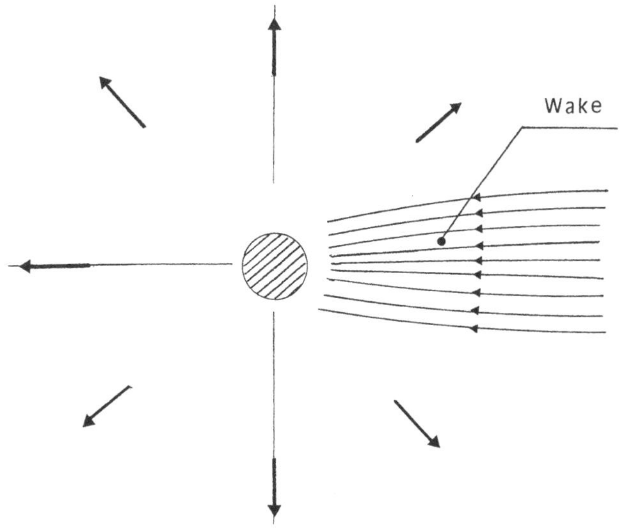

Figure 5.9. Instantaneous streamlines far from a body moving through stationary fluid (unsteady flow). In the text, we mainly consider the frame of reference in which the body is fixed and the fluid moving (steady flow, Figure 5.1b).

[18] A long, but finite-length lifting body also generates trailing vortices which are very persistent and obviously outside the scope of the present discussion.

(an assumption that we found to hold for self-similar jets, with velocity defect replaced by total velocity, since the fluid external to the jets considered was at rest). Because $\overline{u_1 u_2}$ is a quadratic moment of turbulence, it should be proportional to W^{-2}, and so, from (5.112) we obtain $W^{-2} \propto W^{-1} dW/dx_1$. Integrating with respect to x_1 gives $W \propto (x_1 - x_0)^{1/2}$ so that we take

$$W(x_1) = (x_1 - x_0)^{1/2} \tag{5.113}$$

as a measure of the wake width. The resulting similarity variable, $\xi = x_2/(x_1 - x_0)^{1/2}$, is in agreement with experiment, which is the real reason for believing in proportionality of turbulent velocities and velocity defect.

As for jets, we now suppose an eddy viscosity, so that

$$\overline{u_1 u_2} = -\nu_T \frac{\partial V}{\partial x_2} \tag{5.114}$$

to leading order and hence

$$\nu_T = -\frac{1}{2} U_\infty \xi f(\xi) \left(\frac{df}{d\xi}\right)^{-1} \tag{5.115}$$

from (5.110) and (5.112). Finally, assuming ν_T is independent of ξ and therefore constant, we obtain

$$f(\xi) = -Ae^{-U_\infty \xi^2/4\nu_T} \tag{5.116}$$

where A is a constant that is related to the drag force on the body per unit length by $F_D = 2A\rho(\pi \nu_T U_\infty)^{1/2}$, according to (5.104), (5.107), (5.110), and (5.116). This allows the constant A to be determined, given F_D and ν_T.

The self-similar plane wake flow is characterized by

$$V = -\frac{A}{(x_1 - x_0)^{1/2}} e^{-U_\infty \xi^2/4\nu_T} \tag{5.117}$$

$$\overline{U}_2 = -\frac{A}{2(x_1 - x_0)} \xi e^{-U_\infty \xi^2/4\nu_T} \tag{5.118}$$

and

$$\overline{u_1 u_2} = -\frac{AU_\infty}{2(x_1 - x_0)} \xi e^{-U_\infty \xi^2/4\nu_T} \tag{5.119}$$

Rather than spreading as a wedge-shaped region, like the plane jet, the self-similar plane wake has a parabolic shape, $x_2 \propto (x_1 - x_0)^{1/2}$. This reflects the combined effects of convection at the constant speed, U_∞, and wake spreading at a steadily decreasing rate, due to reducing turbulence levels, $u' \propto (x_1 - x_0)^{-1/2}$. The Reynolds number, based on a velocity scale proportional to $(x_1 - x_0)^{-1/2}$ and a length scale proportional to $(x_1 - x_0)^{1/2}$ is constant (and large), as for the axisymmetric jet (the plane wake therefore remains turbulent indefinitely, like jets). Furthermore, the turbulent dissipative scales may participate in the self-similarity of the mean flow and large scales of turbulence.

The Gaussian wake profile, (5.117), resulting from assuming a constant eddy viscosity is in good agreement with measurements behind a circular cylinder.

Fitting of the Gaussian form to the measurements allows determination of ν_T. Constancy of ν_T means that the resulting mean flow is equivalent to a laminar wake flow in which the fluid viscosity is ν_T, rather than the much smaller true viscosity of the fluid in the turbulent flow. The Reynolds number based on ν_T, the maximum velocity deficit, and the distance of the half-velocity point from the center plane of the wake is

$$\mathrm{Re}_T = \left(\frac{\log 2}{\pi}\right)^{1/2} \frac{F_D}{\rho \nu_T U_\infty} \tag{5.120}$$

for the Gaussian distribution, (5.117). In the case of a circular cylinder, the measured values of F_D and ν_T give $\mathrm{Re}_T \approx 15$, which is of the same order of magnitude as the values obtained for jets, although noticeably smaller.

For an axisymmetric far wake, we use (5.65) and

$$U_\infty \frac{\partial V}{\partial x} = -\frac{1}{r} \frac{\partial r \overline{u_r u_x}}{\partial r} \tag{5.121}$$

where $V = \overline{U}_x - U_\infty$, as before. Integration of (5.121) yields constancy of the volume flux deficit

$$Q_w = -\int_0^\infty 2\pi r V \, dr \tag{5.122}$$

expressing the momentum balance with a drag force, $F_D = \rho U_\infty Q_w$. Using (5.65) we conclude that there is no leading-order entrainment in the far wake, as for the case of a plane wake. A near-body source of volume flux Q_w must again be included if one does a momentum audit to calculate the drag force; however, this time it is a *point*, rather than a line source (see Batchelor (G 1967), section 5.12 for details of the corresponding laminar case). Figure 5.9 again illustrates the flow field in the frame of reference of the fluid far from the moving body.

In the self-similar regime, sufficiently far from the body, we suppose similarity variable $\xi = r/W(x)$ and obtain

$$V = W^{-2} f(\xi) \tag{5.123}$$

to maintain Q_w constant. This gives

$$\overline{u_r u_x} = U_\infty W^{-2} \frac{dW}{dx} \xi f(\xi) \tag{5.124}$$

from (5.121). Proportionality of V and the turbulence velocities leads to

$$W(x) = (x - x_0)^{1/3} \tag{5.125}$$

as the wake width dependency on streamwise distance, x.

The eddy viscosity is

$$\nu_T = \frac{\chi}{(x - x_0)^{1/3}} \tag{5.126}$$

where

$$\chi = -\frac{1}{3} U_\infty \xi f(\xi) \left(\frac{df}{d\xi}\right)^{-1} \tag{5.127}$$

Assuming constant χ results in

$$f(\xi) = -Ae^{-U_\infty \xi^2/6\chi} \tag{5.128}$$

and the axisymmetric wake has a Gaussian profile as for the plane wake, according to this model and in agreement with experiment except near the outskirts of the wake. The constant A can be related to the drag force, $F_D = 6\pi\chi\rho A$.

The Reynolds number based on a velocity scale proportional to $(x - x_0)^{-2/3}$, and a length scale, proportional to $(x - x_0)^{1/3}$, *decreases* proportional to $(x - x_0)^{-1/3}$. Thus, as stated earlier, the wake should eventually become laminar, although this may take a very long distance, owing to the rather slow decay of $(x - x_0)^{-1/3}$ and the initially high value of the Reynolds number. Since the Reynolds number is not constant, the smallest scales of turbulence cannot share similarity with the large scales and mean flow. The slowly decreasing Reynolds number should cause η/L to increase slowly with streamwise distance.

The "Reynolds number" based on ν_T is again constant. Using the maximum velocity deficit and half-velocity distance from the axis (calculated from (5.128)), we find

$$\text{Re}_T = \left(\frac{\log 2}{6\chi^3 U_\infty}\right)^{1/2} \frac{F_D}{\pi\rho} \tag{5.129}$$

which is $\text{Re}_T \approx 15$ according to measurements, and thus near the value for the plane wake.

The quantities ρ and U_∞ are fixed by the fluid and ambient flow respectively. The drag force, F_D, can be regarded as determined from empirical values of the drag coefficient of the body (drag coefficients have been tabulated for many body shapes and Reynolds numbers). If the effective "Reynolds number," Re_T, is known, then the far-wake similarity solution is fully determined. Thus, for a plane wake, one can calculate ν_T from (5.120) and A from $A = F_D/2\rho(\pi\nu_T U_\infty)^{1/2}$, while, for an axisymmetric wake, χ is given by (5.129) and A by $A = F_D/6\pi\chi\rho$. As we have seen, the value of Re_T seems to be about 15 for wakes.

It should perhaps be reiterated that the assumptions made in the above calculations, namely self-similarity, that ν_T is a function of x_1 only, and that the turbulent velocities have the same dependence on x_1 as the velocity defect, are empirically based and *only* justified through agreement with experimental results.

The turbulence energetics of self-similar wakes are broadly similar to those of jets, described earlier, and, as before, can be analyzed via measurements of the various terms in the energy equation. Production has a maximum near the inflection point of the mean velocity profile, while dissipation is largest in the wake center. Diffusion acts to transfer energy from the central part of the wake, where the turbulent energy is higher, to its outskirts. The net effect is that the turbulent intensity of the wake decreases with streamwise distance from the body, while showing spreading consistent with self-similarity.

The intermittency of self-similar wakes is also similar to that of jets, increasing away from the wake center, although intermittency makes itself felt nearer the center of a wake than a jet, reflecting the larger excursions of the turbulence boundary about its average position compared to jets. Since these excursions are caused by the convection of bulges of turbulence whose size is determined by the large turbulent eddies, this implies that the large scales of turbulence are bigger in wakes than in jets.

Turbulent jets exhausting into infinite fluid moving at nonzero uniform velocity U_∞ in the same direction as the jet (rather than the quiescent fluid we considered in Section 5.3) are rather similar to wakes, at least far enough downstream. In particular, the self-similar theory of jets in comoving fluid is the same as for wakes, being based on (5.102). However, rather than the expression, (5.107), for Q_w in terms of the drag force, one uses constancy of the jet momentum flux excess

$$Q_m = \int_{-\infty}^{\infty} \overline{U_1}(\overline{U_1} - U_\infty)dx_2 \tag{5.130}$$

which follows from (5.13) and (5.28) for plane jets (with an expression similar to (5.130) in the axisymmetric case). The constant, Q_m, is fixed by its value at the nozzle, while in the far jet we have $Q_w = -Q_m/U_\infty$, allowing determination of the constant Q_w occurring in the self-similar wake-like theory. Clearly, the behavior of the far jet is quite different when there is external flow. For instance, the plane far jet without such flow spreads proportional to $x_1 - x_0$, whereas in comoving uniform flow its width is proportional to $(x_1 - x_0)^{1/2}$, like a plane wake.

It might be asked how the differing similarity behavior of jets exhausting into stationary and comoving fluid can be reconciled as the external velocity, U_∞, is reduced to zero. Clearly, if U_∞ is much smaller than the nozzle exit velocity, the jet will not notice the external flow during its initial development, which will thus take place as if the fluid were stationary. However, as the velocity excess of the jet falls with downstream distance, there will come a time when it becomes comparable with U_∞ and the jet begins to be affected by the external stream. Still further away, $V \ll U_\infty$, as assumed when using the far-wake simplification, (5.102), which was the basis of the wake analysis given above. Thus, there will be a changeover from the one type of behavior to the other when V is of order U_∞, which occurs further and further downstream as $U_\infty \to 0$. If this changeover takes place after establishment of self-similarity, the flow evolves from one type of similarity solution to another, through intermediate behavior which is not self-similar.

The wakes considered above have a nonzero value of F_D and hence of Q_w, but there is an interesting class of bodies for which the *net* force applied to the fluid is zero and so $Q_w = 0$. These are self-propelled bodies (e.g., submarines) moving at constant velocity. Such bodies push themselves through the fluid using a propulsion system that works by applying a force to the fluid to compensate the drag force over the rest of the body. The net force is zero because the body is not accelerating.[19] With zero Q_w, the wake produced is known as momentumless and is quite different in character from the wakes with momentum considered above. Momentumless wakes are also found to become self-similar far from the body, and, although their similarity profiles are *not* the same as wakes with momentum, the principles used to analyze self-similar momentumless wakes are the same. We refer the reader to Tennekes and Lumley (G 1972), section 4.3, for a detailed example. In the same book, the reader will also find (section 4.6) an interesting

[19] Here we neglect gravity, which is often significant for self-propelled bodies. Unless the body is neutrally buoyant, gravity leads to a net static force on the body, which must be counterbalanced by hydrodynamic forces (e.g., the lift on an airplane) if the body is not to accelerate. In that case, the wake is more complicated, generally containing net momentum and organized vortices.

analysis of a thermal plume problem in which a hot spot causes fluid to expand and rise due to buoyancy, creating an upward-going current above the spot, whose self-similar characteristics are considered.

5.5 Turbulent Boundary Layers

As discussed in Section 5.1, a turbulent boundary layer normally comes about when a laminar one becomes unstable and then undergoes transition to turbulence. Broadly speaking, instability arises when a local Reynolds number, based on the laminar boundary-layer thickness rather than on the overall body dimensions, becomes sufficiently large. Thus, increasing either the velocity or layer thickness can lead to transition (as can raising the level of perturbations, explained later in this section). Since the layer thickness is small compared with the size of the body on which the layer develops, the local Reynolds number is considerably smaller than the overall one for the body (we saw earlier that the laminar boundary-layer Reynolds number scales on the square root of the body-scale one). This is why the boundary layer can remain laminar when other parts of the flow have long since undergone transition and become turbulent. However, if the body-scale Reynolds number is increased, the local boundary-layer Reynolds number also rises, and the layer eventually becomes turbulent, at least over part of the body surface. Typically, the boundary-layer thickness, and hence the local Reynolds number, grows with streamwise distance along the surface, thus making the boundary layer increasingly susceptible to transition.

For cylinders and spheres, the boundary layer first becomes turbulent when the body-scale Reynolds number is about 10^5–10^6 (whose square root gives an estimate of the boundary-layer Reynolds number for instability in the hundreds, in rough agreement with stability calculations). Turbulence appears near the separation point, causing separation to move downstream from its laminar location near the lateral extremities of the body. This is because a turbulent boundary layer is more resistant to separation than a laminar one (recall Figure 1.8). A turbulent boundary layer is then established between the points of transition and separation. Transition itself moves upstream at still larger Reynolds numbers, rendering an increasing portion of the boundary layer turbulent. An interesting phenomenon is observed in the case of a cylinder at Reynolds number between about 10^5 and 10^6. The boundary layer separates while still laminar and the resulting shear layer becomes turbulent, thickening rapidly by turbulent mixing and entrainment. The shear layer spreads so quickly that it reattaches to the surface to form a new boundary layer, now turbulent, which separates considerably further downstream to form the wake. The process by which a boundary layer first separates and then reattaches to the surface forms a small pocket of recirculating fluid and also occurs on the suction surface of airfoils at high angles of attack, towards the nose. In the case of the cylinder, the recirculating pocket disappears as the overall Reynolds number increases to about 10^7, when the boundary layer undergoes transition to turbulence before laminar separation occurs. The boundary layer is then laminar from the upstream stagnation point until transition, and turbulent from there until it separates to form the wake. This is typical of bluff bodies at sufficiently high Reynolds number, but streamlined ones, such as an aircraft wing, may avoid separation if the angle of attack is small enough. In that case, the boundary layer is shed from the trailing edge, creating a narrower and less intense wake than when separation happens.

The place at which transition to turbulence occurs is not a precise point in the boundary layer, but rather a zone in which turbulence is sporadic, taking the form of bursts which become more and more frequent as one moves downstream into the fully turbulent zone. Because it is the result of a nonlinear instability, the location of the transition zone can differ in nominally identical flows, depending upon the amplitude and type of perturbations to the laminar flow. For a smooth, flat plate in a uniform stream at zero incidence, transition typically covers about 30% of the distance to the leading edge of the plate. Transition can be made less variable by provoking it deliberately using a spanwise inhomogeneity, such as a wire (see Figure 1.8b). Tripping the layer in this way produces transition earlier than it would otherwise occur, a reflection of the sensitivity of transition to perturbations. Experimentally, one can distinguish between laminar and turbulent flow because turbulence appears as periods of rapid jittering of flow quantities, such as velocity, separating intervals of relatively smooth variation with time, corresponding to laminar flow.

The laminar boundary-layer flow becomes more and more sensitive to perturbations (e.g., disturbances from upstream, body surface vibrations, and inhomogeneities) as its local Reynolds number increases with downstream distance. If care is taken to reduce the perturbations of a flat-plate boundary layer to a minimum, a succession of laminar-flow instabilities can be observed at increasing streamwise distances towards transition (appearance of Tollmien–Schlichting waves and their secondary instabilities). This gives rise to a sequence of unsteady laminar flows of growing complexity, followed by the transition zone in which sporadic turbulent bursts (or spots) occur more and more frequently with increasing streamwise distance. Such a burst originates at a random location and time in the layer and, once triggered, spreads out as it is convected downstream, finally merging with its fellow bursts to form the fully developed turbulent boundary layer. At higher levels of perturbation of the upstream flow, the path taken to transition appears to be somewhat different, with laminar instability occurring upstream of the location predicted by classical stability theory and not involving Tollmien–Schlichting waves. However, the transition zone itself, with its turbulent spots, is found to be similar, although it is located upstream of where it should be according to classical theory. Stability theory has not yet advanced to the point where it can fully describe the transition process. In fact, rather little can currently be said about the transitional zone of a typical boundary layer from a theoretical point of view, and so we will concentrate on fully developed turbulent boundary layers from here on.

The structure of a turbulent boundary layer is quite different from that of a laminar one. For a laminar layer, transverse convection and viscosity provide transfer of momentum across the layer, while viscosity also allows satisfaction of the no-slip condition at the surface. Viscous skin friction at the surface causes the fluid in the layer to move more slowly than fluid in the essentially inviscid flow outside. In a turbulent boundary layer, it is still viscosity which imposes the no-slip condition in the viscous sublayer, but turbulent momentum transfer takes over from viscosity as the dominant mechanism for slowing the fluid further from the surface. Thus, there are two regions making up a turbulent boundary layer: a thin inner layer at the surface (the viscous sublayer), in which both viscous and turbulent momentum transfer across the flow are important, and a much thicker outer layer in which turbulent momentum transfer (via the Reynolds stress term in the mean-flow equations) dominates the mean flow. This is not to say that viscosity plays no role in the outer layer: as usual, it is responsible for turbulent energy dissipation via the smallest scales of turbulence. It will be recalled from Chapter 3 that the size of the smallest scales is such as to bring viscosity into play to dissipate the energy cascade from the larger eddies. As the surface is approached, however, the size of the largest scales decreases (being limited by the proximity of the surface) and, in the sublayer, becomes comparable to that of the smallest scales. When the viscous sublayer is entered, the turbulent Reynolds number, Re_L, is no longer large. This relatively small Reynolds number corresponds to a range of scales, L/η, of $O(1)$. All scales of turbulence are then comparable in size and all are affected by viscosity. It is thus clear that turbulence within the viscous sublayer at a wall is quite different in character from turbulence in the outer part of a boundary layer or away from a wall (e.g., in jets and wakes).

Because turbulent momentum transfer in the outer layer is much more efficient than purely viscous diffusion, turbulent boundary layers are considerably thicker and have higher skin friction than their laminar counterparts, the difference in thickness being a reflection of the existence of the outer layer. Across the transition zone, the overall boundary-layer thickness therefore increases and the boundary-layer structure reorganizes itself into a viscous sublayer and an outer layer. The viscous sublayer is a lot thinner than the original laminar boundary layer, whereas the outer layer is considerably thicker, occupying most of the total boundary-layer thickness, δ. The profile of streamwise mean velocity across a typical turbulent boundary layer is as sketched in Figure 5.1d. Mean shear increases rapidly as the surface is approached from the outer layer, but the viscous sublayer is too thin to be visible in this figure.

As for our earlier discussion of laminar boundary layers, we restrict ourselves to two-dimensional (plane) flow. In general, three-dimensional boundary layers are much more complicated to analyze. To allow for curved surfaces, we use coordinates x and y, with x taken along the surface in the streamwise direction and y normal to the surface, as shown in Figure 5.3.

One of the most important quantities characterizing a turbulent boundary layer is the turbulent friction velocity, u_*. This is defined in terms of the mean frictional stress at the surface

$$\tau_w = \nu\rho \left.\frac{\partial \overline{U_x}}{\partial y}\right|_{y=0} \tag{5.131}$$

which is the average force per unit area acting in the streamwise direction, x, on the body surface. Equation (5.131) follows from taking the mean of the usual expression for the stress in a Newtonian fluid, allowing for the no-slip condition, $U_x = U_y = 0$, at the surface, $y = 0$. The friction velocity is given by

$$u_* = \sqrt{\frac{|\tau_w|}{\rho}} \tag{5.132}$$

and is found to determine many of the properties of the boundary layer. For instance, the turbulent velocities are $O(u_*)$ throughout the layer, both in the viscous sublayer, whose thickness is $O(\nu/u_*)$, and the outer layer, $y = O(\delta)$. The mean velocity in the viscous sublayer is $O(u_*)$, as is the velocity *defect* (relative to the external flow) in the outer layer. However, although the velocity defect is $O(u_*)$ in the outer layer, the mean velocity itself is larger there (by a factor of around twenty or more) and determined by the external flow speed. That is, the external flow speed is considerably greater than u_*, which characterizes the order of magnitude of the *variations* in $\overline{U_x}$ across the outer layer, $y = O(\delta)$. It follows that $\overline{U_x}$ is approximately uniform over the outer layer, away from the surface, although its small deviations, $O(u_*)$, from uniformity are crucial, as we shall see. Note that both the friction velocity, u_*, and the total boundary-layer thickness, δ, change with streamwise distance as the boundary layer develops.

Across the entire boundary layer, $\overline{U_x}$ must change from its value in the external flow to become zero at the surface, in order to satisfy the no-slip condition there. Since the variations of $\overline{U_x}$ in the outer layer, away from the surface, are comparatively small (of $O(u_*)$), most of the overall change in $\overline{U_x}$ occurs near the surface. As the surface is approached from the outer layer, the velocity defect increases, taking on a logarithmic form as a function of y in the inertial layer, $\nu/u_* \ll y \ll \delta$. The inertial layer will be regarded as a zone of overlap between the two extreme asymptotic regions: the viscous sublayer, $y = O(\nu/u_*)$, and the outer layer, $y = O(\delta)$. Moving towards the surface, by the time one enters the viscous sublayer from the inertial layer, $\overline{U_x}$ has undergone a significant fraction of the overall change from its value outside the layer.

The ratio of thicknesses of the viscous sublayer and outer layer is very small, $O(\mathrm{Re}_*^{-1})$, where $\mathrm{Re}_* = u_*\delta/\nu$ is a large boundary-layer Reynolds number. It is the large value of Re_* that leads to clear asymptotic separation between the thin viscous sublayer and outer layer, forming the basis of the theory described in this section. As we shall see later, it is the large value of Re_* that is also responsible for the smallness of u_* compared with flow speeds outside the layer, which in turn leads to length scales for streamwise development long compared with the boundary-layer thickness.

We want to develop approximate turbulent-boundary-layer equations for the mean flow, exploiting the relatively slow streamwise variations of flow statistics to simplify the Navier–Stokes equations by eliminating small terms. However, one needs to be more careful in deriving the approximate equations for a turbulent boundary layer than with the Prandtl equations of the laminar case or the "boundary-layer" approximation for turbulent jets and wakes. The reason for this is that, as noted above, the outer-layer velocity defect of a turbulent boundary layer is a small, but highly significant fraction of the total velocity. To maintain precision in the outer layer, we shall develop a boundary-layer approximation using the velocity defect, rather than the mean velocity itself.

To define the velocity defect precisely, suppose that the flow exterior to the boundary layer is essentially steady, laminar, and inviscid. The absence of significant turbulence in the exterior flow is an important restriction: turbulent boundary layers are observed to behave rather differently if there is a high enough level of turbulence already present in the flow outside the layer. The velocity field outside the boundary layer is imagined to be extrapolated smoothly to the wall, yielding a notional steady, inviscid, incompressible flow, $\mathbf{U}^{(inv)}$, which *does not have a boundary layer.* This notional flow is analogous to that in the outer region[20] of a laminar boundary layer (recall the discussion in Section 5.1) and can be obtained by an inviscid calculation, as for the laminar case. The notional flow field, $\mathbf{U}^{(inv)}$, varies on length scales determined by the body dimensions, even within the turbulent boundary layer, where it is defined by extrapolation. Given $\mathbf{U}^{(inv)}$, the velocity defect field is given by $\mathbf{V} = \overline{\mathbf{U}} - \mathbf{U}^{(inv)}$ and is nonzero in the boundary layer. This definition of $\mathbf{V}$ means that the streamwise velocity defect is really $-V_x$. It might therefore be better to call $\mathbf{V}$ the boundary-layer velocity excess, rather than the defect, but we will nonetheless refer to $\mathbf{V}$ as the velocity defect.

The streamwise component of $\mathbf{U}^{(inv)}$ at the body surface is denoted by

$$U_\infty(x) = U_x^{(inv)}(y=0) \tag{5.133}$$

which, along with u_* and δ, are the main quantities characterizing the boundary layer. Since the boundary layer is thin compared with the body scale, $U_x^{(inv)}$ varies little through the layer, and we have $U_x^{(inv)} = U_\infty$ to a first approximation within the boundary layer. Furthermore, because $V_x = O(u_*)$ in the outer layer and u_*/U_∞ is small, the velocity defect in the outer layer is also small compared with U_∞. Thus, $\overline{U_x} = U_\infty$ to leading order, expressing the approximate uniformity of the streamwise velocity, noted above. Of course, this only applies in the outer layer, $y = O(\delta)$, and $\overline{U_x}$ decreases through the inertial layer and viscous sublayer, to become zero at the surface. Note that U_∞ is positive since x is a streamwise coordinate.

The notional velocity field, $\mathbf{U}^{(inv)}$, satisfies the steady, inviscid equations of an incompressible fluid, whereas $\overline{\mathbf{U}}$ obeys the turbulent mean-flow equations, (5.11) and (5.12). By subtraction, one obtains the velocity defect equations

$$\overline{U}_j \frac{\partial V_i}{\partial x_j} + V_j \frac{\partial U_i^{(inv)}}{\partial x_j} = -\frac{\partial \Gamma}{\partial x_i} - \frac{\partial \overline{u_i u_j}}{\partial x_j} + \nu \frac{\partial^2 \overline{U}_i}{\partial x_j \partial x_j} \tag{5.134}$$

where $\Gamma = (\overline{P} - P^{(inv)})/\rho$ represents the pressure defect, and

$$\frac{\partial V_i}{\partial x_i} = 0 \tag{5.135}$$

Since these equations describe the velocity defect, they can be approximated without fear of loss of precision, even within the outer layer where $V_x = O(u_*)$ is small compared with $\overline{U_x} \approx U_\infty$.

When equation (5.134) is expressed in terms of the curvilinear boundary-layer coordinates,[21] x and y, one obtains

[20] Not to be confused with the outer layer of a turbulent boundary layer, which forms part of the boundary layer itself. A laminar boundary layer has no internal structure and the outer region refers to the flow external to the layer, i.e., at body-scale distances from the surface.

[21] Recall that y is distance from the body surface and x is the streamwise distance along the surface of its nearest point.

$$\frac{R\overline{U_x}}{R+y}\frac{\partial V_x}{\partial x}+\overline{U_y}\frac{\partial V_x}{\partial y}+\left(\frac{R}{R+y}\frac{\partial U_x^{(inv)}}{\partial x}+\frac{U_y^{(inv)}}{R+y}\right)V_x+\left(\frac{\partial U_x^{(inv)}}{\partial y}+\frac{\overline{U_x}}{R+y}\right)V_y=$$
$$-\frac{R}{R+y}\left\{\frac{\partial\Gamma}{\partial x}+\frac{\partial\overline{u_x^2}}{\partial x}+\frac{1}{R}\frac{\partial}{\partial y}((R+y)\overline{u_xu_y})+\frac{\overline{u_xu_y}}{R}\right\}+\nu(\nabla^2\overline{\mathbf{U}})_x \tag{5.136}$$

and

$$\frac{R\overline{U_x}}{R+y}\frac{\partial V_y}{\partial x}+\overline{U_y}\frac{\partial V_y}{\partial y}+\left(\frac{R}{R+y}\frac{\partial U_y^{(inv)}}{\partial x}-\frac{\overline{U_x}+U_x^{(inv)}}{R+y}\right)V_x+\frac{\partial U_y^{(inv)}}{\partial y}V_y=$$
$$-\frac{\partial\Gamma}{\partial y}-\frac{R}{R+y}\left\{\frac{\partial\overline{u_xu_y}}{\partial x}+\frac{1}{R}\frac{\partial}{\partial y}\left((R+y)\overline{u_y^2}\right)-\frac{\overline{u_x^2}}{R}\right\}+\nu(\nabla^2\overline{\mathbf{U}})_y \tag{5.137}$$

where $R(x)$ is the surface radius of curvature, defined as positive for a convex surface. Here, we have spared the reader the messy explicit expressions for the viscous terms, since they are only needed in the very thin viscous sublayer where surface curvature effects are insignificant, even when they are important in the outer layer. Equation (5.135) takes the form

$$\frac{\partial V_x}{\partial x}+\frac{1}{R}\frac{\partial}{\partial y}\left((R+y)V_y\right)=0 \tag{5.138}$$

in terms of x and y.

We aim to derive approximations to (5.136)–(5.138) by dropping small terms, eventually resulting in the leading-order boundary-layer equations for the mean flow. To this end, we introduce a length scale, d, characterizing the body dimensions. Since the boundary layer develops over the body surface, its development generally takes place over length scales of $O(d)$. Thus, as for jets and wakes, d will also be a length scale for streamwise changes in the boundary-layer properties. Typically, the surface radius of curvature will also be of $O(d)$. On the other hand, the boundary layer, of total thickness δ, is assumed thin compared with these length scales, that is, δ/d is supposed small. Another small quantity is the ratio, u_*/U_∞, of the turbulent friction velocity to the external flow speed. As we will see later, δ/d and u_*/U_∞ are of the same order of magnitude, with smallness of both being a consequence of the large Reynolds number, Re_*.

As a first step towards deriving the boundary-layer equations, we note that, because $\delta \ll R$, y is small compared with R within the boundary layer. Thus, we replace $R+y$ by R throughout (5.136)–(5.138), giving

$$\overline{U_x}\frac{\partial V_x}{\partial x}+\overline{U_y}\frac{\partial V_x}{\partial y}+\left(\frac{\partial U_x^{(inv)}}{\partial x}+\frac{U_y^{(inv)}}{R}\right)V_x+\left(\frac{\partial U_x^{(inv)}}{\partial y}+\frac{\overline{U_x}}{R}\right)V_y=$$
$$-\left\{\frac{\partial\Gamma}{\partial x}+\frac{\partial\overline{u_x^2}}{\partial x}+\frac{\partial\overline{u_xu_y}}{\partial y}+\frac{\overline{u_xu_y}}{R}\right\}+\nu(\nabla^2\overline{\mathbf{U}})_x \tag{5.139}$$

$$\overline{U_x}\frac{\partial V_y}{\partial x}+\overline{U_y}\frac{\partial V_y}{\partial y}+\left(\frac{\partial U_y^{(inv)}}{\partial x}-\frac{\overline{U_x}+U_x^{(inv)}}{R}\right)V_x+\frac{\partial U_y^{(inv)}}{\partial y}V_y=$$

$$-\left\{\frac{\partial\Gamma}{\partial y}+\frac{\partial\overline{u_x u_y}}{\partial x}+\frac{\partial\overline{u_y^2}}{\partial y}-\frac{\overline{u_x^2}}{R}\right\}+\nu(\nabla^2\overline{\mathbf{U}})_y \tag{5.140}$$

and

$$\frac{\partial V_x}{\partial x}+\frac{\partial V_y}{\partial y}=0 \tag{5.141}$$

Integrating (5.141) across the boundary layer and using the fact that $\mathbf{V}\to 0$ outside the layer, we obtain the leading-order approximation

$$V_y=\frac{d}{dx}\int_y^\infty V_x dy \tag{5.142}$$

where, as usual in boundary-layer theory, the limit $y\to\infty$ implies leaving the boundary layer and entering the external flow. Because the outer part of the boundary layer, $y=O(\delta)$, occupies most of its total thickness, the main contributions to the integral in (5.142) arise from the outer layer, even when y is near the surface. In the outer layer, $V_x=O(u_*)$ so that (5.142) yields the estimate

$$V_y=O\left(\frac{u_*\delta}{d}\right) \tag{5.143}$$

valid throughout the boundary layer. Another consequence is that, evaluating the relationship $U_y^{(inv)}=\overline{U_y}-V_y$ at $y=0$ using (5.142) and the boundary condition $\overline{U_y}=0$, we obtain

$$U_y^{(inv)}\Big|_{y=0}=-\frac{d}{dx}\int_0^\infty V_x dy=O\left(\frac{u_*\delta}{d}\right) \tag{5.144}$$

giving the normal velocity at the body surface for the notional inviscid flow, $\mathbf{U}^{(inv)}$. Note that this normal velocity is a factor of $O((u_*/U_\infty)(\delta/d))$ less than the external flow speed, $O(U_\infty)$. Thus, since both u_*/U_∞ and δ/d are small, the flow field $\mathbf{U}^{(inv)}$ can be calculated at leading order with the usual inviscid boundary condition of zero normal velocity at the body surface. Equation (5.144) gives a higher-order correction.

The velocity field $\mathbf{U}^{(inv)}$ varies on length scales dictated by the body dimensions and, within the comparatively thin boundary layer, it can be expanded as a Taylor's series in y. To a first approximation

$$U_x^{(inv)}=U_x^{(inv)}\Big|_{y=0}=U_\infty(x) \tag{5.145}$$

while, owing to smallness of $U_y^{(inv)}|_{y=0}$ noted above, a second term may be needed in the expansion of $U_y^{(inv)}$. Thus, we write

$$U_y^{(inv)}=U_y^{(inv)}\Big|_{y=0}+y\frac{\partial U_y^{(inv)}}{\partial y}\Big|_{y=0} \tag{5.146}$$

Here, the derivative can be expressed using the incompressibility condition, $\nabla \cdot \mathbf{U}^{(inv)} = 0$, evaluated at the body surface, giving

$$\left.\frac{\partial U_y^{(inv)}}{\partial y}\right|_{y=0} = -\frac{dU_\infty}{dx} - \frac{U_y^{(inv)}|_{y=0}}{R} \tag{5.147}$$

of which the second term on the right is smaller than the first by a factor of $O((u_*/U_\infty)(\delta/R))$, according to (5.144). Since both u_*/U_∞ and δ/R are small, we can drop the curvature term in (5.147), so that (5.146) becomes

$$U_y^{(inv)} = \left.U_y^{(inv)}\right|_{y=0} - y\frac{dU_\infty}{dx} \tag{5.148}$$

valid throughout the boundary layer. In the outer layer, the last term in (5.148) is $O(U_\infty\delta/d)$, whereas the first term on the right is smaller by a factor of $O(u_*/U_\infty)$, according to (5.144). Thus, we have

$$U_y^{(inv)} = -y\frac{dU_\infty}{dx} \tag{5.149}$$

as a leading-order approximation in the outer layer. Since $\overline{U_y} = U_y^{(inv)} + V_y$ and V_y has order of magnitude (5.143), we can also show that

$$\overline{U_y} = -y\frac{dU_\infty}{dx} \tag{5.150}$$

to a first approximation in the outer layer. The first term on the right of (5.148) is needed nearer the surface.

Turning attention to equation (5.139), we can replace $\partial U_x^{(inv)}/\partial x$ in the third term by dU_∞/dx, according to (5.145), while the quantity $U_y^{(inv)}/R$ is at least $O(\delta/R)$ smaller, from (5.149), and is therefore dropped. That is, the third term of (5.139) is replaced by $(dU_\infty/dx)V_x$. In the fourth term, the derivative $\partial U_x^{(inv)}/\partial y = O(U_\infty/d)$ since the external flow velocity, $U_x^{(inv)}$, is of $O(U_\infty)$ and varies on a body length scale, d. Likewise, $\overline{U_x}/R = O(U_\infty/d)$ because the radius of curvature is $O(d)$ and $\overline{U_x} = O(U_\infty)$ (or less, near the surface). From (5.143), we find that the fourth term of (5.139) is $O(u_* U_\infty \delta/d^2)$. On the other hand, the third term is $O(U_\infty u_*/d)$, that is, larger than the fourth by a factor of $O(d/\delta)$. Thus, we drop the fourth term of (5.139) at leading order.

As regards the right-hand side of (5.139), we recall that the turbulent velocities are $O(u_*)$ throughout the boundary layer, so that the second-order moment, $\overline{u_i u_j} = O(u_*^2)$. Of the Reynolds stress terms, $\partial\overline{u_x u_y}/\partial y = O(u_*^2/\delta)$ within the outer layer and increases towards the surface, where gradients with respect to y become larger, whereas the other two are smaller by a factor of $O(\delta/d)$ and are neglected at leading order. The viscous term is only important in the very thin viscous sublayer where curvature is insignificant and the gradients with respect to y dominate, thus one can write $(\nabla^2\overline{\mathbf{U}})_x = \partial^2\overline{U_x}/\partial y^2$ to a very good approximation. As a final approximation of (5.139), we employ $\overline{U_x} = U_x^{(inv)} + V_x$ and (5.145) in the first term, leading to

$$(U_\infty + V_x)\frac{\partial V_x}{\partial x} + \overline{U_y}\frac{\partial V_x}{\partial y} + \frac{dU_\infty}{dx}V_x = -\frac{\partial \Gamma}{\partial x} - \frac{\partial\overline{u_x u_y}}{\partial y} + \nu\frac{\partial^2\overline{U_x}}{\partial y^2} \tag{5.151}$$

which is valid to leading order throughout the boundary layer and will become the turbulent-boundary-layer equation once the term $\partial\Gamma/\partial x$, the streamwise derivative of the pressure defect Γ, is shown to be negligible and dropped.

To show that this term is indeed negligible, we consider equation (5.140). The Reynolds stress terms other than $\partial\overline{u_y^2}/\partial y$ are neglected and the viscous term expressed using $(\nabla^2\overline{\mathbf{U}})_y = \partial^2\overline{U_y}/\partial y^2$. The result can be written as

$$\frac{\partial\Psi}{\partial y} = -\chi \tag{5.152}$$

where

$$\Psi = \Gamma + \overline{u_y^2} - \nu\frac{\partial\overline{U_y}}{\partial y} \tag{5.153}$$

and χ is the left-hand side of (5.140). Integration of (5.152) across the boundary layer gives

$$\Psi = \int_y^\infty \chi\, dy \tag{5.154}$$

where we have used the fact that Ψ is negligibly small outside the boundary layer because $\Gamma \to 0$ (recall that Γ is proportional to the pressure defect), $\overline{u_y^2} \to 0$ (turbulence is supposed confined to the boundary layer), and viscosity is insignificant outside the layer. The main contributions to the integral in (5.154) arise from the outer layer, even when y is close to the surface. Thus, to estimate Ψ, we determine the order of magnitude of χ, that is, the left-hand side of (5.140), in the outer layer.

In the third term of (5.140), one can use the leading-order outer-layer approximations $\overline{U_x} = U_x^{(inv)} = U_\infty$. Thus, the quantity $(\overline{U_x} + U_x^{(inv)})/R = 2U_\infty/R$ to a first approximation and is of $O(U_\infty/d)$, whereas $\partial U_y^{(inv)}/\partial x = O(\delta U_\infty/d^2)$, according to (5.149), and is smaller by a factor of $O(\delta/d)$. In consequence, we replace the third term of (5.140) by $-(2U_\infty/R)V_x$ in the outer layer, where it is $O(u_* U_\infty/d)$, since $V_x = O(u_*)$. On the other hand, all the other terms on the left of (5.140) can be estimated as $O(u_* U_\infty \delta/d^2)$ in the outer layer, using (5.143), (5.150), and $\partial U_y^{(inv)}/\partial y = O(U_\infty/d)$. Thus, we drop terms on the left of (5.140) other than the third, leading to

$$\Psi = -\frac{2U_\infty}{R}\int_y^\infty V_x dy \tag{5.155}$$

from (5.154), which may be estimated as

$$\Psi = O\left(\frac{u_* U_\infty \delta}{d}\right) \tag{5.156}$$

throughout the boundary layer.

The mean-flow incompressibility condition, $\nabla . \overline{\mathbf{U}} = 0$, can be written as

$$\frac{\partial\overline{U_x}}{\partial x} + \frac{1}{R}\frac{\partial}{\partial y}\left((R+y)\overline{U_y}\right) = 0 \tag{5.157}$$

and $R + y$ replaced by R, giving the leading-order approximation

$$\frac{\partial\overline{U_y}}{\partial y} = -\frac{\partial\overline{U_x}}{\partial x} \tag{5.158}$$

so that (5.153) yields

$$\Gamma = \Psi - \overline{u_y^2} - \nu \frac{\partial \overline{U_x}}{\partial x} \tag{5.159}$$

whose derivative with respect to x gives

$$\frac{\partial \Gamma}{\partial x} = \frac{\partial \Psi}{\partial x} - \frac{\partial \overline{u_y^2}}{\partial x} - \nu \frac{\partial^2 \overline{U_x}}{\partial x^2} \tag{5.160}$$

a result which is used in (5.151). The second and third terms on the right-hand side of (5.160) are respectively small compared with the Reynolds stress and viscous terms of (5.151) and are neglected. The quantity $\partial\Psi/\partial x = O(u_* U_\infty \delta/d^2)$, according to (5.156), and is small compared with the third term of (5.151), which is of $O(u_* U_\infty/d)$ or larger. Thus, one drops the term $\partial\Gamma/\partial x$ in (5.151), yielding the final form

$$\boxed{(U_\infty + V_x)\frac{\partial V_x}{\partial x} + \overline{U_y}\frac{\partial V_x}{\partial y} + \frac{dU_\infty}{dx} V_x = -\frac{\partial \overline{u_x u_y}}{\partial y} + \nu \frac{\partial^2 \overline{U_x}}{\partial y^2}} \tag{5.161}$$

which is the first of the leading-order mean-flow boundary-layer equations. The second is obtained from (5.141), replacing V_y by $\overline{U_y} - U_y^{(inv)}$ and using (5.148). The result is

$$\boxed{\frac{\partial V_x}{\partial x} + \frac{\partial \overline{U_y}}{\partial y} = -\frac{dU_\infty}{dx}} \tag{5.162}$$

Equations (5.161) and (5.162) are the final result of the analysis: the leading-order boundary-layer equations for the mean flow.

The viscous term in (5.161) only becomes important in the viscous sublayer, very near the surface, where the streamwise velocity gradients are much larger than in the outer part of the layer. If desired, one can replace $\partial^2\overline{U_x}/\partial y^2$ by $\partial^2 V_x/\partial y^2$ in this viscous term, because the difference between the two, namely $\partial^2 U_x^{(inv)}/\partial y^2$, retains its order of magnitude from the outer layer and therefore gives a uniformly small contribution to (5.161). If one carries out this replacement, (5.161) and (5.162) no longer contain $\overline{U_x}$, and govern the streamwise development of the boundary-layer velocity defect, V_x, and transverse velocity, $\overline{U_y}$. Equations (5.161) and (5.162), and hence the boundary-layer problem, depend on $U_\infty(x)$, which is imposed from outside the boundary layer, as in the laminar case. They also contain the Reynolds stress, $\overline{u_x u_y}$, which requires closure if one wants a complete system of equations. The boundary conditions for the problem (5.161), (5.162) consist of the no-slip conditions

$$\begin{aligned} V_x &= -U_\infty \\ \overline{U_y} &= 0 \end{aligned} \tag{5.163}$$

at $y = 0$ and the condition

$$V_x \to 0 \tag{5.164}$$

as $y \to \infty$, together with some upstream initial profile, as for the laminar case.

Aside from the closure problem, (5.161), with $\partial^2 \overline{U_x}/\partial y^2$ replaced by $\partial^2 V_x/\partial y^2$, together with (5.162)–(5.164) and the upstream conditions completely determine the turbulent boundary-layer mean flow at leading order and are analogous to (5.1), (5.5), (5.7), and (5.8) for laminar boundary layers. If we define

$$W_x = U_\infty(x) + V_x \tag{5.165}$$

equations (5.161) and (5.162), after replacement of the viscous term by $\nu(\partial^2 V_x/\partial y^2)$, may be rewritten as

$$W_x \frac{\partial W_x}{\partial x} + \overline{U_y} \frac{\partial W_x}{\partial y} = U_\infty \frac{dU_\infty}{dx} - \frac{\partial \overline{u_x u_y}}{\partial y} + \nu \frac{\partial^2 W_x}{\partial y^2} \tag{5.166}$$

and

$$\frac{\partial W_x}{\partial x} + \frac{\partial \overline{U_y}}{\partial y} = 0 \tag{5.167}$$

with boundary conditions

$$W_x = \overline{U_y} = 0 \tag{5.168}$$

at $y = 0$ and

$$W_x \to U_\infty \tag{5.169}$$

as $y \to \infty$. This form of the leading-order turbulent-boundary-layer problem can be compared with the laminar one. It is formally identical, apart from the additional Reynolds stress term, representing the turbulence. It also has the same form as the basic "boundary-layer" approximation of Section 5.2, with W_x instead of $\overline{U_x}$ and an added viscous term. However, it is important to note that, in general, W_x is not the same as $\overline{U_x}$.

The difference between the two is $U_x^{(inv)} - U_\infty$, a quantity that is zero at the surface, but can be significantly nonzero in the outer layer. Adopting a one-term Taylor's expansion of $U_x^{(inv)} - U_\infty$ with respect to y, we have

$$\overline{U_x} - W_x = U_x^{(inv)} - U_\infty = y \left.\frac{\partial U_x^{(inv)}}{\partial y}\right|_{y=0} \tag{5.170}$$

which gives the next order correction to (5.145). The derivative in (5.170) can be expressed using the external vorticity at the surface, that is, the z-component of the curl of $\mathbf{U}^{(inv)}$ at $y = 0$, denoted by $\Omega^{(inv)}$. Recalling that $U_y^{(inv)}|_{y=0}$ is zero at leading order, we write the z-component of the curl, expressed in terms of the curvilinear boundary-layer coordinates, as

$$\Omega^{(inv)} = -\left.\frac{\partial U_x^{(inv)}}{\partial y}\right|_{y=0} - \frac{U_\infty}{R} \tag{5.171}$$

so that (5.170) yields

$$\overline{U_x} - W_x = U_x^{(inv)} - U_\infty = -y\left(\Omega^{(inv)} + \frac{U_\infty}{R}\right) \tag{5.172}$$

showing that the difference between $\overline{U_x}$ and W_x is due to curvature and external flow vorticity. If neither effect is present, that is, the surface curvature and external vorticity are both zero, $\overline{U_x}$ and W_x are the same to the order to which we are working and the refined boundary-layer approximation derived above coincides with the basic approximation of Section 5.2 (with an extra viscous term to describe the viscous sublayer). In this case, we have $\overline{U_x} = U_\infty + V_x$, whereas (5.172) must be added in general.

To determine the levels of curvature and external vorticity needed to make these effects significant in the outer layer, we recall that the velocity defect is $O(u_*)$ there. Thus, the difference between $\overline{U_x}$ and W_x is significant if it is $O(u_*)$ or more. Dividing (5.172) by u_* and setting $y = O(\delta)$, we obtain two nondimensional parameters, $U_\infty \delta / u_* R$ and $\Omega^{(inv)}\delta/u_*$, which respectively measure the importance of curvature and external vorticity in the outer layer. When these parameters are small, the corresponding effects can be neglected, whereas if one or both is $O(1)$, they are significant in the outer layer. Since the velocity defect is larger nearer the surface, whereas (5.172) decreases, the effects of curvature and external vorticity are less important in the inertial layer and less again in the viscous sublayer. Observe that the parameter, $U_\infty \delta/(u_* R)$, measuring the importance of surface curvature, can be $O(1)$, even if δ/R is small, owing to the large value of U_∞/u_*. That is, curvature can be significant even if the boundary layer is thin compared with the radius of curvature. The two parameters, $U_\infty\delta/(u_* R)$ and $\Omega^{(inv)}\delta/u_*$, are implicitly assumed to be $O(1)$ or smaller in the theory developed here.

In the above discussion, significant curvature and external mean vorticity give rise to differences between $\overline{U_x}$ and W_x, which implies that the refined boundary-layer approximation is needed. However, neither curvature nor mean external vorticity appear explicitly in the leading-order boundary-layer equations for the mean flow (i.e., either of the systems (5.161)–(5.164) or (5.166)–(5.169)). As we shall see later, curvature and external vorticity *are* present in the equations for the turbulence, thus modifying $\overline{u_x u_y}$, and hence the Reynolds stress term in the mean-flow equations. Moreover, their importance in the equations for the turbulence is measured by the same parameters, $U_\infty\delta/(u_* R)$ and $\Omega^{(inv)}\delta/u_*$, found above. Thus, surface curvature and external vorticity can modify the mean-flow equations in a nontrivial fashion via the turbulence. In contrast, the external flow velocity, $U_\infty(x)$, occurs directly in the leading-order mean-flow equations given above and is therefore likely to have a more profound effect on the boundary layer.

Streamwise variations of $U_\infty(x)$ are due to an external pressure gradient along the surface, as appears by application of Bernoulli's theorem to a surface streamline of the leading-order external flow. Thus, if $P_\infty(x)$ denotes the pressure field associated with the flow $\mathbf{U}^{(inv)}$, extrapolated to the surface, the quantity $P_\infty + \rho U_\infty^2/2$ is a constant to leading order and so

$$U_\infty \frac{dU_\infty}{dx} = -\frac{1}{\rho}\frac{dP_\infty}{dx} \qquad (5.173)$$

showing the relationship between the derivative of $U_\infty(x)$ and the pressure gradient. The effects of significant pressure gradients on turbulent boundary layers are qualitatively similar to those for laminar ones (see Section 5.1), and will be discussed later in this section (see "Pressure Gradient and Surface Roughness Effects").

Although we have the equivalent system, (5.166)–(5.169), which may appear simpler, we prefer to continue to work with the velocity-defect form, (5.161)–(5.164), of the equations. This choice leads to more straightforward mathematical derivations and, because V_x expresses the (small) outer-layer modifications to $\overline{U_x}$ due to the boundary layer, makes it easier to keep track of orders of magnitude and hence maintain precision of the approximation in the outer layer.

The most important result obtained from the turbulent mean-flow boundary-layer equations is an overall momentum balance, known as the Von Karman equation. From (5.161) and (5.162) we derive

$$\frac{\partial}{\partial x}\left\{(U_\infty + V_x)V_x\right\} + \frac{dU_\infty}{dx} V_x = \frac{\partial}{\partial y}\left(\nu \frac{\partial \overline{U_x}}{\partial y} - \overline{u_x u_y} - \overline{U_y} V_x\right) \tag{5.174}$$

which is integrated with respect to y, across the boundary layer, using (5.131), (5.163), (5.164), and the fact that both turbulence and viscosity are negligible outside the layer, while $\overline{u_x u_y} = 0$ at the surface. The result is

$$\frac{d}{dx}\int_0^\infty (U_\infty + V_x)V_x dy + \frac{dU_\infty}{dx}\int_0^\infty V_x dy = -\frac{\tau_w}{\rho} \tag{5.175}$$

which expresses the leading-order balance of the streamwise component of mean momentum for the layer. The momentum balance is more usually expressed in terms of two thickness scales, δ_m and δ_d, defined by

$$\delta_m = \frac{1}{U_\infty^2}\int_0^\infty \overline{U_x}(U_x^{(inv)} - \overline{U_x})dy = -\frac{1}{U_\infty^2}\int_0^\infty \overline{U_x} V_x dy \tag{5.176}$$

known as the momentum thickness, and

$$\delta_d = \frac{1}{U_\infty}\int_0^\infty (U_x^{(inv)} - \overline{U_x})dy = -\frac{1}{U_\infty}\int_0^\infty V_x dy \tag{5.177}$$

called the displacement thickness, which is generally somewhat larger than δ_m. It can be shown that the small correction, (5.144), to the surface boundary condition may be taken into account in the calculation of $\mathbf{U}^{(inv)}$ by applying the usual inviscid condition of zero normal velocity, not at the real surface, but at a fictitious body surface, $y = \delta_d(x)$. That is, as far as the effect of the boundary layer on the external flow is concerned, the body can be imagined as enlarged by the displacement thickness, hence its name.

At leading order, we may use (5.145) to write $\overline{U_x} = U_\infty + V_x$ in the second integral of (5.176), to obtain

$$\delta_m = -\frac{1}{U_\infty^2}\int_0^\infty (U_\infty + V_x)V_x dy \tag{5.178}$$

and employ (5.173), (5.177), and (5.178) to rewrite (5.175) as

$$\boxed{\frac{d\delta_m}{dx} = \frac{1}{\rho U_\infty^2}\left\{\tau_w + (\delta_d + 2\delta_m)\frac{dP_\infty}{dx}\right\}} \tag{5.179}$$

This is the celebrated Von Karman equation, expressing the momentum balance of the boundary layer, and equally valid in the laminar case. It indicates that the momentum thickness evolves with streamwise distance due to the skin friction and pressure gradient, if there is one.

Equation (5.179) can be reexpressed in the equivalent form

$$\frac{d\delta_m}{dx} = \frac{u_*^2}{U_\infty^2}\left\{1 + \Pi\left(1 + \frac{2\delta_m}{\delta_d}\right)\right\} \tag{5.180}$$

which forms the main theoretical component of the semiempirical theories to be described in the next two subsections. Here, the quantity

$$\Pi = \frac{\delta_d}{\tau_w}\frac{dP_\infty}{dx} = -\frac{\delta_d U_\infty}{u_*^2}\frac{dU_\infty}{dx} \tag{5.181}$$

is a parameter measuring the importance of the pressure gradient. If this parameter is small, the effects of the pressure gradient on the boundary layer are negligible. On the other hand, when $\Pi = O(1)$, the pressure gradient becomes significant in the outer layer and the results depend strongly on the sign of Π, as we shall see later in this section (see "Pressure Gradients and Surface Roughness Effects"). It is implicitly assumed that $\Pi = O(1)$ or smaller in our analysis of turbulent boundary layers, otherwise the scaling properties of the outer layer are quite different. For example, near a separation point the outer layer velocity defect is no longer small compared with the external flow velocity, nor do the turbulent velocities scale on u_*, and so on. *The theory developed here does not apply near separation.*

For a turbulent boundary layer, the main contributions to the integrals in (5.176) and (5.177) arise from the outer layer, where $\overline{U_x} = U_\infty$ to leading order. Thus, we find that

$$\delta_m \approx \delta_d = -\frac{1}{U_\infty}\int_0^\infty V_x dy \tag{5.182}$$

to a first approximation, that is, the momentum and displacement thicknesses are equal to leading order, which is not the case for laminar layers nor for turbulent ones near separation. Since $V_x = O(u_*)$ in the outer layer, one may estimate

$$\delta_m \sim \delta_d = O\left(\frac{u_*}{U_\infty}\delta\right) \tag{5.183}$$

showing that both momentum and displacement thickness are small compared with the overall boundary-layer thickness, δ, by a factor of $O(u_*/U_\infty)$. Their difference

$$\delta_d - \delta_m = \frac{1}{U_\infty^2}\int_0^\infty (\overline{U_x} - U_\infty)V_x dy \tag{5.184}$$

is $O((u_*^2/U_\infty^2)\delta)$ and thus $O(u_*/U_\infty)$ smaller than either δ_m or δ_d taken separately. As we shall see later, although this order of magnitude is correct, the numerical factor multiplying u_*/U_∞ in the expression for $(\delta_d - \delta_m)/\delta_d$ can be quite large, and consequently δ_m and δ_d may differ appreciably (by as much as 40% for a flat plate in a uniform stream at zero incidence, still more for certain other cases, particularly towards separation). For this reason, δ_m and δ_d are usually regarded as distinct. However, from a fundamental point of view, this is not very satisfactory because smallness of u_*/U_∞ has been used in the derivation of the Von Karman momentum balance and so, if one draws a distinction between δ_m and δ_d, one ought also to be consistent and use a higher-order version of the Von Karman equation.

Setting $\delta_d = \delta_m$ in (5.180), we obtain the leading-order form

$$\frac{d\delta_m}{dx} = \frac{u_*^2}{U_\infty^2}(1 + 3\Pi) \tag{5.185}$$

This equation, although formally of the same order of precision as (5.180), is not as accurate in reality, for the reasons given above. However, it is used here to obtain some important order of magnitude estimates. Equation (5.185) indicates that $d\delta_m/dx = O(u_*^2/U_\infty^2)$, while, from (5.183), we obtain $d\delta_m/dx = O(u_*\delta/U_\infty d)$. Equating the two order of magnitudes gives

$$\frac{\delta}{d} = O\left(\frac{u_*}{U_\infty}\right) \tag{5.186}$$

so that the small quantities, δ/d and u_*/U_∞, are of the same order, as stated earlier. Equation (5.186) can be rewritten as

$$d = O\left(\frac{\delta U_\infty}{u_*}\right) \tag{5.187}$$

giving the distance over which boundary-layer development occurs, which is large compared with δ since u_*/U_∞ is small. This is in keeping with the general discussion in the introduction to this chapter. The slow development of the boundary layer is due to small turbulent velocities, $O(u_*)$, compared with the streamwise mean flow, $O(U_\infty)$. The result (5.186) will shortly be confirmed by a more detailed analysis of the outer layer.

A further interesting consequence of (5.186) concerns the relative thicknesses of laminar and turbulent boundary layers. From Section 5.1, it will be recalled that the thickness of a laminar layer over a body of size $O(d)$ in an external stream of velocity $O(U_\infty)$ can be estimated as $O((\nu d/U_\infty)^{1/2})$. The ratio of laminar to turbulent boundary-layer thicknesses under the same conditions is therefore $O(\mathrm{Re}_*^{-1/2})$, from (5.186), where $\mathrm{Re}_* = u_*\delta/\nu$ is the large boundary-layer Reynolds number. Thus, a turbulent boundary layer is considerably thicker (asymptotically in large Re_*) than the corresponding laminar one, having the same body size and external velocity. On the other hand, the viscous sublayer is $O(\mathrm{Re}_*^{-1})$ smaller than the outer layer and hence $O(\mathrm{Re}_*^{-1/2})$ thinner than the laminar one. It follows that the laminar layer is asymptotically intermediate in thickness between the viscous sublayer and outer layer of a comparable turbulent one.

A higher-order momentum balance equation is derived in the appendix, giving

$$\frac{d\delta_m}{dx} = \frac{u_*^2}{U_\infty^2}\left\{1 + \Pi\left(1 + \frac{2\delta_m}{\delta_d}\right) + \frac{1}{u_*^2}\left[\frac{d}{dx}\left(\int_0^\infty \left(\overline{u_x^2} - \overline{u_y^2}\right)dy\right.\right.\right.$$
$$\left.\left.\left. - \left(\Omega^{(inv)} + 3\frac{U_\infty}{R}\right)\int_0^\infty yV_x dy\right) + \frac{2}{R}\frac{d}{dx}\left(U_\infty \int_0^\infty yV_x dy\right)\right]\right\} \tag{5.188}$$

of which the terms in square brackets provide a correction to the Von Karman equation, (5.180). This correction is formally of the same order of magnitude, u_*/U_∞, as $(\delta_d - \delta_m)/\delta_d$, and should therefore be included, in principle, if one distinguishes between δ_m and δ_d, which is necessary to obtain reasonable accuracy. However, as noted above, the difference, $(\delta_d - \delta_m)/\delta_d$, can have a large numerical multiplier in front of u_*/U_∞ and so be more significant than appears from formal orders of magnitude. We shall not be using the higher-order form, (5.188), in what follows, but it provides a reminder that the Von Karman equation is not exact. Notice that, even in the absence of surface curvature and external vorticity, there is a small correction term involving the Reynolds stresses, which arises from the turbulent velocity field. This corresponds to that occurring in the basic boundary-layer approximation at higher order, as in the final term of equation (5.32).

Returning to the leading-order, mean-flow, boundary-layer equation, (5.161)–(5.164), we want next to consider the forms taken by (5.161) in the viscous sublayer and outer layer. That is, although all terms in (5.161) are needed somewhere in the boundary layer, further approximations, dropping different terms, are possible in the different parts of the layer. Let us begin with the viscous sublayer.

The incompressibility condition, $\nabla . \overline{\mathrm{U}} = 0$, yields

$$\frac{\partial \overline{U_y}}{\partial y} = -\frac{\partial \overline{U_x}}{\partial x} \tag{5.189}$$

when we neglect the curvature terms at leading order. The thickness of the sublayer is $O(\nu/u_*)$, while $\overline{U_x} = O(u_*)$ there. Thus, (5.189) gives $\overline{U_y} = O(\nu/d)$. The left-hand side of (5.161) can now be estimated as at most $O(U_\infty^2/d)$, because $V_x = O(U_\infty)$ in the viscous sublayer. On the other hand, both terms on the right of (5.161) are $O(u_*^3/\nu)$, since the turbulent velocities are $O(u_*)$. The ratio of left- to right-hand side is therefore

$$O\left(\frac{U_\infty^2}{u_*^2}\frac{\delta}{d}\,\mathrm{Re}_*^{-1}\right)$$

in the viscous sublayer. Now, although the quantity $(U_\infty^2/u_*^2)(\delta/d)$ is large, it is not as large as the Reynolds number, Re_*. In fact, $d/\delta = O(U_\infty/u_*)$ according to (5.186) and, as we shall see later, the quantity U_∞/u_* is a logarithmic function of Re_* and thus dominated in size by Re_* as $\mathrm{Re}_* \to \infty$. It follows that the left-hand side of (5.161) is negligible in the viscous sublayer, resulting in the leading-order form

$$\frac{\partial}{\partial y}\left(\nu\frac{\partial \overline{U_x}}{\partial y} - \overline{u_x u_y}\right) = 0 \tag{5.190}$$

which can be integrated to obtain

$$\boxed{\nu\frac{\partial \overline{U_x}}{\partial y} - \overline{u_x u_y} = u_*^2} \tag{5.191}$$

where we have employed (5.131), (5.132), and $\overline{u_x u_y} = 0$ at $y = 0$ to determine the constant of integration.

The viscous sublayer mean-flow equation, (5.191), expresses constancy of total shear stress, consisting of a sum of the viscous shear stress, $\nu\rho(\partial\overline{U_x}/\partial y)$, and the Reynolds shear stress, $-\rho\overline{u_x u_y}$. At the wall, the Reynolds stress is zero and the viscous shear stress is $\tau_w = \rho u_*^2$, while as one leaves the viscous sublayer and enters the inertial layer, viscosity becomes unimportant and we have

$$-\overline{u_x u_y} = u_*^2 \tag{5.192}$$

in the inertial layer. Thus, the Reynolds shear stress is approximately constant in the inertial layer and equal to the wall friction. As distance from the wall increases through the viscous sublayer, the Reynolds shear stress rises from zero at the wall to $\tau_w = \rho u_*^2$ outside the sublayer, while the viscous shear stress compensates by falling from τ_w at the wall to zero outside. Notice that (5.191) does not have any pressure gradient, curvature, or external vorticity terms, which suggests that the viscous sublayer is insensitive to such effects.

Turning attention to the outer layer, we drop the viscous term in (5.161), use (5.150) to write

$$\overline{U_y} = -y \frac{dU_\infty}{dx} \tag{5.193}$$

and replace $U_\infty + V_x$ by U_∞, since $V_x = O(u_*)$ is smaller. Thus, we obtain the leading-order outer-layer equation

$$\boxed{U_\infty \frac{\partial V_x}{\partial x} + \frac{dU_\infty}{dx}\left(V_x - y\frac{\partial V_x}{\partial y}\right) = -\frac{\partial \overline{u_x u_y}}{\partial y}} \tag{5.194}$$

which describes the development of V_x in the outer layer, and contains the pressure gradient term, dU_∞/dx. This implies that the outer layer will be sensitive to pressure gradients, as is indeed found to be the case. The first term on the left of (5.194) represents the streamwise development of the outer part of the boundary layer and is determined by the other terms, which respectively describe the effects of a pressure gradient and the cross-stream transfer of momentum by turbulence in the outer layer. The first term is $O(u_* U_\infty/d)$, since $V_x = O(u_*)$, while the Reynolds stress term is $O(u_*^2/\delta)$ because the turbulent velocities are $O(u_*)$ and the outer-layer thickness is $O(\delta)$. Supposing that turbulent momentum transfer plays a significant role in the development of the layer, we may equate these two orders of magnitude, thus recovering (5.186) and showing once again that the small quantities, δ/d and u_*/U_∞, are of the same order of magnitude. This implies that the distance required for streamwise development is given by (5.187), which is large compared with the boundary-layer thickness. If one moves downstream at velocity U_∞, thus following the fluid in the outer layer, the time taken for significant boundary-layer developments is $d/U_\infty = O(\delta/u_*)$. This is the eddy lifetime for the largest turbulent scales of the boundary layer, which have size $O(\delta)$ and associated velocities $O(u_*)$. In this time, the layer thickness changes by an amount of $O(\delta)$, implying an average velocity of the frontier of the layer of $O(u_*)$. Thus, the moving fluid in the outer part of a boundary layer finds that the development of layer thickness takes place at a speed comparable with the turbulent velocities.

The pressure gradient term in (5.194) is important if it is comparable to or larger than the Reynolds stress term. The ratio of the former to the latter is $O((\delta/u_*)(dU_\infty/dx))$, indicating that, like Π, the parameter $(\delta/u_*)(dU_\infty/dx)$ measures the significance of the pressure gradient for the outer layer. In fact, since $\delta_d = O((u_*/U_\infty)\delta)$ from (5.183), we have $\Pi = O((\delta/u_*)(dU_\infty/dx))$, confirming the relationship between the two parameters. A third quantity, $(d/U_\infty)(dU_\infty/dx)$, can also be shown, using (5.186), to be another measure of the importance of the gradient and has a simple interpretation: the pressure gradient is important if $U_\infty(x)$ undergoes significant changes over the distance required for boundary-layer development. For a general body shape, it is natural to suppose that $dU_\infty/dx = O(U_\infty/d)$, which makes $(d/U_\infty)(dU_\infty/dx)$ of $O(1)$. That is, the pressure gradient ought to be significant for general bodies in the outer part of the boundary layer. However, for some cases, such as a flat plate in a uniform flow at zero incidence, dU_∞/dx is small compared with U_∞/d and the pressure gradient can be neglected. For others, such as flow in the vicinity of a sharp edge, the gradient is very strong. In discussing the effects of pressure gradients (see "Pressure Gradient and Surface Roughness Effects"), we use the parameter Π as a measure of the importance of the gradient, rather than one of the other possible parameters. One advantage of Π is that it occurs directly in the Von Karman equation, which is the principal theoretical ingredient of the semiempirical theory of turbulent boundary layers in the next two subsections.

Using (5.186), the dimensionless parameters, $U_\infty\delta/u_*R$ and $\Omega^{(inv)}\delta/u_*$, given earlier as measures of the importance of curvature and external vorticity in the outer layer, can be estimated as $O(d/R)$ and $O(\Omega^{(inv)}d/U_\infty)$. In general, one would expect that the radius of curvature of the body surface would be comparable with the body dimensions, $O(d)$, and hence that curvature effects should be significant. Furthermore, if the external flow, $\mathbf{U}^{(inv)}$, is appreciably rotational, one might expect that its surface vorticity, $\Omega^{(inv)}$, would be $O(U_\infty/d)$. Thus, for a general body in a rotational stream, pressure gradient, curvature, and external vorticity should have significant, but not dominant, effects on the outer part of a turbulent boundary layer, since the corresponding parameters are $O(1)$. Nonetheless, the simplest case is when none of these effects is present, as for the boundary layer on a flat plate in a uniform stream considered in the next subsection.

So far, we have considered the mean-flow equations but we now want to briefly discuss the equations governing the energetics of the turbulence. The turbulent energy equation for a general flow is (5.41). Under the basic "boundary-layer" approximation we obtained (5.43), and the only term that needs reassessment for a boundary layer is the production in the outer layer. By writing the tensor $\partial\overline{U}_i/\partial x_j$ in the curvilinear boundary-layer coordinates, one can show that it has the form

$$\begin{bmatrix} \dfrac{R}{R+y}\dfrac{\partial\overline{U}_x}{\partial x}+\dfrac{\overline{U}_y}{R+y} & \dfrac{\partial\overline{U}_x}{\partial y} \\ \dfrac{R}{R+y}\dfrac{\partial\overline{U}_y}{\partial x}-\dfrac{\overline{U}_x}{R+y} & \dfrac{\partial\overline{U}_y}{\partial y} \end{bmatrix} \approx \begin{bmatrix} \dfrac{dU_\infty}{dx} & \dfrac{\partial V_x}{\partial y}-\dfrac{U_\infty}{R}-\Omega^{(inv)} \\ -\dfrac{U_\infty}{R} & -\dfrac{dU_\infty}{dx} \end{bmatrix} \tag{5.195}$$

so that the leading-order turbulent energy equation is

$$\underbrace{\overline{U}_x\frac{\partial\frac{1}{2}\overline{q^2}}{\partial x}+\overline{U}_y\frac{\partial\frac{1}{2}\overline{q^2}}{\partial y}}_{\text{Convection}} = \underbrace{\frac{dU_\infty}{dx}\left(\overline{u_y^2}-\overline{u_x^2}\right)+\left(\frac{2U_\infty}{R}+\Omega^{(inv)}-\frac{\partial V_x}{\partial y}\right)\overline{u_xu_y}}_{\text{Production}}$$
$$-\underbrace{\overline{\varepsilon}}_{\text{Dissipation}}\underbrace{-\frac{\partial}{\partial y}\left(\overline{u_y\left(\frac{1}{2}q^2+\frac{p}{\rho}\right)}-\overline{\nu u_i\left(\frac{\partial u_i}{\partial y}+\frac{\partial u_y}{\partial x_i}\right)}\right)}_{\text{Turbulent diffusion}} \tag{5.196}$$

showing pressure gradient, surface curvature, and external vorticity effects via the turbulence production terms. Similar behavior is found for the Reynolds stress equation, (4.32), at leading order. The importance of the pressure gradient, surface curvature, and external vorticity for turbulence in the outer layer can be determined by comparing the corresponding terms of (5.196) with the production, $-(\partial V_x/\partial y)\overline{u_x u_y} = O(u_*^3/\delta)$, due to shear alone. Thus, we recover the parameters $(\delta/u_*)(dU_\infty/dx)$, $U_\infty\delta/u_*R$, and $\Omega^{(inv)}\delta/u_*$ that were obtained earlier as measures of the significance of these effects for the outer layer. However, the curvature and external vorticity now appear explicitly, thus modifying the turbulence and hence, indirectly, the mean-flow equations through the Reynolds stress term. The production due to shear, $-(\partial V_x/\partial y)\overline{u_x u_y}$, rises more rapidly as the wall is approached and dominates that due to the pressure gradient, curvature, and external vorticity in the viscous sublayer, again suggesting that the sublayer is less sensitive to such perturbing influences than the outer layer. Notice that, for a flat-plate layer without pressure gradient or significant external vorticity, the production term takes the form $-(\partial \overline{U}_x/\partial y)\overline{u_x u_y}$, the product of the mean shear and Reynolds shear stress, as in (5.43).

The viscous part of the turbulent diffusion term of (5.196) is only important in the viscous sublayer, reflecting the differing nature of turbulence there, with significant viscous effects even at the large scales. This is in contrast with turbulence further from the wall, where the large scales are bigger and less and less strongly affected by viscosity, leading to dissipation via a cascade to smaller scales, rather than the direct action of viscosity on the large ones.

We will return to the turbulent energy equation in the following subsection, where measured values of the various terms for a flat-plate boundary layer will be discussed. For the moment, we simply note that, because of the large mean-flow defect gradients, $\partial V_x/\partial y$, near the wall, turbulent production is much higher there. In particular, the viscous sublayer acts as a strong source of turbulent energy, but also has high turbulent energy dissipation, so that much of the turbulence which is generated there is damped out locally. The net result is that the turbulent kinetic energy, $\overline{q^2}/2$, rises as the wall is approached from the outer layer and is a maximum in the viscous sublayer.

To make further progress, we introduce some experimental results. As has been implied earlier, the viscous sublayer appears to be relatively insensitive to external perturbing effects such as pressure gradients and so we describe it first. On the other hand, the outer layer is affected by significant pressure gradients and its observed properties, with and without a pressure gradient, are therefore discussed separately in the next two subsections. We do not consider the effects of surface curvature or external vorticity any further.

In the viscous sublayer, the viscosity, ν, and turbulent friction velocity, u_*, are observed to completely determine the behavior of the flow. From u_* and ν, only one length scale, namely ν/u_*, can be constructed and this scale characterizes the viscous sublayer. Distance from the wall can be nondimensional using this scale, leading to

$$y_+ = \frac{u_* y}{\nu} \tag{5.197}$$

as an inner coordinate appropriate to the viscous sublayer. Since the flow there is determined by ν and u_* only, any flow quantity, nondimensionalized using ν and u_*, should be a universal function of y_+. For instance,

$$\frac{\overline{U_x}}{u_*} = f(y_+) \tag{5.198}$$

is found to describe the mean flow in the viscous sublayer, where f is a universal function, shown in Figure 5.10 using a logarithmic scale for y_+. Since the no-slip condition applies at the surface, we have

$$f(0) = 0 \tag{5.199}$$

while equations (5.131), (5.132), (5.197), and (5.198) imply

$$\left.\frac{df}{dy_+}\right|_{y_+=0} = 1 \tag{5.200}$$

so that

$$f(y_+) \sim y_+ \tag{5.201}$$

as $y_+ \to 0$. In the opposite limit, it is found that

$$f(y_+) \sim \frac{1}{\kappa} \log y_+ + a \tag{5.202}$$

as $y_+ \to \infty$, which is the celebrated logarithmic law, giving the mean velocity via (5.198). The quantity κ is known as the Von Karman constant and has a value not far from $\kappa = 0.39$. Experiments give a wider spread of values for the other universal constant, a, with $a = 4$ giving reasonable results. Measurements also seem to indicate that the scaled turbulent-velocity moments $\overline{u_i u_j}/u_*^2$ are universal functions of y_+. These moments go to zero at $y_+ = 0$ because of the no-slip condition. They presumably approach constant values as $y_+ \to \infty$, but this is unclear given current experimental data. Note that (5.198) implies that $\overline{U_x} = O(u_*)$ in the viscous sublayer, as supposed previously.

The outer part of the boundary layer is described by the nondimensional variable, $\eta = y/\delta$ (not to be confused with the same symbol, used elsewhere to represent the Kolmogorov scale), where δ is the overall boundary-layer thickness. Figure 5.11 shows the measured $-V_x/u_*$ as a function of y/δ for a turbulent boundary layer on a flat plate in a uniform stream at zero incidence. The velocity defect, $-V_x$ is very close to zero once a certain distance from the surface is exceeded. Of course, it will never be exactly zero, but in practice, there is little leeway in defining δ. If a precise,

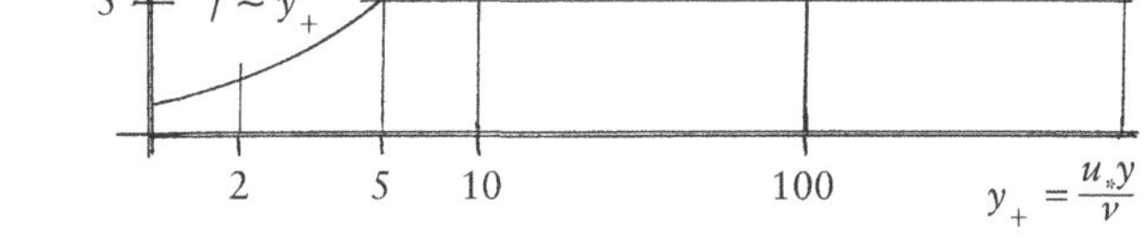

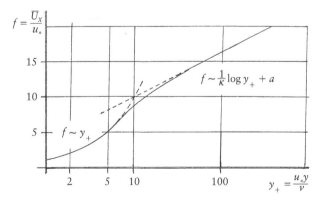

Figure 5.10. Universal representation of the mean velocity in the near-wall region. The vertical scale is linear, the horizontal one is logarithmic.

but necessarily somewhat arbitrary definition of δ is desired, one can take $|V_x| = 0.1u_*$.

FLAT PLATE WITHOUT PRESSURE GRADIENT

In this subsection, we restrict attention to the well-documented case of a flat plate in an infinite, uniform stream at zero incidence. In that case, $U_x^{(inv)} = U_\infty$ is uniform and there is no pressure gradient or other complicating effects, yielding the simplest boundary-layer flow. Provided the boundary-layer Reynolds number is large enough, it is found that the velocity defect, $-V_x$, is only dependent on u_* and δ in the outer layer. Thus, by dimensional analysis,

$$\frac{V_x}{u_*} = F(\eta) \tag{5.203}$$

where $\eta = y/\delta$ and F is another universal function. Figure 5.11 shows the positive quantity, $-F(\eta)$, with a logarithmic scale for η. Note that the velocity deficit, $-V_x$, is $O(u_*)$ in the outer layer, as assumed earlier. This is also true of the turbulent velocities. The universal form, (5.203), of the outer-layer velocity defect is observed to hold when $\text{Re}_* = u_*\delta/\nu$ is larger than about 2,000, but shows strong Reynolds number dependence of $F(\eta)$ at lower values. In fact, there may also be weak variations at still higher Re_* (see Gad-el-Hak and Bandyopadhyay (1994) for a review). However, in keeping with the title of this chapter, we will develop the classical theory of turbulent boundary layers, which treats $F(\eta)$ as universal for flat plates. In so doing, we assume a large enough value of Re_*, that is, the theory is asymptotic as $\text{Re}_* \to \infty$. The Reynolds number rises with increasing streamwise distance, as the boundary layer thickens by entrainment of external fluid.

The mean velocity is described by (5.198) in the viscous sublayer, of thickness $O(\nu/u_*)$, and by (5.203) in the outer layer, $y = O(\delta)$. The relative thickness of the viscous sublayer is $O(\text{Re}_*^{-1})$, and is very small. In between the viscous sublayer and the outer region, that is, for $\nu/u_* \ll y \ll \delta$, lies the inertial layer, in which both (5.198) and (5.203) hold and must agree. This is similar to the asymptotic matching technique in which expansions in some small parameter, valid in two adjoining asymptotic regions, match in their common domain of validity (or overlap region). However, we are matching experimentally determined forms here. The small parameter is Re_*^{-1}, giving asymptotic separation into the inner region, or viscous sublayer, and the outer layer. In the inertial sublayer, we equate $\overline{U}_x/u_*$, given by (5.198) and (5.203), to find that

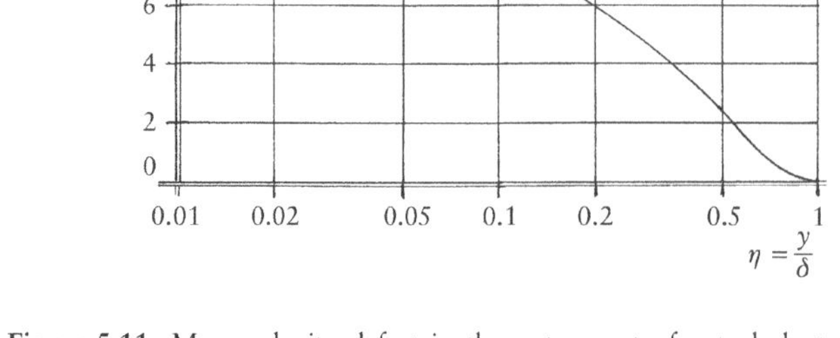

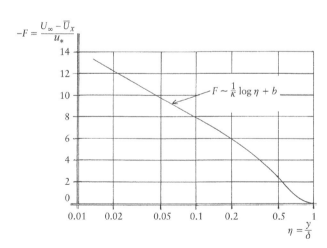

Figure 5.11. Mean-velocity defect in the outer part of a turbulent boundary layer on a flat plate at zero incidence. The vertical scale is linear, the horizontal one is logarithmic.

$$\hat{f}(\mathrm{Re}_* \eta) = \frac{U_\infty}{u_*} + \hat{F}(\eta) \tag{5.204}$$

where $\hat{f}$ denotes the large argument form of f and $\hat{F}$ is the small argument form of F. In deriving (5.204), we have written $V_x = \overline{U}_x - U_x^{(inv)}$ with $U_x^{(inv)} = U_\infty$. The latter holds exactly for the uniform external flow considered here. Differentiating (5.204) with respect to η and multiplying by η gives

$$\mathrm{Re}_* \eta \hat{f}'(\mathrm{Re}_* \eta) = \eta \hat{F}'(\eta) \tag{5.205}$$

where the primes denote derivatives.

We argue that we may vary $\mathrm{Re}_* \eta$ while keeping η fixed, by, for instance, considering different streamwise distances along the boundary layer. Thus, it is apparent that both sides of (5.205) must be constant and

$$y_+ \hat{f}'(y_+) = \frac{1}{\kappa} \tag{5.206}$$

where κ is constant. The integral of this equation yields the log law

$$\hat{f}(y_+) = \frac{1}{\kappa} \log y_+ + a \tag{5.207}$$

given earlier in (5.202), and which is now seen to be necessary for matching of the two expressions (5.198) and (5.203). A further consequence of (5.204) is that

$$\hat{F}(\eta) = \frac{1}{\kappa} \log \eta + b \tag{5.208}$$

where

$$\frac{U_\infty}{u_*} = \frac{1}{\kappa} \log \mathrm{Re}_* + a - b \tag{5.209}$$

completes the matching. Equation (5.208) gives the form of the mean flow in the outer region as $\eta \to 0$, via (5.203). The universal flat-plate constant, b, is hard to determine with great precision from experiments and has a value of about $b = -2$. The matching condition, (5.209), can also be written in the convenient form

$$\frac{U_\infty}{u_*} + \frac{1}{\kappa} \log\left(\frac{U_\infty}{u_*}\right) = \frac{1}{\kappa} \log \mathrm{Re}_\delta + a - b \tag{5.210}$$

where $\mathrm{Re}_\delta = U_\infty \delta / \nu = (U_\infty / u_*) \mathrm{Re}_*$ is a boundary-layer Reynolds number, even larger than Re_*. In equations (5.209) and (5.210), $\kappa \approx 0.39$ is the Von Karman constant and $a - b \approx 6$. These are purely numerical constants characterizing flat-plate boundary layers.

The minimum values of the Reynolds number occur just after the transition zone and are large even there. Moreover, universality of $F(\eta)$ requires still higher Reynolds numbers, that is, one must go further downstream. Because Re_* and Re_δ are large and increasing with streamwise distance (owing to thickening of the boundary layer), U_∞ / u_* is also large and growing with x, according to (5.209) and (5.210). Smallness of u_* / U_∞, repeatedly used earlier, is therefore a consequence of high Reynolds number. As a result, the large Reynolds number forms the basis of the "boundary-layer" approximations derived previously. The logarithmic dependence of U_∞ / u_* on

Re_*, apparent in (5.209), implies that it is of lower order than Re_* as $\mathrm{Re}_* \to \infty$, the ordering being $U_\infty/u_* \ll \mathrm{Re}_* \ll \mathrm{Re}_\delta$. Another way of looking at this is that Re_*^{-1} is an exponentially small function of the small parameter u_*/U_∞, which is reflected in the very small fraction, $O(\mathrm{Re}_*^{-1})$, of the overall boundary-layer thickness occupied by the viscous sublayer. Some numbers may help here: taking the value $\mathrm{Re}_* = 2{,}000$, which is about the lowest allowable by universal $F(\eta)$, we obtain $U_\infty/u_* = 25$ from (5.209), and $\mathrm{Re}_\delta = (U_\infty/u_*)\mathrm{Re}_* = 5 \times 10^4$.

As a consequence of the logarithmic behavior of U_∞/u_* as a function of Re_*, it is relative insensitive to the Reynolds number. Each doubling of Re_* causes an increment of 1.8 to the value of U_∞/u_*, according to (5.209). Since U_∞/u_* is already a large number, it is clear that quite significant changes to Re_* can occur without altering U_∞/u_* very much. As a result, U_∞/u_* increases only slowly with streamwise distance as the boundary layer develops and can be considered as approximately constant over, say, a factor of two in streamwise distance from the leading edge of the plate. Since U_∞ is constant for the flat-plate boundary layer considered here, this is a statement about approximate constancy of the friction velocity, u_*, and hence of the skin friction, $\tau_w = \rho u_*^2$. Despite significant thickening of the boundary layer as a whole, U_∞/u_* changes only gradually with x.

Because u_*/U_∞ is small, and V_x scales on u_* in the outer layer, according to (5.203), the departures of the mean velocity from U_∞ are small there. As one crosses the boundary layer, the streamwise mean velocity, $\overline{U_x}$, slowly decreases from U_∞ at the edge of the boundary layer, to around $U_\infty - 2u_*$ at $y = \delta/2$ and about $U_\infty - 8u_*$ at $y = \delta/10$ (see Figure 5.11). Taking $U_\infty/u_* = 25$, it appears that $\overline{U_x}$ is still some 70% of U_∞ at $y = \delta/10$. Thus, most of the change from U_∞ at the edge of the layer, to zero at the surface, takes place rather close to the wall. The sublayer near the surface in which viscosity is important is characterized by the inner variable y_+, and can be considered as extending at most to, say, $y_+ = 50$ (see Figure 5.10), corresponding to just 2.5% of δ when $\mathrm{Re}_* = 2{,}000$. At this location, $\overline{U_x}$ is about $15u_*$, that is, 60% of U_∞, if $U_\infty/u_* = 25$. The main part of the viscous sublayer lies at smaller values of y_+, around $y_+ = 15$ say, giving $y \approx 0.01\delta$ and $\overline{U_x} \approx 0.4U_\infty$. The variations of $\overline{U_x}$ across the viscous sublayer are therefore a significant fraction of U_∞ unless the Reynolds number is very much greater than the value taken here. For instance, if we arbitrarily specify that $\overline{U_x}$ be less than 5% of U_∞ at $y_+ = 15$, we must have $U_\infty/u_* > 200$, leading to $\mathrm{Re}_\delta = 10^{35}$ from (5.210).[22] Such huge values of the Reynolds number are terrestrially unattainable, even in the atmospheric boundary layer, where the Reynolds numbers are considerably larger than in typical laboratory experiments. This illustrates the fact that, although the scaling for $\overline{U_x}$ in the viscous sublayer is u_*, and therefore formally small compared with U_∞, it is never very small in practice. Figure 5.12 shows a sketch of the typical profile of $\overline{U_x}$ across the whole boundary layer. The reader should note the logarithmic scale for y, necessary in order that the thin viscous sublayer be visible, and the large range of values on the horizontal axis.

[22] This shows how quantities which are logarithmic in the Reynolds number, as many of those in a boundary layer are, can require extremely large values before they take on their asymptotic behavior. This creates difficulties, because much of the semiempirical theory makes assumptions based on asymptotics, whereas experiments, which are the main source of hard information, are obviously limited in Reynolds number.

The log law for $\overline{U_x}$ in the inertial layer is apparent in each of Figures 5.10, 5.11, and 5.12. It extends from about $y_+ = 30$ up to $y = 0.2\delta$, a range of y_+ which grows wider as the Reynolds number increases with x and is quite substantial, even at relatively low Reynolds numbers.

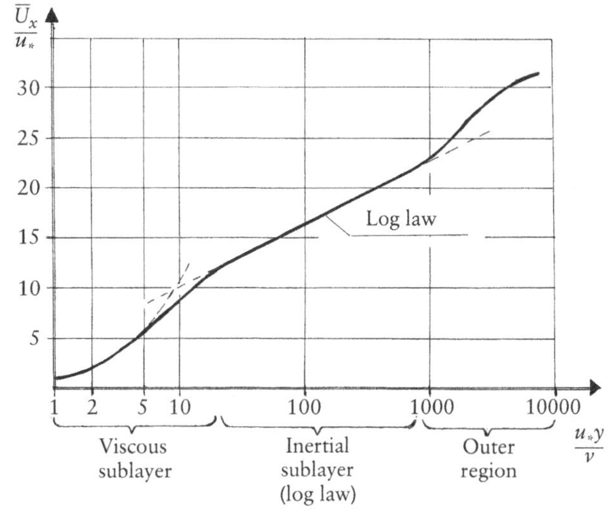

Figure 5.12. A typical velocity distribution across the whole of a turbulent boundary layer. The vertical axis is linear, the horizontal one is logarithmic. Note the wide range of values of y, which increases with Reynolds number.

For the flat-plate boundary layer considered here, U_∞ is constant. Its value is determined by the flow outside the layer and may be considered as given, like the fluid viscosity, ν. If δ were also known, (5.210) allows calculation of u_* and (5.198), (5.203) then give the mean velocity in the boundary layer in the viscous sublayer and outer layer. Thus, there is a single unknown, $\delta(x)$, which completely determines the streamwise mean flow in the boundary layer. An equation describing the development of $\delta(x)$, that is, the thickening of the boundary layer with streamwise distance, is provided by the Von Karman momentum balance equation, (5.179), which becomes

$$\frac{d\delta_m}{dx} = \frac{\tau_w}{\rho U_\infty^2} = \frac{u_*^2}{U_\infty^2} \tag{5.211}$$

since there is no pressure gradient and the momentum thickness changes due to skin friction alone. To exploit (5.211), we first need to relate δ_m and δ.

The momentum and displacement thicknesses are defined by (5.176) and (5.177). Since the viscous sublayer is very thin (exponentially so compared with u_*/U_∞), the integrals in these equations are dominated by contributions from outside the sublayer, where (5.203) applies. Using equation (5.203), we obtain

$$\frac{\delta_m}{\delta} = I_1 \frac{u_*}{U_\infty} - I_2 \left(\frac{u_*}{U_\infty}\right)^2 \tag{5.212}$$

and

$$\frac{\delta_d}{\delta} = I_1 \frac{u_*}{U_\infty} \tag{5.213}$$

where

$$I_1 = -\int_0^\infty F(\eta)\, d\eta \tag{5.214}$$

and

$$I_2 = \int_0^\infty F^2(\eta)\, d\eta \tag{5.215}$$

are purely numerical constants, which have experimentally determined values around $I_1 = 3.5$ and $I_2 = 25$. Equations (5.212) and (5.213) illustrate the earlier discussion of momentum and displacement thicknesses. They are both $O((u_*/U_\infty)\delta)$, and therefore small compared with δ. In fact, with $U_\infty/u_* = 25$, we obtain $\delta_m = 0.1\delta$ and $\delta_d = 0.14\delta$. The difference, $(\delta_d - \delta_m)/\delta = I_2(u_*/U_\infty)^2$, is formally of lower order than either δ_m or δ_d separately, and to leading order $\delta_d \sim \delta_m$. However, because the numerical coefficient, I_2, has quite a large value, the difference is not as small as the order of magnitudes would suggest and more accurate results are obtained if the formally lower-order term in (5.212) is retained, despite the apparent inconsistency of using the leading-order momentum balance, (5.211). As discussed earlier, no doubt the higher-order corrections to (5.211) (i.e., the Reynolds stress term in (5.188)) turn out to be numerically small, even though they are formally of the same order as $\delta_d - \delta_m$. Note that (5.212) and (5.213) imply that $\delta_d > \delta_m$, since $I_2 > 0$.

We use (5.212) to express δ_m in (5.211), and combine the result with the x-derivative of (5.210) to obtain

$$\frac{d\delta}{dx} = \frac{\kappa \dfrac{U_\infty}{u_*} + 1}{\kappa I_1 \left(\dfrac{U_\infty}{u_*}\right)^2 - \kappa I_2 \dfrac{U_\infty}{u_*} + I_2} \tag{5.216}$$

giving the rate of thickening of the boundary layer in terms of U_∞/u_*. Recalling that U_∞/u_* is slowly varying, thanks to its logarithmic dependency on Reynolds number, one can regard it as constant in (5.216), to a first (local in x) approximation. With constant U_∞/u_*, $\delta(x)$ is a linear function of x, according to (5.216) and so the boundary layer grows as a straight-sided wedge (see Figure 5.1d). From (5.212) and (5.213), the same is true of δ_m and δ_d. In reality, the angle of the wedge changes slowly with downstream distance as U_∞/u_* evolves. Notice that, if U_∞/u_* is taken constant, $\overline{U_x}$ is self-similar in the outer layer, according to (5.203) with $\overline{U_x} = U_\infty + V_x$. The viscous sublayer does not partake in this outer-layer local similarity, but instead has a fixed profile of $\overline{U_x}$, from (5.198) with constant u_*. The slow evolution of U_∞/u_* means that $\overline{U_x}$ is not strictly self-similar in the outer layer, nor is its profile fixed in the viscous sublayer. Regardless of the behavior of U_∞/u_*, equation (5.203) implies outer-layer self-similarity of the *velocity defect*.

Taking $U_\infty/u_* = 25$ in (5.216), the angle of the wedge defined by $y = \delta(x)$ is found to be only about 1°, and decreases slowly with streamwise distance. The leading-order form of (5.216), that is, the limit as $u_*/U_\infty \to 0$, is

$$\frac{d\delta}{dx} = \frac{u_*}{I_1 U_\infty} \tag{5.217}$$

which agrees with the previous estimate, (5.186), for the rate of boundary-layer development. These expressions describe the process of boundary-layer thickening by entrainment: turbulence is convected downstream at speed U_∞, while spreading laterally at speed $O(u_*)$. In consequence, the boundary layer develops over the streamwise distance scale (5.187). However, quantitatively, (5.217) is not very precise and it is more accurate to use (5.216). Observe that we are here considering the mean properties of the layer, whereas in any particular realization the frontier between laminar and turbulent fluid is a convoluted and fluctuating surface which only spreads as a wedge in an averaged sense.

From (5.210), we have

$$\frac{U_\infty \delta}{\nu} = \mathrm{Re}_\delta = \frac{U_\infty}{u_*} \exp\left[\kappa\left(\frac{U_\infty}{u_*} - a + b\right)\right] \tag{5.218}$$

which we use to replace δ in (5.216), leading to

$$\frac{\nu}{U_\infty}\frac{d}{dx}\left(\frac{U_\infty}{u_*}\right) = \frac{\exp\left[-\kappa\left(\frac{U_\infty}{u_*} - a + b\right)\right]}{\kappa \mathrm{I}_1\left(\frac{U_\infty}{u_*}\right)^2 - \kappa \mathrm{I}_2 \frac{U_\infty}{u_*} + \mathrm{I}_2} \tag{5.219}$$

a differential equation for U_∞/u_* whose integral gives

$$\mathrm{Re}_x = \left\{\mathrm{I}_1\left(\frac{U_\infty}{u_*}\right)^2 - \left(\mathrm{I}_2 + \frac{2\mathrm{I}_1}{\kappa}\right)\frac{U_\infty}{u_*} + \frac{2}{\kappa}\left(\mathrm{I}_2 + \frac{\mathrm{I}_1}{\kappa}\right)\right\} \exp\left[\kappa\left(\frac{U_\infty}{u_*} - a + b\right)\right] \tag{5.220}$$

where

$$\mathrm{Re}_x = \frac{U_\infty(x - x_0)}{\nu} \tag{5.221}$$

is a very large Reynolds number, much larger than Re_* and Re_δ, which is based on streamwise distance from an unknown origin, $x = x_0$, lying upstream of the transition zone and therefore outside the turbulent boundary layer itself.

Equation (5.220) gives U_∞/u_* implicitly as a function of Re_x, and hence of streamwise position in the boundary layer. From U_∞/u_*, we may calculate Re_δ using (5.218). Thus, (5.218) and (5.220) allow determination of Re_δ and U_∞/u_* as universal functions of Re_x, for flat-plate boundary layers in a uniform external stream. These functional relationships permit the streamwise mean velocity to be computed at any point in the boundary layer from (5.198) and (5.203), thus resolving the boundary-layer problem, at least as far as the mean-flow velocity is concerned. It turns out that, for the large Re_x encountered in turbulent boundary layers, the solution of (5.220) with the values of the numerical constants given previously can be well approximated by the explicit form

$$\frac{U_\infty}{u_*} = 1.2(\log \mathrm{Re}_x - 0.75)^{1.15} \tag{5.222}$$

while, from (5.218) and (5.220), we have

$$\mathrm{Re}_\delta = \mathrm{Re}_x\left(\mathrm{I}_1 \frac{U_\infty}{u_*} + \frac{2}{\kappa}\left(\mathrm{I}_2 + \frac{\mathrm{I}_1}{\kappa}\right)\frac{u_*}{U_\infty} - \mathrm{I}_2 - \frac{2\mathrm{I}_1}{\kappa}\right)^{-1} \tag{5.223}$$

Equations (5.222) and (5.223) allow explicit calculation of Re_δ and U_∞/u_* as functions of Re_x.

The constant of integration, x_0, appearing as an unknown origin via the definition of Re_x, cannot be determined by consideration of the developed turbulent boundary layer alone. In principle, it may be fixed by specifying δ, for instance, at some streamwise location.

The drag on the plate is a parameter of considerable practical interest and can be calculated for a portion of the flat plate via the integral

$$F_D = \int \tau_w \, dx \tag{5.224}$$

giving the drag force per unit length in z. The total drag on a flat plate occupying $0 < x < d$ can be expressed in terms of the nondimensional skin-friction coefficient, c_f, where

$$\frac{1}{2} c_f = \frac{F_D}{d\rho U_\infty^2} = \frac{1}{d}\int_0^d \frac{u_*^2}{U_\infty^2}\, dx \tag{5.225}$$

which is the average value of u_*^2/U_∞^2, including the laminar region near the leading edge (in which skin friction is much lower) and the transition zone. Because of the integral in (5.225), the total skin friction is not a local quantity, and the dependence of $c_f(d)$ on d smears out any rapid variations of local skin friction with streamwise distance, such as occur at transition. As a result, c_f depends noticeably on the existence of the laminar and transition regions over about a decade of downstream distance following transition, although the transition zone is appreciably narrower than this. For sufficiently long plates, the contributions from the laminar and transition zones are negligible and c_f can be obtained by integration using the expressions for U_∞/u_* given above (e.g., (5.222)). Schlichting (1987) gives a number of semiempirical expressions for c_f, as a function of $\mathrm{Re}_d = U_\infty d/\nu$. Note that, for bodies other than flat plates at zero incidence, an additional form drag is needed, due to pressure forces at the surface, as well as the viscous skin friction, which is not described by (5.220) in general.

The reader may ask why we have not used an eddy-viscosity model to determine the transverse structure of the mean-velocity profile, as we did earlier for jets and wakes. In the present case,[23] one would define ν_T by

$$-\overline{u_x u_y} = \nu_T \frac{\partial V_x}{\partial y} \tag{5.226}$$

which may be used in mean-flow equations such as (5.161). The difficulty with this approach is that ν_T is not found to be approximately uniform across the flow, as it is for jets and wakes, and all one has succeeded in doing is to replace one unknown quantity, $-\overline{u_x u_y}$, by another, ν_T.

It will be recalled, from the discussion following (5.191), that $-\overline{u_x u_y}$ increases across the viscous sublayer, from zero at the wall to u_*^2 outside the sublayer and then maintains this value through the inertial layer. It then decreases across the outer layer, as shown in Figure 5.13. Within the inertial layer, we can calculate ν_T using $-\overline{u_x u_y} = u_*^2$ and the log law, (5.203) with (5.208), as

$$\nu_T = \kappa u_* y \tag{5.227}$$

[23] More generally, allowing for the effects of surface curvature and external vorticity, using (5.33) and (5.195) we have

$$-\overline{u_x u_y} = \nu_T \left\{ \frac{\partial V_x}{\partial y} - \frac{2U_\infty}{R} - \Omega^{(inv)} \right\}$$

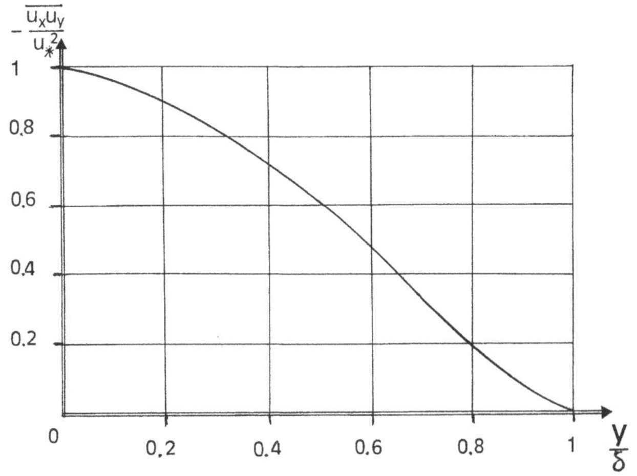

Figure 5.13. Reynolds shear stress in the outer part of a turbulent boundary layer on a flat plate at zero incidence.

showing a linear variation of eddy viscosity with distance from the surface. Interestingly, this result implies that the fictitious "Reynolds number," $u_* y/\nu_T = \kappa^{-1} \approx 2.5$ is a purely numerical constant for inertial layers. In any case, ν_T increases with y in the inertial layer. It reaches a maximum in the outer layer at around $y = 0.3\delta$, and then decreases at larger distances from the wall, owing to the falling value of $-\overline{u_x u_y}$, which wins out over decreasing $\partial V_x/\partial y$ when ν_T is calculated from (5.226). Nowhere is the eddy viscosity approximately independent of y, which effectively precludes the approach used previously for jets and wakes. Closures, such as the k–ε model described in Chapter 8, which are based on an eddy-viscosity approximation, include equations for ν_T. However, these models are heuristic approximations based on fitting to measured data. Near surfaces, they generally end up using the same experimental results, in particular the log law, that we have introduced here. If one is interested in the boundary layer, it is probably better to use experimental results directly, as we have done here. Alternatively, one can employ the heavy computational artillery of numerical simulation based directly on the Navier–Stokes equations. This approach is very computer intensive and is really another means of conducting experiments, having advantages and disadvantages compared with real laboratory measurements, as discussed in Chapter 8. It is currently severely limited in attainable Reynolds numbers by computer power.

Turning next to the properties of the turbulence, rather than the mean flow we have considered until now, Figure 5.14 shows the behavior of the three root-mean-squared components of turbulent velocity, u_x', u_y', and u_z', as a function of position in the viscous sublayer (in fact, this figure is derived from pipe-flow measurements, but the viscous sublayer of boundary layers is believed to yield essentially the same results). The largest is the streamwise fluctuation, u_x', which attains a maximum of about $u_x' = 3u_*$ towards $y_+ = 15$. All three components appear to give fairly universal results when they are scaled on u_* and plotted as a function of y_+, at least up to values of y_+ that are not too large. Likewise, it seems that the scaled turbulent velocities yield approximately universal, decreasing functions of $\eta = y/\delta$ in the outer region. These results suggest that u'_x/u_*, u'_y/u_*, and u'_z/u_* may approach universal limiting values when either y_+ or η is held fixed and $\mathrm{Re}_* \to \infty$. This is reminiscent of matched expansions, in which the limiting values would represent the first terms in two asymptotic expansions: one for the viscous sublayer, the other for the outer layer. For instance, we would have

$$\frac{u'_x}{u_*} \to \phi_x(y_+) \tag{5.228}$$

as $\mathrm{Re}_* \to \infty$ with fixed y_+ and

$$\frac{u'_x}{u_*} \to \Phi_x(\eta) \tag{5.229}$$

when $\mathrm{Re}_* \to \infty$ at fixed η. If the expansions are to match through the inertial layer like the mean flow, it is difficult to avoid the conclusion that $\phi_x(y_+)$ and $\Phi_x(\eta)$ should share the same constant limits as $y_+ \to \infty$ and $\eta \to 0$, respectively. The same is true of the other components of turbulent velocity, and, from Figure 5.14, the common limiting values would appear to be about $u'_x/u_* = 2, u'_y/u_* = 1$, and $u'_z/u_* = 1.5$.

As the Reynolds number increases, the separation of scales between the viscous sublayer, described by y_+, and the outer layer, described by η, widens and an intermediate inertial layer plateau, corresponding to the common values given above, should appear if one plots $u'_x/u_*, u'_y/u_*$, or u'_z/u_* as a function of y. Thus, as $\mathrm{Re}_* \to \infty$, the leading-order forms, for instance, $\phi_x(y_+)$ and $\Phi_x(\eta)$, should apply to the left and right of the plateau, whereas the plateau takes on the asymptotic common value. In the case of u'_y/u_* or u'_z/u_*, whose leading-order forms as $\mathrm{Re}_* \to \infty$ appear to be increasing functions of y_+, but decreasing with η, a maximum occurs in the plateau range, whose position, being of intermediate nature, will scale neither with inner, nor outer variables. The same is true of $-\overline{u_x u_y}/u_*^2$, which as we saw earlier, has a common asymptotic value of 1 in the inertial layer. This contrasts with u'_x/u_*, which has its maximum within the viscous sublayer at an asymptotically constant value of y_+.

Although the above description of the high Reynolds number asymptotic properties of the turbulent velocities is no doubt qualitatively correct, it is quantitatively questionable. One of the difficulties is that the appearance of well-defined plateaux is found to require considerably larger Reynolds numbers than are necessary for the existence of a log law in the mean-velocity profile. Furthermore, even when such plateaux are clearly observed, it is found that there remains significant Reynolds number dependence in the results for turbulent velocities, scaled on u_* as functions of y_+ (see Gad-el-Hak and Bandyopadhyay (1994)) and no

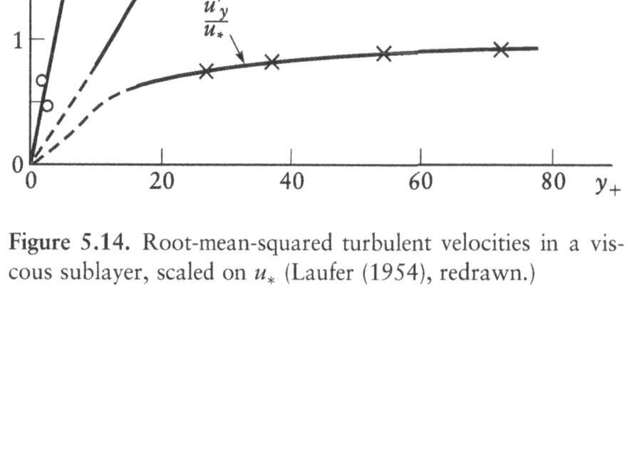

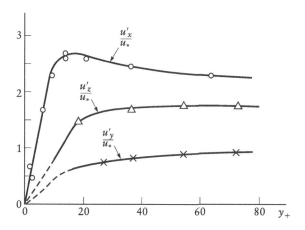

Figure 5.14. Root-mean-squared turbulent velocities in a viscous sublayer, scaled on u_* (Laufer (1954), redrawn.)

doubt of η. This might be interpreted as meaning that higher-order terms in the inner and outer asymptotic expansions are still important at the Reynolds numbers considered, in which case much larger values might be needed to attain the final asymptotic state.[24] Alternatively, the scalings used could be inappropriate, there might be an intermediate and as yet unidentified asymptotic region, and so forth. Since there is no real analytical basis for the theory, other than experimental observations, it is difficult to tell.

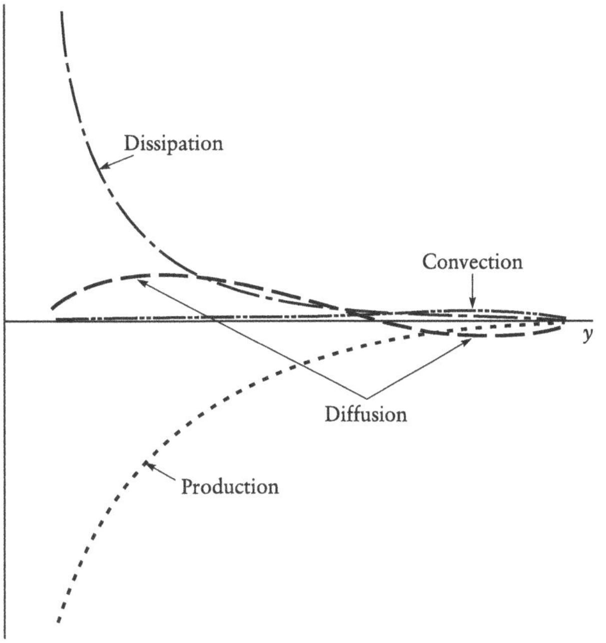

Figure 5.15. Terms in the turbulent energy balance for a flat-plate boundary layer at zero incidence. The sign of the production term has been switched so the different contributions sum to zero. The curve labeled diffusion incorporates both the pressure and cubic velocity diffusive terms of (5.43). (Klebanoff (1955), redrawn.)

The energetics of zero-pressure gradient, flat-plate boundary-layer turbulence are illustrated in Figure 5.15, which shows different terms in (5.43), with the sign of production switched so that they sum to zero. Note that (5.43) applies directly to the zero-gradient flat plate, although, in general, the turbulent energy equation needs refinement by inclusion of pressure gradient, wall curvature, and external vorticity production terms, as we saw earlier. Both turbulence production (which, it will be recalled from (5.43), has the form $-(\partial \overline{U_x}/\partial y)\overline{u_x u_y}$ according to the boundary-layer approximation for the zero-pressure-gradient flat plate considered here) and dissipation increase rapidly as the wall is approached and there is very nearly equilibrium between the two near the wall. That is, there is a balance, with production and dissipation of turbulent energy proceeding at nearly equal rates, but with a small excess of production over dissipation. In the inertial sublayer, we may use (5.192) and the log law for $\overline{U_x}$ to evaluate the production term, $-(\partial \overline{U_x}/\partial y)\overline{u_x u_y}$ under the boundary-layer approximation, gives $u_*^3/\kappa y$, increasing proportional to y^{-1} as the wall is approached. Owing to equilibrium between production and dissipation, the latter also follows the same inertial-range form. Both the mean shear, $\partial \overline{U_x}/\partial y$, and Reynolds shear stress, $-\overline{u_x u_y}$, factors appearing in the production depart from inertial-range laws once the viscous sublayer is entered and the production peaks at around $y_+ = 10$ (which is not visible in Figure 5.15 since only the outer region is shown) falling to zero at the surface because $\overline{u_x u_y} = 0$ there. Thus, as was mentioned earlier, the largest production of turbulent energy occurs within the viscous sublayer, although this may give

[24] The expansion of u_x'/u_*, etc., might be conjectured to proceed in powers of u_*/U_∞, which is only slowly (logarithmically) decreasing with increasing Reynolds number. In that case, higher-order terms could persist up to extremely large values of Re_*.

a somewhat misleading impression since the dissipation is also a maximum in that region, mopping up much of the turbulence production in situ. Nonetheless, the diffusion term causes transfer of the small excess of production over dissipation from the wall half of the layer towards its less energetic outskirts, a result similar to that found earlier for wakes and jets. The effects of mean convection are fairly small everywhere, which implies that the boundary layer is approximately in local energy balance.

Boundary-layer intermittency is similar to that of jets and wakes. Bulges in the turbulence interface, reflecting large eddies in the outer layer, are convected downstream and cause entrainment of laminar fluid from outside. On the average, this results in the growth of the boundary layer with streamwise distance in the locally linear fashion which was described earlier. The wedge angle of the layer is considerably smaller than for jets, a result of the comparative weakness of boundary-layer turbulence at comparable mean-flow velocities.

PRESSURE GRADIENT AND SURFACE ROUGHNESS EFFECTS

In this subsection we assume that the surface is flat[25] and the external flow irrotational, so that there is no curvature or external vorticity. Qualitatively, the effects of pressure gradients are similar in turbulent and laminar boundary layers (recall the discussion of laminar layers in Section 5.1). The sign of the pressure gradient is crucial, and gradients can be classified as either adverse if they tend to decelerate the fluid in the layer or favorable if they act to accelerate the flow. In terms of the parameter Π, defined by (5.181), $\Pi > 0$ yields an adverse gradient, while $\Pi < 0$ corresponds to a favorable one. We saw earlier that $|\Pi|$ measures the importance of the pressure gradient in the outer layer, whereas the viscous sublayer is relatively insensitive to perturbing effects. Thus, one would expect $|\Pi| = O(1)$ to yield significant effects of the pressure gradient in the outer layer, whereas, for small $|\Pi|$, such effects should be unimportant.

An adverse gradient, $dU_\infty/dx < 0$, implies $\overline{U_y} > 0$ in the outer layer, according to (5.193). That is, the mean flow is directed outwards from the surface, which tends to cause the boundary layer to thicken with streamwise distance. This effect is reinforced by transverse turbulent momentum transfer, which tries to thicken the layer no matter what the sign of the pressure gradient. As a result, the boundary-layer thickness increases rather quickly. This is also apparent from the Von Karman equation, (5.180), in which the term with Π as factor represents the pressure gradient. If $\Pi > 0$, the skin friction and pressure gradient terms of (5.180) act in consort, causing the momentum thickness to increase more quickly than in the absence of a gradient. From the definition, (5.181), of $\Pi = (\delta_d/\tau_w)(dP_\infty/dx)$, we see that the factor δ_d/τ_w, which multiplies the pressure gradient, will tend to increase due to two effects. Firstly, as the layer thickens, δ_d will grow, and secondly, deceleration of the fluid due to the pressure gradient leads to lower shear across the near-wall region, and hence to reducing skin friction, τ_w. Both effects cause Π to increase at constant dP_∞/dx, thus amplifying the pressure gradient term in (5.180). A uniform adverse

[25] Although later in this subsection we will consider rough surfaces, the size of the roughness elements will be taken to be small compared with the layer thickness and, at that scale, the surface will be assumed flat.

pressure gradient therefore has an increasing influence as the boundary layer develops and is self-reinforcing in the sense that it brings about thickening of the layer and decreased skin friction, which leads to a larger value of Π, and hence greater influence of the pressure gradient. A constant adverse pressure gradient results in increasingly rapid thickening of the boundary layer.

As well as causing the boundary layer to thicken, an adverse gradient tends to slow the fluid in the layer. If it is sufficiently strong and maintained for sufficiently long, the fluid near the surface can be brought to rest, resulting in separation of the boundary layer and reverse flow downstream. Because the fluid outside the layer continues to move forwards, turbulent momentum transfer with the external flow resists the decelerating effects of the adverse gradient. The mean flow near separation is no doubt similar to that shown in Figure 5.4 for a laminar layer, although, since we are now dealing with turbulent flows, there are fluctuations about the mean from realization to realization, so a precise definition of the separation point and the mean flow in its vicinity is not especially meaningful. However, the skin friction, τ_w, changes sign at separation, owing to the reverse flow at the surface, and the zero of τ_w is conventionally taken as a definition of the separation point in a turbulent boundary layer. Using (5.131), this definition can be shown to coincide with the location at which a dividing streamline of the mean flow branches out from the surface, although the reader is cautioned that mean-flow streamlines are not physically meaningful in highly fluctuating flows such as those near separation of a turbulent boundary layer.

Separation always occurs for bluff bodies at high Reynolds numbers, where the separated boundary layer forms the edges of the near wake. Turbulent boundary layers are more resistant to separation than laminar ones, as witnessed by the fact that the separation point moves downstream when boundary-layer transition occurs (cf. Figure 1.8). Immediately behind the body, at sufficiently high Reynolds number, the near wake consists of relatively slowly moving fluid bounded by thin shear layers that are the product of boundary-layer separation. Since the near wake is comparatively slowly moving, it has approximately uniform pressure, which is maintained across the shear layer because the layer cannot support a significant pressure difference. The flow outside the wake region is rapidly moving and, using a rough-and-ready Bernoulli's theorem argument, the pressure there is lower than at the nose of the body. Thus, the pressure over the rear part of the body is lower than at the front, which is the origin of the significant form drag that is observed when the boundary layer separates.

The flow downstream of boundary-layer separation is outside the scope of the present discussion. Thus, if separation occurs, we restrict attention to the attached layer upstream of the separation zone. When separation is approached, the boundary layer thickens and the skin friction drops, as noted earlier. Figure 5.16 illustrates the qualitative changes in the mean flow and Reynolds stress profiles in the outer layer as separation is approached (the profiles have been normalized and are shown as functions of $\eta = y/\delta$; bear in mind that δ is increasing). As separation is approached, the shear stress near the wall drops, because u_* falls, and the location of maximum shear stress moves away from the wall. At the same time, the mean velocity takes on a characteristic inflectional form with growing velocity defects in the outer part of the layer. By the time separation is reached (bottom right), the shear stress and mean-flow

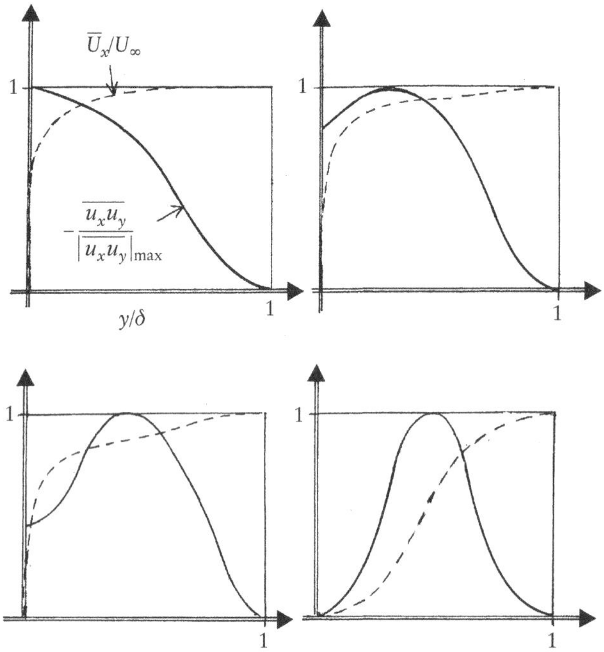

Figure 5.16. Sketches of the profiles of $-\overline{u_x u_y}$ and $\overline{U}$ in the outer part of a turbulent boundary layer at various locations moving toward separation from left to right and downwards. Separation is attained at the bottom right.

profiles resemble those of a shear layer rather than a normal boundary layer. Clearly, a turbulent boundary layer near separation has quite different properties, and becomes more like a shear flow than a wall-bounded flow. This is reasonable, since, following separation, the vorticity shed from the boundary layer does indeed form a shear layer.

The case of a boundary layer with a favorable pressure gradient is quite different. Equation (5.193) now indicates that $\overline{U_y} < 0$, so the mean flow tends to cause thinning of the layer and acts in opposition to the ever present thickening effects of turbulent momentum transfer across the layer. This is also apparent from (5.180), where the skin friction and pressure gradient terms are now of opposite signs. The favorable gradient therefore slows down the thickening of the layer, or causes it to become thinner if it is sufficiently strong. The gradient also accelerates the fluid, increasing the shear near the wall and hence the skin friction. If the layer becomes thinner, the combined effects of falling δ_d and increasing τ_w, cause the factor δ_d/τ_w in (5.181) to decrease, so that Π drops at constant dP_∞/dx. Thus, in contrast with an adverse gradient, whose results are self-reinforcing, a favorable gradient tends to bring about changes in the boundary layer which decrease its own importance. That is, a strong enough favorable pressure gradient has self-attenuating effects.

As $|\Pi|$ is increased from small values, at which the pressure gradient can be neglected, the effects of the pressure gradient on the outer layer become stronger, while the viscous sublayer is not greatly modified at this stage, except in so far as the value of u_*, and hence the sublayer scalings, is altered by the presence of the gradient. In particular, (5.198) should still hold in the viscous sublayer, so that the log law resulting from (5.202) applies over an inertial layer of small y/δ. With increasing $|\Pi|$, the outer layer is strongly modified by the pressure gradient, while the range of validity of the log law, expressed in terms of $\eta = y/\delta$, is found to decrease. Given the self-attenuating effects of a favorable gradient, large negative values of Π are hard to attain. However, as separation is approached with an adverse gradient, the skin friction drops toward zero (i.e., $\tau_w, u_* \to 0$) and $\Pi \to \infty$, according to (5.181). We implicitly assume that $|\Pi| = O(1)$ in most of the remainder of this chapter, excluding the neighborhood of separation unless otherwise stated.

In developing a quantitative theory of boundary layers with pressure gradients, as we will now attempt to do, it is best to make clear from the start that we are on much less solid ground than for the flat plate of the previous subsection. The difficulty is that no equivalent of the outer-layer defect law, (5.203), has been found for general $U_\infty(x)$. This is not surprising, since the boundary layer at any streamwise location depends on the entire streamwise history of $U_\infty(x)$ up to that point, that is it has memory and integrates the effects of varying $U_\infty(x)$ in some complicated nonlinear way. The same is true of laminar layers, of course, but one can always integrate the Prandtl equations numerically with the specified $U_\infty(x)$, whereas, in the turbulent case, one is stymied by the closure problem.

To make progress, we suppose that it is possible to adjust $U_\infty(x)$ so as to produce a class of special boundary layers with pressure gradients, known as self-preserving layers,[26] which are such that their outer layers have self-similar velocity defects (cf. Townsend (G 1976), chapter 7, for a detailed discussion of self-preserving layers from a somewhat different perspective than that adopted here). That is, the velocity defect in the outer layer is given by (5.203), where $F(\eta)$ does not vary with streamwise position and $\eta = y/\delta(x)$ is the similarity variable. The flat-plate boundary layer considered in the previous subsection is one member of this class, but we assume that there are others having pressure gradients. The gradients are not uniform, but carefully chosen so that the outer layer has a self-similar velocity defect profile. If such self-preserving layers exist, as presumed in what follows, they are analogous to the Falkner–Skan laminar boundary layers, which are self-similar solutions of the Prandtl equations with power laws for $U_\infty(x)$ as a function of x.

Since the viscous sublayer is insensitive to the pressure gradient, we continue to use (5.198) there. Matching of (5.198) and (5.203) requires that $F(\eta)$ have the logarithmic form (5.208) as $\eta \to 0$, where b is determined from (5.209) and is independent of x, since $F(\eta)$ is, but varies from one self-preserving layer to another. In particular, the constant b will generally not have the same value as in a layer with zero pressure gradient. The same is true of the constants I_1 and I_2, given by (5.214) and (5.215).

Since the Reynolds number is large, (5.209) implies that U_∞/u_* is also large, and of lower order than Re_*, owing to the logarithm. Furthermore, as for the flat plate, the logarithmic dependence implies that U_∞/u_* will be a slowly varying function of x, and can be locally approximated as constant for the self-preserving layer. We assume that the same is true of Π, which is plausible because Π is a parameter measuring the influence of the pressure gradient on the outer layer. If Π were to undergo rapid changes, this would no doubt be reflected in the velocity defect profile, $F(\eta)$, which is disallowed by the self-similar nature of the outer part of the self-

[26] If the wall were roughened, a case we examine later, outer-region self-similarity could also be obtained by allowing the roughness properties to vary appropriately with streamwise distance.

preserving layer. On the other hand, Π may, and probably does, vary slowly with streamwise distance in the self-preserving layer, owing to the changing relationship between the viscous sublayer and outer layer with Reynolds number, as reflected in the gradually evolving U_∞/u_*. In any case, we shall suppose that Π, like U_∞/u_*, is locally constant.

To derive the equivalent of (5.216) for a self-preserving layer with a pressure gradient, we proceed as follows. From (5.181) and (5.213), we have

$$\frac{dU_\infty}{dx} = -\frac{u_*}{I_1\delta}\,\Pi \tag{5.230}$$

which can be combined with the x-derivative of (5.210) to obtain

$$\frac{d}{dx}\left(\frac{u_*}{U_\infty}\right) = \frac{u_*}{U_\infty\delta\left(\kappa\dfrac{U_\infty}{u_*}+1\right)}\left\{\frac{u_*}{U_\infty}\,\Pi - \frac{d\delta}{dx}\right\} \tag{5.231}$$

Using (5.212), (5.213), and (5.231) in (5.180), we find

$$\frac{d\delta}{dx} = \frac{1+\kappa\dfrac{U_\infty}{u_*}+\Pi\left\{2+\kappa\left(\dfrac{3U_\infty}{u_*}-\dfrac{2\mathrm{I}_2}{\mathrm{I}_1}\right)\right\}}{\kappa\mathrm{I}_1\left(\dfrac{U_\infty}{u_*}\right)^2-\kappa\mathrm{I}_2\dfrac{U_\infty}{u_*}+\mathrm{I}_2} \tag{5.232}$$

generalizing (5.216), to which it reduces when $\Pi = 0$. Equation (5.232) describes the thickening or thinning of a self-preserving layer, which depends on U_∞/u_*, Π, the unknown constants I_1 and I_2, whose values are different with a pressure gradient, and the Von Karman constant, $\kappa \approx 0.39$.

Since U_∞/u_* and Π vary slowly with x and can therefore be approximated as locally constant, $d\delta/dx$ is also slowly varying. Thus, a self-preserving boundary layer develops locally as a straight-sided wedge, of angle determined from (5.232), which varies only slowly as the layer develops. From (5.212) and (5.213), the same holds for δ_m and δ_d, and indeed, locally, the entire outer layer evolves linearly with x for a self-preserving boundary layer. This is not generally true of non-self-preserving layers and it ought to be recalled that $U_\infty(x)$ needs to be carefully adjusted to obtain such self-preserving layers. According to (5.232), the layer thickens or thins depending on whether $\Pi > \Pi_0$ or $\Pi < \Pi_0$, where

$$\Pi_0 = -\frac{\kappa\dfrac{U_\infty}{u_*}+1}{\kappa\left(3\dfrac{U_\infty}{u_*}-2\dfrac{I_2}{I_1}\right)+2} \tag{5.233}$$

which can be shown to lie in the range $-1 < \Pi_0 < -\frac{1}{3}$ (assuming δ_m and δ_d are positive, so that $I_1 > 0$ and $I_2/I_1 < U_\infty/u_*$, according to (5.212) and (5.213)). Thus, thinning always occurs if $\Pi < -1$ and never happens when $\Pi > -\frac{1}{3}$. In particular, as noted earlier, an adverse gradient, $\Pi > 0$, leads to thickening of the layer, whereas a favorable gradient can produce thinning if it is sufficiently strong. The high Reynolds number limit, $U_\infty/u_* \to \infty$, of (5.233) gives $\Pi_0 = -\frac{1}{3}$ as the borderline between thinning and thickening of a self-preserving layer. Incidentally, this indicates that pressure gradients as small as $|\Pi| = \frac{1}{3}$ may have important effects.

We next want to use a local analysis, taking constant values of U_∞/u_* and Π, to determine approximately the form of $U_\infty(x)$ needed to maintain a self-preserving layer. Locally, $\delta(x)$ is either constant if (5.232) is zero, or

$$\delta(x) = (x - x_0)\,\delta' \tag{5.234}$$

where $\delta' = d\delta/dx$ is the constant given by (5.232) and x_0 is an unknown origin. If $\delta' > 0$, the layer thickens, spreading out linearly from $x = x_0$, which lies upstream of the location considered. On the other hand, when $\delta' < 0$, the layer thins, shrinking towards the point $x = x_0$, which lies downstream. In that case, it tends to zero thickness as $x = x_0$ is approached, which is clearly unphysical. Moreover, as we shall see, the external velocity, $U_\infty(x)$, needed to maintain the self-preserving layer approaches infinity there. Such a singularity calls into question the approximations used in deriving these results, but suggests that self-preserving layers that become thinner with streamwise distance cannot be maintained beyond a finite range and require increasingly large pressure gradients as that limiting range is approached.

Using (5.230) and (5.234), we obtain an equation for $U_\infty(x)$, which can be integrated, taking constant U_∞/u_* and Π, as before. Thus, we find the local approximations

$$U_\infty \propto |x - x_0|^{-u_*\Pi/I_1 U_\infty \delta'} \tag{5.235}$$

if $\delta' \neq 0$, or

$$U_\infty \propto \exp\left[-\frac{u_*\Pi}{I_1 U_\infty \delta}\,x\right] \tag{5.236}$$

in the special case $\delta' = 0$, that is, when $\Pi = \Pi_0$. The signs of proportionality in (5.235) and (5.236) indicate that there are multiplicative constants of integration. In reality, both these multiplicative factors and the power exponent in (5.235) are slowly varying with x, rather than constants. The power law form, (5.235), of $U_\infty(x)$ for self-preserving layers brings to mind the Falkner–Skan self-similar laminar layers, which also have power laws for the external velocity (see Batchelor (G 1967), section 5.9). Since δ', given by (5.232), is slowly varying with x, the special exponential form, (5.236), representing the borderline between layer thinning and thickening, will not generally persist.

With an adverse gradient, that is, $\Pi > 0$, we saw earlier that $\delta' > 0$, so that the exponent in (5.235) is negative, giving a pressure gradient which decreases with increasing streamwise distance. As discussed before, an adverse gradient has self-reinforcing effects on the boundary layer and therefore needs to be continuously decreased with streamwise distance to maintain self-preservation. For favorable gradients, sufficiently strong as to produce thinning, $\delta' < 0$ and the exponent is again negative, but now $x = x_0$ lies downstream and the result is an increasing pressure gradient with an infinite singularity at $x = x_0$, as noted above. This reflects the self-attenuating effects of a favorable gradient, which needs to be continuously increased to maintain a self-preserving layer.

As $U_\infty/u_* \to \infty$, the limit of (5.232) is

$$\frac{d\delta}{dx} = \frac{u_*}{U_\infty I_1}(1 + 3\Pi) \tag{5.237}$$

where the factor $1 + 3\Pi$ will be recognized from the leading-order Von Karman equation, (5.185), and from the corresponding limiting value $\Pi_0 = -\frac{1}{3}$, noted above. In this limit, the power-law exponent in (5.235) becomes $-\Pi/(1 + 3\Pi)$. As usual, these leading-order, high Reynolds number limiting expressions are probably not very precise because the increase of U_∞/u_* with increasing Reynolds number is only logarithmic and therefore the limit is approached rather slowly.

The self-preserving layers considered above are a rather special class of boundary-layer flows, but have the virtue of allowing some analytical progress. Based on empirical evidence, Coles (1956) proposed that more general $U_\infty(x)$ result in a single-parameter family of outer layers having the velocity defect law (5.203) with

$$F(\eta) = \frac{1}{\kappa} \log \eta + b\left(1 - \frac{1}{2} w(\eta)\right) \tag{5.238}$$

in the outer layer. The so-called wake function, $w(\eta)$, is supposed universal and gives an outer-layer correction to the log law. Coles found that

$$w(\eta) = 1 - \cos \pi\eta \tag{5.239}$$

fitted the experimental data well. The single parameter b describes the variations between different boundary layers and different streamwise locations within a given layer. It will be noted that $w(\eta) \to 0$ as $\eta \to 0$, so that (5.238) takes on the form (5.208) near the surface, as it must to match to viscous sublayer. Thus, the parameter b may be obtained from the matching relation, (5.209) or (5.210) and can vary with streamwise location. Using (5.238) to evaluate the integrals in (5.214) and (5.215), we obtain

$$I_1 = \frac{1}{\kappa} - \frac{1}{2} b \tag{5.240}$$

and

$$I_2 = \frac{2}{\kappa^2} + \frac{3}{8} b^2 - 1.59 \frac{b}{\kappa} \tag{5.241}$$

which lead to $I_1 = 3.6$, $I_2 = 23$ in the zero gradient case, $b = -2$, values that are not far from those given earlier.

In adverse gradients $b < -2$, while $b > -2$ corresponds to a favorable one. It is, however, doubtful whether (5.238) applies with significantly favorable pressure gradients, so we restrict attention to adverse ones in what follows. Figure 5.17 shows the function $b + 2 - F(\eta)$, according to (5.238) with (5.239), for a number of values of $b \leq -2$, of which $b = -2$ represents the zero gradient layer and increasingly negative b illustrate the effects of adverse gradients. These graphs show the scaled outer-layer velocity defect, $-F(\eta)$, shifted vertically by the amount $b + 2$, so that their log-law asymptotes coincide. The reader may imagine displacing the curves upwards to give the value zero at $\eta = 1$, leading to graphs of the velocity defects themselves. It will be noted that, as remarked earlier, the region of applicability of the log law decreases as the effect of the pressure gradient becomes stronger.

We may combine Coles' outer-layer defect form, (5.238), and the viscous sublayer expression, (5.198), to obtain Coles' composite form

$$\frac{\overline{U_x}}{u_*} = f(y_+) - \frac{1}{2}bw(\eta) \qquad (5.242)$$

for the mean velocity, valid throughout the boundary layer. It reduces to (5.198) in the viscous sublayer, since $w(\eta) \to 0$ as $\eta \to 0$, and to (5.203) with (5.238) in the outer layer, by virtue of (5.202) and the matching condition, (5.209).

The limit $b \to -\infty$ corresponds to very strong adverse gradient effects and has been conjectured to correspond to separation. In that case, $u_* \to 0$ (to represent separation) and (5.203) with (5.238) show that

$$V_x \sim u_* b\left(1 - \frac{1}{2}w(\eta)\right) \qquad (5.243)$$

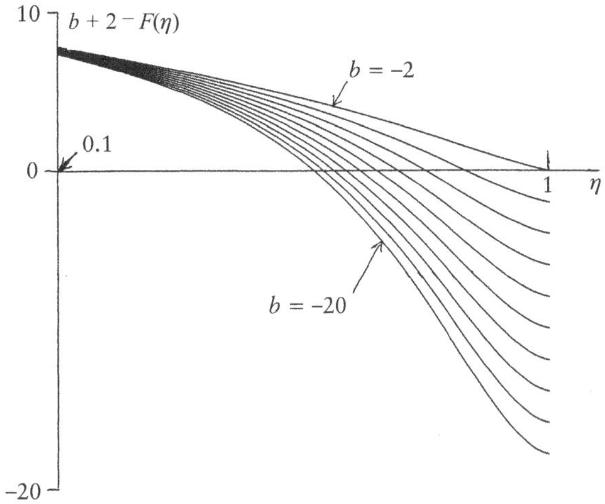

Figure 5.17. Coles' outer-layer velocity defect function at equally spaced values of the parameter $b \leq -2$ representing adverse pressure gradients. To obtain the velocity defect, each curve should be shifted vertically to make the defect zero at $\eta = 1$. The vertical scale is linear, the horizontal one logarithmic.

The logarithmic part has now disappeared, implying that there are no longer large velocity gradients near the surface. If we apply the no-slip condition, $V_x = -U_\infty$ at $y = 0$, to (5.243), we have $b \sim -U_\infty/u_*$, leading to

$$\overline{U_x} = U_\infty + V_x = \frac{1}{2}U_\infty w(\eta) = \frac{1}{2}U_\infty(1 - \cos \pi\eta) \qquad (5.244)$$

which indicates a sinusoidal velocity distribution across the layer at separation. We discussed the approach to separation earlier, and indeed, as sketched in Figure 5.16, the mean-velocity profile appears to take on an inflectional form, qualitatively similar to (5.244), and resembling that of the nascent shear layer to which the boundary layer gives rise downstream.

Since Coles' velocity defect form, (5.238), is intended for boundary layers with general $U_\infty(x)$, it should also apply to self-preserving layers, for which the value of b is constant. Thus, self-preserving layers form a one parameter family whose velocity defect functions, $F(\eta, b)$, are representative of more general boundary layers with pressure gradients. Figure 5.17 can therefore also be interpreted as representing the velocity defect in the outer parts of different self-preserving layers with adverse gradients.

Given the two quantities $\delta(x)$ and $b(x)$, for a general $U_\infty(x)$, one can calculate $u_*(x)$ from the matching condition, (5.210). The boundary-layer mean flow can then be determined using Coles' expression, (5.242). One equation describing the streamwise development of the two unknown quantities, $\delta(x)$ and $b(x)$, can be obtained from the Von Karman equation, (5.180), in which Π contains the pressure gradient via (5.181), and δ_m, δ_d may be determined from (5.212), (5.213) with (5.240), (5.241). However, we are missing an equation. Thus, Coles' expressions are not sufficient to allow calculations of boundary-layer development with general

$U_\infty(x)$. It should also be borne in mind that they are based on a finite amount of experimental data and may not apply to all $U_\infty(x)$, particularly extreme cases.

We now want to change the subject and briefly consider another perturbing effect on turbulent boundary layers, namely *surface roughness*. Whereas, we have so far supposed a smooth, flat surface and considered the influence of pressure gradients, we now assume the gradient to be negligible, but that the surface is rough. The reader may imagine a flat plate, whose texture is similar to sandpaper with grains small compared with the total boundary-layer thickness, δ. Provided the roughness height, k, is much less than the viscous sublayer thickness in the absence of roughness, the effects are negligible, but once k becomes comparable with the sublayer thickness, roughness begins to influence the sublayer, and hence the flow as a whole. It may be remarked that, because a laminar boundary layer has no internal structure, the corresponding criterion compares the roughness height and total boundary-layer thickness. Recalling that, everything else being equal, a laminar layer is considerably thicker than a viscous sublayer, it is apparent that turbulent boundary layers are more sensitive to roughness than their laminar counterparts. Furthermore, once k has become sufficient to perturb a laminar layer significantly, the entire boundary layer flow is directly modified because it must pass around the grains of roughness. In contrast, a turbulent layer can be significantly altered by roughness heights much smaller than its overall thickness, which then only have a direct effect on the flow near the surface, but thereby indirectly change the outer layer.

As for a smooth surface let the mean frictional force per unit area be $\tau_w = \rho u_*^2$, thus defining the friction velocity, u_*. It is found that the outer-layer velocity defect law

$$\frac{V_x}{u_*} = F(\eta) \tag{5.245}$$

continues to hold, with the same $F(\eta)$ as for a smooth surface. That is, roughness only influences the velocity defect in the outer layer through modification of u_* and δ (the latter appearing via $\eta = y/\delta$ in (5.245)). This is reasonable because k is assumed sufficiently small that the outer layer is altered only indirectly, through changes to the near-surface flow. As $\eta \to 0$, (5.208) and (5.245) imply that

$$\frac{\overline{U_x}}{u_*} \approx \frac{1}{\kappa} \log \eta + \frac{U_\infty}{u_*} + b \tag{5.246}$$

showing log-law behavior that must match to the flow nearer the surface, where roughness has a direct effect. Here, b has its usual flat-plate, zero-gradient value, as do the constants I_1 and I_2 in (5.212) and (5.213).

The importance of roughness is measured by the parameter $u_* k/\nu$, which is a Reynolds number based on the roughness size. If this parameter is small, the wall may be considered smooth, while, as it grows in magnitude the influence of roughness increases. The flow around the grains then depends on u_*, k, ν, and the geometrical form of the roughness. At distances from the surface large compared with k, $\overline{U_x}$ approaches the logarithmic form

$$\frac{\overline{U_x}}{u_*} \approx \frac{1}{\kappa} \log\left(\frac{y}{k}\right) + \tilde{a} \tag{5.247}$$

to match with (5.246). By dimensional analysis, based on the parameters u_*, k, ν, we see that the nondimensional quantity $\tilde{a}$ should depend only on u_*k/ν and the geometrical form of the roughness. That is, the function $\tilde{a}(u_*k/\nu)$ may depend on the type of roughness used (e.g., close-packed hemispheres, etc.). Equality of (5.246) and (5.247) yields the matching condition

$$\frac{U_\infty}{u_*} = \frac{1}{\kappa} \log\left(\frac{\delta}{k}\right) + \tilde{a} - b \tag{5.248}$$

which expresses u_* in terms of δ for a rough plate. It is analogous to the smooth-plate result (5.209), to which it reduces when $\tilde{a} = a + (1/\kappa)\log(u_*k/\nu)$, the form taken by the function $\tilde{a}(u_*k/\nu)$ in the smooth limit, $u_*k/\nu \to 0$. Equation (5.248) may be combined with the Von Karman equation, (5.211), and equation (5.212), allowing determination of the streamwise development of δ and U_∞/u_* if $\tilde{a}(u_*k/\nu)$ is known.

The quantity $\tilde{a}(u_*k/\nu)$ is found to approach a limiting value, $\tilde{a}_\infty$, for sufficiently large u_*k/ν and the surface is then said to be fully rough. The flow around the roughness grains is then of high Reynolds number and it is plausible that its overall properties, such as $\tilde{a}$, should no longer depend on the fluid viscosity and hence on u_*k/ν. The constants k and $\tilde{a}_\infty$ are geometrical properties of the roughness and completely characterize a fully rough surface in so far as the streamwise development and outer layer are concerned. Observe that, according to (5.248), u_* increases with k, everything else being equal. That is, the skin friction grows with the roughness height for a fully rough surface, which is as one might intuitively expect, but need not hold for surfaces which are not fully rough.

We have deliberately avoided giving a precise definition of the roughness height, k, and a variety of definitions might be used for a given rough surface, differing by numerical factors. Changing the definition of k also modifies $\tilde{a}$, in such a way that the combination $ke^{-\kappa\tilde{a}}$, occurring implicitly in (5.247), remains the same. It is convenient to normalize k so that $\tilde{a}_\infty$ has the same value for all fully rough surfaces and a fixed value of about $\tilde{a}_\infty = 8$ corresponds to the conventional equivalent sand-grain roughness height, k. The single parameter k then suffices in the fully rough limit of large u_*k/ν. For smaller values of u_*k/ν, one also needs the function $\tilde{a}(u_*k/\nu)$.

Assuming a fully rough plate, so that $\tilde{a} = \tilde{a}_\infty$ is constant, and that the roughness properties do not vary with x, equations (5.211) and (5.212) with the x-derivative of (5.248) lead to

$$\frac{d\delta}{dx} = \frac{1}{I_1 \dfrac{U_\infty}{u_*} - \dfrac{I_1}{\kappa} - I_2 + 2\dfrac{I_2}{\kappa}\dfrac{u_*}{U_\infty}} \tag{5.249}$$

which is analogous to the smooth plate form, (5.216), and describes the development of the boundary-layer thickness. According to (5.248), the quantity U_∞/u_* is large and slowly varying, thanks to the assumed large value of δ/k, rather than largeness of the Reynolds number, which plays the same role for a smooth plate. Thus, the boundary layer again develops locally as a straight-sided wedge, whose angle varies gradually with streamwise distance. As $U_\infty/u_* \to \infty$, (5.249) reduces to (5.217), as for a smooth plate. Thus, the distance for development of the boundary layer, $d = O(U_\infty \delta / u_*)$ has the same order as in the smooth case, although the values of U_∞ / u_* and δ, and hence d, will generally be altered by roughness. Equation (5.248) can be rewritten as

$$\delta = k \exp\left\{\kappa\left(\frac{U_\infty}{u_*} - \tilde{a}_\infty + b\right)\right\} \tag{5.250}$$

which can be used in (5.249) to obtain a differential equation for U_∞ / u_*.

In summary, the effects of roughness on the skin friction and flow properties at distances from the wall large compared with k are expressed by $\tilde{a}(u_* k/\nu)$, experimentally determined. When $u_* k/\nu$ is sufficiently low, the wall is effectively smooth, while at large enough $u_* k/\nu$, $\tilde{a} \to \tilde{a}_\infty$ yields a fully rough wall. In between, k is comparable with the viscous sublayer thickness and the function $\tilde{a}(u_* k/\nu)$ depends on the type of roughness.

EXAMPLES

This subsection briefly describes two examples of flows, which although they are not pure boundary layers, show significant effects of walls and hence share some of the characteristics of boundary layers. The first of these is the plane channel flow introduced in the previous chapter, to which discussion the reader should refer for basic properties and notation.

At the entrance to the channel, we imagine a uniform flow profile. This will produce boundary layers on the two walls, $x_2 = 0$ and $x_2 = 2D$, which develop and thicken downstream until they meet at the middle of the channel. The flow then approaches the fully developed regime in which the mean flow is in the x_1-direction and only the mean pressure, $\overline{P}$, depends on the streamwise coordinate. We consider fully developed flow in what follows.

There are viscous sublayers near the walls, in which the velocity is given by (5.198) in terms of u_*. The central part of the channel is similar to the outer part of a boundary layer and the analog of the velocity defect relation, (5.203), is

$$\frac{\overline{U}_1 - \overline{U}_{\max}}{u_*} = F\left(\frac{x_2}{2D}\right) \tag{5.251}$$

where $\overline{U}_{\max}$ is the maximum mean velocity, $\overline{U}_1(x_2 = D)$, and F is a universal function for developed plane channel flow (i.e., it does not depend on the Reynolds number). To match to the viscous sublayers, F must have logarithmic behavior as we approach either of the walls. For instance,

$$F\left(\frac{x_2}{2D}\right) \sim \frac{1}{\kappa} \log\left(\frac{x_2}{2D}\right) + c \tag{5.252}$$

as $x_2 \to 0$, where c is a universal constant. Matching of (5.198) and (5.252) leads to the friction law

$$\frac{\overline{U}_{\max}}{u_*} + \frac{1}{\kappa} \ln\left(\frac{\overline{U}_{\max}}{u_*}\right) = \frac{1}{\kappa} \ln\left(\frac{2\overline{U}_{\max} D}{\nu}\right) + a - c \tag{5.253}$$

which relates $\overline{U}_{\max}/u_*$ to the large overall Reynolds number, $\overline{U}_{\max} D/\nu$. This result indicates that u_* is small compared with $\overline{U}_{\max}$, so, according to (5.251), the mean velocity in the central region of the channel is approximately uniform, as for the outer part of a turbulent boundary layer. As the walls are approached, the log-law zone is entered and there are increasingly large mean-velocity gradients, until the thin viscous sublayer is reached. The near-wall region is much the same as for a boundary layer.

In the channel problem, the mean-wall-pressure gradient is usually regarded as the controlling parameter and determines the friction velocity via

$$\rho u_*^2 = \tau_w = -D\,\frac{dP_w}{dx_1} \tag{5.254}$$

From u_*, one may calculate $\overline{U}_{\max}$ using

$$\frac{\overline{U}_{\max}}{u_*} = \frac{1}{\kappa}\ln\left(\frac{2u_*D}{\nu}\right) + a - c \tag{5.255}$$

showing that $\overline{U}_{\max}$ is equal to u_* times a large, slowly increasing, logarithmic function of the high Reynolds number, u_*D/ν. The volume flux in the channel can be calculated by neglecting the contribution of the viscous sublayers, just as for δ_m and δ_d in a boundary layer, giving

$$\frac{Q_v}{2D} = \int_0^{2D} \frac{\overline{U}_1}{2D}\,dx_2 = \overline{U}_{\max}\left(1 - I_1\,\frac{u_*}{\overline{U}_{\max}}\right) \tag{5.256}$$

from (5.251), where

$$I_1 = -\int_0^1 F(\eta)d\eta \tag{5.257}$$

Thus, the volume flux is determined from the pressure gradient via (5.254)–(5.256). Both c and I_1 are universal constants specific to plane channel flow. Compared with the assumed uniform flow at the channel entrance, $U_1 = Q_v/2D$, the developed mean flow, $\overline{U_1}$, is slower moving near the walls, but has the somewhat higher value, $\overline{U}_{\max}$, at the channel center. During the development phase, the fluid is slowed in the vicinity of the walls and accelerated in the center of the channel, conserving the volume flux through the channel. A technique similar to that used above works for a circular pipe, a case that is left as an exercise for the reader.

Finally, we want to briefly discuss the wall-jet flow, illustrated in Figure 5.18, in which a plane jet exhausts tangentially to a plane wall in ambient fluid at rest. We can divide the flow into two regions: a viscous sublayer at the wall and an outer region. Like a boundary layer, the viscous sublayer is found to scale on the friction velocity, u_*, where $\tau_w = \rho u_*^2$ is the mean wall shear stress. However, the outer region is quite different from a boundary layer, and rather similar to a jet in its overall properties, although it is a jet that is modified significantly by the presence of the wall. Whereas for a boundary layer, the turbulent velocities in the outer region are $O(u_*)$, here they are found to be much larger than u_*. This increased turbulence level is no doubt due to the existence of an inflection point in the mean-velocity profile away from the surface, making the flow more unstable and producing higher levels of turbulence. As noted earlier, inflection points correspond to maxima of shear and tend to act as production centers for turbulence. For the wall jet, there are thus two production centers, one

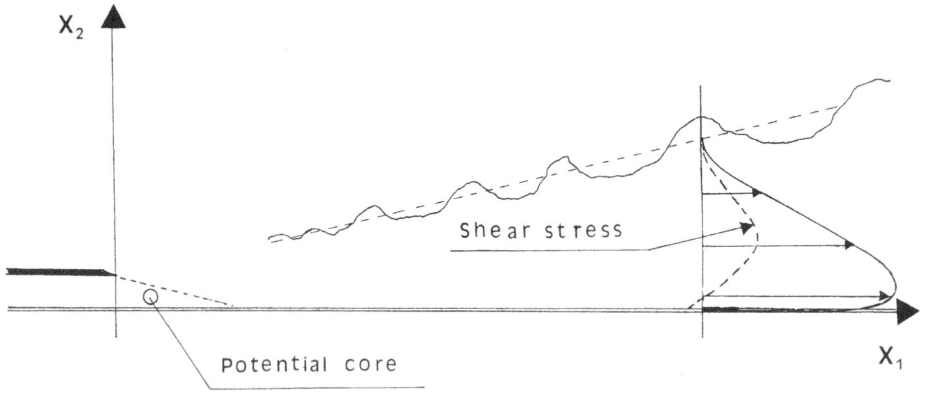

Figure 5.18. Illustration of a plane wall jet.

near the wall due to the high shear there, and similar in nature to that of a boundary layer, the other near around the inflection point in the jet flow away from the wall.

In the case of boundary layers, we used a refined boundary-layer approximation, expressed in terms of the velocity defect. This was necessary to maintain precision in the outer region because the streamwise mean-flow velocity there has only small departures from the external flow, but is not required for the wall jet. Thus, we have equations (5.13) and (5.28), with the viscous term

$$\nu \frac{\partial^2 \overline{U}_1}{\partial x_2^2} \tag{5.258}$$

necessary within the viscous sublayer. Using (5.13), (5.28) can be rewritten as (5.45). Including the viscous term, and integrating across the wall jet yields

$$\frac{dQ_m}{dx_1} = -u_*^2 \tag{5.259}$$

where Q_m, given by

$$Q_m = \int_0^\infty \overline{U}_1^2 dx_2 \tag{5.260}$$

is the jet momentum flux, which decreases with range due to skin friction, according to (5.259).

In Section 5.2 (equation (5.19)), we saw that jets spread over a streamwise distance scale

$$d = O\left(\frac{U_0^2 \Delta}{\overline{u_1 u_2}}\right) \tag{5.261}$$

where U_0, Δ, and $\overline{u_1 u_2}$ are measures of the streamwise velocity, width, and Reynolds shear stress in the jet. Over this distance, the momentum flux decreases by $O(U_0^2 \Delta u_*^2 / \overline{u_1 u_2})$, according to (5.259), which is a fraction $O(u_*^2/\overline{u_1 u_2})$ of the total momentum flux, $O(U_0^2 \Delta)$. Since the turbulent velocities in the outer part of the flow (the jet) are considerably larger than u_*, we would expect that $\overline{u_1 u_2}$ be large compared with u_*^2, and this indeed turns out to be the case. The result is that the jet only loses a small fraction of its momentum flux in the distance required for entrainment and spreading. The momentum flux is approximately constant as the jet spreads, but decreases slowly over longer ranges.

It is found that the wall jet becomes *approximately* self-similar sufficiently far downstream of the nozzle, having the velocity profile

$$\overline{U}_1 = (x_1 - x_0)^{-1/2} f(\xi) \tag{5.262}$$

where $\xi = x_2/(x_1 - x_0)$ is the similarly variable expressing linear spreading of the jet and the factor $(x_1 - x_0)^{-1/2}$ gives a constant momentum flux. This behavior is similar to a jet without a wall; however, the profile, $f(\xi)$, is quite different and has the form sketched in Figure 5.18. The wall jet has a wedge angle which is noticeably less than that of the free jet, because the entrainment process at the outside of the jet is inhibited by the effects of the wall on the largest scales of the jet flow. The decreased jet width means that the maximum velocity is initially higher, to maintain the jet momentum flux at its nozzle value. However, over sufficiently long distances, the decrease of momentum flux due to skin friction causes the maximum velocity to decrease faster than the factor, $(x_1 - x_0)^{-1/2}$, in (5.262) would imply. That is, equation (5.262) holds only locally in x.

The wall jet is not correctly described by the eddy-viscosity approximation (5.34): as noted in Chapter 4, the maximum of $\overline{U}_1$ does not coincide with the zero of $-\overline{u_1 u_2}$ and the eddy viscosity, ν_T, which would be needed to agree with measurements therefore has an infinite singularity and a region of ξ in which it is negative! Even if this difficulty is ignored, the eddy-viscosity approximation is not very helpful because ν_T varies greatly across the flow.

The viscous sublayer, $y_+ = O(1)$, is described by (5.198), as for a boundary layer, but, as one leaves this zone, the behavior begins to depart markedly from that of the boundary layer. The much higher turbulence levels in the outer region percolate down almost to $y_+ = O(1)$. A log-law zone *is* found outside the viscous sublayer, but the effective value of κ appears to be changed to about 0.5, rather than the value $\kappa = 0.39$ that we noted before for boundary layers. Outside the log-law zone, the velocity profile in the outer region bears no resemblance to a boundary layer.

The wall jet is a somewhat extreme example of the effects of turbulence in the external stream of a boundary layer, here originating from the jet flow. As one might expect, the first effect of external turbulence, as its intensity is raised, is to modify the outer part of the boundary layer. The outer layer is perturbed by increased turbulent mixing until it can no longer be clearly distinguished from the external turbulence. This leaves the viscous sublayer and log-law zones, of which the latter is more and more deeply modified by increasing external turbulence. The growing turbulent momentum transfer in the log-law zone tends to make $\overline{U}_1$ more uniform, thereby reducing the slope of $\overline{U}_1$ as a function of $\log y$ and increasing the apparent value of κ in the log law.

NEAR-WALL ORGANIZED STRUCTURES

Before bringing this lengthy chapter to a close, we want to discuss one final aspect of wall-bounded flows, namely the near-wall structure and dynamics of turbulence in *individual realizations* of the flow. Thus, rather than considering simple statistical measures, such as the mean velocity or root-mean-squared fluctuations, as we have until now, attention will be focused on the detailed properties of the time-dependent flow. Work has been going on in this area since the 1950s, mainly experimental studies in the laboratory, but more recently computer simulations of the Navier–Stokes equations have allowed controlled numerical experiments to be carried out, while the development of more sophisticated flow visualization techniques has provided more incisive tools for laboratory analysis of individual realizations (see, e.g., Figures 1.4 and 5.19). The basic idea behind these studies is that turbulence in the near-wall region possesses certain types of flow structures that recur and can be identified, in the process illuminating the physical mechanisms by which turbulence is generated in that zone. It should be noted that, unfortunately, a precise and generally accepted definition of what constitutes such an organized structure is lacking, lending a subjective element to an area that is already largely descriptive. Nonetheless, certain features of the near-wall zone are widely accepted and, in this subsection, we will briefly attempt to throw light on some of these, inevitably (given the nature of the material) coloring the description with our own perceptions. We make no pretense at a complete treatment of the field here, pointing the reader to the review by Robinson (1991) for detailed discussion of and references to the abundant literature.

Figure 5.19. Visualization of a turbulent boundary layer using smoke injection and illumination by thin planes of laser light perpendicular to the flow direction, which runs into the page. (Courtesy of F. Ladhari.)

Many studies have concerned the flat-plate boundary layer without pressure gradient, which is the case we concentrate on. One of the best-established features [27] is that turbulence in the viscous sublayer ($y_+ < 50$, say) possesses structures (often referred to as streaks) that are elongated in the streamwise direction, a tendency that becomes more marked the nearer to the surface one looks. This streamwise elongation is apparent in a number of flow quantities, both from laboratory and numerical experiments, and is associated with longer correlation lengths in the streamwise direction. Figure 5.20 shows a plan view of the instantaneous streamwise fluctuating velocity, u_x, in a numerical boundary-layer simulation. The lighter zones in the figure are strikingly elongated and represent fluid which is instantaneously moving significantly more slowly than the average in the stream-

[27] First explicitly noted by Grant (1958), intensively studied by the team at Stanford (see, e.g., Kline et al. (1967), Kim, Kline, and Reynolds (1971), and the review by Robinson (1991), referred to above) as part of a general attack on the problem of near-wall organized structures.

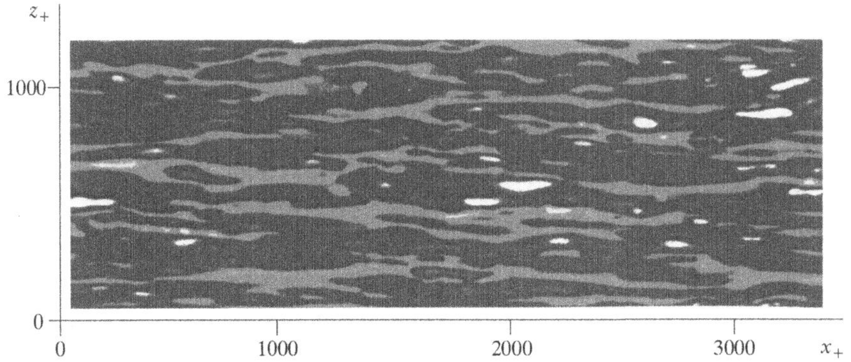

Figure 5.20. Instantaneous streamwise component, u_x, of fluctuating velocity in the plane $y_+ = 3$, illustrating streamwise elongation of turbulent structures. The lighter zones indicate where u_x is negative and $-u_x$ exceeds a certain threshold value. (Robinson (1991), reproduced with permission.)

wise direction, that is, $u_x < 0$ with $-u_x$ above a certain threshold value. This tends to be associated with motion outwards from the wall, that is, $u_y > 0$, which is understandable, since fluid nearer the wall moves more slowly on the average and, when a fluid particle moves away from the wall, it tends to keep this slower streamwise speed. Thus, broadly speaking, there are elongated streaks of slow-moving fluid, which came from nearer the wall, alternating with faster-moving fluid originating from further away. Typical dimensions of the low-speed streaks (in wall units, i.e., multiples of ν/u_*) are of the order of hundreds in the spanwise direction, tens in the wall-normal direction, and thousands in the streamwise direction, thus resembling ribbons. As time goes on, these structures are convected downstream, leading, at a fixed point, to alternation of low- and high-speed fluid. Needless to say, the time-averaged velocity is determined by the mean flow, that is, the negative and positive values of u_x and u_y cancel.

As noted above, low-speed streaks tend to involve outward moving fluid, ejected from the wall by the passage of the flow structure which created the streak. The precise nature of the associated flow structures has been the subject of considerable debate and a number of different theoretical models, most of them involving streamwise vortices of one form or other, have been proposed. In our view, there is considerable confusion surrounding the nature and role of near-wall vortices, which we will attempt to shed some light on later, but for the moment we continue to describe the observed behavior. As well as ejections, there are motions toward the wall, which tend to be associated with the high-speed fluid between streaks, and which are often referred to as sweeps. Ejections and sweeps alternate in time at a fixed point, appearing as a quasi-cyclic process. Following Wallace, Eckelmann, and Brodkey (1972), it is traditional to represent results in a u_x–u_y plane, with the quadrants numbered in the usual fashion, anticlockwise beginning with $u_x > 0, u_y > 0$. The second quadrant, $u_x < 0, u_y > 0$, represents ejections, while the fourth, $u_x > 0, u_y < 0$, corresponds to sweeps. The remaining two quadrants represent rarer and less intense events that are of neither of the above types.

As we saw earlier in this chapter, under the boundary-layer approximation, turbulence affects the mean flow via the Reynolds stress component $-\overline{u_x u_y}$, which also appears in the turbulent energy production, $-\overline{u_x u_y}(\partial \overline{U_x}/\partial y)$ for a flat-plate boundary layer without pressure gradient. Using conditional averaging, one can determine the relative contributions made to the average $\overline{u_x u_y}$ by the different quadrants of the u_x–u_y, plane, thus ranking them in importance as far as the mean flow and turbulence production is concerned. Both quadrants two and four, representing ejections and sweeps, yield negative $u_x u_y$ and their contributions are responsible for the overall negative value of $\overline{u_x u_y}$, exceeding those of the other two quadrants, which are smaller owing to the greater scarcity and lesser intensity of the corresponding events. Ejections and sweeps are found to vary in relative importance depending on distance from the surface, with ejections contributing more for higher values of y_+ and sweeps more important nearer the surface (with the changeover occurring at about $y_+ = 15$). It is reasonable that ejections become more significant further away from the surface, since there is more room from which the ejection can originate. This leads to the idea that, in the outer part of the viscous sublayer (and perhaps the inertial sublayer), turbulence is mainly produced and carried away from the surface by ejections, which consequently play a key role in the near-wall turbulence dynamics.

As discussed earlier, the outer part of the boundary layer is different in character to the near-wall zone, resembling free shear flows such as jets and wakes. The main outer-layer structure apparent in individual realizations is the frontier between turbulent and laminar flow (see Figure 5.1d), whose position fluctuates due to bulges of turbulence that are convected downstream, leading to turbulent intermittency. Preferential elongation of turbulence structures in the streamwise direction is not observed in this region. If anything, flow visualizations, such as Figure 1.4, show structures which slope downstream and outwards from the wall at an angle of roughly 45°. The laminar flow may be thought of as having to pass around a somewhat porous (due to entrainment), corrugated, and unsteadily deforming body, whose surface is the boundary of the turbulent zone. This suggests, as has been observed, that regions of high shear, similar to boundary layers over the surfaces of solid bodies, might form over the upstream part of a bulge in the turbulence frontier, with wakelike regions behind the bulges.

It is time to make some attempts at explanation, beginning with the question: Why are turbulence structures found to be elongated in the streamwise direction in the viscous sublayer, but not further away? The obvious candidate is the high mean shear, $s = \partial \overline{U_x}/\partial y$, which grows like y^{-1} as y decreases through the inertial sublayer and reaches a maximum at the wall. The basic idea is that turbulent structures will be sheared out by mean-flow convection and hence end up aligned in the streamwise direction, an idea which we believe to be fundamentally correct, but which merits further examination. Turbulence is subject to the mean shear s, which has the dimensions of an inverse time, with s^{-1} giving the time scale required for mean shear to act. Turbulence is also subject to self-interactions, evolving on a time scale of $O(L/u')$ in the absence of mean shear. The ratio, sL/u' is a nondimensional measure of the importance of mean shear. If it is sufficiently large, one can neglect the self-interaction of turbulence to a first approximation, only taking into account the effects of mean shear, whereas, when it is small, the mean shear can be neglected. The quantity L is a measure of the size of the large scales of turbulence and is found to increase with distance from the wall, roughly proportional to y. Taking $L = y$ and using the measured profiles of $\overline{U_x}$ and u' leads to estimates for sL/u' as a function of distance from the wall, which is found to decrease through the viscous sublayer, from around six at the wall to a constant value of about 1.5 in the inertial layer. This suggests that mean shear should be more important in the part of the viscous sublayer near the wall, hence the tendency to streamwise elongation there. Having attained a minimum in the inertial sublayer, sL/u' increases with distance from the wall in the outer region due to the decreasing intensity of turbulence, which might suggest that mean shear again be more significant toward the outskirts of the layer. However, since (for reasons that are unclear) streamwise elongation is not observed in that part of the boundary layer, we concentrate on the near-wall zone.

If the mean shear is sufficiently large, its effects on the turbulence dominate self-interaction. This is the regime of applicability of linear theory, in which the nonlinear term in the equations, (4.31), for the fluctuations is neglected. Furthermore, in order to simplify the problem and render it mathematically tractable using spectral analysis, we consider the case of homogeneous turbulence in an unbounded domain subjected to the uniform mean shear, $\overline{U_x} = sy$, where s is a constant. As a final simplification, viscous effects are neglected, so that linear theory becomes rapid-distortion theory. Obviously, this is only a very crude model of a viscous sublayer, but one which shows some of the features of near-wall turbulence. The flow was already used as an example illustrating the effects of mean shear on turbulence in Section 4.5 (see "Homogeneous Turbulence Subject to Uniform Shear").

The quantitative consequences of rapid-distortion theory for homogeneous turbulence subjected to simple mean shear can be worked out using spectral analysis, that is, Fourier decomposition of the velocity field, a technique introduced in Chapter 6 (although there we will only examine the simplest case of turbulence in the absence of mean shear). It is found that the turbulence does indeed become elongated in the streamwise direction at large times, due to shearing out by the mean flow and preferential amplification of certain Fourier components of the initial velocity field. Furthermore, the streamwise component, u_x, of the fluctuating velocity is predicted to grow with time, dominating the other components at large times. This is in accord with the significantly larger value of u_x' in the viscous sublayer apparent in Figure 5.14. The crudeness of the model used should be borne in mind, as should the fact that, whereas a boundary layer is a steady flow, the modeled flow is statistically unsteady. Nonetheless, some of the correct trends are apparent, which suggests that mean shearing is responsible for structuring the turbulence in the viscous sublayer, as well as preferring streamwise velocity fluctuations. In particular, the presence of the wall would seem to be unnecessary, except as the means of producing the strong shear, leading in turn to highly anisotropic turbulence in the near-wall region.

As remarked earlier, many models that have been proposed to explain the near-wall behavior of turbulence involve streamwise vortices. The first point to note is that the mean vorticity is spanwise and dominates the fluctuating vorticity in the viscous sublayer (the parameter sL/u' provides a measure of their relative magnitudes and is large in the region in which streamwise elongation is most apparent, as observed earlier). Thus, it is difficult to see how streamwise vortices of the total flow, mean plus fluctuation, could arise. Furthermore, of the fluctuating velocity components, u_x' is the largest, and the only one which does not appear in the expression, $\omega_x = \partial u_z/\partial y - \partial u_y/\partial z$, for the streamwise fluctuating vorticity. In consequence, even the fluctuating vorticity is mainly in the spanwise direction, again suggesting that streamwise vortices are unlikely. This is not to say that the flow does not have streamwise turbulence structures, structures that are associated, as always in incompressible flow, with vorticity. Indeed, we saw earlier that rapid-distortion theory predicts the appearance of such flow structures, but they are not dominated by streamwise, but rather by cross-stream, vorticity. In our view, the concept of streamwise vortices has produced significant confusion in the literature on near-wall turbulence and care needs to be taken to clearly define terms.

Attempts have been made to *reduce* wall skin friction by deliberately introducing specially chosen types of roughness, known as riblets, usually taking the form of grooves aligned with the streamwise direction whose dimensions are comparable with the thickness of the viscous sublayer. If the riblet geometry is appropriately matched to the flow, this has been found to reduce the skin friction somewhat (by the order of 10%). This is thought to work because the streaks are channeled by the grooves, becoming more stable and less subject to ejections, thus reducing the mean momentum transfer and hence the drag. Other means for drag reduction have also been suggested, including placing airfoil-like devices known as LEBUs (large-eddy break-up devices) in the outer part of the boundary layer, wall suction, polymer injection, and so on (see, e.g., Savill, Truong, and Ryhming 1988). Many of these techniques have been found to have beneficial effects on skin friction in the laboratory and some are under industrial development.

Finally, we note that one can also study the structures of turbulence present in single realizations of jets, mixing layers, and wakes. These flows are rather different because their mean-velocity profiles have inflection points that are associated with Kelvin–Helmholtz type instabilities and act as turbulence production centers in the heart of the flow, rather than at a wall. Jets in particular are efficient amplifiers of perturbations of all sorts and may be persuaded to produce specific organized structures by appropriate matching of external perturbations (e.g., acoustic in origin) to the natural modes of instability of the jet. The latter can be calculated using an inviscid, linear model (i.e., rapid-distortion theory) and many features of jets and mixing layers can be explained (even quantitatively in some cases) in terms of such modes and their nonlinear interactions. We refer the interested reader to, for instance, Crow and Champagne (1971), and Ho and Huerre (1984), among many others in the extensive literature on this subject.

5.6 Conclusions

The turbulent flows considered in this chapter are of considerable fundamental and practical interest. Even given the space devoted to them here, we have far from exhausted all aspects of these classes of flows. For instance, to name but a few, we have not discussed three-dimensional boundary layers, ones with suction/blowing at the surface, and effects of compressibility (when the Mach number is not negligible).

The constant interplay between theoretical analysis and experiment that is characteristic of much of the theory of turbulence has been apparent in this chapter. This is, in part, a reflection of the fundamental closure problem: experimental data is often referred to (nowadays including numerical simulations of the full Navier–

Stokes equations) in order to overcome the lack of a reliable closure approximation or as a means of verifying and calibrating some proposed theoretical model, generally heuristic in origin.

In the next chapter, we turn attention to the spectral theory of *homogeneous* turbulence, which provides a simpler case than the inhomogeneous flows of this chapter, allowing the development of more incisive analytical tools for the study of turbulent flows.

Appendix: A Higher-Order Boundary-Layer Momentum Balance Equation

In this appendix, we give a brief derivation of equation (5.188). A mean momentum audit of the boundary-layer flow yields

$$\oint \left(\overline{U_i U_j} - \frac{1}{\rho} \overline{\sigma_{ij}} \right) n_j ds = 0 \tag{5.A1}$$

where the integral is taken over any closed curve in the x_1–x_2 plane within the flow, n_i is a unit vector, normal to the curve, and σ_{ij} is the fluid stress tensor. Likewise, for the notional, inviscid flow, $\mathbf{U}^{(inv)}$, we have

$$\oint \left(U_i^{(inv)} U_j^{(inv)} + \frac{1}{\rho} P^{(inv)} \delta_{ij} \right) n_j ds = 0 \tag{5.A2}$$

and the difference between (5.A1) and (5.A2) is

$$\oint T_{ij} n_j ds = 0 \tag{5.A3}$$

where

$$T_{ij} = \overline{U_i U_j} - U_i^{(inv)} U_j^{(inv)} - \frac{\overline{\sigma_{ij}} + P^{(inv)} \delta_{ij}}{\rho} \tag{5.A4}$$

which goes to zero outside the boundary layer, where the mean flow, $\overline{\mathbf{U}}$, approaches the external flow, $\mathbf{U}^{(inv)}$, and viscosity is insignificant.

We choose the bounding curve to consists of two straight lines of constant x, which are perpendicular to the surface, and two curves of constant y, namely $y = 0$ (the surface itself) and $y = \infty$ (meaning outside the boundary layer). In the limit as the two lines of constant x approach one another, the component of (5.A3) tangential to the surface can be shown to give

$$\frac{d}{dx} \int_0^\infty T_{xx} dy + \frac{1}{R} \int_0^\infty T_{xy} dy - U_\infty V_y \Big|_{y=0} + \frac{\tau_w}{\rho} = 0 \tag{5.A5}$$

where we have used the no-slip condition $\mathbf{U}|_{y=0} = 0$ and the definition $U_\infty = U_x^{(inv)}|_{y=0}$, of $U_\infty(x)$. Integration of (5.138) across the layer, from $y = 0$ to $y = \infty$, yields

$$V_y|_{y=0} = \frac{d}{dx} \int_0^\infty V_x dy \tag{5.A6}$$

which we use to replace $V_y|_{y=0}$ in (5.A5).

From (5.A4), we obtain

$$T_{xx} = (\overline{U_x} + U_x^{(inv)})V_x + \overline{u_x^2} + \Gamma - \frac{1}{\rho}\,\overline{\tau_{xx}} \tag{5.A7}$$

and

$$T_{xy} = \overline{U_x}V_y + U_y^{(inv)}V_x + \overline{u_x u_y} - \frac{1}{\rho}\,\overline{\tau_{xy}} \tag{5.A8}$$

where $\Gamma = (\overline{P} - P^{(inv)})/\rho$, as in the main text, and τ_{ij} is the viscous part of the stress tensor. Equation (5.A5) with (5.A7) and (5.A8) is an exact expression of the boundary-layer streamwise mean momentum balance. It can also be obtained from (5.136), using the incompressibility conditions $\nabla . \overline{\mathbf{U}} = \nabla . \mathbf{U}^{(inv)} = 0$, expressed in terms of x and y, multiplication by $(R + y)/R$, and integration with respect to y.

At leading order, one neglects the second term in (5.A5), sets $U_x^{(inv)} = U_\infty$ in (5.A7), and drops the final three terms in that equation to obtain

$$\frac{d}{dx}\int_0^\infty \overline{U_x}V_x dy + \frac{dU_\infty}{dx}\int_0^\infty V_x dy + \frac{\tau_w}{\rho} = 0 \tag{5.A9}$$

giving the Von Karman equation, (5.179), when the definitions, (5.176) and (5.177), of δ_m and δ_d, and the relationship (5.173) are used.

At the next order, the viscous stress contributions to the integrals in (5.A5), arising from $\overline{\tau_{xx}}$ and $\overline{\tau_{xy}}$ in (5.A7) and (5.A8), remain negligible, because the mean viscous stresses are $O(\mathrm{Re}_*^{-1})$ smaller than the Reynolds stresses in the outer layer, rising to the same order in the viscous sublayer, which is, however, very thin, $O(\mathrm{Re}_*^{-1})$ smaller than the boundary layer as a whole. Thus, the importance of the viscous stresses in (5.A5) is measured by Re_*^{-1}, which is much smaller than the correction terms considered here (in fact, exponentially smaller). We therefore drop the viscous stresses in (5.A7) and (5.A8), so that (5.A5) can be rewritten, using (5.A6)–(5.A8), as

$$\begin{aligned}&\frac{d}{dx}\int_0^\infty \overline{U_x}V_x dy + \frac{dU_\infty}{dx}\int_0^\infty V_x dy + \frac{\tau_w}{\rho} = \\ &\frac{d}{dx}\int_0^\infty \left\{\left(U_\infty - U_x^{(inv)}\right)V_x - \Gamma - \overline{u_x^2}\right\}dy - \frac{1}{R}\int_0^\infty \left\{\overline{U_x}V_y + U_y^{(inv)}V_x + \overline{u_x u_y}\right\}dy\end{aligned} \tag{5.A10}$$

where the right-hand side provides a correction to the leading-order equation, (5.A9).

The main contributions to the right-hand side of (5.A10) arise from the outer layer and we approximate it using leading-order, outer-layer expressions. Thus, we use (5.172) to express $U_\infty - U_x^{(inv)}$, and (5.153), without the viscous term, with (5.155), to show that

$$\frac{\partial}{\partial y}\left(\Gamma + \overline{u_y^2}\right) = \frac{2U_\infty}{R}\,V_x \tag{5.A11}$$

in the outer layer. Thus, from integration by parts,

$$\int_0^\infty \left(\Gamma + \overline{u_y^2}\right) dy = -\int_0^\infty y \frac{\partial}{\partial y}\left(\Gamma + \overline{u_y^2}\right) dy = -\frac{2U_\infty}{R}\int_0^\infty yV_x dy \tag{5.A12}$$

leading to

$$\int_0^\infty \left\{\left(U_\infty - U_x^{(inv)}\right)V_x - \Gamma - \overline{u_x^2}\right\} dy = \int_0^\infty \left\{\overline{u_y^2} - \overline{u_x^2} + \left(\Omega^{(inv)} + 3\frac{U_\infty}{R}\right) yV_x\right\} dy \tag{5.A13}$$

The other term in (5.A10) is treated as follows: $\overline{U_x}$ is replaced by U_∞ and, integrating by parts,

$$\int_0^\infty V_y dy = -\int_0^\infty y \frac{\partial V_y}{\partial y} dy = \frac{d}{dx}\int_0^\infty yV_x dy \tag{5.A14}$$

where we have employed (5.141). The quantity $U_y^{(inv)}$ is replaced by $-y(dU_\infty/dx)$, by virtue of (5.149), and the integral of $\overline{u_x u_y}$ evaluated using the leading-order outer-layer equation (5.194):

$$\frac{\partial \overline{u_x u_y}}{\partial y} = y\frac{dU_\infty}{dx}\frac{\partial V_x}{\partial y} - \frac{\partial}{\partial x}(U_\infty V_x) \tag{5.A15}$$

Thus, integrating by parts,

$$\begin{aligned}\int_0^\infty \overline{u_x u_y}\, dy &= -\int_0^\infty y \frac{\partial \overline{u_x u_y}}{\partial y} dy = \frac{d}{dx}\left(U_\infty \int_0^\infty yV_x dy\right) - \frac{dU_\infty}{dx}\int_0^\infty y^2 \frac{\partial V_x}{\partial y} dy \\ &= \frac{d}{dx}\left(U_\infty \int_0^\infty yV_x dy\right) + 2\frac{dU_\infty}{dx}\int_0^\infty yV_x dy\end{aligned} \tag{5.A16}$$

The result of the above approximations is

$$\int_0^\infty \left\{\overline{U_x}V_y + U_y^{(inv)}V_x + \overline{u_x u_y}\right\} dy = 2\frac{d}{dx}\left(U_\infty \int_0^\infty yV_x dy\right) \tag{5.A17}$$

Combining (5.A10), (5.A13), and (5.A17), we obtain

$$\begin{aligned}&\frac{d}{dx}\int_0^\infty \overline{U_x}V_x dy + \frac{dU_\infty}{dx}\int_0^\infty V_x dy + \frac{\tau_w}{\rho} = \\ &\frac{d}{dx}\left\{\int_0^\infty \left(\overline{u_y^2} - \overline{u_x^2}\right) dy + \left(\Omega^{(inv)} - \frac{U_\infty}{R}\right)\int_0^\infty yV_x dy\right\} - \frac{2}{R}\frac{d}{dx}\left(U_\infty \int_0^\infty yV_x dy\right)\end{aligned} \tag{5.A18}$$

which yields (5.188) when the definitions of δ_m and δ_d are introduced, and (5.173), (5.181) used. We note that (5.173), although approximate, still holds at the order to which we are working.

References

Bradbury, L. J. S., 1965. The structure of a self-preserving turbulent plane jet. *J. Fluid Mech.*, **23**, 31–64.

Coles, D., 1956. The law of the wake in the turbulent boundary layer. *J. Fluid Mech.*, **1**, 191–226.

Crow, S. C., Champagne, F. H., 1971. Orderly structure in jet turbulence. *J. Fluid Mech.*, **48**, 547–91.

Gad-el-Hak, M., Bandyopadhyay, P. R., 1994. Reynolds number effects in wall-bounded turbulent flows. *J. ASME*, **47**(8), 307–65.

Grant, H. L., 1958. The large eddies of turbulent motion. *J. Fluid Mech.*, **4**, 149–90.

Ho, C. M., Huerre, P., 1984. Perturbed free shear layers. *Ann. Rev. Fluid Mech.*, **16**, 365–424.

Hussain, F., Husain, H. S., 1989. Elliptic jets. Part 1: Characteristics of unexcited and excited jets. *J. Fluid Mech.*, **208**, 257–320.

Kim, H. T., Kline, S. J., Reynolds, W. C., 1971. The production of turbulence near a smooth wall in a turbulent boundary layer. *J. Fluid Mech.*, **50**, 133–60.

Klebanoff, P. S., 1955. Characteristics of turbulence in a boundary layer with zero pressure gradient. *NACA*, report no. 1247.

Kline, S. J., Reynolds, W. C., Schraub, F. A., Runstadler, P. W., 1967. The structure of turbulent boundary layers. *J. Fluid Mech.*, **30**, 741–73.

Laufer, J., 1954. The structure of turbulence in fully developed pipe flow. *NACA*, report no. 1174.

Robinson, S. K., 1991. The kinematics of turbulent boundary layer structure. *NASA* technical memorandum no. 103859, NASA Ames Research Center, April 1991.

Savill, A. M., Truong, T. V., Ryhming, I. L., 1988. Turbulent drag reduction by passive means: a review and report on the first European drag reduction meeting. *J. Theor. Appl. Mech.*, **7**, 353–78.

Schlichting, H., 1987. *Boundary Layer Theory*. McGraw-Hill, New York.

Smith, F. T., 1979. Laminar flow of an incompressible fluid past a bluff body: the separation, reattachment, eddy properties and drag. *J. Fluid Mech.*, **92**, 171–205.

Smith, F. T., 1982. On the high Reynolds number theory of laminar flows. *IMA J. Appl. Math.*, **28**, 207–81.

Sunyach, M., Mathieu, J., 1969. Intermittence cinématique et thermique aux frontières de la zone de mélange d'un jet plan. *C.R. Acad. Sci. Paris*, **268**, 781–4.

Van Dyke, M., 1975. *Pertubation Methods in Fluid Mechanics*. Parabolic Press, Stanford, CA.

Wallace, J. M., Eckelmann, H., Brodkey, R. S., 1972. The wall region in turbulent shear flow. *J. Fluid Mech.*, **54**, 39–48.

Wygnanski, I., Fiedler, H. E., 1969. Some measurements in the self-preserving jet. *J. Fluid Mech.*, **38**, 577–612.

CHAPTER SIX

Spectral Analysis of Homogeneous Turbulence

In this chapter, we apply Fourier analysis to turbulence. The idea is to decompose the turbulent fluctuations into sinusoidal components and study the distribution of turbulent energy among the different wavelengths, representing different scales of turbulence, and its evolution with time. As we shall see, if the turbulence is homogeneous, this idea works out quite well and we obtain equations governing the velocity spectra, which can be readily interpreted physically in terms of the transfer of energy between different scales of turbulence and dissipation of turbulent energy by viscosity. Of course, the equations obtained suffer from the usual closure problem, owing to the nonlinear terms in the Navier–Stokes equation, but have the advantage over the moment equations in physical space that the nonlocally determined pressure term can be easily expressed in terms of the velocities. Additional closure assumptions are needed if one wants a closed set of spectral equations. However, we do not want to introduce closure approximations here, preferring to interpret the spectral equations in physical terms.

The simplest case, and the one considered here, is when there are no mean velocity gradients, which means that the turbulence interacts only with itself, decaying as it evolves in time due to the absence of a source of turbulent energy.

We refer the reader at the outset to the classic text by Batchelor (G 1953), which perhaps contributed more than any other to the development of spectral analysis of turbulence, even if it now appears a little dated. Monin and Yaglom (G 1975) also contains a vast amount of valuable material on homogeneous turbulence, much of which is directly relevant to this and the next chapter.

Fourier analysis is one of the most powerful mathematical methods and is used in a wide variety of scientific problems. It allows the decomposition of general functions into sinusoidal components, from which they may be reconstructed using the Fourier inversion theorem. Thus, given a function, $F(x)$, its Fourier transform is

$$\tilde{F}(k) = \frac{1}{2\pi}\int_{-\infty}^{\infty} F(x)e^{-ikx}\,dx \tag{6.1}$$

and the inversion theorem gives

$$F(x) = \int_{-\infty}^{\infty} \tilde{F}(k)e^{ikx}\,dk \tag{6.2}$$

thereby expressing $F(x)$ as a linear combination of sinusoidal components, e^{ikx}, with complex amplitudes, $\tilde{F}(k)$. Fourier analysis is often employed to reduce linear differential or integral equations to algebraic equations, a powerful technique that allows the solution of many difficult and interesting scientific problems. However, its use in the theory of turbulence, although in part also intended to turn differential operators into algebraic factors, is more subtle because turbulent quantities are random functions.

The reader who has studied the theory of signal processing with random noise will already be acquainted with the basic ideas of Fourier analysis of random functions, such as correlations, spectra, and their relationships (see, e.g., Bendat and Piersol 1971). In these applications, the signal is usually thought of as a function of time, whereas here we are mainly concerned with the use of *spatial* Fourier transforms, with time playing only a secondary role (correlations and spectra evolve with time). However, the principles involved are the same.

One can decompose the velocity into mean and fluctuating parts as

$$U_i = \overline{U_i} + u_i \tag{6.3}$$

and define the velocity correlation function by

$$R_{ij}(\mathbf{x}, \mathbf{x}', t) = \overline{u_i(\mathbf{x}, t)u_j(\mathbf{x}', t)} \tag{6.4}$$

In the case of homogeneous turbulence, that is, when the statistical properties of u_i remain the same under: an arbitrary constant displacement in space, R_{ij} is solely a function of $\mathbf{r} = \mathbf{x} - \mathbf{x}'$:

$$R_{ij}(\mathbf{x} - \mathbf{x}', t) = \overline{u_i(\mathbf{x}, t)u_j(\mathbf{x}', t)} \tag{6.5}$$

We assume that the velocity correlations tend to zero sufficiently rapidly as the distance, $|\mathbf{x} - \mathbf{x}'|$, between the points increases:

$$R_{ij}(\mathbf{r}, t) \to 0 \tag{6.6}$$

as $r = |\mathbf{r}| \to \infty$. Decorrelation of the turbulent fluctuating velocities is a consequence of their assumed asymptotic statistical independence, as discussed in Chapter 3. Such asymptotic independence is also implicitly used for higher-order velocity moments later in the development of spectral analysis and, for simplicity's sake, is supposed from the start. Notice that R_{ij} is a two-point average: in fact, this chapter can be thought of as implementing two-point analysis for homogeneous turbulence without mean flow.

Let us define the spectral functions, Φ_{ij}, as the three-dimensional Fourier transform of R_{ij}:

$$\Phi_{ij}(\mathbf{k}, t) = \frac{1}{(2\pi)^3} \int R_{ij}(\mathbf{r}, t) e^{-i\mathbf{k}.\mathbf{r}} d^3\mathbf{r} \tag{6.7}$$

where the volume integral is over all $\mathbf{r}$. The spectral matrix, $\Phi_{ij}(\mathbf{k}, t)$, depends on the wavenumber vector, $\mathbf{k}$, and on t as the turbulence evolves with time. The inverse transform is

$$R_{ij}(\mathbf{r}, t) = \int \Phi_{ij}(\mathbf{k}, t) e^{i\mathbf{k}.\mathbf{r}} d^3\mathbf{k} \tag{6.8}$$

which is an integral over all three-dimensional wavenumber space. Setting $\mathbf{r} = 0$ and $j = i$ in (6.8) gives

$$\frac{1}{2}\overline{u_i u_i} = \frac{1}{2}R_{ii}(0, t) = \frac{1}{2}\int \Phi_{ii}(\mathbf{k}, t)d^3\mathbf{k} \tag{6.9}$$

showing that the turbulent kinetic energy can be expressed as an integral over **k**. The function $\Phi_{ii}(\mathbf{k}, t)/2$ can therefore be interpreted as the distribution of turbulent energy over different wavenumbers. Notice that, here as elsewhere, an average such as $\overline{u_i u_i}$, in which we do not specify spatial positions implies the same position for all quantities appearing inside the average (and it does not matter which position, by homogeneity).

Some elementary properties of R_{ij} and Φ_{ij} follow from their definitions. Thus,

$$R_{ij}(\mathbf{x} - \mathbf{x}', t) = \overline{u_i(\mathbf{x}, t)u_j(\mathbf{x}', t)} = \overline{u_j(\mathbf{x}', t)u_i(\mathbf{x}, t)} = R_{ji}(\mathbf{x}' - \mathbf{x}, t) \tag{6.10}$$

that is, $R_{ij}(\mathbf{r}) = R_{ji}(-\mathbf{r})$. When this result is used in (6.7), we obtain

$$\Phi_{ij}(\mathbf{k}, t) = \Phi_{ji}(-\mathbf{k}, t) \tag{6.11}$$

while, since $R_{ij}(\mathbf{r})$ is real, the complex conjugate of (6.7) yields

$$\Phi^*_{ij}(\mathbf{k}, t) = \Phi_{ij}(-\mathbf{k}, t) \tag{6.12}$$

Combining (6.11) and (6.12) results in

$$\Phi^*_{ij}(\mathbf{k}, t) = \Phi_{ji}(\mathbf{k}, t) \tag{6.13}$$

which is the statement that Φ_{ij} is a Hermitian matrix. It follows that each of the diagonal terms of Φ_{ij} is real and hence that $\Phi_{ii}(\mathbf{k}, t)/2$ is real, as it should be for a quantity we interpret as the distribution of energy with wavenumber. The real and imaginary parts of Φ_{ij} can be related to the even and odd parts of $R_{ij}(\mathbf{r}, t)$ with respect to **r**, an exercise we leave for the reader (the even part is $(R_{ij}(\mathbf{r}, t) + R_{ij}(-\mathbf{r}, t))/2$, while the odd part is $(R_{ij}(\mathbf{r}, t) - R_{ij}(-\mathbf{r}, t))/2$).

The magnitude of the wavenumber vector, $k = |\mathbf{k}|$, has the dimensions of $(\text{length})^{-1}$ and so k^{-1} is a length. One often interprets the length scale $\ell = O(k^{-1})$ as giving the spatial scale represented by wavenumber **k**. For instance, if L is an integral scale of turbulence, wavenumbers of order L^{-1} represent the large scales of turbulence, which contain most of the turbulent kinetic energy. The spectral function, $\Phi_{ii}(\mathbf{k})$, which, as we noted above, describes the distribution of turbulent energy among different wavenumbers, has its largest values for $|\mathbf{k}| = O(L^{-1})$ and decays to zero as $|\mathbf{k}| \to \infty$. The dominant contributions to (6.9) arise from $|\mathbf{k}| = O(L^{-1})$ because $\Phi_{ii}(\mathbf{k})$ drops off rapidly outside that range. Note that volume integrals, such as (6.9), emphasize the higher wavenumbers more than is apparent by simply considering the magnitude of the integrand. If **k**-space spherical polar coordinates, k, θ, ϕ, are used, the volume element is $k^2 \sin\theta \, dk d\theta d\phi$, and the factor of k^2 shows the increasing volume weighting with k. In the present case, Φ_{ii} decreases sufficiently rapidly that the dominant contributions to (6.9) nonetheless come from $k = O(L^{-1})$. This is not to say that wavenumbers larger than L^{-1} have no importance: as we have seen in Chapter 3, there are spatial scales extending from the largest, $O(L)$, to the smallest, $O(\eta)$, where η is the Kolmogorov scale at which viscous dissipation is most effective. The spectrum of turbulence has correspondingly significant contributions at all wavenumbers between $O(L^{-1})$ and $O(\eta^{-1})$, as illustrated in Figure 6.1. The figure shows the spectrum $E = 2\pi k^2 \Phi_{ii}$ as a function of k, where the factor $2\pi k^2$ represents the k^2 volumetric weighting discussed above. The log–log plot means that

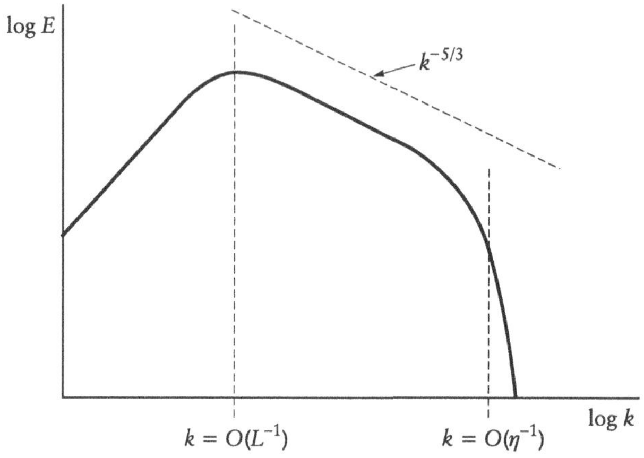

Figure 6.1. Sketch of the energy spectrum of turbulence in traditional log–log format, which brings out power laws as straight lines. The figure is drawn for a sufficiently high Reynolds number that there is a $k^{-5/3}$ inertial range, a feature we will address later in this chapter and in more detail in the next.

power laws, often found in turbulent spectra, appear as straight lines. Notice that the relationship, $\ell \leftrightarrow k^{-1}$, between spatial length scale and wavenumber is reciprocal: small wavenumbers correspond to large spatial scales and vice versa.

Although the identification of wavenumber k with spatial scales of order k^{-1} can be very useful in qualitative interpretation of spectral properties, both theoretically and experimentally, it should not be pushed too far. The relationship $\ell = O(k^{-1})$ represents an order of magnitude and one cannot, for instance, meaningfully distinguish between length scales k^{-1} and, say, $2\pi/k$ (which is the wavelength of the Fourier component $e^{i\mathbf{k}.\mathbf{x}}$). The Fourier transform of some spatially localized function of spatial width $O(\ell)$ extends over all k, even if its largest values occur at $k = O(\ell^{-1})$. In the same way, turbulent spatial scales of $O(\ell)$ will mostly contribute to the spectrum for k of $O(\ell^{-1})$, but will have effects at all k. In short, the correspondence $\ell \leftrightarrow k^{-1}$ is often used, but should be treated as an order-of-magnitude, interpretative relationship.

In a similar vein, it should be borne in mind that spectra do not, in general, contain full statistical information about turbulence. This is perhaps obvious, after all second-order moments like R_{ij} are usually insufficient to fully define a random process such as $u_i(\mathbf{x}, t)$. However, it is easy to forget this fact when faced with a spectrum. It will be recalled from Chapter 2 that the full statistics of Gaussian variables are determined by their mean (which is zero here) and second-order moments, but the turbulent fluctuating velocities are not usually Gaussian, although they can be approximately so.

In Chapter 3, we observed that turbulence contains a wide range of spatial scales. The largest scales are of size $O(L)$, where L is a correlation scale and is determined by the physical processes that created the turbulence and does not depend on the fluid viscosity. The dynamics of the large scales are essentially independent of viscosity at high Reynolds numbers. These large scales are intrinsically unstable and tend to progressively give up their energy to smaller scales by a process which is inviscid and depends on the nonlinear convective terms of the Navier–Stokes equation (see Chapter 3 and Section 4.4). Eventually, as the spatial scale decreases, a point is reached at which the increasing importance of viscosity stops the cascade of energy to smaller scales. Viscous energy dissipation acts to destroy the kinetic energy that the small structures inherited from their larger parents. This intervention of viscosity takes place around the Kolmogorov scale, η, which therefore gives the order of size of the smallest scales of turbulence. Thus, we have spatial scales ranging all the way from $O(\eta)$ up to $O(L)$. Since η decreases with the viscosity, the ratio, L/η, grows larger as viscosity decreases. The lack of importance of viscosity at the largest scales is measured by the turbulent Reynolds number, $u'L/\nu$, which must be large for neglect of viscosity at scales $O(L)$ and hence for the cascade mechanism to operate effectively. The larger the turbulent Reynolds number, the larger L/η and the wider the range of scales of turbulence.

This description can be given a more quantitative basis using spectral analysis, as we shall see in this chapter. In particular, the turbulent energy spectrum gives the distribution of energy among the different spatial scales. The spectrum extends from $k = O(L^{-1})$ up to $k = O(\eta^{-1})$, going to zero rapidly above $k = O(\eta^{-1})$ and also below $k = O(L^{-1})$. If the viscosity, or equivalently, the turbulent Reynolds number is changed, the position of the viscous "cutoff" at $k = O(\eta^{-1})$ is altered to accommodate this. We shall obtain equations governing the temporal evolution of the spectrum that contain terms corresponding to the energy transfer from large to small scales (the cascade) and to viscous dissipation at the smallest scales ($k = O(\eta^{-1})$). This provides a quantitative underpinning for the Kolmogorov theory of the small scales, described in the next chapter.

In addition to its magnitude, the wavenumber vector, $\mathbf{k}$, has a direction and, in general, the energy distribution $\Phi_{ii}(\mathbf{k}, t)/2$ depends on both. However, for isotropic turbulence, that is, whose statistical properties have no preferred direction, we would expect the energy distribution $\Phi_{ii}(\mathbf{k}, t)/2$ to be independent of the direction of $\mathbf{k}$, possessing spherical symmetry in $\mathbf{k}$-space. In that case, it is conventional to define the energy spectrum, $E(k, t)$, via

$$E(k, t) = 2\pi k^2 \Phi_{ii}(\mathbf{k}, t) \tag{6.14}$$

so that, according to (6.9),

$$\frac{1}{2}\overline{u_i u_i} = \int_0^\infty E(k, t) dk \tag{6.15}$$

where we have integrated over the volume, $4\pi k^2 dk$, between two infinitesimally separated spheres, $|\mathbf{k}| = k$ and $|\mathbf{k}| = k + dk$. Equation (6.15) shows that $E(k, t)$ represents the distribution of energy over $k = |\mathbf{k}|$, with the k^2 factor in (6.14) increasing with k to account for the greater volume of $\mathbf{k}$-space occupied by the larger wavenumbers. Figure 6.1 illustrates a typical spectrum of isotropic turbulence. Note the maximum at energy-containing scales, $k = O(L^{-1})$, and the rapid falloff above $k = O(\eta^{-1})$. The consequences of isotropy will be investigated in detail in Section 6.4, but, until then, we develop the theory for the general anisotropic case. In practice, turbulence is usually anisotropic, although it tends to be less so when one considers the small scales, as we shall see in the next chapter.

In order to define the spectral function $\Phi_{ij}(\mathbf{k})$, we have assumed homogeneity, an ideal property that most turbulent flows do not possess: at least not the exact, three-dimensional type of homogeneity assumed above (although grid-generated turbulence is a good approximation to such homogeneity, many grid spacings downstream of the grid). However, many turbulent flows have restricted or approximate homogeneity properties. For instance, statistically two-dimensional turbulent flow is homogeneous in the direction perpendicular to the planes of the flow (here, the mean flow is two dimensional, but the turbulence in individual realizations is not, of course). A boundary layer changes slowly in directions other than perpendicular to the wall and is thus approximately homogeneous in these directions, even though it is grossly inhomogeneous in the wall-normal direction. Furthermore, as described in the next chapter, it is found that turbulence is approximately homogeneous at small scales for quite general flows at high Reynolds number. With homogeneity only in certain directions, the correlation, $R_{ij}(\mathbf{x}, \mathbf{x}', t) = \overline{u_i(\mathbf{x}, t)u_j(\mathbf{x}', t)}$, is a function of the components of the separation, $\mathbf{r} = \mathbf{x} - \mathbf{x}'$, in the directions of homogeneity. In the other directions, it is dependent on the corresponding components of both $\mathbf{x}$ and $\mathbf{x}'$. One can define spectral functions by Fourier transforming in the homogeneous directions alone, while leaving the spatial dependence on the inhomogeneous coordinates in the spectral functions. Local spectra can also be usefully defined for weakly homogeneous turbulence, that is, when its statistical properties vary over many correlation lengths. The correlation functions, $R_{ij}(\mathbf{x}, \mathbf{x}', t)$, are then regarded as functions of separation, $\mathbf{r} = \mathbf{x} - \mathbf{x}'$, and mean position, $(\mathbf{x} + \mathbf{x}')/2$. They are Fourier transformed with respect to $\mathbf{r}$, leading to spectral functions of $\mathbf{k}$ *and* position. This definition of local spectral functions applies to general inhomogeneous turbulence, but is really only useful when the inhomogeneity is weak, in which case many of the results derived for homogeneous turbulence in this and later chapters can be carried over, at least in an approximate sense. For simplicity's sake, except for a brief description of one-dimensional spectra in Section 6.5, we will only be concerned with full homogeneity in all three dimensions. Of course, strict homogeneity in any direction implies that the flow is infinite in that direction and so full three-dimensional homogeneity means no boundaries at all, as we assume in what follows.

In analyses of the spectral properties of turbulence, one often finds that direct use is made of the Fourier transforms of flow quantities, such as $u_i(\mathbf{x}, t)$. Thus, we find the transform pair

$$\tilde{u}_i(\mathbf{k}, t) = \frac{1}{(2\pi)^3} \int u_i(\mathbf{x}, t) e^{-i\mathbf{k}.\mathbf{x}} d^3\mathbf{x} \tag{6.16}$$

$$u_i(\mathbf{x}, t) = \int \tilde{u}_i(\mathbf{k}, t) e^{-i\mathbf{k}.\mathbf{x}} d^3\mathbf{k} \tag{6.17}$$

At first sight, it would appear to be more straightforward to derive the equations for the spectra in this direct manner, rather than going via the correlation function. There is, however, a technical difficulty in using (6.16), namely lack of convergence of the integral! The random function $u_i(\mathbf{x}, t)$ extends over all space and does not decay to zero as $|\mathbf{x}| \to \infty$ in the manner required for the existence of a classical Fourier transform. The resolution of this difficulty is not entirely straightforward and is complicated by the fact that $u_i(\mathbf{x}, t)$ is a random function. As we shall see in the next section, one can define a Fourier transform of $u_i(\mathbf{x}, t)$ over a finite volume of space and then consider what happens as that volume grows very large, thus approaching the infinite case. The resulting transformed function is a *random function* of $\mathbf{k}$ (and t) with somewhat pathological properties. While it is often more direct to use such an approach, the novice spectral analyst may find it easier to proceed via the somewhat less direct, but more conventional route of setting up equations for the correlation function and taking their transforms to obtain the spectral equations. For this reason, and because the correlation equations are of interest in themselves, we will later rederive the spectral evolution equations using this approach.

Although homogeneity is necessary for developing the spectral theory, isotropy is not, and we do not introduce the additional assumption of isotropic turbulence until Section 6.4. Aside from increased generality, it seems to us to be easier to grasp this way, because one does not need to come to grips with two essentially different sets of difficulties at once. Thus, spectral theory is initially developed without confronting the reader with the special properties of the various tensors when isotropy is assumed. As in all developments of the subject, we assume the reader is reasonably conversant with Cartesian tensors and subscript notation (see Jeffreys (1931)).

6.1 Direct Fourier Transforms

We want first to explain the way in which the spectrum arises using the direct Fourier transform of $u_i(\mathbf{x}, t)$. Since we cannot use (6.16) directly because the integral diverges at infinity, an obvious remedy is to define a finite-range transform

$$\tilde{u}_i^{(X)}(\mathbf{k}, t) = \frac{1}{(2\pi)^3} \int_{V_X} u_i(\mathbf{x}, t) e^{-i\mathbf{k}.\mathbf{x}} d^3\mathbf{x} \tag{6.18}$$

where V_X is the cubic volume defined by $-X < x_1 < X, -X < x_2 < X, -X < x_3 < X$, and $2X$ defines the sides of the cube. Equation (6.18) yields a complex-valued, random function $\tilde{u}_i^{(X)}(\mathbf{k}, t)$, a finite-range transform of $u_i(\mathbf{x}, t)$. The complex conjugate of (6.18) with index j in place of i and integration variable $\mathbf{x}'$ in place of $\mathbf{x}$ gives

$$\tilde{u}_j^{(X)^*}(\mathbf{k}, t) = \frac{1}{(2\pi)^3} \int_{V_X} u_j(\mathbf{x}', t) e^{i\mathbf{k}.\mathbf{x}'} d^3\mathbf{x}' \tag{6.19}$$

whose product with (6.18) is averaged, yielding the spectral moment

$$\overline{\tilde{u}_i^{(X)} \tilde{u}_j^{(X)^*}} = \frac{1}{(2\pi)^6} \int_{V_X} \int_{V_X} R_{ij}(\mathbf{x} - \mathbf{x}', t) e^{-i\mathbf{k}.(\mathbf{x}-\mathbf{x}')} d^3\mathbf{x}' d^3\mathbf{x} \tag{6.20}$$

where we have used the definition, (6.5), of the correlation function, R_{ij}, as a function of $\mathbf{x} - \mathbf{x}'$, owing to homogeneity.

We now want to consider what becomes of (6.20) as the cubic volume, V_X, goes to infinity, that is, as $X \to \infty$. The velocity autocorrelation, R_{ij}, is a function of $\mathbf{x} - \mathbf{x}'$ and, as always, we assume decorrelation as $|\mathbf{x} - \mathbf{x}'| \to \infty$, that is, we suppose that $R_{ij}(\mathbf{x} - \mathbf{x}', t)$ goes to zero rapidly as the two points $\mathbf{x}$ and $\mathbf{x}'$ are separated. The distance scale over which $R_{ij}(\mathbf{x} - \mathbf{x}', t)$ has significantly nonzero values is $|\mathbf{x} - \mathbf{x}'| = O(L)$, L being a correlation length. Let $\mathbf{r} = \mathbf{x} - \mathbf{x}'$, the vector separation of two points, replace $\mathbf{x}'$ as integration variable in (6.20) leading to

$$\overline{\tilde{u}_i^{(X)} \tilde{u}_j^{(X)^*}} = \frac{1}{(2\pi)^6} \int_{V_X} \int_{V_X^{(\mathbf{x})}} R_{ij}(\mathbf{r}, t) e^{-i\mathbf{k}.\mathbf{r}} d^3\mathbf{r} d^3\mathbf{x} \tag{6.21}$$

where the cubic volume, $V_X^{(\mathbf{x})}$, depends on $\mathbf{x}$ and is defined by $x_1 - X < r_1 < x_1 + X, x_2 - X < r_2 < x_2 + X, x_3 - X < r_3 < x_3 + X$. The cube, $V_X^{(\mathbf{x})}$, has side $2X$ and is centered on $\mathbf{r} = \mathbf{x}$.

The integral in (6.21) is only significantly different from zero when $|\mathbf{r}| = O(L)$ and this region lies well inside $V_X^{(\mathbf{x})}$ provided that $\mathbf{x}$ is many correlation lengths from the boundary of V_X. Thus, as $X \to \infty$, for most $\mathbf{x}$ in V_X, we can consider the integration over $\mathbf{r}$ as extending to infinity. Using the definition, (6.7), of the spectral function Φ_{ij}, (6.21) then becomes

$$\overline{\tilde{u}_i^{(X)}\tilde{u}_j^{(X)*}} \sim \frac{1}{(2\pi)^3}\int_{V_X} \Phi_{ij}(\mathbf{k}, t)d^3\mathbf{x} \tag{6.22}$$

where the approximation of the integral over $\mathbf{r}$ by $\Phi_{ij}(\mathbf{k}, t)$ applies everywhere except for $\mathbf{x}$ of the order of a correlation length from the boundary of V_X. Since $\Phi_{ij}(\mathbf{k}, t)$ is independent of $\mathbf{x}$, we have

$$\overline{\tilde{u}_i^{(X)}\tilde{u}_j^{(X)*}} \sim \left(\frac{X}{\pi}\right)^3 \Phi_{ij} \tag{6.23}$$

As $X \to \infty$, the above approximation becomes better and better because the fraction of the volume V_X within a correlation length of the boundary decreases. In other words,

$$\Phi_{ij}(\mathbf{k}, t) = \lim_{X\to\infty}\left\{\left(\frac{\pi}{X}\right)^3 \overline{\tilde{u}_i^{(X)}(\mathbf{k}, t)\tilde{u}_j^{(X)*}(\mathbf{k}, t)}\right\} \tag{6.24}$$

gives the spectral function as a limit of an average involving finite-range Fourier transforms, rather than from its original definition as the Fourier transform of the velocity correlation function. Note that, using (6.24), we have

$$\Phi_{ii}(\mathbf{k}, t) = \lim_{X\to\infty}\left\{\left(\frac{\pi}{X}\right)^3 \overline{\tilde{u}_i^{(X)}(\mathbf{k}, t)\tilde{u}_i^{(X)*}(\mathbf{k}, t)}\right\} \tag{6.25}$$

confirming that $\Phi_{ii}(\mathbf{k}, t)/2$ is real and positive, which is just as well for a quantity we interpret as the distribution of turbulent energy with wavenumber. It also follows from (6.24) that the matrix Φ_{ij} is positive definite, which is a stronger result than $\Phi_{ii} > 0$.

To interpret (6.24) and (6.25), let us imagine that we are given velocity measurements of a homogeneous turbulent flow and want to calculate the spectral function $\Phi_{ij}(\mathbf{k}, t)$. One could always return to the definition (6.7) and first calculate the velocity correlation, then take the Fourier transform, but suppose one wanted to use equation (6.25). As we have seen, X must be chosen much larger than a correlation length if we want to obtain an accurate answer. One can compute the finite-range Fourier transform, (6.18), for any given realization, $u_i(\mathbf{x}, t)$, to obtain $\tilde{u}_i^{(X)}(\mathbf{k}, t)$, but the result will fluctuate randomly from one realization to another. The mean value of the transform, $\tilde{u}_i^{(X)}(\mathbf{k}, t)$, will be zero, of course, as follows by taking the average of (6.18), but what of its fluctuations from one realization to another? The squared amplitude of these fluctuations is measured by the variance, which is given by equation (6.23) with $i = j$ (and, for once, no implied sum). The amplitude of fluctuations of $\tilde{u}_i^{(X)}(\mathbf{k}, t)$ from one realization to another thus increases proportional to $X^{3/2}$. Since $\tilde{u}_i^{(X)}(\mathbf{k}, t)$ has zero mean, this also gives its typical size in any one realization. The scaled spectral product

$$\left(\frac{\pi}{X}\right)^3 \tilde{u}_i^{(X)}(\mathbf{k},t)\tilde{u}_i^{(X)^*}(\mathbf{k},t) \tag{6.26}$$

will also fluctuate from realization to realization with variance that *does not decrease* as $X \to \infty$, but its mean value, taken over many realizations, will provide an estimate of Φ_{ii}. The fact that the variance of (6.26) does not decrease with growing X means that the averaging over realizations is essential: a single realization will give an inaccurate result for the spectra which varies randomly from realization to realization.

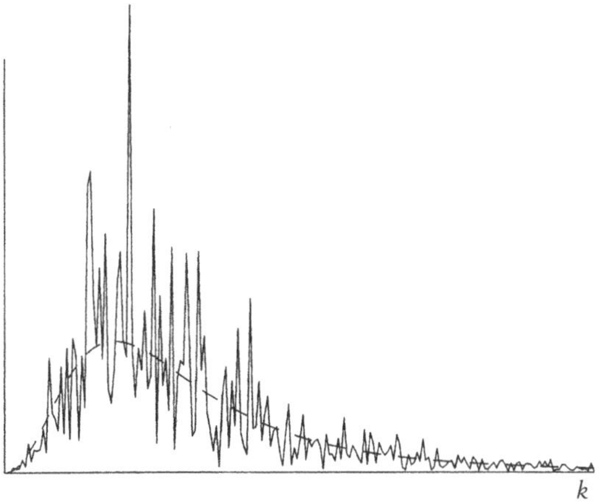

Figure 6.2. A single realization of the spectral product, (6.26) (solid line), in the construction of the spectral function Φ_{ii} (dashed line) by averaging a large number of such realizations.

Suppose that we plot the scaled spectral product, (6.26), whose average is Φ_{ii}, as a function of $\mathbf{k}$ for a single realization. To represent it as a graph we choose a particular direction of $\mathbf{k}$ and imagine plotting (6.26) as a function of $k = |\mathbf{k}|$. A sketch of a typical result is shown in Figure 6.2, in which the dashed line gives the average value, Φ_{ii}. There are rapid variations with wavenumber, whose scale in $\mathbf{k}$ is $O(X^{-1})$ and which get finer the larger the transform volume is made. As X is increased, the function becomes more and more "furry" and varies randomly from one realization to another. Clearly, the result does not converge as the size of the volume increases unless averaging is performed. When averaged over many realizations, however, the function leads to the smooth spectral function Φ_{ii} indicated by the dashed line in Figure 6.2. Those readers who have seen a spectral analyzer construct a (usually temporal) spectrum by averaging in real time will appreciate the elegance of this process.

The above description of a single realization as an increasingly furry function when $X \to \infty$ applies equally well to the transform, $\tilde{u}_i^{(X)}$, itself. It has zero mean, so there is no point in averaging it, and, as we have seen, its standard deviation *grows* like $X^{3/2}$. Clearly such a quantity is quite pathological and does not converge in any classical sense as $X \to \infty$. Nonetheless, as we shall see, in many respects it can be treated like a normal, infinite-range Fourier transform.

As mentioned earlier, a fundamental property of the Fourier transform is its ability to convert differential to algebraic operators. Let $F(\mathbf{x})$ be zero mean, with homogeneous statistics, and decorrelation at large separations. Define

$$\tilde{F}^{(X)}(\mathbf{k}) = \frac{1}{(2\pi)^3}\int_{V_X} F(\mathbf{x})e^{-i\mathbf{k}.\mathbf{x}}d^3\mathbf{x} \tag{6.27}$$

Using the divergence theorem, we can express the transform of $\partial F/\partial x_i$ as follows:

$$\begin{aligned} \frac{1}{(2\pi)^3}\int_{V_X}\frac{\partial F}{\partial x_i}e^{-i\mathbf{k}.\mathbf{x}}d^3\mathbf{x} &= \frac{1}{(2\pi)^3}\int_{V_X}\left\{ik_iFe^{-i\mathbf{k}.\mathbf{x}} + \frac{\partial}{\partial x_i}(Fe^{-i\mathbf{k}.\mathbf{x}})\right\}d^3\mathbf{x} \\ &= ik_i\tilde{F}^{(X)}(\mathbf{k}) + \frac{1}{(2\pi)^3}\int_{S_X}Fe^{-i\mathbf{k}.\mathbf{x}}dS_i \end{aligned} \tag{6.28}$$

where S_X is the boundary of the cubic volume, V_X. We have already seen that, as $X \to \infty$, the random variable $\tilde{F}^{(X)}$ has standard deviation $O(X^{3/2})$ and is thus typically of $O(X^{3/2})$ itself. The standard deviation of the surface integral can be estimated asymptotically using much the same method as we employed for $\tilde{u}_i^{(X)}$, but this time considering large squares of side $2X$ rather than cubes. The surface integral has standard deviation $O(X)$ and is typically small compared with the term in $\tilde{F}^{(X)}$ as $X \to \infty$. If X is large enough, we can neglect the surface integral in (6.28), giving

$$\frac{\partial \tilde{F}^{(X)}}{\partial x_i} = ik_i \tilde{F}^{(X)} \tag{6.29}$$

which is the classical result for infinite-range Fourier transforms. Thus, the derivative $\partial/\partial x_i$ becomes ik_i under transformation.

Another useful property of Fourier transforms is that a product of functions becomes a convolution over wavenumber under transformation. We want to show that this also applies to the finite-range transforms. Let $F(\mathbf{x})$ and $G(\mathbf{x})$ be two functions, and $\tilde{F}^{(X)}(\mathbf{k})$, $\tilde{G}^{(X)}(\mathbf{k})$ their finite-range transforms. By the Fourier inversion theorem, we have

$$F(\mathbf{x}) = \int \tilde{F}^{(X)}(\mathbf{k}) e^{i\mathbf{k}.\mathbf{x}} d^3\mathbf{k} \tag{6.30}$$

for $\mathbf{x}$ inside V_X and

$$0 = \int \tilde{F}^{(X)}(\mathbf{k}) e^{i\mathbf{k}.\mathbf{x}} d^3\mathbf{k} \tag{6.31}$$

outside V_X. In (6.30) and (6.31), the integrals are taken over all $\mathbf{k}$. The product of (6.30), (6.31), with the corresponding results for $G(\mathbf{x})$ can be shown to yield

$$F(\mathbf{x})G(\mathbf{x}) = \int \left(\tilde{F}^{(X)} \otimes \tilde{G}^{(X)}\right) e^{i\mathbf{k}.\mathbf{x}} d^3\mathbf{k} \tag{6.32}$$

for $\mathbf{x}$ inside V_X and

$$0 = \int \left(\tilde{F}^{(X)} \otimes \tilde{G}^{(X)}\right) e^{i\mathbf{k}.\mathbf{x}} d^3\mathbf{k} \tag{6.33}$$

outside V_X. Here, the convolution, $\tilde{F}^{(X)} \otimes \tilde{G}^{(X)}$, of $\tilde{F}^{(X)}$ and $\tilde{G}^{(X)}$ is defined by

$$\left(\tilde{F}^{(X)} \otimes \tilde{G}^{(X)}\right)(\mathbf{k}) = \int \tilde{F}^{(X)}(\mathbf{k}')\tilde{G}^{(X)}(\mathbf{k} - \mathbf{k}') d^3\mathbf{k}' \tag{6.34}$$

Equations (6.32) and (6.33) are simply the inverse Fourier transform of the equality

$$\widetilde{FG}^{(X)} = \tilde{F}^{(X)} \otimes \tilde{G}^{(X)} \tag{6.35}$$

that is, the finite-range transform of a product is the convolution of the finite-range transforms. Thus, the convolution theorem carries over exactly to finite-range transforms. Notice that in the convolution, (6.34), the arguments of $\tilde{F}^{(X)}$ and $\tilde{G}^{(X)}$ sum to $\mathbf{k}$. This is one way of remembering the definition of convolution: it is the integral over products whose arguments sum to $\mathbf{k}$.

There is a final, very important property of the transform of a homogeneous random function which we want to derive. This arises when we consider quantities such as $\tilde{u}_i^{(X)}(\mathbf{k}, t)\tilde{u}_j^{(X)*}(\mathbf{k}', t)$, where $\mathbf{k}$ and $\mathbf{k}'$ may have *different* values. We have already treated the case $\mathbf{k}' = \mathbf{k}$, leading to (6.24), and much the same method can be applied here. Taking the product of (6.18) and (6.19) with $\mathbf{k}$ replaced by $\mathbf{k}'$ in (6.19), leads to

$$\overline{\tilde{u}_i^{(X)}(\mathbf{k}, t)\tilde{u}_j^{(X)*}(\mathbf{k}', t)} = \frac{1}{(2\pi)^6}\int_{V_X}\int_{V_X^{(\mathbf{x})}} R_{ij}(\mathbf{r}, t)e^{-i\mathbf{k}'.\mathbf{r}}e^{-i(\mathbf{k}-\mathbf{k}').\mathbf{x}}d^3\mathbf{r}d^3\mathbf{x} \tag{6.36}$$

where we have switched from the integration variable $\mathbf{x}'$ to $\mathbf{r} = \mathbf{x} - \mathbf{x}'$, as before. Following the previous line of reasoning, we argue that we can extend the integration over $\mathbf{r}$ to infinity provided that $\mathbf{x}$ is many correlation lengths from the boundary of V_X. Using (6.7), this leads to

$$\overline{\tilde{u}_i^{(X)}(\mathbf{k}, t)\tilde{u}_j^{(X)*}(\mathbf{k}', t)} \sim \Phi_{ij}(\mathbf{k}', t)\Xi_X(\mathbf{k} - \mathbf{k}') \tag{6.37}$$

where

$$\Xi_X(\mathbf{k} - \mathbf{k}') = \frac{\sin(k_1 - k_1')X\sin(k_2 - k_2')X\sin(k_3 - k_3')X}{\pi^3(k_1 - k_1')(k_2 - k_2')(k_3 - k_3')} \tag{6.38}$$

The detailed form of this function of $\mathbf{k} - \mathbf{k}'$ is not important. Its value at $\mathbf{k} = \mathbf{k}'$ is $(X/\pi)^3$, which is its maximum and reproduces (6.23). The function becomes increasingly sharply peaked about $\mathbf{k} = \mathbf{k}'$ as $X \to \infty$ (the width of the peak is $O(X^{-1})$) and its maximum value increases while maintaining the constant integral (see Figure 6.3)

$$\int \Xi_X(\mathbf{q})d^3\mathbf{q} = 1 \tag{6.39}$$

It will be seen that Ξ_X has all the properties of a Dirac function:

$$\Xi_X(\mathbf{k} - \mathbf{k}') \to \delta(\mathbf{k} - \mathbf{k}') \tag{6.40}$$

as $X \to \infty$. Finally, combining this with (6.37) gives the very important result

$$\overline{\tilde{u}_i^{(X)}(\mathbf{k}, t)\tilde{u}_j^{(X)*}(\mathbf{k}', t)} = \Phi_{ij}(\mathbf{k}, t)\delta(\mathbf{k} - \mathbf{k}') \tag{6.41}$$

in the infinite X limit (see for instance, Lighthill (1958), for a thorough discussion of generalized functions, such as the Dirac function). Here we have replaced $\Phi_{ij}(\mathbf{k}', t)$ by $\Phi_{ij}(\mathbf{k}, t)$ since the Dirac function is, in any case, effectively zero when $\mathbf{k}' \neq \mathbf{k}$.

Another form of the fundamental cross-spectral relationship (6.41) is often used, which follows from (6.18) and the fact that u_i is real. The complex conjugate of (6.18) yields

$$\tilde{u}_j^{(X)*}(\mathbf{k}', t) = \tilde{u}_j^{(X)}(-\mathbf{k}', t) \tag{6.42}$$

and so (6.41) can also be written as

$$\overline{u_i^{(X)}(\mathbf{k}, t)\tilde{u}_j^{(X)}(\mathbf{k}', t)} = \Phi_{ij}(\mathbf{k}, t)\delta(\mathbf{k} + \mathbf{k}') \tag{6.43}$$

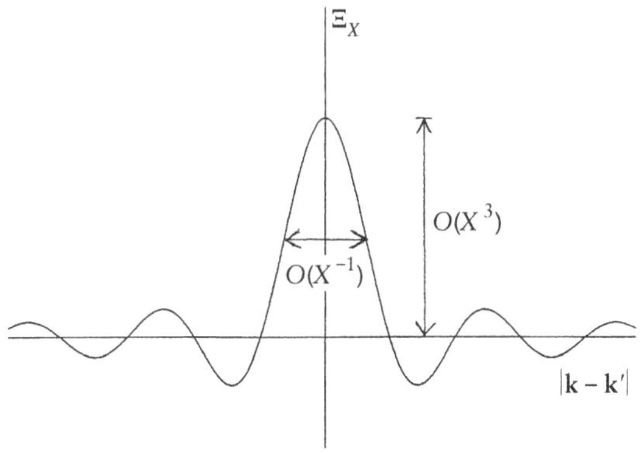

Figure 6.3. Illustration of the limiting process by which the spectral correlations approach a Dirac function as $X \to \infty$. The height increases, while the width decreases with increasing X. This limiting process should be imagined as taking place in three dimensions.

in the infinite X limit. If desired, one may take the sum and difference of (6.41) and (6.43). This allows the calculation of the spectral correlation of $\tilde{u}_i^{(X)}(\mathbf{k}, t)$ with both the real and imaginary parts of $\tilde{u}_j^{(X)}(\mathbf{k}', t)$ and hence the individual correlations of any combination of their real and imaginary parts at two different values of wavenumber can be expressed. It is then apparent that, unless $\mathbf{k} = \pm\mathbf{k}'$, $\tilde{u}_i^{(X)}(\mathbf{k}, t)$ and $\tilde{u}_j^{(X)}(\mathbf{k}', t)$ are uncorrelated in the limit $X \to \infty$.

In summary, the finite, but large-range transforms of velocity, $\tilde{u}_i^{(X)}$, are random, furry functions of wavenumber $\mathbf{k}$, of zero mean and having variance that increases in proportion to $8X^3$, the volume of the transform region V_X. Qualitatively, the variance can be thought of as increasing in this way because the amount of turbulent energy in the transform volume increases proportional to the volume. The result is that the transform, $\tilde{u}_i^{(X)}(\mathbf{k}, t)$, has values that, while fluctuating from realization to realization, increase as the square root of the transform volume. Another way of looking at this is that, when $X \to \infty$, the volume of integration in (6.18) can be split into a large number of subcubes, each of which is itself large compared with the correlation length of the flow, giving essentially independent[1] contributions to $\tilde{u}_i^{(X)}(\mathbf{k}, t)$, which therefore consists of a sum of a large number of (almost) statistically independent random variables. Keeping a fixed size of subcube, large compared with the correlation scale, the number of terms in the sum is proportional to the transform volume and thus we expect the variance of $\tilde{u}_i^{(X)}(\mathbf{k}, t)$ to increase in the manner found above. This way of looking at the problem has the further advantage that it suggests that $\tilde{u}_i^{(X)}(\mathbf{k}, t)$ should be asymptotically Gaussian, by the central limit theorem, a suggestion that is confirmed by more careful analysis (see Lumley (1970), section 3.16). Notice, however, that this is for a *single* value of the wavevector and does not imply that the random function $\tilde{u}_i^{(X)}(\mathbf{k}, t)$ can be asymptotically taken to be *jointly* Gaussian at multiple values of $\mathbf{k}$. As we saw above, two values of wavenumber, $\mathbf{k}$ and $\mathbf{k}'$, give transforms that become uncorrelated as the transform volume goes to infinity unless $\mathbf{k} = \pm\mathbf{k}'$. However, decorrelation does *not* mean statistical independence of different wavenumbers, as it would if $\tilde{u}_i^{(X)}(\mathbf{k}, t)$ really became a Gaussian random process (i.e., with joint Gaussian statistics at multiple values of $\mathbf{k}$). Indeed, as we will see later, third-order spectral moments of the form $\overline{\tilde{u}_i(\mathbf{k}, t)\tilde{u}_l(\mathbf{p}, t)\tilde{u}_m(\mathbf{q}, t)}$, which would be zero if $\tilde{u}_i(\mathbf{k}, t)$ were a Gaussian process, play an important role in the spectral dynamics and, like the second-order moments considered above, show Dirac function behavior, this time when $\mathbf{k} + \mathbf{p} + \mathbf{q} = 0$, that is, when the three wavevectors form a closed triangle. Thus, there are triads of unequal wavenumbers with significantly nonzero third-order moments, which means that, although $\tilde{u}_i^{(X)}(\mathbf{k}, t)$ is asymptotically Gaussian at a single wavenumber, it cannot be treated as a Gaussian process.

Spectral correlation functions, such as $\overline{\tilde{u}_i^{(X)}(\mathbf{k}, t)\tilde{u}_j^{(X)}(\mathbf{k}', t)}$, are *averages* of products of transforms and are thus no longer random quantities. They are expressed in terms of the cross-spectra, Φ_{ij}, using (6.41) or (6.43). At large, but finite X, equations (6.37) and (6.38) show that there are residual correlations over a small wavenumber range of size $O(X^{-1})$, which corresponds to the scale of "furriness" in Figure 6.2. However, one or other of the two relations, (6.41) and (6.43), are all one usually needs to recall from the above analysis.

[1] As noted earlier, we assume asymptotic statistical independence at large separations.

In practice, one takes the Fourier transform of, for instance, the Navier–Stokes equation as if the infinite transform existed in the usual sense. In the process, derivatives are replaced by factors of $i\mathbf{k}$ and products by convolutions. When one wants to derive equations for the spectra, products of two transforms are introduced and averaged to produce spectra using (6.41) or (6.43). We shall carry out this procedure for the Navier–Stokes equations in the next section and, in the process, we will drop the (X) in the notation for the transforms, so that the velocity transform is written $\tilde{u}_i(\mathbf{k}, t)$: it is, of course, implied that what we really mean are finite, but large volume transforms. The reader will, no doubt, find it easier not to worry overly about the basis of the procedure, an attitude shared by many research workers in the field.

There is a school of thought that the direct use of transforms is merely an alternative method for obtaining results that are more surely derived using correlations in physical space, followed by Fourier transformation. We believe this is too extreme: the physical meaning is often clearer when expressed in terms of the direct transforms of the fluctuations. Nonetheless, it *is* safer to rederive a result using correlations in physical space to reassure oneself that no mathematical faux pas has been committed in what is, after all, a quite difficult area of analysis involving random functions and a variety of limiting processes.

6.2 Transformation of the Navier–Stokes Equations

We assume that there is no mean flow ($\overline{U}_i = 0$) and so the homogeneous turbulence decays because there is dissipation but no energy input. As it does so, there is transfer of energy to smaller scales via the cascade and viscous dissipation at the smallest scales. Steady, homogeneous turbulence without a mean flow is impossible (see the turbulent energy equation, (4.37); since there is no mean flow, the production term is zero, so the turbulent kinetic energy decays under the effects of viscous dissipation).

The Navier–Stokes equations for an incompressible fluid can be written as

$$\frac{\partial u_i}{\partial t} + \frac{\partial}{\partial x_j}(u_i u_j) = -\frac{1}{\rho}\frac{\partial p}{\partial x_i} + \nu \frac{\partial^2 u_i}{\partial x_j \partial x_j} \tag{6.44}$$

with the incompressibility condition

$$\frac{\partial u_i}{\partial x_i} = 0 \tag{6.45}$$

As usual, one can derive an equation for the pressure by taking the divergence of (6.44), using (6.45). Thus,

$$\nabla^2 p = -\rho \frac{\partial^2}{\partial x_i \partial x_j}(u_i u_j) \tag{6.46}$$

Taking the Fourier transform of (6.45) gives

$$k_i \tilde{u}_i = 0 \tag{6.47}$$

showing that the velocity transform is perpendicular to the wavenumber, $\mathbf{k}$. This is the expression of incompressibility in spectral terms and would not apply if we were studying a compressible flow. We do not consider compressible flows in this book, but one of the novelties of spectral analysis applied to such flows is that $\tilde{u}_i$ has a radial component in **k**-space, in addition to the two components perpendicular to **k** that are present for incompressible flows.

Since $\tilde{u}_i(\mathbf{k}, t)$ is a function of **k**, one can think of it as a (complex) vector field in **k**-space. The fact that $\tilde{u}_i$ is complex makes it harder to grasp, but it is not mathematically dissimilar in nature to the physical flow field $u_i(\mathbf{x}, t)$. Both the real and imaginary parts of $\tilde{u}_i$ are perpendicular to the vector **k**, which, of course, changes direction from one point in **k**-space to another. At any given point in **k**-space, the real and imaginary parts of $\tilde{u}_i$ lie in the plane perpendicular to **k** (i.e., the tangent plane to the sphere of radius $|\mathbf{k}|$). As we shall see below, this condition of orthogonality allows expression of the pressure in terms of velocity. This leads to the appearance of an operator expressing projection onto the plane perpendicular to **k** in the evolution equation of $\tilde{u}_i$.

The Fourier transform of (6.44) can be expressed, using a convolution for the nonlinear term, as

$$\frac{\partial \tilde{u}_i}{\partial t} + ik_j \tilde{u}_i \otimes \tilde{u}_j = -\frac{ik_i}{\rho}\tilde{p} - \nu k^2 \tilde{u}_i \tag{6.48}$$

while, either from (6.46) or multiplication of (6.48) by k_i and use of (6.47),

$$\tilde{p} = -\rho \frac{k_i k_j}{k^2} \tilde{u}_i \otimes \tilde{u}_j \tag{6.49}$$

We may eliminate $\tilde{p}$ from (6.48) using (6.49) to find

$$\frac{\partial \tilde{u}_i}{\partial t} = \underbrace{-i\Delta_{il} k_m \tilde{u}_l \otimes \tilde{u}_m}_{\text{Nonlinear interactions}} \quad \underbrace{-\nu k^2 \tilde{u}_i}_{\text{Viscous dissipation}} \tag{6.50}$$

where

$$\Delta_{il} = \delta_{il} - \frac{k_i k_l}{k^2} \tag{6.51}$$

Equation (6.50) describes the time evolution of the transformed velocity field, $\tilde{u}_i$, in **k**-space. As indicated by the annotation, the first term in (6.50) represents nonlinear interactions between different wavenumbers in **k**-space, while the second describes viscous dissipation of the given wavenumber.

The quantity, Δ_{il}, can be thought of as a projection operator: projecting any given vector onto the plane perpendicular to **k**. To see this, let **a** be any vector, then

$$\Delta_{il} a_l = a_i - \frac{k_i}{k}\frac{k_l}{k} a_l \tag{6.52}$$

The second term gives the component of **a** in the direction of **k**, which, when subtracted from **a**, leaves the projection perpendicular to **k**. The operator, Δ_{il}, removes the radial component of a vector in **k**-space. The effect of this operator in (6.50) is to make the right-hand side orthogonal to **k**. Δ ensures this for the first term, while the second is orthogonal by virtue of (6.47). This means that, as $\tilde{u}_i$ evolves according to (6.50), it continues to satisfy the incompressibility condition, (6.47). This elegant expression of incompressibility further results in an equation, (6.50), for $\tilde{u}_i$ alone: the pressure has been eliminated. Recall that, in physical space, the pressure is nonlocally related to the velocity field by the Green's function solution of (6.46). This spatial nonlocality is one of the difficulties associated with treatments of turbulence in physical space, as we noted in Chapter 4, where two-point pressure–velocity correlations cropped up in equations for the single-point moments. Spectral methods allow us to deal with two-point averages in homogeneous turbulence and have no difficulty with the pressure, but there is spectral nonlocality, arising from the convolution term in (6.50).

In fact, all the difficulty of the problem is contained in the nonlinear interaction term, which involves the convolution

$$\tilde{u}_l \otimes \tilde{u}_m = \int \tilde{u}_l(\mathbf{k}', t)\tilde{u}_m(\mathbf{k} - \mathbf{k}')d^3\mathbf{k}' \tag{6.53}$$

This convolution is nonlocal and nonlinear. Any given wavenumber $\mathbf{k}$ has contributions from all other pairs of wavenumbers that sum to $\mathbf{k}$. These are often referred to as triad interactions because there are three wavenumbers, $\mathbf{k}$, $\mathbf{k}'$, and $\mathbf{k} - \mathbf{k}'$, involved. Such action of two wavenumbers on a third arises because the Navier–Stokes equation exhibits *quadratic* nonlinearity.

Notice that, in deriving equation (6.50), we have made no use of the assumption of statistical homogeneity, nor even that u_i be the fluctuating part of a turbulent flow. Equation (6.50) is simply the formal consequence of applying Fourier transformation to the Navier–Stokes equations, (6.44) and (6.45), in an infinite flow domain. In fact, equations very similar to (6.50) are often used to solve the Navier–Stokes equations numerically (spectral methods). When this is done, one obviously needs to discretize the wavenumber, and this is most naturally done by considering a spatially periodic flow from the start. Thus, the actual flow one wants to treat, which is imagined to occur inside some large rectangular box, is made to be periodic via infinitely many repeated copies of the box. If the box is large enough, one expects that the flow will take a long time to notice the fact that it has been constrained to be periodic. Adopting spatial periodicity means that one can express the flow variables as Fourier series, which are analogous to taking Fourier transforms in an infinite, aperiodic flow. The wavenumbers occurring in the series are discrete, rather than continuously variable, but the derivation of the analog of (6.50) is very similar to the aperiodic case considered above. The main difference is that the convolution integral, (6.53), is replaced by a discrete sum over wavenumbers. We will return to the question of numerical simulation of turbulent flows in Chapter 8. As well as numerical applications, such formulations are often used in analytical treatments of spectral theory, since they provide another way of overcoming the problem of nonconvergence of the infinite Fourier transforms.

To derive an evolution equation for the spectral tensor, Φ_{ij}, we first rewrite (6.50) with j in place of i and $\mathbf{k}'$ in place of $\mathbf{k}$. Thus

$$\frac{\partial \tilde{u}_j'}{\partial t} = -i\Delta_{jl}'k_m'(\tilde{u}_l \otimes \tilde{u}_m)' - \nu k'^2\tilde{u}_j' \tag{6.54}$$

where the primes indicate quantities which are to be evaluated at $\mathbf{k}'$, rather than $\mathbf{k}$. From (6.50) and (6.54), we obtain

$$\frac{\partial}{\partial t}(\tilde{u}_i\tilde{u}_j') = -i\left[\Delta_{il}k_m(\tilde{u}_l \otimes \tilde{u}_m)\tilde{u}_j' + \Delta_{jl}'k_m'(\tilde{u}_l \otimes \tilde{u}_m)'\tilde{u}_i\right] - \nu(k^2 + k'^2)\tilde{u}_i\tilde{u}_j' \tag{6.55}$$

and we want to take the average of this equation. There is no problem in evaluating the average of the terms which are quadratic in $\tilde{u}$, using (6.43), but the convolution terms are cubic, arising from the quadratic nonlinearity of the Navier–Stokes equation. They are not expressible in terms of the cross-spectra, Φ_{ij}, and the closure problem rears its ugly head!

To determine the average of the cubic terms, we need to evaluate quantities such as

$$\overline{\tilde{u}_i(\mathbf{k}, t)\tilde{u}_l(\mathbf{p}, t)\tilde{u}_m(\mathbf{q}, t)} \tag{6.56}$$

and this requires further analysis, similar to that used to derive (6.43). To this end, let us introduce the third-order velocity moments

$$\chi_{ilm}(\mathbf{r}, \mathbf{r}', t) = \overline{u_i(\mathbf{x}, t)u_l(\mathbf{x} - \mathbf{r}, t)u_m(\mathbf{x} - \mathbf{r}', t)} \tag{6.57}$$

where homogeneity has been used to infer that χ_{ilm} is solely a function of the separations, $\mathbf{r}$ and $\mathbf{r}'$. The twofold Fourier transform of χ_{ilm} is denoted by

$$\tilde{\tilde{\chi}}_{ilm}(\mathbf{k}, \mathbf{k}', t) = \frac{1}{(2\pi)^6}\int\int \chi_{ilm}(\mathbf{r}, \mathbf{r}', t)e^{-i(\mathbf{k}.\mathbf{r}+\mathbf{k}'.\mathbf{r}')}d^3\mathbf{r}d^3\mathbf{r}' \tag{6.58}$$

and is a spectral function representing third-order moments involving three points in space.

To calculate (6.56), we use (6.18) for each term in the product, replacing i by l, $\mathbf{k}$ by $\mathbf{p}$ and $\mathbf{x}$ by $\mathbf{x}'$ for the second term and similarly for the third. The result is a threefold volume integral over $\mathbf{x}$, $\mathbf{x}'$, and $\mathbf{x}''$, whose average introduces the quantity χ_{ilm}. When X is much larger than the correlation scale, L, and $\mathbf{x}$ is many correlation lengths from the boundary of V_X, the integrals over $\mathbf{x}$ and $\mathbf{x}'$ (or equivalently, $\mathbf{r} = \mathbf{x} - \mathbf{x}'$ and $\mathbf{r}' = \mathbf{x} - \mathbf{x}''$) can be extended to infinity, because χ_{ilm} decays rapidly to zero as $|\mathbf{r}| = |\mathbf{x} - \mathbf{x}'|$ or $|\mathbf{r}'| = |\mathbf{x} - \mathbf{x}''| \to \infty$. One finds

$$\overline{\tilde{u}_i(\mathbf{k}, t)\tilde{u}_l(\mathbf{p}, t)\tilde{u}_m(\mathbf{q}, t)} = \tilde{\tilde{\chi}}_{ilm}(-\mathbf{p}, -\mathbf{q}, t)\Xi_X(\mathbf{k} + \mathbf{p} + \mathbf{q}) \tag{6.59}$$

where Ξ_X is the function defined by (6.38) and approaches a Dirac function as $X \to \infty$. Thus,

$$\overline{\tilde{u}_i(\mathbf{k}, t)\tilde{u}_l(\mathbf{p}, t)\tilde{u}_m(\mathbf{q}, t)} = \tilde{\tilde{\chi}}_{ilm}(-\mathbf{p}, -\mathbf{q}, t)\delta(\mathbf{k} + \mathbf{p} + \mathbf{q}) \tag{6.60}$$

in the limit of infinite transform volume. Equation (6.60) is the equivalent of (6.43) for the third-order spectral moment.

Recalling that our aim in deriving (6.60) was to calculate the average of the cubic terms in (6.55), we write

$$\overline{u_i(\tilde{u}_l \otimes \tilde{u}_m)'} = \int \overline{\tilde{u}_i(\mathbf{k}, t)\tilde{u}_l(\mathbf{p}, t)\tilde{u}_m(\mathbf{k}' - \mathbf{p}, t)}d^3\mathbf{p} = \delta(\mathbf{k} + \mathbf{k}')\int \tilde{\tilde{\chi}}_{ilm}(-\mathbf{p}, \mathbf{p} - \mathbf{k}', t)d^3\mathbf{p} \tag{6.61}$$

where we have used equation (6.60) to express the third-order spectral moment. Since the right-hand side of (6.61) contains the factor $\delta(\mathbf{k} + \mathbf{k}')$, it is zero unless $\mathbf{k}' = -\mathbf{k}$. Thus, one can replace $\tilde{\tilde{\chi}}_{ilm}(-\mathbf{p}, \mathbf{p} - \mathbf{k}', t)$ by $\tilde{\tilde{\chi}}_{ilm}(-\mathbf{p}, \mathbf{p} + \mathbf{k}, t)$ and then define the new spectral function

$$\Theta_{ilm}(\mathbf{k}, t) = \int \tilde{\tilde{\chi}}_{ilm}(-\mathbf{p}, \mathbf{p} + \mathbf{k}, t) d^3\mathbf{p} \tag{6.62}$$

to rewrite (6.61) in the form

$$\overline{\tilde{u}_i(\tilde{u}_l \otimes \tilde{u}_m)'} = \Theta_{ilm}(\mathbf{k}, t)\delta(\mathbf{k} + \mathbf{k}') \tag{6.63}$$

We are now ready to derive the evolution equation for the spectral tensor, Φ_{ij}, by averaging (6.55). According to (6.43),

$$\overline{\tilde{u}_i \tilde{u}_j'} = \Phi_{ij}(\mathbf{k}, t)\delta(\mathbf{k} + \mathbf{k}') \tag{6.64}$$

while the cubic terms are evaluated using (6.63) (and a similar equation obtained by interchanging $\mathbf{k}$ and $\mathbf{k}'$), resulting in

$$\begin{aligned} &\Delta_{il} k_m \overline{(\tilde{u}_l \otimes u_m)\tilde{u}_j'} + \Delta_{jl}' k_m' \overline{(\tilde{u}_l \otimes \tilde{u}_m)'\tilde{u}_i} \\ &\qquad = \delta(\mathbf{k} + \mathbf{k}')\left\{\Delta_{il} k_m \Theta_{jlm}(\mathbf{k}', t) + \Delta_{jl}' k_m' \Theta_{ilm}(\mathbf{k}, t)\right\} \end{aligned} \tag{6.65}$$

which can be further simplified, since the Dirac function is zero unless $\mathbf{k}' = -\mathbf{k}$. Thus, putting $\mathbf{k}' = -\mathbf{k}$ in the terms other than the Dirac function

$$\begin{aligned} &\Delta_{il} k_m \overline{(\tilde{u}_l \otimes u_m)\tilde{u}_j'} + \Delta_{jl}' k_m' \overline{(\tilde{u}_l \otimes \tilde{u}_m)'\tilde{u}_i} \\ &\qquad = \delta(\mathbf{k} + \mathbf{k}') k_m \left\{\Delta_{il} \Theta_{jlm}(-\mathbf{k}, t) - \Delta_{jl} \Theta_{ilm}(\mathbf{k}, t)\right\} \end{aligned} \tag{6.66}$$

Using (6.64) and (6.66) in the average of (6.55), we find that all terms contain a factor of $\delta(\mathbf{k} + \mathbf{k}')$. This allows us to write $k^2 + k'^2 = 2k^2$ so that

$$\delta(\mathbf{k} + \mathbf{k}')\left[\left(\frac{\partial}{\partial t} + 2\nu k^2\right)\Phi_{ij}(\mathbf{k}, t) - ik_m\{\Delta_{jl}\Theta_{ilm}(\mathbf{k}, t) - \Delta_{il}\Theta_{jlm}(-\mathbf{k}, t)\}\right] = 0 \tag{6.67}$$

The term in square brackets is a function of $\mathbf{k}$ only and so, integrating over any region of $\mathbf{k}'$ including $\mathbf{k}' = -\mathbf{k}$, gives the final result

$$\frac{\partial \Phi_{ij}}{\partial t} = T_{ij} - 2\nu k^2 \Phi_{ij} \tag{6.68}$$

where

$$T_{ij}(\mathbf{k}, t) = ik_m\{\Delta_{jl}\Theta_{ilm}(\mathbf{k}, t) - \Delta_{il}\Theta_{jlm}(-\mathbf{k}, t)\} \tag{6.69}$$

Equation (6.68) is Craya's equation without mean flow. It is a key result of this chapter and describes the temporal evolution of the spectral tensor, $\Phi_{ij}(\mathbf{k}, t)$. Notice that it contains a relatively straightforward viscous dissipation term and a cubic term, T_{ij}, which represents energy transfer between different wavenumbers and the given wavenumber $\mathbf{k}$ (the triad interaction referred to earlier). If this transfer term was absent, the solution of (6.68) would be of simple exponential form, $e^{-2\nu k^2 t}$, with each wavenumber decaying independently of all others. This is indeed what happens in the final stages of decay of turbulence in the absence of mean flow, once the amplitude of the turbulent fluctuations has become sufficiently small that one can neglect the nonlinear (convective) terms in the Navier–Stokes equation (one can only forget about the nonlinear terms in this manner if the turbulent Reynolds number, $\mathrm{Re}_L = u'L/\nu$, is sufficiently small). However, the more interesting phase of turbulent decay happens earlier, when the nonlinear terms lead to an energy cascade from large to small spatial scales (or from small to large $|\mathbf{k}|$, when expressed in spectral terms), with viscous dissipation at the smallest (Kolmogorov) scale, η.

Unless we are nearing the final phase of decay (when the magnitude of the turbulent Reynolds number, Re_L, has dropped to $O(1)$ or less), there is a clear separation between wavenumbers, $k = O(L^{-1})$, which represent the large, energy-containing turbulent scales, and dissipative wavenumbers, $k = O(\eta^{-1})$. The cascade idea implies that there should be energy transfer from small to large $|\mathbf{k}|$, with viscous dissipation mainly at large values, $k = O(\eta^{-1})$. Thus, we would expect the nonlinear term, T_{ij} in (6.68), which expresses spectral transfer by the cascade, to dominate over the dissipative term at low wavenumbers. As Re_L increases, the ratio, L/η, separating the energy containing and dissipative scales grows larger. Provided that Re_L is sufficiently large, a range of intermediate wavenumbers, $L^{-1} \ll k \ll \eta^{-1}$, appears. This is known as the inertial range and has characteristic asymptotic properties (asymptotic in the limit $\mathrm{Re}_L \to \infty$). In particular, based on the discussion of the Kolmogorov theory in the next chapter, one expects universal statistical properties there, for example, the well-known power-law spectrum $E \propto k^{-5/3}$. But here we are getting ahead of ourselves.

The third-order spectral function, $\Theta_{ilm}(\mathbf{k}, t)$, which is defined by (6.62), can be reexpressed using the definition, (6.58), of $\tilde{\chi}_{ilm}$ as

$$\Theta_{ilm}(\mathbf{k}, t) = \frac{1}{(2\pi)^3} \int \left\{ \frac{1}{(2\pi)^3} \int \int \chi_{ilm}(\mathbf{r}, \mathbf{r}', t) e^{i\mathbf{p}.\mathbf{r}} d^3\mathbf{r}\, e^{-i\mathbf{p}.\mathbf{r}'} d^3\mathbf{p} \right\} e^{-i\mathbf{k}.\mathbf{r}'} d^3\mathbf{r}' \tag{6.70}$$

and the contents of the curly brackets will be recognized as a Fourier transform, followed by its inverse transform. Equation (6.70) can therefore be rewritten as

$$\Theta_{ilm}(\mathbf{k}, t) = \frac{1}{(2\pi)^3} \int \chi_{ilm}(\mathbf{r}', \mathbf{r}', t) e^{-i\mathbf{k}.\mathbf{r}'} d^3\mathbf{r}' \tag{6.71}$$

or, if we define the *two-point*, cubic moment

$$Q_{ilm}(\mathbf{r}, t) = \chi_{ilm}(\mathbf{r}, \mathbf{r}, t) = \overline{u_i(\mathbf{x}, t) u_l u_m(\mathbf{x} - \mathbf{r}, t)} \tag{6.72}$$

then

$$\Theta_{ilm}(\mathbf{k}, t) = \frac{1}{(2\pi)^3} \int Q_{ilm}(\mathbf{r}, t) e^{-i\mathbf{k}.\mathbf{r}} d^3\mathbf{r} \tag{6.73}$$

shows that Θ_{ilm} is the Fourier transform of Q_{ilm}. It is, of course, a simplification to have reexpressed the spectral transfer quantities, Θ_{ilm}, in terms of two-point rather than three-point statistics, but this does not remove the underlying closure problem. Although calculations for homogeneous turbulence are definitely easier in spectral space, no amount of transformation changes the fact that the equations for the second-order moments (correlations or spectra) contain third-order moments, and so on up the hierarchy of increasingly higher-order moment equations. This means that closure schemes are needed in the end if definite calculations are to be carried out based on the equations, such as (6.68), for the spectra.

At this point, we should clearly define what is conventionally meant by the expressions linear and nonlinear terms. As discussed in Chapter 4, linearity is defined in terms of the equations governing the fluctuating velocities, e.g., (6.50), and refers to the fluctuations, so that a term which is a product of a mean and a fluctuating quantity would be called linear. Once equations for the correlations or spectra are formulated, such linear terms lead to quantities which are quadratic in the fluctuations, while the nonlinear (quadratic) terms in the original equations yield cubic moments. One continues to refer to those terms that *originated* from linear terms in the equations for the fluctuations as linear and the others as nonlinear. For instance, the time-derivative and viscous terms in (6.68) are linear, whereas the transfer term, T_{ij}, is nonlinear.

A variety of spectral closures have been proposed, ranging in sophistication from simple neglect of the nonlinear terms to those, such as the cumbersomely named EDQNM and others that explicitly allow for triad interactions. In the absence of mean flow, linear theory (often called rapid-distortion theory, or RDT, when viscosity is neglected as well as nonlinearity) predicts simple exponential decay of the spectra due to viscosity. Linear theory is a rather crude approximation, since it does not allow for the action of turbulence on itself, an effect that appears as nonlinear interactions of wavenumber triads in the spectral theory of homogeneous turbulence, as we saw above. However, once the theory is extended to include, e.g., mean-flow gradients, an extension we do not attempt in this book, a surprising amount of physics is captured, including the distorting effects on the turbulence of physical mechanisms such as mean flow, stratification, and rotation.[2]

The other extreme of sophistication in spectral closure schemes is exemplified by the EDQNM (Eddy-Damped Quasi-Normal Markovian) model, which is based on heuristic assumptions leading to expressions for T_{ij} in terms of Φ_{ij}, thus closing equation (6.68). The details of nonlinear spectral closures lie beyond the scope of this volume, but we give a brief overview here. In quasi-normal models such as EDQNM, the next order of spectral evolution equations above (6.68) are used, that is, evolution equations for the cubic moments (such as Θ_{ilm}) are formulated, which introduces fourth-order spectral moments owing to the closure problem. The method of closure involves supposing that these fourth-order moments are given in terms of products of the second-order spectra Φ_{ij}, as if the velocity field had Gaussian statistics (cf. (2.59)). This closure is known as the quasi-normal approximation and, after some modifications, intended to make it more realistic, yields the EDQNM model. The resulting expression for T_{ij} in terms of Φ_{ij} consists of integrals over all wavenumber triads involving the given $\mathbf{k}$, so (6.68) becomes a nonlinear integro-differential equation that describes the time evolution of Φ_{ij}. Thus, triadic interactions, represented by the wavenumber integrals, appear explicitly in the model, hence the term triadic closure, which is sometimes used to qualify EDQNM and other models of a similar degree of sophistication. Like all closures, heuristic approximations are involved that

[2] Turbulence in a rotating fluid can be considered as a special case of a mean flow which it is easier to analyze using a rotating frame of reference, in which the mean flow disappears, but Coriolis terms arise, as we saw in Section 4.5 ("Rotating Turbulence"). Physical mechanisms other than mean flow can be introduced, such as density stratification in a gravitational field, which also modify the turbulence, but, like mean velocity gradients, will not be considered in this book.

are not derived from the equations of motion, but are instead introduced as additional hypotheses about the statistics of turbulence that are supposed to be approximately satisfied by real turbulent flows. Simpler closures than EDQNM have been proposed, as have other triadic closures, for instance, the Direct Interaction Approximation (or DIA) and its relatives. Each such closure leads to a heuristic expression for the nonlinear spectral transfer and hence completes the spectral evolution equation. This equation can then, in principle, be integrated forward in time, allowing quantitative prediction of the spectral evolution of turbulence, according to the particular closure used. However, we do not want to introduce closure approximations here, preferring to develop the theory towards the applications, such as Kolmogorov's theory of the small scales, described in the next chapter.

Based on (6.72) and (6.73), we can derive some elementary properties of Θ_{ilm}. Since, from (6.72),

$$Q_{ilm} = Q_{iml} \tag{6.74}$$

we have

$$\Theta_{ilm}(\mathbf{k}, t) = \Theta_{iml}(\mathbf{k}, t) \tag{6.75}$$

while, because Q_{ilm} is real,

$$\Theta^*_{ilm}(\mathbf{k}, t) = \Theta_{ilm}(-\mathbf{k}, t) \tag{6.76}$$

is the complex conjugate of (6.73). Using (6.76) it follows, from (6.69), that T_{ij} has the properties

$$T^*_{ij}(\mathbf{k}, t) = T_{ji}(\mathbf{k}, t) = T_{ij}(-\mathbf{k}, t) \tag{6.77}$$

This is necessary, because Φ_{ij} itself has these properties, which should be conserved under time evolution by (6.68).

Some important constraints on the spectral functions Φ_{ij} and Θ_{ilm} are imposed by the incompressibility condition, (6.47). Multiplying (6.64) by k_i and integrating with respect to $\mathbf{k}'$ over any region including $\mathbf{k}' = -\mathbf{k}$ yields

$$k_i \Phi_{ij} = 0 \tag{6.78}$$

Since Φ_{ij} is Hermitian, we also have

$$\Phi_{ij} k_j = (k_j \Phi_{ji})^* = 0 \tag{6.79}$$

Likewise, (6.63) implies that

$$k_i \Theta_{ilm} = 0 \tag{6.80}$$

which can be used, with (6.69) and the definition of the projection operator, Δ, to find

$$k_i T_{ij} = T_{ij} k_j = 0 \tag{6.81}$$

Equation (6.81) shows that, if Φ_{ij} satisfies the incompressibility constraints, (6.78) and (6.79), at $t = 0$, it will continue to satisfy them at later times, given time evolution according to (6.68).

So far, we have implicitly used a rectangular Cartesian coordinate system to define the components of vectors and tensors and this is by far the best approach for development of the general theory. However, it is often useful to adopt spherical polar coordinates k, θ, ϕ in **k**-space as shown in Figure 6.4. The axis of the spherical polar system is arbitrary in the absence of an externally imposed preferred direction (effects of mean shear, gravity, or rotation can select an unambiguous axis, but these are not present here). As for an orthogonal curvilinear coordinate system in physical space, one defines local unit vectors $\mathbf{e}_k$, $\mathbf{e}_\theta$, $\mathbf{e}_\phi$, aligned with the coordinate directions, and which vary from point to point in **k**-space. Any vector ot tensor field in **k**-space can then be expressed using its components with respect to these vectors. For instance, **k** takes the form $(k, 0, 0)$. The incompressibility condition, (6.78), and the Hermitian nature of Φ_{ij} imply that is has components

$$\Phi_{ij} = \begin{bmatrix} 0 & 0 & 0 \\ 0 & \Phi_{\theta\theta} & \Phi_{\theta\phi} \\ 0 & \Phi^*_{\theta\phi} & \Phi_{\phi\phi} \end{bmatrix} \tag{6.82}$$

in the spherical polar system, where $\Phi_{\theta\theta}$ and $\Phi_{\phi\phi}$ are both real. There are thus four independent real components of Φ_{ij}: $\Phi_{\theta\theta}$, $\Phi_{\phi\phi}$, and the real and imaginary parts of $\Phi_{\theta\phi}$. This use of polar coordinates in **k**-space to describe spectral quantities is often known as the Craya–Herring representation. We will see later that the number of independent components of Φ_{ij} is further reduced if one assumes isotropy of the turbulence. We shall nonetheless continue to use Cartesian components in what follows.

From (6.68), we can obtain the evolution equation of the spectral energy density, $\Phi_{ii}(\mathbf{k}, t)/2$, as

$$\frac{\partial}{\partial t}\left(\frac{1}{2}\Phi_{ii}\right) = \frac{1}{2}T_{ii} - \nu k^2 \Phi_{ii} \tag{6.83}$$

of which the first term represents transfer of energy from other wavenumbers by triadic interactions and the second gives the viscous dissipation, which is important at large k, comparable with η^{-1}. If (6.83) is integrated over all **k**, and (6.9) is used, one obtains the turbulent energy equation for homogeneous turbulence without a mean flow discussed in Chapter 4. It can be shown that the nonlinear transfer term integrates to zero:

$$\int T_{ii} d^3\mathbf{k} = 0 \tag{6.84}$$

indicating that it redistributes energy among the wavenumbers, but does not create or dissipate turbulent energy. This leaves the viscous dissipation term, whose integral over **k** is shown to be given by

$$\nu \int k^2 \Phi_{ii} d^3\mathbf{k} = \nu \overline{\frac{\partial u_i}{\partial x_j}\frac{\partial u_i}{\partial x_j}} = \bar{\varepsilon} \tag{6.85}$$

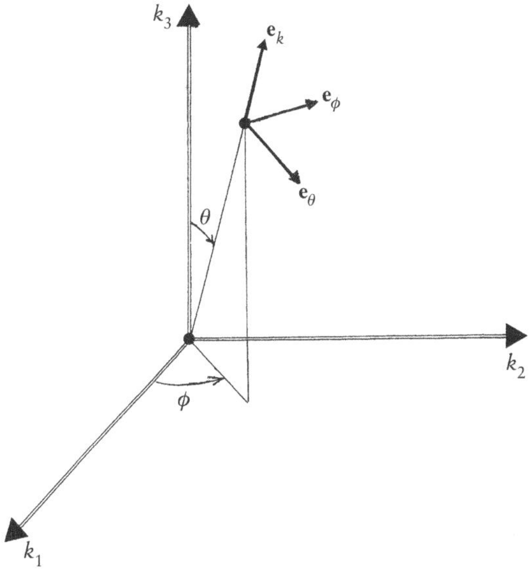

Figure 6.4. Spherical polar coordinates in **k**-space defining the Craya–Herring components for spectral vectors and tensors. The axis of the polar system has been arbitrarily taken to coincide with that for k_3.

in the appendix (equation (6.A22)). Thus, when integrated over all **k**, (6.83) becomes

$$\frac{d}{dt}\left(\frac{1}{2}\overline{u_i u_i}\right) = -\bar{\varepsilon} \tag{6.86}$$

which is, indeed, the turbulent energy equation (4.37), in the case where there are no mean velocity gradients. Overall, the turbulence decays continuously due to dissipation, but there is energy redistribution between wavenumbers, represented by the transfer term in (6.83). Note that (6.84) is necessary to recover (4.37).

In the special case of isotropic turbulence, if we multiply equation (6.83) by $4\pi k^2$ and use the definition (6.14), we obtain Lin's equation

$$\frac{\partial E}{\partial t} = \underbrace{T}_{\text{Transfer}} - \underbrace{2\nu k^2 E}_{\text{Dissipation}} \tag{6.87}$$

where $T(k,t) = 2\pi k^2 T_{ii}(\mathbf{k},t)$ is a function of $k = |\mathbf{k}|$ alone, owing to isotropy. This important equation describes the time evolution of the spectrum $E(k,t)$. The first term on the right represents spectral transfer of energy from all other wavenumbers to the given wavenumber k, while the final term describes its viscous dissipation. Figure 6.5 shows a sketch of the behavior of the two terms on the right-hand side of (6.87). The nonlinear transfer term, T_{ii}, integrates to zero according to (6.84) and therefore has both negative and positive regions in **k**-space, as consequently does $T(k,t)$ as a function of k. In fact, from (6.84) and its definition, $T(k,t)$ has zero integral from $k = 0$ to $k = \infty$, representing transfer of energy among wavenumbers, rather than production or dissipation. Within the range, $k = O(L^{-1})$, of energy-containing wavenumbers, energy is lost to higher wavenumbers (smaller scales) via the cascade, and so $T(k,t)$ is dominantly negative there. Nonetheless, as we will see in the next chapter, near $k = 0$ there is a small subrange in which $T(k,t)$ is slightly positive, corresponding to transfer of energy from around the spectral peak toward smaller wavenumbers, that is, scales larger than L. Since this subrange represents only a minor part of the energy-containing range and the associated values of $T(k,t)$ are small, it is not visible in Figure 6.5. Over most of $k = O(L^{-1})$, and in particular near the spectral peak, T is negative, giving transfer of energy to larger k by the cascade (with a lesser amount going to the small k subrange). Energy supply from the large to the small scales means that T changes sign and becomes positive at higher values of k, thereby offsetting viscous dissipation, which rises as the Kolmogorov scale is approached. The dissipation is always positive, but is relatively unimportant for the energy-containing scales, reflecting their essentially inviscid nature. It increases gradually with k and peaks in the dissipative range, $k = O(\eta^{-1})$. The net result is that the energy spectrum, $E(k,t)$, decays with time (except in the small k subrange), reflecting the absence of turbulence pro-

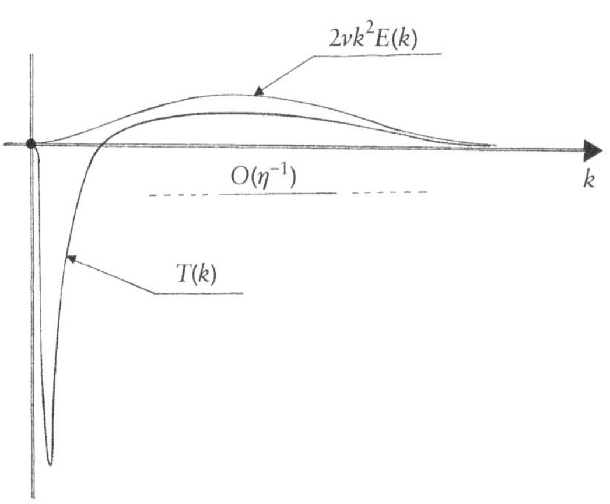

Figure 6.5. Energy transfer and dissipation terms of (6.87) in spectral space for isotropic turbulence.

duction to offset dissipation, there being no mean flow. We would expect the anisotropic case to be qualitatively similar, although all terms of (6.83) are then functions of the direction, as well as the magnitude of $\mathbf{k}$. Anisotropy allows transfer of energy between different directions, in addition to between different values of $|\mathbf{k}|$. Without mean flow, it is found that the transfer term tends to cause initially anisotropic turbulence to approach isotropy.

6.3 Spectral Equations via Correlations in Physical Space

In the previous section, we took direct Fourier transforms of the Navier–Stokes equations and obtained the spectral equations, (6.68), from those. As we noted earlier, it is also possible to proceed via the evolution equations for the velocity correlations in physical space and only afterwards take Fourier transforms to obtain the spectral equations. This is what we shall do here. In the process we obtain the equations governing the correlations in physical space, which are of some interest in themselves.

In this section, we shall denote turbulent quantities evaluated at point $\mathbf{x}'$ by a prime, while unprimed quantities refer to point $\mathbf{x}$. Thus, u_i is $u_i(\mathbf{x}, t)$ and u_j' is $u_j(\mathbf{x}', t)$. The Navier–Stokes equation, (6.44), can be written as

$$\frac{\partial u_i}{\partial t} + \frac{\partial}{\partial x_m}(u_i u_m) = -\frac{1}{\rho}\frac{\partial p}{\partial x_i} + \nu \frac{\partial^2 u_i}{\partial x_m \partial x_m} \tag{6.88}$$

at position $\mathbf{x}$ and

$$\frac{\partial u_j'}{\partial t} + \frac{\partial}{\partial x_m'}(u_j' u_m') = -\frac{1}{\rho}\frac{\partial p'}{\partial x_j'} + \nu \frac{\partial^2 u_j'}{\partial x_m' \partial x_m'} \tag{6.89}$$

at position $\mathbf{x}'$. Multiplying (6.88) by u_j', (6.89) by u_i, adding and averaging yields

$$\begin{aligned} &\frac{\partial}{\partial t}\overline{(u_i u_j')} + \frac{\partial}{\partial x_m}\overline{(u_i u_m u_j')} + \frac{\partial}{\partial x_m'}\overline{(u_i u_j' u_m')} = \\ &-\frac{1}{\rho}\left\{\frac{\partial}{\partial x_i}\overline{(p u_j')} + \frac{\partial}{\partial x_j'}\overline{(p' u_i)}\right\} + \nu\left\{\frac{\partial^2}{\partial x_m \partial x_m}\overline{(u_i u_j')} + \frac{\partial^2}{\partial x_m' \partial x_m'}\overline{(u_i u_j')}\right\} \end{aligned} \tag{6.90}$$

where primed quantities have been taken inside unprimed derivatives, because they do not depend on $\mathbf{x}$, and vice versa.

We now use homogeneity and the definitions ((6.5) and (6.72)) of R_{ij} and Q_{ilm}, and introduce the pressure–velocity correlation

$$\Pi_i(\mathbf{r}, t) = \overline{u_i(\mathbf{x}, t) p(\mathbf{x} - \mathbf{r}, t)} \tag{6.91}$$

to obtain

$$\frac{\partial}{\partial t}\left(R_{ij}(\mathbf{x}-\mathbf{x}',t)\right)+\frac{\partial}{\partial x_m}\left(Q_{jim}(\mathbf{x}'-\mathbf{x},t)\right)+\frac{\partial}{\partial x'_m}\left(Q_{ijm}(\mathbf{x}-\mathbf{x}',t)\right)=$$
$$-\frac{1}{\rho}\left\{\frac{\partial}{\partial x_i}\left(\Pi_j(\mathbf{x}'-\mathbf{x},t)\right)+\frac{\partial}{\partial x'_j}\left(\Pi_i(\mathbf{x}-\mathbf{x}',t)\right)\right\}$$
$$+\nu\left\{\frac{\partial^2}{\partial x_m\partial x_m}\left(R_{ij}(\mathbf{x}-\mathbf{x}',t)\right)+\frac{\partial^2}{\partial x'_m\partial x'_m}\left(R_{ij}(\mathbf{x}-\mathbf{x}',t)\right)\right\} \tag{6.92}$$

or, if we introduce $\mathbf{r}=\mathbf{x}-\mathbf{x}'$,

$$\frac{\partial R_{ij}}{\partial t}+\frac{\partial}{\partial r_m}\left(Q_{jim}(-\mathbf{r},t)-Q_{ijm}(\mathbf{r},t)\right)=$$
$$-\frac{1}{\rho}\left\{\frac{\partial}{\partial r_i}\left(\Pi_j(-\mathbf{r},t)\right)-\frac{\partial}{\partial r_j}\left(\Pi_i(\mathbf{r},t)\right)\right\}+2\nu\frac{\partial^2 R_{ij}}{\partial r_m\partial r_m} \tag{6.93}$$

which is the evolution equation for $R_{ij}(\mathbf{r},t)$. Note that it contains both the cubic velocity correlations, Q, and the pressure–velocity correlations Π. The latter are nonlocally determined from the velocity in physical space, which is its major disadvantage, compared with a spectral formulation.

From the incompressibility condition, (6.45), at $\mathbf{x}$ and $\mathbf{x}'$, we can derive

$$\frac{\partial R_{ij}}{\partial r_i}=\frac{\partial R_{ij}}{\partial r_j}=\frac{\partial \Pi_i}{\partial r_i}=0 \tag{6.94}$$

while (6.46) gives

$$\frac{\partial^2\Pi_i}{\partial r_m\partial r_m}=-\rho\frac{\partial^2 Q_{ilm}}{\partial r_l\partial r_m} \tag{6.95}$$

a Poisson equation that can be used to determine Π_i nonlocally from Q_{ilm} via the usual Green's function solution. Equations (6.93)–(6.95) describe the correlations in physical space.

We now take the Fourier transform of (6.93)–(6.95) with respect to $\mathbf{r}$, that is, we multiply by $e^{-i\mathbf{k}.\mathbf{r}}/(2\pi)^3$ and integrate over all $\mathbf{r}$. As usual, the derivatives with respect to $\mathbf{r}$ become factors of $i\mathbf{k}$, giving

$$\frac{\partial\Phi_{ij}}{\partial t}+ik_m\left\{\Theta_{jim}(-\mathbf{k},t)-\Theta_{ijm}(\mathbf{k},t)\right\}=-\frac{i}{\rho}\left\{k_i\tilde{\Pi}_j(-\mathbf{k},t)-k_j\tilde{\Pi}_i(\mathbf{k},t)\right\}-2\nu k^2\Phi_{ij} \tag{6.96}$$

$$k_i\Phi_{ij}=\Phi_{ij}k_j=k_i\tilde{\Pi}_i=0 \tag{6.97}$$

and

$$\tilde{\Pi}_i=-\frac{\rho k_l k_m}{k^2}\Theta_{ilm} \tag{6.98}$$

The last of these equations determines $\tilde{\Pi}_i$ locally in spectral space and so, given the definition (6.51), of the projector, Δ, we can easily express (6.96) as Craya's equation, (6.68) with (6.69).

We have now rederived the evolution equation for the spectra by Fourier transformation of the equations for the correlations in physical space. Although the direct transform method appears to involve somewhat more work, one can interpret equations, such as (6.50), in terms of triad interactions of Fourier components and (6.47) in terms of orthogonality, whereas (6.93)–(6.95) are more opaque. Furthermore, when one extends spectral analysis to problems involving, for instance, stratification or rotation, there are waves present (internal waves in stratified fluids and inertial waves with rotation). These waves are manifest in the linear operators that crop up using direct transforms of the fluctuating turbulent quantities, and indeed waves are often treated by such direct transform methods. The normal modes (e.g., plane-wave solutions) obtained by such methods provide a rather convenient basis for expressing the spectral equations, since the linear operators in the equations then take on rather simple forms. It is harder to extract such wavelike physics using the correlation approach. On the other hand, it can be argued that the correlation method is more mathematically robust, for the reasons given in Section 6.1. Which is chosen for a given problem is largely a matter of taste.

6.4 Consequences of Isotropy

In many works on turbulence, in the absence of a mean flow, homogeneous turbulence is treated as isotropic from the beginning. While it is true that such turbulence is observed to tend toward isotropy, because there is no preferred direction external to the turbulence, initially anisotropic, homogeneous turbulence can nonetheless exist. The conditions of homogeneity and isotropy are really independent and, despite simplifying the results a little by reducing the number of independent components of the various tensors, the introduction of isotropy at the same time as spectral analysis confuses the issues and means that one has to cope with technical difficulties at two levels at once. For this reason, and the fact that isotropic turbulence is a further idealization that is rarely precisely achieved in practice, we have chosen to postpone consideration of isotropy until now.

Suppose then that the turbulence is isotropic. This means that it has no preferred direction, that is, its statistical properties are unchanged if we rotate the flow or reflect the flow in any plane. The result, for *one-point* quadratic moments, is easily seen to be that all off-diagonal terms of $\overline{u_i u_j}$ are zero, while the diagonal terms are all equal to $u'^2 = \overline{u_i u_i}/3$. We used these and other simplifications of form due to symmetries in Chapter 4. Similar simplifications occur for the correlations (i.e., two-point quadratic moments) and spectra, but are not as obvious. We begin by considering the consequences for the spectral tensor, $\Phi_{ij}(\mathbf{k}, t)$. The vector wavenumber, $\mathbf{k}$, gives a preferred direction and we rotate axes so that $\mathbf{k}$ lies in the x_1-direction. Even for anisotropic turbulence, the incompressibility conditions, (6.78) and (6.79), then imply that $\Phi_{1j} = \Phi_{i1} = 0$, which leaves the nonzero components Φ_{22}, Φ_{33}, Φ_{23}, and Φ_{32}.

Symmetry under reflection in the plane $x_2 = 0$, say, results in $\Phi_{23} = \Phi_{32} = 0$. To see this, note that reflection of the flow field causes any given flow realization, $u_i(\mathbf{x}, t)$, to become

$$\begin{bmatrix} u_1(x_1, -x_2, x_3, t) \\ -u_2(x_1, -x_2, x_3, t) \\ u_3(x_1, -x_2, x_3, t) \end{bmatrix} \tag{6.99}$$

and so its Fourier transform, $\tilde{u}_i(\mathbf{k}, t)$, becomes $(\tilde{u}_1, -\tilde{u}_2, \tilde{u}_3)$, since $\mathbf{k}$ is in the x_1-direction. The fact that the flow is symmetric under reflection means that $(\tilde{u}_1, \tilde{u}_2, \tilde{u}_3)$ and its reflection $(\tilde{u}_1, -\tilde{u}_2, \tilde{u}_3)$, are equally probable within the ensemble of realizations and therefore produce canceling contributions when the average in (6.23) is taken for $i = 2, j = 3$ or $i = 3, j = 2$. It follows that $\Phi_{23} = \Phi_{32} = 0$. This leaves the components Φ_{22} and Φ_{33}, which must be equal by virtue of rotational symmetry. Thus, the only nonzero components, Φ_{22} and Φ_{33}, are equal. They are also real and positive, since Φ_{ij} is positive definite and Hermitian.

We can express the above results in the form

$$\Phi_{ij} = A\left(\delta_{ij} - \frac{k_i k_j}{k^2}\right) \tag{6.100}$$

which indeed has nonzero components $\Phi_{22} = \Phi_{33} = A$ when the x_1-axis is aligned with $\mathbf{k}$, but is also valid in any other rectangular Cartesian coordinates because both the left- and right-hand sides are tensors. From (6.100), we deduce that $\Phi_{ii}/2 = A(\mathbf{k}, t)$, which therefore gives the distribution of turbulent energy over $\mathbf{k}$. By isotropy, this distribution is independent of the direction of $\mathbf{k}$ and so $A(\mathbf{k}, t)$ depends on $k = |\mathbf{k}|$ only. It is more conventional to use the quantity

$$E(k, t) = 4\pi k^2 A(\mathbf{k}, t) \tag{6.101}$$

which is called the energy spectrum and, like A, is real and positive. The turbulent energy is given by

$$\frac{1}{2}\overline{u_i u_i} = \int A(\mathbf{k}, t) d^3\mathbf{k} = \int_0^\infty E(k, t) dk \tag{6.102}$$

where the volume integral has been performed using spherical elementary volumes, $4\pi k^2 dk$. Equation (6.102) shows that $E(k, t)$ gives the distribution of turbulent energy over different values of $k = |\mathbf{k}|$, there being no directionality for isotropic turbulence. The spectral tensor has the form

$$\Phi_{ij} = \frac{E(k, t)}{4\pi k^2}\left(\delta_{ij} - \frac{k_i k_j}{k^2}\right) \tag{6.103}$$

according to (6.100) and (6.101). Figure 6.1 shows a typical spectrum of isotropic turbulence, having its maximum at $k = O(L^{-1})$ and extending up to $k = O(\eta^{-1})$.

We can determine the form of $R_{ij}(\mathbf{r})$ by taking the inverse Fourier transform, (6.8), of (6.103). To do this, we use the mathematical identity

$$\int F(k) e^{i\mathbf{k}.\mathbf{r}} d^3\mathbf{k} = \frac{4\pi}{r}\int_0^\infty kF(k) \sin kr\, dk \tag{6.104}$$

where $r = |\mathbf{r}|$. This identity is easily established using spherical polar coordinates for $\mathbf{k}$, with axis in the direction of $\mathbf{r}$ ($\mathbf{k}.\mathbf{r} = kr\cos\theta$, $d^3\mathbf{k} = 2\pi k^2 \sin\theta\, d\theta\, dk$).

Employing (6.103) and (6.104) in (6.8), one finds

$$R_{ij}(\mathbf{r}, t) = \delta_{ij}\int \frac{E(k, t)}{4\pi k^2} e^{i\mathbf{k}.\mathbf{r}} d^3\mathbf{k} + \frac{\partial^2}{\partial r_i \partial r_j}\left\{\int \frac{E(k, t)}{4\pi k^4} e^{i\mathbf{k}.\mathbf{r}} d^3\mathbf{k}\right\} = B(r, t)\delta_{ij} - C(r, t)\frac{r_i r_j}{r^2} \tag{6.105}$$

where

$$B(r,t) = \int_0^\infty E(k,t)\left\{\frac{\sin kr}{kr} + \frac{kr\cos kr - \sin kr}{k^3r^3}\right\}dk \tag{6.106}$$

and

$$C(r,t) = \int_0^\infty E(k,t)\left\{\frac{\sin kr}{kr} + 3\,\frac{kr\cos kr - \sin kr}{k^3r^3}\right\}dk \tag{6.107}$$

Following Batchelor (G 1953), it is conventional to write R_{ij} in terms of nondimensional functions $f(r,t)$ and $g(r,t)$, rather than B and C, thus

$$R_{ij}(\mathbf{r},t) = u'^2\left(g(r,t)\delta_{ij} + (f(r,t) - g(r,t))\,\frac{r_ir_j}{r^2}\right) \tag{6.108}$$

where $u'^2 = \overline{u_iu_i}/3$ is the mean-squared turbulence velocity, which can be expressed in terms of the spectrum E via (6.102) as

$$u'^2(t) = \frac{2}{3}\int_0^\infty E(k,t)dk \tag{6.109}$$

For isotropic turbulence, u'^2 is the mean-squared value of the velocity component in any direction, for instance, $\overline{u_1^2} = \overline{u_2^2} = \overline{u_3^2} = u'^2$. The quantity f is given by

$$u'^2 f(r,t) = B - C = 2\int_0^\infty E(k,t)\,\frac{\sin kr - kr\cos kr}{k^3r^3}\,dk \tag{6.110}$$

while $u'^2 g = B$ is expressed by (6.106) or, equivalently, after some mathematical manipulation,

$$g(r,t) = f + \frac{1}{2}\,r\,\frac{\partial f}{\partial r} \tag{6.111}$$

which can also be derived more directly from (6.108) and the incompressibility condition, (6.94), for R_{ij}.

Equation (6.108) takes a particularly simple form if one uses coordinates in which one of the axes is parallel to the vector $\mathbf{r} = \mathbf{x} - \mathbf{x}'$. In that case, the matrix R_{ij} is diagonal with components $u'^2 f$ and $u'^2 g$ parallel and perpendicular to $\mathbf{r}$. The quantities f and g can be interpreted as illustrated in Figure 6.6. Denote the velocity components along the line joining $\mathbf{x}'$ with $\mathbf{x}$ by u_p and any one of the two components perpendicular to that line by u_n. The average of the product of u_p at $\mathbf{x}$ and u_p at $\mathbf{x}'$ normalized by u'^2, gives f, that is, $\overline{u_pu_p'} = f(r,t)u'^2 = B - C$. Likewise, $\overline{u_nu_n'} = g(r,t)u'^2 = B$. Thus, f describes the velocity correlations parallel to the line joining $\mathbf{x}'$ and $\mathbf{x}$, while g corresponds to velocity correlations normal to this line (also called longitudinal and transverse correlations). Moreover, the transverse correlations can be determined from the longitudinal one using (6.111).

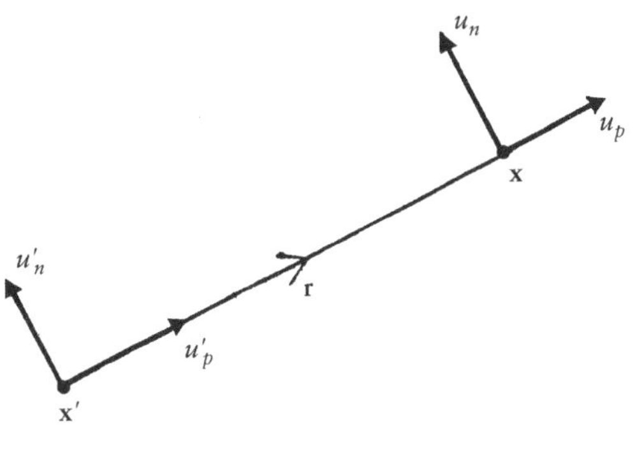

Figure 6.6. Velocity components parallel and normal to the line joining two points, giving correlations in isotropic turbulence.

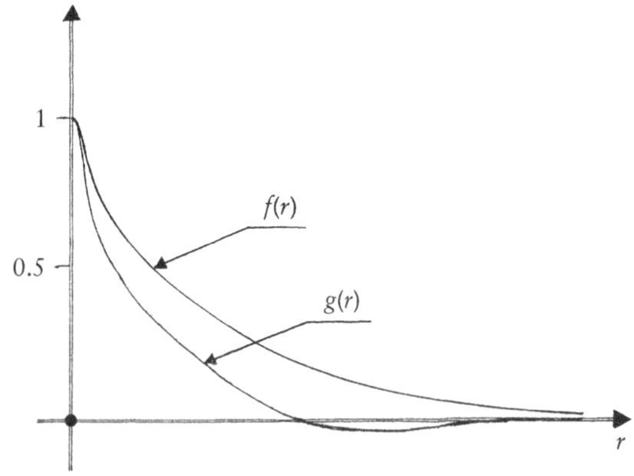

Figure 6.7. Typical longitudinal and transverse velocity correlations for isotropic turbulence.

If we take the limit $r \to \infty$, we know that $R_{ij} \to 0$, so that the large separation limits $f(r, t), g(r, t) \to 0$ are apparent from (6.108). The longitudinal and transverse correlations go rapidly to zero as the separation becomes large and the turbulence decorrelates. As $r \to 0$, (6.110) and (6.111) give $f = g = 1$ using (6.109). This also follows from the interpretation of f and g as longitudinal and transverse correlations, since the mean-squared velocity component in any direction is u'^2, that is, $\overline{u_p^2} = \overline{u_n^2} = u'^2$, when the points $\mathbf{x}$ and $\mathbf{x}'$ coalesce. Figure 6.7 shows a typical plot of the correlation functions, f and g, as a function of r.

From (6.111) we can show that

$$\int_0^\infty rg(r, t)dr = 0 \tag{6.112}$$

using integration by parts. This result implies the g must have both negative and positive values as a function of r.

One can use f and g to define turbulent integral scales

$$L_p = \int_0^\infty f(r)dr \tag{6.113}$$

$$L_n = \int_0^\infty g(r)dr \tag{6.114}$$

and from (6.111) one deduces that $L_n = L_p/2$ by integration by parts. These relations provide particular quantitative definitions of the integral scales, but it is probably better to consider L as an order of magnitude rather than a definite numerical value, because that is the way it is usually employed. One can give numerous different quantitative definitions of the integral scale, but it is difficult to choose between them for this reason.

Equation (6.110) gives f in terms of E, but one can also go the other way around. Taking the Fourier transform of (6.108) and applying (6.7) and the equivalent of (6.104) for integrals over $\mathbf{r}$, after some algebraic manipulations using (6.111), we recover (6.103) with

$$E(k, t) = \frac{u'^2}{\pi} \int_0^\infty f(r, t)kr(\sin kr - kr\cos kr)dr \tag{6.115}$$

which expresses the energy spectrum in terms of the longitudinal correlations function, $f(r, t)$. One can also define another correlation function, $R(r, t) = R_{ii}/2 = u'^2(f + 2g)/2$, which includes both transverse and longitudinal correlations with a "natural" weighting for isotropic turbulence (there are two transverse directions and one longitudinal direction for any given $\mathbf{r}$). Using (6.106) with $u'^2 g = B$, (6.110), (6.111), and (6.115), it can be shown that

$$E(k,t) = \frac{2}{\pi}\int_0^\infty R(r,t) kr \sin kr \, dr \tag{6.116}$$

and

$$R(r,t) = \int_0^\infty E(k,t)\, \frac{\sin kr}{kr}\, dk \tag{6.117}$$

One can continue to define different functions of r and k. The main point is that isotropy reduces the number of independent real functions from four for the anisotropic case to one, which might be E, R, f, or g. Whichever function is chosen, its values depend on the specific turbulent flow considered, and all the others can be calculated from this single one.

Recall that the Taylor microscales of turbulence are defined so that mean-squared velocity gradients can be calculated by expressions of the form u'^2/λ^2, where λ is a microscale. One can define a variety of such scales, but the most commonly used are λ_f and λ_g, which are directly related to the correlation functions, f or g. Let us define λ_f by the relationship

$$\overline{\left(\frac{\partial u_1}{\partial x_1}\right)^2} = \frac{2u'^2}{\lambda_f^2} \tag{6.118}$$

and, similarly, λ_g via

$$\overline{\left(\frac{\partial u_1}{\partial x_2}\right)^2} = \frac{2u'^2}{\lambda_g^2} \tag{6.119}$$

Since we are considering isotropic turbulence here, we could equally well take any velocity components in (6.118), provided the derivative was taken in the same direction, that is, λ_f is a longitudinal microscale. Likewise, λ_g could be defined using the derivative of any of the velocity components with respect to a coordinate perpendicular to that component: λ_g is a transverse microscale.

Taking the x_1 and x_1' derivatives of (6.5) with $i = j = 1$ and setting $\mathbf{x}' = \mathbf{x}$ yields

$$\overline{\left(\frac{\partial u_1}{\partial x_1}\right)^2} = -\left.\frac{\partial^2 R_{11}}{\partial r_1^2}\right|_{r=0} \tag{6.120}$$

and so, from (6.108) and (6.118), we have

$$\left.\frac{\partial^2 f}{\partial r^2}\right|_{r=0} = -\frac{2}{\lambda_f^2} \tag{6.121}$$

Thus the Taylor's series of f for small r begins

$$f(r) = 1 - \frac{r^2}{\lambda_f^2} + \cdots \tag{6.122}$$

and it can be likewise shown that

$$g(r) = 1 - \frac{r^2}{\lambda_g^2} + \cdots \tag{6.123}$$

while from (6.111), we deduce that these series are related, and hence that

$$\lambda_f^2 = 2\lambda_g^2 \tag{6.124}$$

More, generally, one can write

$$\overline{\frac{\partial u_i}{\partial x_l}\frac{\partial u_j}{\partial x_m}} = -\left.\frac{\partial^2 R_{ij}}{\partial r_l \partial r_m}\right|_{r=0} \tag{6.125}$$

and, using (6.108), (6.122), (6.123), and (6.124), this leads to

$$\overline{\frac{\partial u_i}{\partial x_l}\frac{\partial u_j}{\partial x_m}} = \frac{2u'^2}{\lambda_g^2}\left(\delta_{ij}\delta_{lm} - \frac{1}{4}(\delta_{il}\delta_{jm} + \delta_{im}\delta_{jl})\right) \tag{6.126}$$

for homogeneous, isotropic turbulence. The turbulent energy dissipation is given by (6.85), which becomes $\bar{\varepsilon} = 15\nu u'^2/\lambda_g^2$ when we use (6.126). This, and a number of other useful results for mean-squared velocity gradients and dissipation are described in the appendix.

The point has been made a number of times that the velocity gradients, unlike the turbulent velocities themselves, are dominated by the small scales. This is related to the importance of the small scales in dissipation, because the dissipation is determined by the mean-squared velocity gradients. Moreover, since, as described in the next chapter, the small scales tend to be approximately homogeneous and isotropic, even if the large scales are not, the above results should have quite general approximate validity.

The mean-squared turbulent velocity, u'^2, is determined by an integral, (6.109), over all the spectrum. The main contributions to this integral come from small wavenumbers, $k = O(L^{-1})$, representing the large scales of turbulence. On the other hand, the mean-squared velocity gradients are dominated by the smallest scales (large $k = O(\eta^{-1})$). The microscales (6.118) and (6.119) are thus defined as a mixture of quantities appropriate to the large and the smallest scales and are intermediate. For this reason, it is not clear what, if any, physical processes are associated with the length scales λ. The microscales are best thought of as convenient quantities for estimating velocity gradients and dissipation rates. They are certainly not the smallest scales in turbulence, which are $O(\eta)$ in size.

The functions $f(r)$ and $g(r)$ reflect the existence of these smallest scales. The Taylor's series, (6.122) and (6.123), only apply when r is smaller than the Kolmogorov scale, η, which itself is small compared with the integral scale, L, when the turbulent Reynolds number, $\mathrm{Re}_L = u'L/\nu$, is large. Indeed, in the next chapter, we shall see that $\eta/L = O(\mathrm{Re}_L^{-3/4})$ as $\mathrm{Re}_L \to \infty$.

For large Re_L, one can distinguish three asymptotic ranges of r. When $r = O(\eta)$, corresponding to the smallest dissipative scales of turbulence, the plots of $f(r)$ and $g(r)$, which have zero gradient at $r = 0$, turn downwards. Provided that the Reynolds number is high enough, there is an inertial range of separations, $\eta \ll r \ll L$, for which $f(r)$ and $g(r)$ are the sum of a constant and a term proportional to $r^{2/3}$. If this power law were continued all the way to $r = 0$, it would lead to infinite derivatives at $r = 0$: in fact, one enters the dissipative range, and the first derivative is zero at $r = 0$, while the second derivative is $O(\lambda^{-2}) = O(\mathrm{Re}_L L^{-2})$ and large, but finite. Nonzero viscosity makes Re_L finite and is responsible for the existence of the dissipative scale, $r = O(\eta)$, which shrinks in size relative to L as Re_L increases. Finally, when $r = O(L)$, $f(r)$ and $g(r)$ make the transition from the $r^{2/3}$ form of the inertial range (if an inertial range exists), via curves whose specific form depends on the particular turbulent flow considered, to zero at $r = \infty$.

These asymptotic ranges of r (asymptotic in the limit $\mathrm{Re}_L \to \infty$) are reflected in the energy spectrum, $E(k)$, with k related to r according to $k = O(r^{-1})$ (see Figure 6.1). In particular, if Re_L is sufficiently large that an inertial range exists, its wavelength equivalent lies in $L^{-1} \ll k \ll \eta^{-1}$ and $E(k)$ is proportional to $k^{-5/3}$ there. This form is known as the Kolmogorov spectrum. The relationship between $r^{2/3}$ for correlations and $k^{-5/3}$ for spectra in the inertial range of scales is fairly subtle. We shall return to this topic in the next chapter.

Let us now derive the isotropic form of the spectral evolution equation, (6.68). We have seen that Φ_{ij} is determined from $E(k, t)$ by (6.103) and, if we substitute into (6.68), we conclude that T_{ij} must have the form

$$T_{ij} = \frac{T(k,t)}{4\pi k^2}\left(\delta_{ij} - \frac{k_i k_j}{k^2}\right) \tag{6.127}$$

and that (6.87), that is, Lin's equation

$$\frac{\partial E}{\partial t} = T - 2\nu k^2 E \tag{6.128}$$

governs the evolution of $E(k, t)$. As noted earlier, the transfer term, T, causes energy to change hands across the spectrum, but does not create or destroy turbulent energy. This fact is expressed by

$$\int_0^\infty T(k,t)dk = 0 \tag{6.129}$$

which follows from (6.84) and

$$T_{ii} = \frac{T}{2\pi k^2} \tag{6.130}$$

One can go further and consider the effects of isotropy on Q_{ilm}, Θ_{ilm}, and hence the expression, (6.69), for T_{ij}. This allows T to be written in terms of a single cubic correlation function, as we shall see later. Figure 6.5 illustrates the transfer and dissipation terms which are present on the right of (6.128). As discussed earlier, the transfer term goes from negative values to positive values as k increases, expressing transfer from large to small scales (with a small range of positive values near $k = 0$ as well). The dissipation is always positive and peaks at k^{-1} of the order of the Kolmogorov scale. This is all a consequence of the cascade process, of which (6.128) provides a quantitative expression.

We next consider the consequences of isotropy for the time evolution of the correlations, $R_{ij}(\mathbf{r}, t)$, which is governed by (6.92). The pressure–velocity correlation, $\Pi_i(\mathbf{r}, t)$, which appears in this equation, is zero for isotropic turbulence, as we now show. A priori, isotropy implies no preferred direction, but two-point correlations, such as $\Pi_i(\mathbf{r}, t)$, introduce $\mathbf{r}$, the displacement between the two points, as a preferred direction. Clearly then the vector $\Pi_i(\mathbf{r}, t)$ must be in the direction of $\mathbf{r}$, that is,

$$\Pi_i(\mathbf{r}, t) = \Pi(r, t)\,\frac{r_i}{r} \tag{6.131}$$

According to (6.94), the divergence of $\Pi_i(\mathbf{r}, t)$ is zero:

$$\frac{\partial \Pi_i}{\partial r_i} = \frac{1}{r^2}\frac{\partial \Pi r^2}{\partial r} = 0 \tag{6.132}$$

so that

$$\Pi(r, t) = \frac{D}{r^2} \tag{6.133}$$

but this make $\Pi_i(\mathbf{r}, t)$ infinitely singular at $\mathbf{r} = 0$ unless $D = 0$. We conclude that $D = 0$ and so $\Pi_i(\mathbf{r}, t) = 0$ for isotropic turbulence.

Next consider the cubic correlation function $Q_{ilm}(\mathbf{r}, t)$, which is defined by equation (6.72):

$$Q_{ilm}(\mathbf{r}, t) = \overline{u_i(\mathbf{x}, t)u_l u_m(\mathbf{x} - \mathbf{r}, t)} \tag{6.134}$$

where u_l and u_m are both evaluated at $\mathbf{x} - \mathbf{r}$. Rotate the coordinates axes so that $\mathbf{r}$ lies in the x_1-direction, that is, so that the two points involved in (6.134) are separated only in the x_1-direction. By considering reflections in the planes $x_2 = 0$ and $x_3 = 0$, one can show that the only components of Q_{ilm} that are nonzero are Q_{111}, Q_{122}, Q_{133}, Q_{212}, Q_{221}, Q_{313}, and Q_{331} (see illustration in Figure 6.8). Furthermore, rotational symmetry about the x_1-axis and the relation $Q_{ilm} = Q_{iml}$, which is evident from (6.134), imply the equalities $Q_{122} = Q_{133}$ and $Q_{212} = Q_{221} = Q_{313} = Q_{331}$. This leaves three independent components and the result can be written in the tensor form

$$Q_{ilm}(\mathbf{r}, t) = A(r, t)r_i r_l r_m + B(r, t)(r_l\delta_{im} + r_m\delta_{il}) + C(r, t)r_i\delta_{lm} \tag{6.135}$$

where A, B, and C should not be confused with the same symbols used earlier for other functions of k and r. Of course, since (6.135) is an equality between tensors, it holds in all coordinate systems, since it applies in the one we have used.

Incompressibility imposes further constraints. Writing (6.134) in the equivalent form

$$Q_{ilm}(\mathbf{r}, t) = \overline{u_i(\mathbf{x} + \mathbf{r}, t)u_l u_m(\mathbf{x}, t)} \tag{6.136}$$

we have

$$\frac{\partial Q_{ilm}}{\partial r_i} = \overline{\frac{\partial u_i}{\partial x_i}(\mathbf{x} + \mathbf{r}, t)u_l u_m(\mathbf{x}, t)} = 0 \tag{6.137}$$

Applying this condition to (6.135) leads to

$$5A + r\frac{\partial A}{\partial r} + \frac{2}{r}\frac{\partial B}{\partial r} = 0 \tag{6.138}$$

and

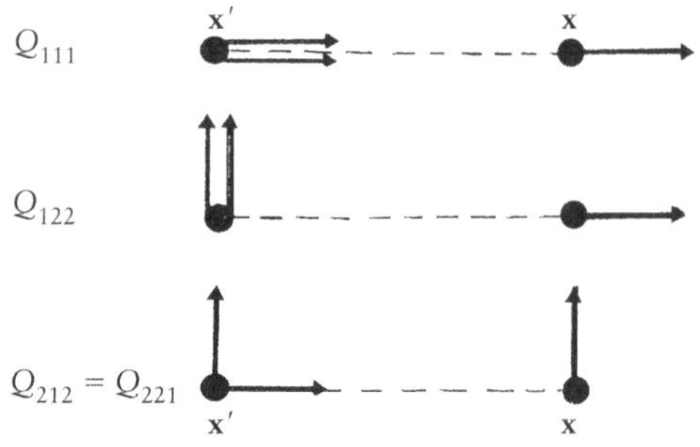

Figure 6.8. Illustration of the nonzero components of $Q_{ilm}(\mathbf{r}, t)$ for isotropic turbulence.

$$2B + 3C + r\frac{\partial C}{\partial r} = 0 \tag{6.139}$$

Introducing $K(r, t)$ via

$$u'^3 K = r^{-3} Q_{ilm} r_i r_l r_m = Ar^3 + (2B + C)r \tag{6.140}$$

which is the quantity, Q_{111}, in the special coordinate system with x_1-axis parallel to $\mathbf{r}$, one can solve (6.138)–(6.140) for A, B, and C to obtain finally,

$$Q_{ilm} = u'^3\left\{\frac{1}{2}\left(K - r\frac{\partial K}{\partial r}\right)\frac{r_i r_l r_m}{r^3} + \frac{1}{4}\left(\frac{2K}{r} + \frac{\partial K}{\partial r}\right)(r_l\delta_{im} + r_m\delta_{il}) - \frac{1}{2}K\frac{r_i}{r}\delta_{lm}\right\} \tag{6.141}$$

This equation gives the required cubic correlations in terms of a single function, $K(r, t)$, which is the triple, longitudinal correlation function. We remark that a similar result can derived in Fourier space, where it can be shown that Θ_{ilm} has the form $k_m\Delta_{il} - k_l\Delta_{im}$ multiplying a function of $k = |\mathbf{k}|$, which can be expressed in terms of $T(k, t)$ using (6.69) and (6.127).

Setting $i = j$ in (6.93), we have

$$\frac{\partial R_{ii}}{\partial t} + \frac{\partial}{\partial r_m}(Q_{iim}(-\mathbf{r}, t) - Q_{iim}(\mathbf{r}, t)) = 2\nu\frac{\partial^2 R_{ii}}{\partial r_m \partial r_m} = \frac{2\nu}{r^2}\frac{\partial}{\partial r}\left(r^2\frac{\partial R_{ii}}{\partial r}\right) \tag{6.142}$$

where we have used $\Pi_i = 0$. From (6.141), we find

$$Q_{iim}(\mathbf{r}, t) = \frac{u'^3}{2r^4}\frac{\partial(Kr^4)}{\partial r}r_m \tag{6.143}$$

so that

$$Q_{iim}(-\mathbf{r}, t) - Q_{iim}(\mathbf{r}, t) = -\frac{u'^3}{r^4}\frac{\partial(Kr^4)}{\partial r}r_m \tag{6.144}$$

and hence, taking the divergence,

$$\frac{\partial}{\partial r_m}(Q_{iim}(-\mathbf{r}, t) - Q_{iim}(\mathbf{r}, t)) = -\frac{u'^3}{r^2}\frac{\partial}{\partial r}\left(\frac{1}{r}\frac{\partial(Kr^4)}{\partial r}\right) \tag{6.145}$$

Applying (6.145) and

$$R_{ii} = u'^2(f + 2g) = \frac{u'^2}{r^2}\frac{\partial}{\partial r}(r^3 f) \tag{6.146}$$

which follows from (6.108) and (6.111), to (6.142) leads to the equation

$$\frac{\partial}{\partial r}\left\{r^3\frac{\partial(u'^2 f)}{\partial t} - \frac{1}{r}\frac{\partial}{\partial r}\left(r^4\left(u'^3 K + 2\nu u'^2\frac{\partial f}{\partial r}\right)\right)\right\} = 0 \tag{6.147}$$

whose first integral with respect to r is the Karman–Howarth equation

$$\frac{\partial(u'^2 f)}{\partial t} = \frac{1}{r^4}\frac{\partial}{\partial r}\left(r^4\left(u'^3 K + 2\nu u'^2\frac{\partial f}{\partial r}\right)\right) \tag{6.148}$$

This equation describes the evolution of $f(r, t)$ with time and is the physical space equivalent of the spectral evolution equation, (6.128). Just as the latter has the unknown spectral transfer term, $T(k, t)$, as a consequence of the closure problem, (6.148) contains the unknown cubic correlation function $K(r, t)$, which sits alongside f and g in the theory. The spectral transfer term can be related to K via

$$T = \frac{ku'^3}{\pi}\int_0^\infty \frac{\sin kr}{r}\frac{\partial}{\partial r}\left(\frac{1}{r}\frac{\partial}{\partial r}(r^4 K)\right)dr \tag{6.149}$$

whose derivation may be found in Batchelor (G 1953). This expression can be rewritten using (6.143) and integration by parts to give

$$T = \frac{2k}{\pi}\int_0^\infty (\sin kr - kr\cos kr)S(r, t)dr \tag{6.150}$$

where

$$S(r, t) = \frac{Q_{iim}r_m}{r} = \overline{u_i(\mathbf{x}, t)u_i u_m(\mathbf{x} - \mathbf{r}, t)}\,\frac{r_m}{r} \tag{6.151}$$

is a weighted sum of longitudinal and transverse cubic velocity moments at two points.

6.5 One-Dimensional Transforms, Time Spectra, and Some Experimental Results

As we mentioned in the introduction, some flows possess homogeneity in only one or two directions. For such flows, it is natural to define one- and two-dimensional spectra. Furthermore, experimental measurements are often made along what amounts to a line within the flow, and again it is appropriate to consider one-dimensional spectra. Here, we give only a brief description.

Suppose homogeneity in the x_1-direction so that the velocity correlation function is of the form $R_{ij}(x_1 - x_1', x_2, x_3, x_2', x_3', t) = \overline{u_i(\mathbf{x}, t)u_j(\mathbf{x}', t)}$. The one-dimensional spectrum is defined by

$$E_{ij}^{[1]}(k_1, x_2, x_3, t) = \frac{1}{2\pi}\int_{-\infty}^{\infty} R_{ij}(r_1, x_2, x_3, x_2, x_3, t)e^{-ik_1 r_1}dr_1 \tag{6.152}$$

in which it may be noted that the correlation is taken at two points having the same values of x_2 and x_3, but separated in x_1. The superscript 1 in $E_{ij}^{[1]}$ indicates that the x_1-direction is used for transformation. More generally, one can define one-dimensional spectra involving correlations at different values of x_2 and x_3, but we will not consider them here.

The inverse transform of (6.152) can be used to obtain $R_{ij}(x_1 - x_1', x_2, x_3, x_2', x_3', t)$ and if evaluated at $x_1' = x_1$ gives

$$\overline{u_i u_j} = \int_{-\infty}^{\infty} E_{ij}^{[1]}(k_1, x_2, x_3, t)dk_1 \tag{6.153}$$

which becomes

$$\frac{1}{2}\overline{u_i u_i} = \int_{-\infty}^{\infty}\frac{1}{2}E_{ii}^{[1]}(k_1, x_2, x_3, t)dk_1 \tag{6.154}$$

when we set $i = j$ and sum. This shows that the distribution of turbulent energy over the one-dimensional wavenumber, k_1, is $E_{ii}^{[1]}(k_1, x_2, x_3, t)/2$. One can easily show that

$$E_{ij}^{[1]}(k_1, x_2, x_3, t) = E_{ji}^{[1]}(-k_1, x_2, x_3, t) \tag{6.155}$$

and

$$E_{ij}^{[1]*}(k_1, x_2, x_3, t) = E_{ji}^{[1]}(k_1, x_2, x_3, t) \tag{6.156}$$

where the star indicates complex conjugation, as usual. From (6.155), it follows that

$$E_{ii}^{[1]}(k_1, x_2, x_3, t) = E_{ii}^{[1]}(-k_1, x_2, x_3, t) \tag{6.157}$$

so that the distribution of energy is equally in positive and negative wavenumbers (a similar result holds for the three-dimensional spectral function $\Phi_{ii}(\mathbf{k}, t)$). Thus, one can also write

$$\frac{1}{2}\overline{u_i u_i} = \int_0^\infty E_{ii}^{[1]}(k_1, x_2, x_3, t)dk_1 \tag{6.158}$$

and hence restrict attention to positive k_1, if required. It may be remarked that, whereas the three-dimensional energy spectrum, $E = 2\pi k^2 \Phi_{ii}$, of isotropic turbulence is zero at $k = 0$, one-dimensional spectra, such as $E_{11}^{[1]}$, are not usually zero when $k_1 = 0$.

Integral scales and Taylor microscales can be defined using the one-dimensional spectrum. For instance,

$$L_{11}^{[1]} = \frac{1}{\overline{u_1^2}}\int_0^\infty R_{11}(r_1, x_2, x_3, x_2, x_3, t)dr_1 = \pi\,\frac{E_{11}^{[1]}(k_1 = 0)}{\overline{u_1^2}} \tag{6.159}$$

and

$$\frac{1}{\lambda_{11}^{[1]^2}} = \frac{1}{2\overline{u_1^2}}\overline{\left(\frac{\partial u_1}{\partial x_1}\right)^2} = \frac{1}{\overline{u_1^2}}\int_0^\infty k_1^2 E_{11}^{[1]}(k_1, x_2, x_3, t)dk_1 \tag{6.160}$$

where the second equality in (6.160) follows by differentiation of the expression for $\overline{u_i(\mathbf{x}, t)u_j(\mathbf{x}', t)}$ obtained from the inverse transform of (6.152).

Turbulence which is homogeneous in all three dimensions is also homogeneous in any single direction, for instance x_1, and one can obtain the one-dimensional spectrum from $\Phi_{ij}(\mathbf{k}, t)$ via

$$E_{ij}^{[1]}(k_1, t) = \int_{-\infty}^{\infty}\int_{-\infty}^{\infty} \Phi_{ij}(k_1, k_2, k_3, t)dk_2 dk_3 \tag{6.161}$$

which does not depend on x_2 or x_3 because the turbulence is now supposed homogeneous in those directions. Of course, one loses information in going from Φ_{ij} to $E_{ij}^{[1]}$ by integration. This is a reflection of the fact that the one-dimensional spectrum, defined by (6.152), involves only points having the same values of x_2 or x_3, whereas Φ_{ij} contains information about correlations between different x_2 and x_3. Taking $i = j$ in (6.161) and dividing by two yields an expression for the one-dimensional distribution of turbulent energy as an integral over the three-dimensional energy distribution, with contributions from all values of k_2 and k_3.

If one further assumes isotropy, Φ_{ij} has the form (6.103) so that the integral in (6.161) can be reexpressed after a little mathematical manipulation as

$$E_{11}^{[1]}(k_1) = \frac{1}{2}\int_{k_1}^{\infty} \frac{E(k)}{k}\left(1 - \frac{k_1^2}{k^2}\right)dk \tag{6.162}$$

and

$$E_{22}^{[1]}(k_1) = E_{33}^{[1]}(k_1) = \frac{1}{4}\int_{k_1}^{\infty} \frac{E(k)}{k}\left(1 + \frac{k_1^2}{k^2}\right)dk \tag{6.163}$$

while all off-diagonal components of $E_{ij}^{[1]}$ are zero. From (6.162) and (6.163), one can show that

$$E(k) = k^2 \left.\frac{d^2 E_{11}^{[1]}}{dk_1^2}\right|_{k_1=k} - k \left.\frac{dE_{11}^{[1]}}{dk_1}\right|_{k_1=k} \tag{6.164}$$

and

$$2E_{22}^{[1]} = E_{11}^{[1]} - k_1 \frac{dE_{11}^{[1]}}{dk_1} \tag{6.165}$$

Of course, for the isotropic turbulence considered here, the one-dimensional spectra are independent of the choice of x_1-direction. The reader should also note that a different notation for the one-dimensional spectra is often used in the literature, in which $E_1(k_1) = 2E_{11}^{[1]}(k_1)$ and $E_2(k_1) = 2E_{22}^{[1]}(k_1)$, rather than $E_{11}^{[1]}(k_1)$ and $E_{22}^{[1]}(k_1)$, represent the longitudinal and transverse one-dimensional spectra of isotropic turbulence.

Apart from spatial spectra for homogeneous flows, *steady* flows allow the definition of spectra in time, often called frequency spectra. For this purpose, we consider velocity correlations $R_{ij}(\mathbf{x}, t - t') = \overline{u_i(\mathbf{x}, t)u_j(\mathbf{x}, t')}$ at the same spatial point and different times. The frequency spectral functions can then be obtained from

$$\Psi_{ij}(\omega, \mathbf{x}) = \frac{1}{2\pi}\int_{-\infty}^{\infty} R_{ij}(\mathbf{x}, \tau)e^{i\omega\tau}d\tau \tag{6.166}$$

where ω is the frequency. Properties include

$$\Psi_{ij}(\omega, \mathbf{x}) = \Psi_{ji}(-\omega, \mathbf{x}) \tag{6.167}$$

$$\Psi_{ij}^*(\omega, \mathbf{x}) = \Psi_{ji}(\omega, \mathbf{x}) \tag{6.168}$$

and

$$\frac{1}{2}\,\overline{u_i u_i} = \frac{1}{2}\int_{-\infty}^{\infty} \Psi_{ii}(\omega, \mathbf{x})d\omega \tag{6.169}$$

of which the latter indicates that the positive quantity $\Psi_{ii}/2$ gives the distribution of turbulent energy over frequency. As for wavenumber, one can use $\Psi_{ii}(\omega, \mathbf{x}) = \Psi_{ii}(-\omega, \mathbf{x})$ to restrict attention to $\omega > 0$. If desired, the one-point temporal correlations can be recovered from Ψ_{ij} using the inverse transform of (6.166).

The frequency spectra take their largest values for $\omega = O(T^{-1})$, where T is an Eulerian correlation time scale, and the integral in (6.169) is dominated by such frequencies. However, the frequency spectra have significant contributions up to values of ω corresponding to the smallest time scales, where viscosity is important. They are related to the spatial spectra: small spatial structures (large wavenumbers) contribute to the high-frequency parts of the time spectra because they are convected past a fixed point in a short time, as discussed in Chapter 3. More generally, the reader is encouraged to refer back to that chapter for detailed discussion of the different time scales of turbulence, in particular the distinction between Eulerian and intrinsic time scales, as well as their relationship with turbulent spatial structure.

Frequency spectra are rather convenient experimentally because they require only measurements at a single sensor. Furthermore, many flows are steady, even if they are not homogeneous, and there is thus a wider domain of applicability. However, the theoretical foundations of homogeneous turbulence are much more extensive. For flows that are both steady and, at least partially or approximately, homogeneous, one can define spectra in both wavenumber and frequency using velocity correlations at different points and times. These provide the widest application of spectral methods, but apply strictly to the smallest class of flows.

Experimentally, measurement of flow velocities at several points simultaneously is far from easy. Single-point measurements are much easier, but one must make rather restrictive assumptions about the flow to be able to use the *temporal* data from a single sensor to obtain information about the *spatial* structure of turbulence. Specifically, one can assume that the spatial structures one is interested in do not change significantly during the time taken for their passage across the probe. Thus, one supposes that the intrinsic time scale for turbulence dynamics is long compared with this passage time (Eulerian time scale). If the mean flow, $\overline{U_i}$, is dominantly in the x_1-direction, this assumption can be expressed by

$$u_i = u_i(x_1 - \overline{U_1}t, x_2, x_3) \tag{6.170}$$

that is, that the velocity fluctuations are merely convected at constant speed $\overline{U_1}$. Such special behavior of the turbulence is never exactly realized, of course, and this approximation is called the Taylor hypothesis. One can imagine superimposing a sufficiently large uniform velocity on a turbulent flow to produce a close approximation to (6.170). The Taylor hypothesis is approximately verified in many experiments on parallel or nearly parallel flows.

Homogeneity in x_1 and steadiness become synonymous if (6.170) holds exactly, and the two-point, two-time correlations take the form

$$\overline{u_i(x_1, x_2, x_3, t)u_j(x_1', x_2, x_3, t')} = R_{ij}(x_1 - x_1' - (t - t')\overline{U_1}, x_2, x_3) \tag{6.171}$$

Favre and co-workers (see, e.g., Favre, Gaviglio, and Dumas (1957, 1958, 1962) and also Hinze (G 1975) for discussion and further references) experimentally studied the behavior of such space–time correlation functions in great detail, producing many important and interesting results. Amongst these, they found that R_{ij} is relatively slowly varying if $x_1 - x_1' - (t - t')\overline{U_1}$ is held constant. Thus, in practice, although (6.171) is not exact, because real turbulence evolves as it is convected, it is approximately true in many flows.

For steady flow in which the Taylor hypothesis is assumed to hold, (6.171) implies a direct relationship between the velocity correlations at two points and a single time (i.e., with $t' = t$) and those at two times and a single point (i.e., $x_1' = x_1$). More precisely, the two-time correlations at delay $\tau = t - t'$ coincide with the two-point correlations at separation $x_1 - x_1' = -\overline{U_1}\tau$, a relation which could be more accurate than the more general (6.171). Using this relation in (6.152) and (6.166) yields

$$E_{ij}^{[1]}(k_1, x_2, x_3) = |\overline{U_1}|\Psi_{ij}(\overline{U_1}k_1, x_2, x_3) \tag{6.172}$$

which connects the frequency and one-dimensional wavenumber spectra according to the Taylor hypothesis. Equation (6.172) reflects the relationship, $\omega = \overline{U_1}k_1$, between frequency and one-dimensional wavenumber for Fourier components convected at speed $\overline{U_1}$. The spectrum at large ω (or k) is determined by the correlations at small $t - t'$ (or $x_1 - x_1'$). Since Favre and coworkers found that the correspondence between two-point and two-time correlations that leads to (6.172) gets better as the time delay/spatial separation is reduced, one expects (6.172) to improve at high k_1, i.e., for the small scales.

A single sensor suffices to obtain $\Psi_{ij}(\omega)$ experimentally and $E_{ij}^{[1]}$ can then be calculated using (6.172). If one further assumes isotropy, the three-dimensional spectrum is given by (6.164), while (6.165) is one means which can be used to check the degree of isotropy of the turbulence. Of course, it would be better to obtain three-dimensional spatial spectra directly from Fourier transformation of two-point correlations obtained by simultaneous measurements at pairs of points, but this is harder and requires a larger amount of data. Most experimental results have been obtained using the Taylor hypothesis.

Grid-generated turbulence is a good approximation to homogeneity 40 or 50 mesh spacings downstream of the grid and is accompanied by a very nearly uniform mean flow $\overline{U_1}$. This is a little surprising given the inhomogeneous nature of the grid, but has been verified in a large number of experiments over the last fifty years, involving many different grid geometries (single and double rows of round and square bars, flat strips, differing mesh sizes and bar aspect ratios, etc.). The turbulence is also approximately isotropic. Since there is little mean shear, turbulent energy production is negligible and energy is simply transferred from large to small scales, where it is dissipated. Owing to the decay of turbulence with downstream distance, the flow is obviously inhomogeneous over sufficiently large downstream distances, but this is of little signficance locally (i.e., within a correlation length, representing the largest scales of the turbulence).

Grid flow is steady and to obtain something close to the decaying homogeneous turbulence without mean flow that we considered earlier, one needs to consider a frame of reference moving with the uniform mean flow $\overline{U_1}$. Thus, time in the theory is related to downstream distance in the laboratory by $t = x_1/\overline{U_1}$. Measurements in the laboratory have turbulence convected at speed $\overline{U_1}$ past a stationary sensor and, adopting the Taylor hypothesis, one can obtain the one-dimensional spectra from the frequency spectra at a single sensor using (6.172). The resulting one-dimensional spectra change with downstream distance of the sensor, which is equivalent to temporal evolution in the theory.

We do not want to go into details of the many experimental studies of grid-generated turbulence. Townsend, Dryden, and others made significant early contributions, while Hinze (G 1975) and Monin and Yaglom (G 1975) cite a considerable number of experimental results on grid turbulence. The measurements by Comte-Bellot and Corrsin (1966) are perhaps amongst the most complete, although the Reynolds number is not as high as in, for instance, Kistler and Vrebalovich (1966), who observed a clear inertial range. Comte-Bellot and Corrsin considered a variety of grids in a wind tunnel of sufficient length that the turbulence had entered its final stage of decay (Re_L small) by the end of the measurement section. One is, nonetheless, most interested in the regime where Re_L is large, because this represents active turbulence.

A basic measure of isotropy is provided by the mean-squared velocity fluctuations $\overline{u_1^2}$ in the streamwise direction and $\overline{u_2^2}, \overline{u_3^2}$ perpendicular to the mean flow. If no special precautions are taken, one typically finds that $\overline{u_2^2}$ and $\overline{u_3^2}$ are very nearly equal, but noticeably smaller than $\overline{u_1^2}$. Comte-Bellot and Corrsin (1966) introduced a slight contraction into the wind tunnel upstream of the measurement section to make all three components more nearly equal, which presumably leads to improved isotropy.

Overall, it is found that the decay of grid turbulence is well described by power laws such as

$$u'^2 \propto t^{-1.3} \tag{6.173}$$

$$L \propto t^{0.35} \tag{6.174}$$

and

$$\lambda \propto t^{0.5} \tag{6.175}$$

provided that Re_L is large. The origin of x_1 used in $t = x_1/\overline{U_1}$ is found to be noticeably different from the grid location, presumably because of near-grid effects before the turbulence has time to fully settle down to the power law behavior. It should be remarked that, although power laws like (6.173), (6.174) are obtained for all grid geometries, the exponents appear to vary significantly from grid to grid. Furthermore, owing to the unknown origin, it is in fact quite difficult to determine the precise values of the exponents in any particular experiment. That is, one may obtain an acceptable data fit using a range of origins, each of which leads to somewhat different exponents, even in the case of a single experiment. Thus, there is no really fundamental significance to the precise exponents given above, with the exception of (6.175), which is always found to hold. Note that the root-mean-squared velocity fluctuations and correlation lengths, whose behavior is considered here, are properties of the *large scales* of the turbulence. The decay of homogeneous, isotropic turbulence, and its relationship with possible large-scale self-similarity of turbulence at long times, will be discussed in the next chapter.

From (6.173) and (6.174) we deduce that

$$\mathrm{Re}_L \propto t^{-0.3} \tag{6.176}$$

indicating that the Reynolds number decreases as the turbulence decays. This is a consequence of two conflicting effects: the integral scale increases, which tends to make $\mathrm{Re}_L = u'L/\nu$ grow, but the turbulent velocities fall faster, so Re_L decreases overall. If one constructs a Reynolds number, $\mathrm{Re}_\lambda = \lambda u'/\nu$, based on the Taylor microscale

$$\mathrm{Re}_\lambda \propto t^{-0.15} \tag{6.177}$$

so that Re_λ also decreases, but rather more slowly. Typical laboratory values of Re_λ are of order 100 (in the atmosphere, Re_λ can go as high as 2,000, leading to a rather wide inertial range).

Equation (6.86) describes the decay of turbulent energy. The overall kinetic energy, $\overline{u_i u_i}/2 = 3u'^2/2$, is dominated by the large scales of turbulence ($k = O(L^{-1})$ in (6.15)); however, the right-hand side of (6.86) (viscous dissipation) mainly arises from the smallest scales ($k = O(\eta^{-1})$ in (6.86)) when the Reynolds number is large. The observations show that high-Reynolds-number turbulence decays significantly over a time which scales on the quantity L/u'. This time is characteristic of the evolution of large turbulence scales and increases with time, leading to the algebraic decay, (6.173). The small-scale, viscous dissipation on the right-hand side of (6.86) adjusts itself according to the supply of energy from the large scales.

The distribution of energy among the different scales is represented by the spectrum, of which a log–log plot at sufficiently high Reynolds number indicates that there is an inertial range, as sketched in Figure 6.1 (see, for instance, the measurements of Kistler and Vrebalovich (1966)). Such experimental spectra are usually obtained from measurements at a single sensor, employing the Taylor hypothesis and assuming isotropy as described earlier in this section. We have already noted the existence of the peak at $k = O(L^{-1})$ and dissipative cutoff at $k = O(\eta^{-1})$. The straight line portion over the inertial range in Figure 6.1, which lies in $L^{-1} \ll k \ll \eta^{-1}$, has a slope of $-5/3$, representing the power law form

$$E(k) \propto k^{-5/3} \tag{6.178}$$

which is the Kolmogorov inertial-range spectrum and is also found in more general flows than grid turbulence, as will be discussed in the next chapter. As time goes by, the spectral peak moves to smaller k, since it is determined by $k = O(L^{-1})$ and L is increasing, while the dissipative range, $k = O(\eta^{-1})$, also moves towards smaller k. The separation between the two, $L/\eta = O(\mathrm{Re}_L^{3/4})$ according to Kolmogorov's theory, decreases with Re_L until the inertial range disappears. Later, when Re_L falls to $O(1)$, there is no longer a cascade nor a wide continuum of different scales, because viscosity acts directly on the large scales and there is no separation of energy-containing and dissipative scales. Turbulence proper dies at this point, but apparently random, decaying motions of the fluid continue. Finally, when $\mathrm{Re}_L \ll 1$, the "turbulence" is essentially dead; it decays passively (each wavenumber independently as $e^{-\nu k^2 t}$) under the action of viscosity.

The transfer term, $T(k, t)$, in (6.128), has been determined by at least two different methods. Both the left-hand side and dissipation terms can be calculated from measurements and hence T obtained. Secondly, it is possible to measure $S(r, t)$, given by

(6.151), and hence calculate $T(k,t)$ from (6.150). The results are as sketched in Figure 6.5 and show transfer from large to small scales, confirming the cascade idea.

6.6 Conclusions

Although homogeneous turbulence is a rather idealized case, it allows us to define spectra that quantify the energy contributions of different spatial scales, whose wide continuum of sizes (between η and L) is one of the fundamental characteristics of high-Reynolds-number turbulence. As we saw in Chapter 4, mean shear can lead to production of turbulent energy, but if, as in this chapter, there is no mean velocity gradient, turbulence decays under viscous dissipation at the smallest scales following energy transfer from large ones via the cascade. This is borne out by the spectral evolution equations developed in this chapter, which show nonlinear transfer of energy from small to large wavenumbers with viscous dissipation at $k = O(\eta^{-1})$. The spectral equations of homogeneous, isotropic turbulence without mean flow, which have been developed in this chapter, are employed in the next chapter, which includes the Kolmogorov theory of the small scales. Using spectral theory, we develop the theory for the relatively simple case of homogeneous, isotropic turbulence without mean flow, but, in studying such special flows, our main aim is to elucidate general properties of turbulence that we hope and expect will allow us to understand less idealized turbulent flows.

The nonlinear transfer terms in the spectral evolution equation are expressed in terms of cubic moments, leading to a closure problem. The nonlinear transfer terms can be thought of as representing the action of turbulence on itself, whereas the only linear ones occurring in the absence of mean flow are due to viscosity. As noted earlier, spectral analysis may be extended to allow for uniform mean velocity gradients or stratification, together with other effects compatible with homogeneity of the turbulence, but such extensions go beyond the scope of this book.

Finally, we should perhaps stress some limitations of spectral analysis even for homogeneous flows. The spectrum is equivalent in information content to the correlation function, since they form a Fourier transform pair. The correlation functions are second-order velocity moments, but these hardly represent a full statistical picture of a turbulent flow. Small-scale intermittency, discussed in Chapter 7, is but one feature that is not captured by the spectrum. It should not, therefore, be surprising that alternative methods of analysis, such as fractals or probability distribution functions, have been used in the study of turbulence. The fact that there are numerous applications of spectral methods should not blind the reader to their limitations.

Appendix: Miscellaneous Expressions for the Mean Dissipation in Homogeneous, Isotropic Turbulence

As a source of reference to the reader, this appendix gives some useful equivalent expressions for the mean energy dissipation of homogeneous, isotropic turbulence. Some of the results also apply without the assumption of isotropy, and we shall point these out as we go.

The definition of $\bar{\varepsilon}$ is

$$\overline{\varepsilon} = \frac{1}{2}\nu\overline{\left(\frac{\partial u_i}{\partial x_j} + \frac{\partial u_j}{\partial u_i}\right)\left(\frac{\partial u_i}{\partial x_j} + \frac{\partial u_j}{\partial u_i}\right)} \tag{6.A1}$$

which is the basic expression for the turbulent energy dissipation per unit mass and applies whether or not the turbulence is homogeneous and isotropic. For homogeneous turbulence:

$$\overline{\varepsilon} = \nu\overline{\omega_i\omega_i} \tag{6.A2}$$

where ω_i is the fluctuating vorticity. Further supposing isotropy:

$$\overline{\varepsilon} = \frac{15}{2}\nu\overline{\left(\frac{\partial u_1}{\partial x_2}\right)^2} \tag{6.A3}$$

$$\overline{\varepsilon} = 15\nu\overline{\left(\frac{\partial u_1}{\partial x_1}\right)^2} \tag{6.A4}$$

where x_1 and x_2 are arbitrary orthogonal Cartesian coordinates,

$$\overline{\varepsilon} = 15\nu\frac{u'^2}{\lambda_g^2} \tag{6.A5}$$

$$\overline{\varepsilon} = 30\nu\frac{u'^2}{\lambda_f^2} \tag{6.A6}$$

where λ_g and λ_f are the Taylor microscales,

$$\overline{\varepsilon} = 30\nu\int_0^\infty k_1^2 E_{11}^{[1]}(k_1)dk_1 \tag{6.A7}$$

where k_1 is the wavenumber in the x_1-direction, and finally

$$\overline{\varepsilon} = 2\nu\int_0^\infty k^2 E(k)dk \tag{6.A8}$$

where k is the total wavenumber.

Using $\omega = \nabla \times \mathbf{u}$ expressed in terms of components and the identity

$$\overline{\left(\frac{\partial u_i}{\partial x_j} + \frac{\partial u_j}{\partial u_i}\right)\left(\frac{\partial u_i}{\partial x_j} + \frac{\partial u_j}{\partial u_i}\right)} - \overline{\left(\frac{\partial u_i}{\partial x_j} - \frac{\partial u_j}{\partial u_i}\right)\left(\frac{\partial u_i}{\partial x_j} - \frac{\partial u_j}{\partial u_i}\right)} = 4\overline{\frac{\partial u_i}{\partial x_j}\frac{\partial u_j}{\partial u_i}} \tag{6.A9}$$

(6.A1) becomes

$$\overline{\varepsilon} = \nu\overline{\omega_i\omega_i} + 2\nu\overline{\frac{\partial u_i}{\partial x_j}\frac{\partial u_j}{\partial x_i}} \tag{6.A10}$$

which gives (6.A2) when we use

$$\overline{\frac{\partial u_i}{\partial x_j}\frac{\partial u_j}{\partial x_i}} = \frac{\partial^2}{\partial x_i \partial x_j}\overline{u_i u_j} = 0 \tag{6.A11}$$

of which the first equality holds because the flow is assumed incompressible and the second follows from homogeneity. Isotropy is not needed for (6.A2). It also holds to a good approximation without homogeneity in high-Reynolds-number turbulence, away from viscous sublayers at boundaries, as discussed in Section 4.2 (see "The Turbulent Energy Dissipation Rate").

Equation (6.A3) can be derived using (6.126), rewritten as

$$\overline{\frac{\partial u_i}{\partial x_l}\frac{\partial u_j}{\partial x_m}} = \overline{\left(\frac{\partial u_1}{\partial x_2}\right)^2}\left(\delta_{ij}\delta_{lm} - \frac{1}{4}(\delta_{il}\delta_{jm} + \delta_{im}\delta_{jl})\right) \tag{6.A12}$$

where the constant has been evaluated using $i = j = 1$, $l = m = 2$. Thus,

$$\overline{\omega_i\omega_i} = \overline{\left(\frac{\partial u_3}{\partial x_2} - \frac{\partial u_2}{\partial x_3}\right)^2} + \overline{\left(\frac{\partial u_1}{\partial x_3} - \frac{\partial u_3}{\partial x_1}\right)^2} + \overline{\left(\frac{\partial u_2}{\partial x_1} - \frac{\partial u_1}{\partial x_2}\right)^2} = \frac{15}{2}\overline{\left(\frac{\partial u_1}{\partial x_2}\right)^2} \tag{6.A13}$$

via which, (6.A3) follows from (6.A2). Equation (6.A4) results from (6.A12) with $i = j = l = m = 1$ and (6.A3). Strictly speaking, (6.A3) and (6.A4) require homogeneity and isotropy. However, assuming that the finest scales of turbulence, which dominate the dissipation, are approximately isotropic, they should hold approximately in general high-Reynolds-number turbulence, away from viscous layers.

Equations (6.A5) and (6.A6) are a consequence of the definitions of the Taylor microscales:

$$\frac{1}{\lambda_f^2} = \frac{1}{2u'^2}\overline{\left(\frac{\partial u_1}{\partial x_1}\right)^2} \tag{6.A14}$$

and

$$\frac{1}{\lambda_g^2} = \frac{1}{2u'^2}\overline{\left(\frac{\partial u_1}{\partial x_2}\right)^2} \tag{6.A15}$$

together with (6.A3) and (6.A4).

Equation (6.A7) results from (6.A6) and

$$\frac{1}{\lambda_f^2} = \frac{1}{u'^2}\int_0^\infty k_1^2 E_{11}^{[1]} dk_1 \tag{6.A16}$$

which follows from (6.160), (6.A14) and $\overline{u_1^2} = u'^2$.

Finally, to prove (6.A8), we use (6.164) to show that

$$\int_0^\infty k^2 E(k) dk = \int_0^\infty k_1^2\left[k_1^2\frac{d^2E_{11}^{[1]}}{dk_1^2} - k_1\frac{dE_{11}^{[1]}}{dk_1}\right]dk_1 \tag{6.A17}$$

from which integrations by parts yield

$$\int_0^\infty k^2 E(k) dk = 15\int_0^\infty k_1^2 E_{11}^{[1]} dk_1 \tag{6.A18}$$

which provides the link between (6.A7) and (6.A8).

An alternative derivation of (6.A8) can be given, which has the advantage that it provides a spectral expression for the dissipation which is valid in homogeneous turbulence, even if it is not isotropic. Taking the second derivative of (6.8) and evaluating the result at $\mathbf{r} = 0$ gives

$$\left.\frac{\partial^2 R_{ij}}{\partial r_k \partial r_l}\right|_{\mathbf{r}=0} = -\int k_k k_l \Phi_{ij}(\mathbf{k}, t) d^3\mathbf{k} \tag{6.A19}$$

The left-hand side of (6.A19) can also be expressed by taking derivatives of (6.5) and setting $\mathbf{x} = \mathbf{x}'$. Thus,

$$\left.\frac{\partial^2 R_{ij}}{\partial r_k \partial r_l}\right|_{\mathbf{r}=0} = -\overline{\frac{\partial u_i}{\partial x_k} \frac{\partial u_j}{\partial x_l}} \tag{6.A20}$$

Equating the right-hand sides of (6.A19) and (6.A20) gives

$$\overline{\frac{\partial u_i}{\partial x_k} \frac{\partial u_j}{\partial x_l}} = \int k_k k_l \Phi_{ij}(\mathbf{k}, t) d^3\mathbf{k} \tag{6.A21}$$

which is a useful general expression for the second-order moments of the velocity derivatives which does not depend on isotropy. Setting $i = j$ and $k = l$ in (6.A21) and employing the first equality of (4.42), we have

$$\overline{\varepsilon} = \nu \int k^2 \Phi_{ii}(\mathbf{k}, t) d^3\mathbf{k} \tag{6.A22}$$

which, using (6.14), reduces to (6.A8) in the isotropic case, but applies to homogeneous turbulence in general.

References

Bendat, J. S., Piersol, A. G., 1971. *Random Data: Analysis and Measurement Procedures.* Wiley, New York.

Comte-Bellot, G., Corrsin, S., 1966. The use of a contraction to improve the isotropy of grid-generated turbulence. *J. Fluid Mech.*, **25**(4), 657–82.

Favre, A. J., Gaviglio, J.-J., Dumas, R. J., 1957. Space-time double correlations and spectra in a turbulent boundary layer. *J. Fluid Mech.*, **2**, 313–42.

Favre, A. J., Gaviglio, J.-J., Dumas, R. J., 1958. Further space-time correlations of velocity in a turbulent boundary layer. *J. Fluid Mech.*, **3**, 344–56.

Favre, A. J., Gaviglio, J.-J., Dumas, R. J., 1962. Correlations spatiotemporelles en écoulements turbulents. In *Mécanique de la Turbulence*, ed. A. J. Favre, 419–45.

Jeffreys, H., 1931. *Cartesian Tensors.* Cambridge University Press, Cambridge.

Kistler, A. L., Vrebalovich, T., 1966. Grid turbulence at large Reynolds numbers. *J. Fluid Mech.*, **26**, 37–47.

Lighthill, M. J., 1958. *An Introduction to Fourier Analysis and Generalised Functions.* Cambridge University Press, Cambridge.

Lumley, J. L., 1970. *Stochastic Tools in Turbulence.* Academic Press, New York.

CHAPTER SEVEN

Kolmogorov's and Other Theories Based on Spectral Analysis

This chapter covers a variety of different subjects, of which the most important is the Kolmogorov theory of inertial and dissipative range scales in its original (Kolmogorov 1941a, b, c) version. The Kolmogorov theory is discussed in Section 7.3 and is one of the foundation stones of the theory of turbulence, leading to the famous $E \propto k^{-5/3}$ inertial-range energy spectrum, which is now a benchmark for measurements and theoretical models alike. However, the original theory is not without its problems, as we will see in Section 7.5, where it is reformulated to avoid Landau's objection to the original version, and the possibility of intermittency corrections is introduced. To illustrate intermittency effects, we use the β-model here, not because it gives a fully satisfactory account of intermittency, but because it is relatively simple to describe, understand, and analyze.

Aside from the theory of the small scales (high wavenumbers), we also discuss spectral behavior at small wavenumbers (in the course of Section 7.1), the final phase of viscous decay (Section 7.2), in which the turbulent Reynolds number is small, and the consequences of large-scale self-similarity for high-Reynolds-number gridlike turbulence (Section 7.4). Throughout the chapter, the case of high-Reynolds-number turbulence will form our main preoccupation, and we will suppose developed turbulence, that is, that the cascade has been allowed the time to reach equilibrium over the full range of scales from the large, energy-containing scales, down to the small, dissipative ones. We further assume homogeneous, isotropic turbulence without mean flow[1] in the analyses of this chapter, although these assumptions are not really essential to all the conclusions reached, in particular Kolmogorov's theory of the small scales is observed to be quite robust, as we will discuss later.

7.1 Properties of the Energy Spectrum and Velocity Correlations

We first want to recall some relevant results from the spectral analysis of homogeneous, isotropic turbulence, which were derived in the previous chapter. All second-

[1] As we saw in earlier chapters, homogeneity without mean flow implies decay of turbulent energy. It is possible to make such a flow steady by introducing random, statistically homogeneous body forces, whose spectrum is peaked at the energy-containing scales of the turbulence. However, we do not adopt such an approach here.

order velocity correlation functions at separation $r = |\mathbf{r}|$ can be determined from the single one

$$R(r, t) = \frac{1}{2}\overline{u_i(\mathbf{x}, t)u_i(\mathbf{x} + \mathbf{r}, t)} \tag{7.1}$$

thanks to isotropy. Furthermore, the energy spectrum, $E(k, t)$, may be expressed in terms of $R(r, t)$ via (6.116) as

$$E(k, t) = \frac{2}{\pi}\int_0^\infty R(r, t)kr \sin kr \, dr \tag{7.2}$$

while (6.117) gives the inverse relationship

$$R(r, t) = \int_0^\infty E(k, t)\frac{\sin kr}{kr} dk \tag{7.3}$$

Equations (7.2) and (7.3) relate velocity correlations in physical space to the energy spectrum in spectral space.

The total turbulent energy per unit mass is given by any one of the equivalent expressions

$$\frac{3}{2}u'^2 = \frac{1}{2}\overline{u_i u_i} = R(0, t) = \int_0^\infty E(k, t)dk \tag{7.4}$$

where the first equality follows from the definition, $u'^2 = \overline{u_i u_i}/3$, of u', the second is obtained from (7.1) with $\mathbf{r} = 0$, and the third from (7.3) with[2] $r = 0$. The integral in (7.4) shows that $E(k, t)$ gives the contributions to the total energy of different wavenumbers.

The mean viscous dissipation per unit mass is given by equation (6.A8) of the appendix to Chapter 6 as

$$\overline{\varepsilon} = 2\nu\int_0^\infty k^2 E(k, t)dk \tag{7.5}$$

which, like the integral, (7.4), for the energy, expresses the distribution of dissipation over wavenumber. The k^2 weighting tends to emphasize the high wavenumbers, compared with the energy integral in (7.4). Bearing in mind the rough correspondence, $\ell \approx k^{-1}$, between length scales in physical space and wavenumbers in spectral space, this is the spectral expression of the importance of the small scales for dissipation.

The velocity correlation function, $R(r, t)$, has the value $3u'^2/2$ at $r = 0$, according to (7.4), and tends to zero as $r \to \infty$, because the velocity field decorrelates with increasing separation. Its behavior as a function of r is much as shown[3] in Figure 3.2. As explained in that chapter, the behavior of $R(r, t)$ at different separations reflects

[2] Note that the function $(\sin kr)/(kr)$ is perfectly well behaved at $kr = 0$ if it is given its limiting value of 1 there.

[3] Strictly speaking, that figure is for a longitudinal correlation function, whereas $2R = R_{ii} = u'^2(f + 2g)$ contains both longitudinal and transverse velocity correlations. Furthermore, $R(r, t)$ is usually slightly negative above a certain value of r.

the corresponding length scales of turbulence, with the large scales of turbulence represented by $r = O(L)$. The correlation length, L, is defined via the decorrelation of the velocity field. Thus, $R(r, t)$ is only significantly nonzero for $r \leq O(L)$ and rapidly approaching zero for $r \gg L$. This means that integrals over r, such as (7.2), are dominated by separations, $r \leq O(L)$, at which $R(r, t) = O(u'^2)$. Separations much greater than L are of negligible importance in this context.

Small separations, $r \ll L$, mirror the fine scales of turbulence, with the smallest scales, $r = O(\eta)$, dissipating turbulent energy, and intermediate sizes, $\eta \ll r \ll L$, providing the cascade. As discussed in Chapter 3, velocity correlations are related to velocity differences. Thus, combining the central equality of (7.4) with (7.1), we find

$$(\Delta u_r)^2 = \overline{|\mathbf{u}(\mathbf{x} + \mathbf{r}, t) - \mathbf{u}(\mathbf{x}, t)|^2} = 4(R(0, t) - R(r, t)) \tag{7.6}$$

which defines Δu_r, a measure of typical velocity differences at separation r. In particular, the manner in which $R(r, t)$ approaches $R(0, t)$ as r goes to zero determines typical velocity differences at small separations. From Chapter 3, we recall that the velocity field is "furry" when examined on scales $r \gg \eta$, corresponding to velocity differences, Δu_r, that decrease more slowly than r. It is only when the smallest scales, $r = O(\eta)$, are reached that the velocity field finally appears as smooth, and $\Delta u_r \propto r$ for $r \ll \eta$. It follows from (7.6) that $R(0, t) - R(r, t)$ is proportional to r^2 at such very small r, but decreases less rapidly for larger separations (see Figure 3.2). As Re_L increases, so does the ratio, L/η, of the largest to the smallest scales of turbulence. If Re_L is sufficiently large, a range of separations exists between the dissipative scales, $r = O(\eta)$, and large scales, $r = O(L)$, in which $\Delta u_r \propto r^{1/3}$ and $R(0, t) - R(r, t) \propto r^{2/3}$. This is known as the inertial range and represents intermediate turbulence scales. These three ranges of r show the characteristic asymptotic structure of developed turbulence as $\mathrm{Re}_L \to \infty$.

Figure 7.1 shows a typical energy spectrum, having an inertial range of wavenumbers, as a log–log plot. The spectrum, $E(k, t)$, has three different ranges of k corresponding, via $k \approx r^{-1}$, to those described above for $R(r, t)$. $E(k, t)$ is zero at $k = 0$, according to (7.2). It increases to a peak in the range $k = O(L^{-1})$ and then decreases, continuing to fall through the inertial range, which lies in $L^{-1} \ll k \ll \eta^{-1}$, separating the large scale and dissipative zones. The inertial range appears as a straight portion in the figure, representing the power law

$$E(k, t) = ak^{-n} \tag{7.7}$$

where the observed exponent is very close to the value, $n = 5/3$, which is predicted by Kolmogorov's theory, as we will see later. Inertial-range wavenumbers satisfy $L^{-1} \ll k \ll \eta^{-1}$, so that they

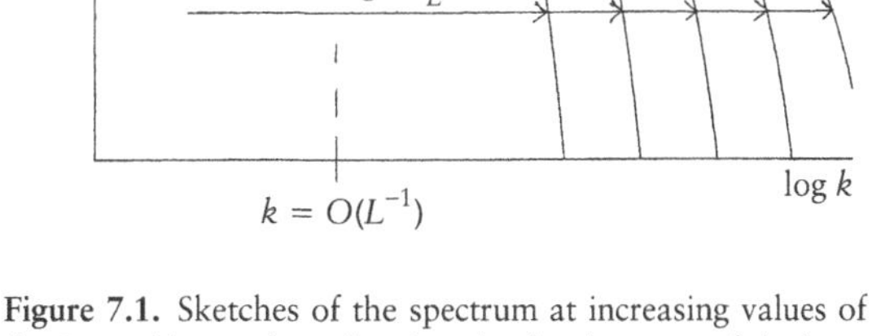

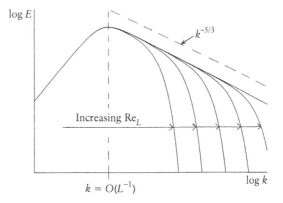

Figure 7.1. Sketches of the spectrum at increasing values of the Reynolds number, showing the development of the inertial range between the energy-containing and dissipative wavenumbers (log–log plots).

separate the large scale and dissipative zones. For future use, we prefer to allow a more general value, $1 < n < 2$, in (7.7), but, for the time being, the reader may assume $n = 5/3$ whenever n occurs in what follows. Finally, the spectrum falls away from the power law of the inertial range, dropping off increasingly rapidly at wavenumbers corresponding to the dissipative scales, $k = O(\eta^{-1})$, and above.

Figure 7.1 shows sketched log–log plots of the spectrum for different values of the Reynolds number. The inertial range, in which the power law, (7.7), holds, is not apparent unless Re_L is sufficiently large. As Re_L increases, the ratio, L/η, of the largest to the smallest scales grows. Since the spectral peak occurs at $k = O(L^{-1})$ and the dissipative falloff takes place for $k = O(\eta^{-1})$, the gap between them widens. This creates increasing room between the two wavenumber ranges (i.e., in $L^{-1} \ll k \ll \eta^{-1}$), which is progressively filled by the inertial range. The appearance of the power law is gradual, and it is difficult to identify exactly when it comes into existence, but it becomes more and more evident with increasing Re_L. As a very rough rule of thumb, Re_L of order 1,000 or more gives a noticeable inertial range in the spectrum,[4] which widens as Re_L increases further. The asymptotic structure in the limit $\mathrm{Re}_L \to \infty$ consists of an energy-containing wavenumber range, a dissipative range, and the inertial range between the two. The reader who would like to think in terms of matched asymptotic expansions could consider the inertial range as a zone of overlap between two asymptotic regions, $k = O(L^{-1})$ and $k = O(\eta^{-1})$ in spectral space, or $r = O(\eta)$ and $r = O(L)$ in physical space.[5] The lack of an inertial range, because Re_L is not large enough, does not mean that there is no cascade. The continuum of different scales, characteristic of turbulence in general, is still present, maintained by essentially the same physical processes as at higher Reynolds numbers: there is just insufficient separation of large and small scales to allow the intermediate asymptotic structure to become apparent.

Suppose that Re_L is sufficiently large that a wide inertial range is present. The integrals in (7.4) and (7.5) express the turbulent energy and dissipation rate in terms of the energy spectrum. In (7.4), the energy is given as the integral of the spectrum, which has a peak in the range $k = O(L^{-1})$, and then decreases through the inertial range, according to (7.7). The integral in (7.4) is dominated by contributions from $k = O(L^{-1})$, rather than the tail of the spectrum, because $n > 1$ makes the integral of (7.7) convergent at $k = \infty$ (the existence of the dissipative range further reduces the contribution from the far tail). Most of the turbulent energy thus comes from $k = O(L^{-1})$. This is the spectral expression of the physical space idea that the large scales of turbulence contain most of the kinetic energy. However, when $E(k, t)$ is multiplied by k^2 to form the integrand of (7.5), the situation changes radically: since $n < 2$, the integrand now increases through the inertial range until the rapid dissipative scale dropoff in the spectrum is reached. Thus, the integral giving the dissipation is dominated by wavenumbers in the dissipative range, $k = O(\eta^{-1})$, as we might expect, since these correspond to the smallest scales of turbulence. As the inertial range widens with increasing Reynolds number, the

[4] However, precise meaning, other than as an order of magnitude, has not been attributed to L, nor therefore to Re_L. Here, one might take the longitudinal integral length, L_p, defined by (6.112).

[5] We do not mean to suggest that a fully rigorous theory, matched expansions or otherwise, exists. This is purely conceptual.

turbulent energy and dissipation come mostly from the other two asymptotic regions, $k = O(L^{-1})$ and $k = O(\eta^{-1})$, respectively. The contributions of the inertial range to either is small, and its role is neither to supply the energy, nor to dissipate it, but simply to transfer energy to higher wavenumbers through the cascade.

The relationship between $R(r, t)$ and $E(k, t)$ can be investigated using (7.3). It is convenient to introduce the final equality of (7.4) to obtain

$$R(0, t) - R(r, t) = \int_0^\infty E(k, t)\left\{1 - \frac{\sin kr}{kr}\right\}dk \tag{7.8}$$

which shows weighted contributions of the spectrum from all wavenumbers. The spectral weighting is the term in brackets; it is a function of kr and is shown in Figure 7.2. For $k \ll r^{-1}$ the weighting function is small, of the form $(kr)^2/6$ as $kr \to 0$, which suppresses spectral contributions from wavenumbers smaller than $O(r^{-1})$. In the opposite limit, $k \gg r^{-1}$, the weighting function is close to 1 and the decay of $E(k, t)$ with increasing k makes the contribution from wavenumbers larger than $O(r^{-1})$ small. Thus, the integral in (7.8) comes mostly from $k = O(r^{-1})$, which is a mathematical expression of the correspondence, $k \approx r^{-1}$, between wavenumbers in spectral space and separations in physical space. The behavior of $R(0, t) - R(r, t)$, and hence Δu_r, is determined by that of $E(k, t)$ for $k = O(r^{-1})$.

With a wide enough inertial range, we can apply (7.7) in (7.8) for separations corresponding to the inertial range wavenumbers. Thus,

$$R(0, t) - R(r, t) = a\int_0^\infty k^{-n}\left\{1 - \frac{\sin kr}{kr}\right\}dk \tag{7.9}$$

which can be evaluated in terms of the gamma function (see, e.g., Abramowitz and Stegun (1970), chapter 6, for a definition and properties of the gamma function, $\Gamma(z)$), leading to

$$R(0,t) - R(r, t) = \frac{a\Gamma(2-n)\sin\frac{1}{2}n\pi}{n(n-1)} r^{n-1} \tag{7.10}$$

showing the power-law behavior of $R(0, t) - R(r, t)$ at inertial range separations. We see that the power law of the inertial range spectrum implies that of the correlation function and velocity difference. The converse is also true, that is, given a power law, (7.10), for $R(0, t) - R(r, t)$, the spectrum, $E(k, t)$, is (7.7), a result which can be derived from (7.2) and which we leave as an exercise for the reader.[6] Note that the departures of the spectrum from the power-law form, (7.7), outside the iner-

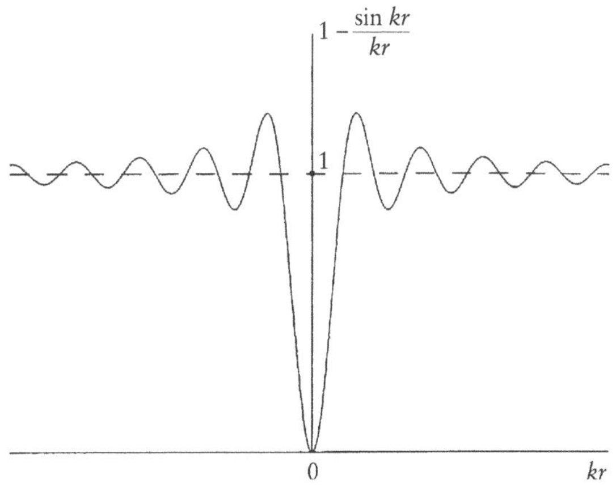

Figure 7.2. The spectral weighting function, $1 - (\sin kr)/(kr)$, occurring in the calculation of the correlation function from the spectrum, using (7.8).

[6] Hint: equation (7.2) should be integrated twice by parts, leading to an integral containing $d^2(rR)/dr^2$, so that it converges when (7.10) is applied.

tial range of wavenumbers have been ignored here. This means that (7.10) will not hold at dissipative or energy-containing scales, that is, $r = O(\eta)$ or $r = O(L)$.

To bring this section to a close, we want to discuss the behavior of $E(k, t)$ at very small wavenumbers, $k \ll L^{-1}$. We have already noted that the spectrum is zero at $k = 0$, from (7.2). When $k \ll L^{-1}$, kr is small over the range $r = O(L)$, for which the integrand of (7.2) is significantly nonzero. Thus, we expand $\sin kr$ as a power series for small argument, leading to

$$E(k, t) = c_2 k^2 + c_4 k^4 + \cdots \tag{7.11}$$

where

$$c_2 = \frac{2}{\pi} \int_0^\infty r^2 R(r, t) dr \tag{7.12}$$

and

$$c_4 = -\frac{1}{3\pi} \int_0^\infty r^4 R(r, t) dr \tag{7.13}$$

The reason for bothering with the k^4 term in (7.11) is that one can show that $c_2 = 0$, provided the velocity correlations tend to zero sufficiently rapidly as $r \to \infty$. The argument runs as follows: we combine the expression, (6.103), for Φ_{ij} with (7.11) to find

$$\Phi_{ij}(\mathbf{k}, t) = \frac{1}{4\pi} \left\{ c_2 \left(\delta_{ij} - \frac{k_i k_j}{k^2} \right) + c_4 \left(k^2 \delta_{ij} - k_i k_j \right) + \cdots \right\} \tag{7.14}$$

as $k \to 0$. If $c_2 \neq 0$, one obtains different limiting values for $\Phi_{ij}(0, t)$ depending on the direction from which $\mathbf{k} = 0$ is approached in $\mathbf{k}$-space. However, (6.7), evaluated at $\mathbf{k} = 0$, gives

$$\Phi_{ij}(0, t) = \frac{1}{(2\pi)^3} \int R_{ij}(\mathbf{r}, t) d^3\mathbf{r} \tag{7.15}$$

unambiguously, provided that $R_{ij} \to 0$ as $|\mathbf{r}| \to \infty$ sufficiently rapidly that there are no problems of convergence of the integral. In that case, we conclude that $c_2 = 0$ and the expansion, (7.11), reads

$$E(k, t) = c_4 k^4 + \cdots \tag{7.16}$$

for $k \ll L^{-1}$.

Doubts have been expressed (see, e.g., Saffman (1967)) about the convergence of integrals such as (7.15) and it has been suggested that c_2 might be nonzero in some flows, hence

$$E(k, t) = c_2 k^2 + \cdots \tag{7.17}$$

at small k. It can then be shown that the components of the correlation tensor, $R_{ij}(\mathbf{r}, t)$, decay to zero like $|\mathbf{r}|^{-3}$ at large $|\mathbf{r}|$, that is, rather slowly, leading to logarithmic divergence of (7.15). Furthermore, the expansion of $E(k, t)$ about $k = 0$ need not continue in the power series form, (7.11), beyond the first term if (7.13) has problems of convergence. That is, the straightforward power series expansion of $\sin kr$ used above does not give higher terms correctly, and a more sophisticated analysis is necessary if one wants to go beyond the k^2 term in (7.17). More generally, having admitted the possibility of a discontinuity in the spectral tensor at $\mathbf{k} = 0$, reflecting comparatively slow decay of the velocity correlations with distance, there is no obvious reason why $E(k, t)$ should not behave differently from either k^2 or k^4 as $k \to 0$. For instance, $E(k, t) \sim c_m k^m$, where m need not be an integer, and which includes both (7.16) and (7.17) as special cases, or even a low-wavenumber spectral form that is not a power of k. Given such a spectrum, one may calculate the corresponding correlations using (6.8) and (6.103). In summary, if the velocity correlations approach zero sufficiently rapidly as $|\mathbf{r}| \to \infty$, one has (7.16) as $k \to 0$, whereas other low-wavenumber spectral forms, in particular (7.17), are possible if the correlations decay less rapidly. Grid turbulence, many grid spacings downstream of the grid, which is the archetypal experimental approximation to the homogeneous turbulence we consider here, is believed to have correlations that decay sufficiently rapidly that $m = 4$, the case usually supposed in theoretical work.

7.2 Spectral Dynamics and the Final Phase of Passive Decay

The time evolution of the spectrum, $E(k, t)$, is governed by Lin's equation

$$\frac{\partial E}{\partial t} = T - 2\nu k^2 E \tag{7.18}$$

where

$$T(k, t) = \frac{2k}{\pi} \int_0^\infty S(r, t)(\sin kr - kr\cos kr)dr \tag{7.19}$$

represents nonlinear transfer of energy from other wavenumbers, while the second term on the right-hand side of (7.18) is the dissipation at wavenumber k. The quantity, $S(r, t)$, occurring in (7.19) is the cubic velocity correlation

$$S(r, t) = \frac{r_j}{r}\overline{u_i(\mathbf{x}, t)u_j(\mathbf{x}, t)u_i(\mathbf{x} + \mathbf{r}, t)} \tag{7.20}$$

We note that, from (7.19), $T = O(k^4)$ as $k \to 0$, provided that $S(r, t)$ is sufficiently rapidly decaying at large r, as is generally believed to be the case. The behavior of the low-wavenumber part of the spectrum, following time evolution according to (7.18), depends on how quickly the initial spectrum drops to zero as $k \to 0$ compared with k^4, representing the nonlinear transfer term (the viscous term is negligible at such low wavenumbers). If the initial spectrum has the form (7.17) as $k \to 0$, the spectrum continues to have that form at later times, with the same value of c_2, which is therefore constant. In fact, a similar result holds for any spectrum of the form $E \sim c_m k^m$ as $k \to 0$ with $m < 4$. On the other hand, although an initial spectrum of the form (7.16) persists at later times, the coefficient, $c_4(t)$, will generally evolve with time.[7] Finally, initial spectra that go to zero faster than k^4 as $k \to 0$ (e.g.,

[7] It should be remarked that such variation of c_4 with time is not in accord with a result of Loitsianskii that the integral in (7.13) be constant. However, this result was subsequently shown to be erroneous by Batchelor and Proudman (1956).

$E(k, 0) \propto k^m$ with $m > 4$) are overshadowed by nonlinear transfer as they evolve according to (7.18) and at later times take on the low-wavenumber form (7.16) with a time-varying c_4. Thus, we see that the low-wavenumber form of the spectrum is determined by the initial conditions. The evolution, or lack of it, of the low-wavenumber part of the spectrum due to the nonlinear spectral transfer term has implications for spectral evolution at large times, as we shall see in Section 7.4.

The overall decay of the turbulence is described by the turbulent energy equation

$$\frac{d\frac{1}{2}\overline{u_i u_i}}{dt} = -\overline{\varepsilon} \tag{7.21}$$

which follows from (4.37), without mean flow. Integrating (7.18) from $k = 0$ to $k = \infty$ and using (7.4) and (7.5), we recover (7.21), provided that

$$\int_0^\infty T(k, t) dk = 0 \tag{7.22}$$

which expresses the fact that nonlinear transfer of energy between different wavenumbers does not change the overall energy, but rather, redistributes it differently. As discussed in the previous chapter, the transfer is mainly from large to small scales, that is, from small to high k, through the cascade process.

The existence of the nonlinear transfer term, $T(k, t)$, in Lin's equation results in a closure problem. Without an additional closure hypothesis to determine the cubic moments, (7.18) is not complete, and we cannot "solve" such an incomplete equation to obtain $E(k, t)$. However, there is one limiting case, that of very weak, passively decaying turbulence, in which the nonlinear transfer term becomes negligible, and we can solve (7.18).

Provided that Re_L is sufficiently small, the nonlinear, convective term in the Navier–Stokes equation is small compared with the viscous one, and $T(k, t)$, which is a reflection of that nonlinearity, is correspondingly small compared with the dissipative term in (7.18). As the turbulent energy decays, according to (7.21), it is found that Re_L falls and eventually the turbulence enters the final phase of its decay, in which Re_L is sufficiently small and we can neglect nonlinear transfer. Dropping the term $T(k, t)$ in (7.18), its solution is

$$E(k, t) = E_\infty(k) e^{-2\nu k^2 t} \tag{7.23}$$

Now, let us suppose that the behavior of $E(k, t)$ as $k \to 0$ is given by either (7.16) or (7.17), so that

$$E_\infty(k) \sim c_m k^m \tag{7.24}$$

where $m = 2$ or $m = 4$. As t increases, the exponential factor in (7.23) suppresses the higher wavenumbers more and more, leaving only small k, for which (7.24) holds, as $t \to \infty$. Thus, in the limit, we have

$$E(k, t) \sim c_m k^m e^{-2\nu k^2 t} \tag{7.25}$$

which describes the final period of passive viscous decay.

From (7.3) and (7.25), one can calculate the correlation function in this final phase. For $m = 2$, one obtains

$$R(r,t) \sim \frac{1}{8}c_2\left(\frac{\pi}{2(\nu t)^3}\right)^{1/2} e^{-\xi^2} \tag{7.26}$$

while, with $m = 4$,

$$R(r,t) \sim \frac{1}{32}c_4\left(\frac{\pi}{2(\nu t)^5}\right)^{1/2} \left(3 - 2\xi^2\right)e^{-\xi^2} \tag{7.27}$$

In both cases, the result is self-similar, with the usual similarity variable

$$\xi = \frac{r}{(8\nu t)^{1/2}} \tag{7.28}$$

describing viscous diffusion. According to (7.26)–(7.28), the correlation length, $L \propto (\nu t)^{1/2}$, while, using (7.4), the turbulent velocity is

$$u' = \left(\frac{2}{3}R(0,t)\right)^{1/2} \propto t^{-(m+1)/4} \tag{7.29}$$

It follows that $\mathrm{Re}_L \propto t^{-(m-1)/4}$ decreases with time, so that the above formulation, assuming small Re_L, is self-consistent.

The above results describe the final phase in the decay of homogeneous, isotropic turbulence without mean flow, and have been verified experimentally, using $m = 4$, for grid turbulence at sufficiently large distances downstream of the grid that Re_L has become small. This agreement with the theory for $m = 4$ suggests that the spectrum of grid turbulence is described by (7.16) at low wavenumbers. Although inhomogeneous turbulence does not allow the use of spectral theory, the basic idea of a final phase of viscous decay in which linear theory applies no doubt carries over. In this final phase, the turbulence is essentially dead, there is no cascade, and the effects of viscosity dominate, owing to the small Reynolds number. The opposite limit, $\mathrm{Re}_L \to \infty$, is more interesting and gives active turbulence with a cascade. In that case, we cannot "solve" (7.18), because it is incomplete unless a closure hypothesis is invoked. Nonetheless, one can extract useful information.

7.3 Kolmogorov's Theory of the Small Scales

We now tackle the subject of the behavior of the small scales, which constitutes the most important topic of this chapter. The theory developed here is due to Kolmogorov and is one of the key elements of the theory of turbulence.

Integrating (7.18) from $k = 0$ up to some fixed value of k, we obtain

$$\frac{d}{dt}\int_0^k E(k',t)dk' = -\aleph(k,t) - 2\nu\int_0^k k'^2 E(k',t)dk' \tag{7.30}$$

where

$$\aleph(k,t) = -\int_0^k T(k',t)dk' = \int_k^\infty T(k',t)dk' \tag{7.31}$$

We can interpret (7.30) as the statement that the energy in wavenumbers below k changes due to two effects: transfer of energy to higher wavenumbers via a spectral energy flux, $\aleph$, and viscous dissipation at each individual wavenumber. Thus $\aleph(k,t)$

is interpreted as the flux of energy through wavenumber k. Likewise, integration from fixed k up to $k = \infty$ gives

$$\frac{d}{dt}\int_k^\infty E(k', t)dk' = \aleph(k, t) - 2\nu \int_k^\infty k'^2 E(k', t)dk' \tag{7.32}$$

which says that positive spectral flux increases the energy at higher wavenumbers, offset by dissipation from each wavenumber.

At high Reynolds numbers, a cascade occurs and, given time, the spectrum is thought to approach a statistical equilibrium in which the time evolution of all wavenumbers takes place at a rate determined by the decay of the large scales. In equilibrium, small scales are continuously being produced at the expense of the large ones, and so on, through the cascade to smaller scales, until viscosity intervenes in the dissipative range. The entire continuum of scales is controlled by the slowest step in the cascade, namely the supply from the large, energy-containing scales. To quantify such an equilibrium cascade, suppose that the Reynolds number is sufficiently large that there is a wide inertial range, and that k is chosen so that most of the energy comes from below k, while the bulk of the dissipation arises from wavenumbers above k. We then neglect the left-hand side of (7.32) and take the lower limit to zero in the integral on the right-hand side. The result may be written as

$$\aleph = \bar{\varepsilon} \tag{7.33}$$

using (7.5). This equation is easily interpreted: since there is negligible dissipation in the inertial range and the fraction of the total energy stored in such scales is small, the energy flux given to the larger scales of the inertial range by progressive decay of the large scales is rapidly transferred through the cascade, essentially unmodified, to be dissipated at the smallest scales. The spectral energy flux is thus approximately the same at all inertial-range wavenumbers and equal to the dissipation rate, as indicated by (7.33). It is important that the inertial range be in developed equilibrium, otherwise a significant fraction of the spectral flux coming from the large scales may go into creating smaller scales, rather than contributing to the dissipation, that is, the left-hand side of (7.32) need not be negligible, which is the case for developing turbulence.

Kolmogorov proposed that the properties of turbulence in the developed inertial range are *only* dependent on the spectral energy flux, which equals $\bar{\varepsilon}$, as we saw in (7.33). The idea behind this is that, as the cascade proceeds through successively smaller scales, it "scrambles" information about the specific large-scale turbulent flow, leaving only the spectral energy flux, equal to $\bar{\varepsilon}$, as a parameter. By dimensional analysis, the only possible form for the inertial-range spectrum is then the famous Kolmogorov power law

$$E(k, t) = C\bar{\varepsilon}^{2/3}k^{-5/3} \tag{7.34}$$

where C is a purely numerical constant, universal in Kolmogorov's theory. As remarked earlier, the existence of a clear $k^{-5/3}$ part of the spectrum requires rather large values of Re_L: the range of validity of (7.34) grows with increasing Re_L, as the available room for the inertial range, $L^{-1} \ll k \ll \eta^{-1}$, gets larger. The inertial-range power law, (7.34), has been verified experimentally many times (see Monin and Yaglom (G 1975), section 23.4, for some results and discussion). Values of about

$C \approx 1.5$ are obtained and, although there is some scatter in the observed values of C, we will take $C = 1.5$ when a definite value of the Kolmogorov constant is required in later work.

As we shall see in Section 7.5, there are reasons for doubting Kolmogorov's assumption that the inertial range is controlled solely by $\bar{\varepsilon}$, and hence universality of C. Although C *is* found to have similar values in a wide variety of flows, including inhomogeneous ones and ones with mean velocity gradients, it can vary somewhat from place to place within a single inhomogeneous flow and between different flows, even among the homogeneous flows without mean velocity considered here. Furthermore, the reader may reasonably ask how one can explain the observed $k^{-5/3}$ spectral form if one does not suppose $\bar{\varepsilon}$ as controlling parameter, since the basis of the dimensional analysis has gone. This issue is addressed in Section 7.5 using a revised hypothesis of inertial-range self-similarity, which leads to the $k^{-5/3}$ spectrum without universality of C. However, historically the theory was developed as described here, that is, assuming that the inertial range is controlled by $\bar{\varepsilon}$ and this assumption serves well for most practical purposes. Indeed, the reason for alerting the reader to the difficulty at this stage is so that it does not come as a surprise later.

Two further remarks concerning (7.34) should be made immediately. Firstly, rather than the three-dimensional spectrum, $E(k, t)$, it is usually the one-dimensional spectrum, $E_{11}^{[1]}(k_1, t)$, that is measured from temporal data with the help of the Taylor's hypothesis, as explained in Chapter 6. Given (7.34), one can show that

$$E_{11}^{[1]}(k_1, t) = \frac{9}{55} C \bar{\varepsilon}^{2/3} k^{-5/3} \tag{7.35}$$

using (6.162). Thus, the one-dimensional spectrum also obeys a $-5/3$ power law in the inertial range. Secondly, from (7.34), one may evaluate the constant, a, occurring in (7.7), and hence determine the inertial range form of the correlation function using (7.10). Thus, taking $n = 5/3$, we have

$$R(0, t) - R(r, t) = \frac{9C\Gamma\left(\frac{1}{3}\right)}{20} \bar{\varepsilon}^{2/3} r^{2/3} \tag{7.36}$$

which is the Kolmogorov inertial-range velocity correlation, from which one may calculate Δu_r, using (7.6). The result, that velocity differences at separation r follow a one-third power law, $\Delta u_r \propto r^{1/3}$, in the inertial range is an important consequence of Kolmogorov's theory.

Above the inertial range lie the dissipative wavenumbers, and since these come about as a result of continuous transfer of energy from the larger, inertial-range scales, it is reasonable to suppose that the dissipative scales too only depend on the particular turbulent flow through $\bar{\varepsilon}$. However, the fluid viscosity becomes important now and gives a second dimensional parameter at dissipative range scales. From the parameters, $\bar{\varepsilon}$ and ν, one can form a length scale

$$\eta = \left(\frac{\nu^3}{\bar{\varepsilon}}\right)^{1/4} \tag{7.37}$$

giving a quantitative definition of the Kolmogorov scale, which determines the size of the smallest, dissipative scales of turbulence. It adjusts itself to the energy flux coming from the large scales and the viscosity of the fluid to make the mean dissipation rate, (4.27), equal to the mean energy flux. The Kolmogorov scale becomes smaller if the viscosity is decreased or the energy flux, equal to $\bar{\varepsilon}$, is raised, thus increasing the velocity gradients at the smallest scales in accord with (4.27). Again, by dimensional analysis, one finds

$$E(k,t) = \bar{\varepsilon}^{1/4}\nu^{5/4}F(k\eta) \tag{7.38}$$

as the dissipative range spectrum, where F is a universal function, that is, the same in any flow, according to the theory. Given the doubts noted above concerning Kolmogorov's assumption that small-scale statistical properties are controlled by $\bar{\varepsilon}$ alone, universality of $F(\kappa)$ is questionable, like that of C, an issue that will be discussed in Section 7.5. However, for the moment we will assume universality of $F(\kappa)$ and investigate the consequences.

As $k\eta \to 0$, we reenter the inertial range, and, to match with (7.34), we must have

$$F(\kappa) \sim C\kappa^{-5/3} \tag{7.39}$$

in the limit of small $\kappa = k\eta$. In consequence, (7.38) includes the inertial-range power law, (7.35), as a limiting case. At the other extreme, $\kappa \to \infty$, $F(\kappa)$ drops off rapidly to zero. The dissipation can be determined using (7.38) in (7.5). Thus, one can show that the function $F(\kappa)$ must satisfy

$$\int_0^\infty \kappa^2 F(\kappa)d\kappa = \frac{1}{2} \tag{7.40}$$

Naturally, (7.38) would only be expected to hold for sufficiently large Re_L in the wavenumber range $k \gg L^{-1}$. There is, however, evidence that (7.38), which allows for dissipation, may even apply at values of Re_L that are too low for an inertial range to be apparent, but large enough that there is still significant asymptotic separation between η and L (see Figure 7.3). This is reasonable, since the cascade is still present at such values of Re_L, and both F and $\bar{\varepsilon}$ are essentially dissipative range quantities. In any case, if there is an inertial range, (7.38) also describes the $k^{-5/3}$ spectrum there.

The prediction of a universal spectral form, (7.38), has been compared favorably with experimental data and some results for one-dimensional spectra (which, by (6.162) and (7.38), have universal expressions like (7.38) in which the function of $\kappa = k\eta$ differs from that for the three-dimensional spectrum) are shown in Figure 7.3. Notice that the spectrum in Figure 7.3 begins to fall away from the $k^{-5/3}$ power law at about $k\eta = 0.1$ and by $k\eta = 1$ has dropped off very significantly (by a factor of about 500). The rapid decay at large values of $k\eta$ is a reflection of smoothness, rather than "furriness," of realizations of the velocity field at spatial separations small compared with η^{-1}. The fact that viscous dissipation has a noticeable effect at wavenumbers lower than η^{-1} indicates that viscosity modifies the behavior of scales in physical space that are somewhat larger than η. The reader should therefore bear in mind that viscous effects can be important at smallish values of κ, corresponding to dissipative scales that are somewhat larger than, though proportional to η.

An approximate expression:

$$F(\kappa) = C\kappa^{-5/3}\exp\left\{-\frac{3}{2}C\kappa^{4/3}\right\} \tag{7.41}$$

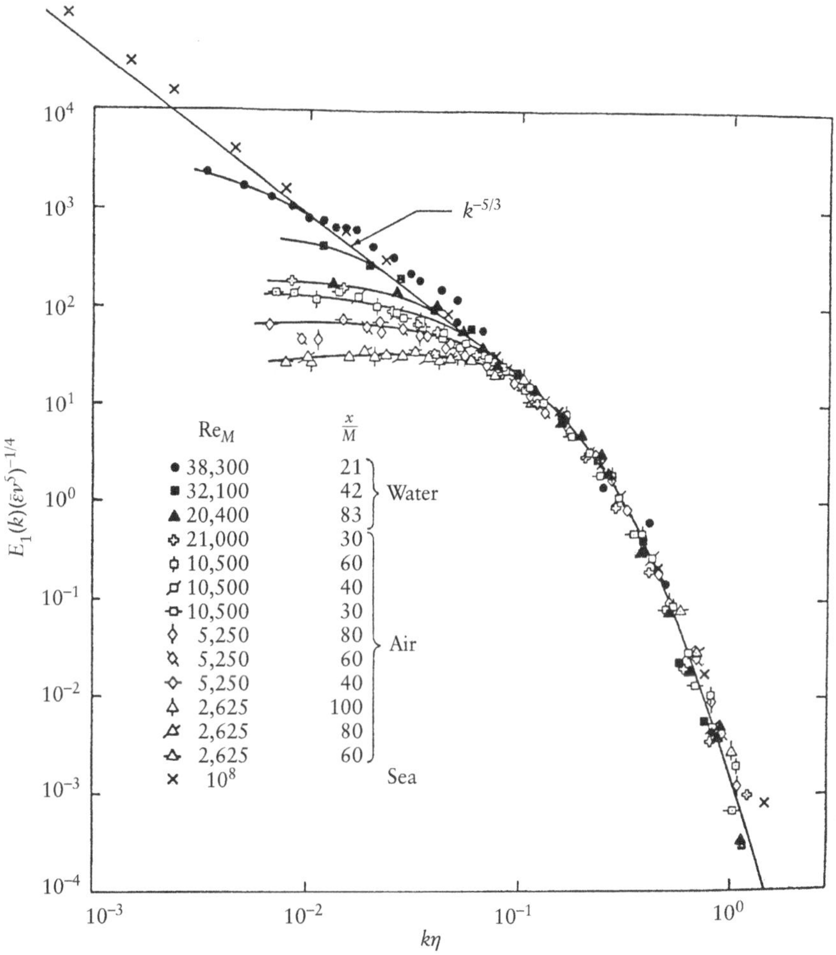

Figure 7.3. Measured one-dimensional spectra in different flows, scaled in Kolmogorov fashion, and plotted in log–log form (the quantity $E_1 = 2E_{11}^{[1]}$ is a longitudinal spectrum). The labeling "Sea" is for a tidal current, while all others are for grid turbulence (M is the mesh spacing, x the distance downstream of the grid). The Reynolds number varies between different flows, hence the location of the spectral peak changes when expressed in terms of $k\eta$. Inertial ranges are only visible at the higher Reynolds numbers. (Gibson and Schwarz (1963), including results of Stewart and Townsend (1951), Grant, Stewart, and Moilliet (1962).)

which is due to Pao (who derived it using a closure hypothesis) fits experimental data reasonably well, has the correct $\kappa \to 0$ behavior, (7.39), to match the inertial range, and satisfies (7.40). This data fit can be useful when performing calculations involving dissipative range wavenumbers.

One may use (7.38) in (7.8) to determine the correlation function at dissipative and inertial range separations. Thus,

$$R(0,t) - R(r,t) = \bar{\varepsilon}^{1/2}\nu^{1/2}\int_0^\infty F(\kappa)\left\{1 - \frac{\sin\kappa\frac{r}{\eta}}{\kappa\frac{r}{\eta}}\right\}d\kappa \tag{7.42}$$

provides the correlation function for $r \ll L$, and hence the velocity difference, as a function of r/η. Needless to say, this result matches with (7.36) in the limit of large r/η.

From the parameters, $\bar{\varepsilon}$ and ν, which are supposed to determine the behavior in the dissipative range, one can construct Kolmogorov time and velocity scales

$$t_\eta = \left(\frac{\nu}{\bar{\varepsilon}}\right)^{1/2} \tag{7.43}$$

and

$$u_\eta = (\nu\bar{\varepsilon})^{1/4} \tag{7.44}$$

whose product is the Kolmogorov spatial scale. These scales must be interpreted with care. In fact, t_η and u_η give scalings for the *evolution time* and *velocity differences* in the dissipative range. As explained in Chapter 3, temporal measurements at a fixed point in space do not generally yield evolution times, but rather reflect the spatial structure as it is convected past the measurement probe. In the absence of mean flow, as here, the velocity of convection of the small scales is dominated by the large ones, and is typically of $O(u')$. The spatial size η therefore leads to the shortest Eulerian time scale, $O(\eta/u')$, which is considerably shorter (by a factor of $O(\mathrm{Re}_L^{-1/4})$) than the lifetime, $O(t_\eta)$, of the smallest eddies. The Reynolds number based on the velocity difference scale, u_η, and length scale, η, is 1, which is another way of saying that viscosity becomes important at such length scales.

If one imagines following a given infinitesimal particle of fluid in its motion, the time taken for changes in the dissipative scales is their evolution time, t_η. In its strongest form, Kolmogorov's theory then implies that all statistical properties of the small-scale velocity differences in the neighborhood of the particle are approximately universal, that is, independent of the particular flow considered, provided that one scales velocity differences, distances, and times using u_η, η, and t_η respectively. Such universality of the small scales presumes, of course, that Re_L is sufficiently large, and that the length and time scales considered correspond to the inertial or dissipative ranges. In particular, it is supposed to hold at dissipative length and time scales, that is, space and time separations of order η and t_η, asymptotically as $\mathrm{Re}_L \to \infty$. We refer the reader to Monin and Yaglom (G 1975, section 21) for a more detailed discussion of statistical universality of the small scales.

The energy flux via the cascade, $\bar{\varepsilon}$ per unit mass, is provided by progressive decay of the large scales. The energy available is $O(u'^2)$ per unit mass, while the lifetime of the large eddies is $O(L/u')$. Thus, we have the estimate

$$\bar{\varepsilon} = O\left(\frac{u'^3}{L}\right) \tag{7.45}$$

for the dissipation rate. Equivalently, one can employ (7.21), using a decay time of $O(L/u')$ to estimate the time derivative. In any case, putting (7.45) in (7.37) gives

$$\frac{\eta}{L} = O\left(\mathrm{Re}_L^{-3/4}\right) \tag{7.46}$$

showing that, as stated earlier, the Kolmogorov scale is asymptotically small compared with the correlation length as $\mathrm{Re}_L \to \infty$. This asymptotic separation of scales forms the basis of the theory described here.

Let $k \gg L^{-1}$ lie in the inertial or dissipative ranges. We can determine the energy dissipation from above wavenumber k, using (7.38), with the semiempirical form (7.41), in (7.5), taking the integral over wavenumbers above k. Subtracting the result from the total dissipation, we find that a proportion

$$1 - \exp\left(-\frac{3}{2}C(k\eta)^{4/3}\right) \tag{7.47}$$

of the dissipation comes from *below* k. As regards the energy from wavenumbers higher than k, since the spectrum falls off at dissipative scales, an upper bound is obtained using the inertial-range power law, (7.34), in the integral of (7.4), restricted to wavenumbers above k. This upper bound represents a proportion

$$C\left(\frac{\bar{\varepsilon}L}{u'^3}\right)^{2/3}(kL)^{-2/3} \tag{7.48}$$

of the total energy, $3u'^2/2$.

Negligible dissipation and energy content are fundamental properties that distinguish the inertial range from the energy-containing and dissipative zones that border it. That is, for k to lie in the inertial range, most of the energy should lie below k and most of the dissipation above (so that we can replace (7.32) by (7.33)). In other words, we need (7.47) and (7.48) to be small simultaneously. If we choose, somewhat arbitrarily, the requirement that (7.47) and (7.48) should be less than 0.2 (i.e., at least 80% of the energy below, 80% of the dissipation above), the result is

$$(5C)^{3/2}\frac{\bar{\varepsilon}}{u'^3} < k < 0.24C^{-3/4}\eta^{-1} \tag{7.49}$$

which describes a range of inertial wavenumbers, provided that the lower bound for k is indeed less than the upper bound. Using the definition, (7.37), of η, this leads to

$$\mathrm{Re}_L > 168C^3\frac{\bar{\varepsilon}L}{u'^3} \approx 570\frac{\bar{\varepsilon}L}{u'^3} \tag{7.50}$$

which, since $\bar{\varepsilon}L/u'^3 = O(1)$ according to (7.45), gives a rough idea of the *minimum* Reynolds number required for an inertial range. Using (6.A5) of the appendix to Chapter 6 to express (7.50) in terms of the Reynolds number, Re_λ, based on the Taylor microscale, λ_g, rather than Re_L, we find

$$\mathrm{Re}_\lambda = \frac{u'\lambda_g}{\nu} = \left(\frac{15u'^3}{\bar{\varepsilon}L}\mathrm{Re}_L\right)^{1/2} > 50C^{3/2} \approx 90 \tag{7.51}$$

as a rule of thumb for the onset of an inertial range. Of course, the inertial range does not "switch on" suddenly as the Reynolds number is increased, but rather appears gradually, and the above criteria are intended as estimates of the threshold for existence of the inertial range. A wide power-law range for the spectrum will therefore require higher Reynolds numbers than these. Furthermore, the threshold value of 0.2 chosen here is rather arbitrary: the number appearing in (7.51) is roughly proportional to the threshold taken to the power of $-3/2$ (taking a more stringent threshold of 0.1, instead of 0.2, we obtain $\mathrm{Re}_\lambda > 146C^{3/2} \approx 270$). We emphasize once again that the dissipative range theory, (7.38) et seq., may well apply even if there is no inertial range, provided that there is a sufficient separation between η and L.

We saw earlier that, as a consequence of the small dissipation and energy content of the inertial range, presuming one exists, the spectral energy flux through that range is approximately constant and given by $\aleph = \bar{\varepsilon}$. Constancy of the spectral flux is asymptotically approached in the limit $\mathrm{Re}_L \to \infty$, but is only approximate at finite, but large Re_L. If the energy flux were really constant, the transfer term, $T(k, t)$, in (7.18) would be zero in the inertial range, according to (7.31). It is interesting to evaluate the transfer term at the next higher approximation in the inertial range by rewriting (7.18) as an equation for T and using the Kolmogorov $k^{-5/3}$ spectral form, (7.34), to calculate the other terms. The time derivative of the spectrum provides the negative term $(2/3)C\bar{\varepsilon}^{-1/3}(d\bar{\varepsilon}/dt)k^{-5/3}$, which decreases in size with increasing k, while the viscous term, $2\nu C\bar{\varepsilon}^{2/3}k^{1/3}$, is a positive and increasing function of k. The two cancel to produce a zero of T when $k = k_T$, where

$$k_T^2 = -\frac{1}{3\nu\bar{\varepsilon}}\frac{d\bar{\varepsilon}}{dt}$$

can be estimated as follows. The time scale for evolution of the turbulence, and hence its rate of energy dissipation, $\bar{\varepsilon}$, is determined by that, $O(L/u')$, of the large scales. Thus, one obtains $k_T = O(\mathrm{Re}_L^{1/2}L^{-1})$, which lies well above the energy-containing wavenumbers, $k = O(L^{-1})$, and below the dissipative ones, given by $k = O(\mathrm{Re}_L^{3/4}L^{-1})$, according to (7.46). That is, k_T is located in the inertial range and therefore within the range of applicability of the above determination of $T(k, t)$. For inertial-range wavenumbers below k_T, $T(k, t)$ is negative, whereas it is positive above k_T. In fact, the inertial-range $T(k, t)$ is easily shown to be proportional to $(k/k_T)^{1/3} - (k/k_T)^{-5/3}$, a function of k/k_T that the reader is encouraged to sketch and compare with Figure 6.5.

The spectral flux, $\aleph(k, t)$, is defined by (7.31) as integrals of $T(k, t)$, of which we examine the first. The flux through the inertial range arises mainly from the energy-containing range, $k = O(L^{-1})$, but it continues to increase slowly up to a maximum at $k = k_T$, decreasing thereafter due to extraction of energy to counterbalance increasing dissipation. Although the integral in (7.31) extends outside the inertial range, where the above evaluation of $T(k, t)$ ceases to apply, one may evaluate $\aleph(k, t)$ to within an integration constant. Thus, we find that the variations of $\aleph(k, t)$ with respect to k are described by $-3\nu C\bar{\varepsilon}^{2/3}k_T^{4/3}((k/k_T)^{4/3}/2 + (k/k_T)^{-2/3})$ within the inertial range. Order-of-magnitude estimation, using (7.45), (7.46), and the estimate $k_T = O(\mathrm{Re}_L^{1/2}L^{-1})$ derived above, shows that this expression for the variations of spectral flux only becomes comparable to $\bar{\varepsilon}$ when k decreases to $O(L^{-1})$ or increases to $O(\eta^{-1})$, that is, outside the inertial range. This is as it should be, confirming that variations of spectral flux are indeed small in the inertial range.

So far, we have discussed the consequences of Kolmogorov's theory for correlations and spectra, which are derived from second-order moments of velocity differences. Moments, $\chi_p(r) = \overline{|\mathbf{u}(\mathbf{x}+\mathbf{r}) - \mathbf{u}(\mathbf{x})|^p}$, of orders other than two may also be defined and are often called the structure functions of turbulence. From dimensional analysis, using $\bar{\varepsilon}$ and r, we deduce that $\chi_p \propto \bar{\varepsilon}^{p/3}r^{p/3}$ in the inertial range. This result is obviously a consequence of velocity differences scaling like $\bar{\varepsilon}^{1/3}r^{1/3}$, that is, the one-third law noted earlier. The structure functions will play an important role in the discussion of inertial-range intermittency in Section 7.5.

Kolmogorov's theory of the small scales of turbulence has been described above for homogeneous, isotropic turbulence with no mean flow. The theory presupposes a large Reynolds number, so that one can talk sensibly about a turbulent energy cascade to smaller scales, and that this cascade has been allowed to develop to equilibrium over a full range of scales down to the Kolmogorov scale. The small scales are then assumed to (statistically) equilibrate and be controlled by the average energy flux through the inertial range, which originates at the large scales and equals the mean dissipation rate, $\bar{\varepsilon}$. Thus, the theory predicts universal statistical properties for the velocity differences at small separations, for instance, correlations and spectra, determined solely by $\bar{\varepsilon}$ and ν (with the latter playing a role only at dissipative scales).

More general turbulent flows have *large* scales that can have quite different properties from those of the idealized, homogeneous, isotropic turbulence considered above. Moreover, as we shall see in Section 7.4, even homogeneous, isotropic turbulence from different flows may have different large-scale behavior. However, Kolmogorov's $k^{-5/3}$ spectrum is remarkably robust and is found to apply to the *small* scales of many turbulent flows that are inhomogeneous, anisotropic, or have mean flow, provided the turbulence is developed, the Reynolds number is large enough, and the scales considered sufficiently small. Strictly speaking, spectra are only precisely defined for homogeneous flows, but one may instead use the predictions for velocity correlations or velocity differences (or generalize the definition of spectra to the small scales of inhomogeneous flows). For instance, as we have seen, velocity differences at sufficiently small separations are proportional to $r^{1/3}$ according to Kolmogorov's theory. As one moves from place to place within an inhomogeneous turbulent flow, statistical properties vary over distances determined by the flow as a whole, which are large compared with the inertial and dissipative range scales that concern us here. Thus, it is reasonable to suppose that the small scales behave as if the flow were homogeneous. It is also plausible, although perhaps less so, that large-scale directionality is lost by "scrambling" in the cascade to smaller scales, which suggests that small enough scales may be approximately isotropic. These remarks are consistent with the basic idea of a cascade that loses information about the particular turbulent flow, so that only $\bar{\varepsilon}$ remains at sufficiently fine scales. In summary, the small scales of turbulence may be approximately statistically homogeneous and isotropic, even if the large scales are not. They may also be only weakly affected by mean velocity gradients and other additional effects, such as density stratification or rotation, except in so far as these change the value of $\bar{\varepsilon}$ through modification of the large scales. For instance, the direct effects of mean-flow gradients should diminish as one considers smaller and smaller scales because the velocity gradients associated with the turbulent velocity field increase (recall the discussion of "furriness" in Chapter 3). Naturally, for inhomogeneous flows, $\bar{\varepsilon}$ becomes a function of position, varying over the volume of the flow, but is nonetheless effectively uniform when viewed on small enough length scales. The above remarks concerning the wide applicability of the Kolmogorov spectrum are well illustrated by Figure 7.4, which shows spectra measured in three different "real-life" flows, which collapse to a

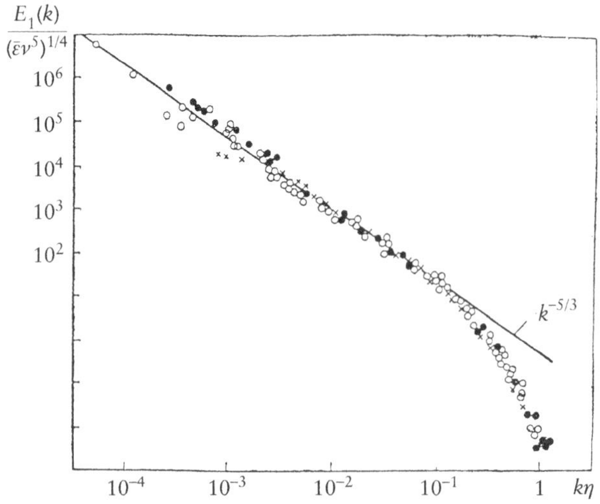

Figure 7.4. Measured longitudinal, one-dimensional spectra, scaled in Kolmogorov fashion, and plotted in log–log form. Data from three flows are superimposed: a boundary layer, a tidal flow in the sea, and the atmosphere near the surface of the sea. The Reynolds numbers are large enough that inertial ranges are clearly seen. (Sandborn and Marshall (1965), redrawn, including results of Grant et al. (1962), Pond, Stewart, and Burling (1963).)

universal curve in the inertial and dissipative ranges when plotted using Kolmogorov scalings.

Care must nonetheless be exercised in applying Kolmogorov's theory to general flows. For instance, within the viscous sublayer of a turbulent boundary layer, the turbulent Reynolds number is comparatively low, the range of scales is not very wide, and strong mean shear leads to anisotropy and gross inhomogeneity. For these reasons, we would not expect Kolmogorov's theory to apply there. On the other hand, the outer part of a turbulent boundary layer has lower shear and a higher turbulent Reynolds number, Re_L, leading to a wider range of scales, and Kolmogorov's theory may hold at scales sufficiently small compared to the distance from the wall. Additional influences, such as strong rotation or stratification, can also significantly modify the behavior of inertial-range, or even dissipative, scales, as well as the large scales, if they are sufficiently strong. Lack of universality and intermittency are also complicating factors, which will be discussed in Section 7.5.

We refer the reader to Monin and Yaglom (G 1975, Section 21) for further discussion of Kolmogorov's theory as applied to general turbulent flows, and to the Kolmogorov anniversary edition of *Proceedings of the Royal Society* (1991) for articles on a variety of different aspects of Kolmogorov's work, viewed from a modern perspective. The topics covered include small-scale intermittency, whose existence modifies the description of those scales, even for homogeneous, isotropic turbulence without mean flow, and can be important for the high-order velocity moments and dissipative scales, as we shall see in Section 7.5.

7.4 Self-Similar Evolution of the Large Scales of Homogeneous, Isotropic Turbulence

Whereas the small scales may be insensitive to the particular turbulent flow considered, apart from its value of $\bar{\varepsilon}$, the large scales *do* depend on the details of the flow and the discussion of the behavior of the large scales, given in this section, should be understood to be specific to homogeneous, isotropic turbulence without mean velocity gradients, to which grid turbulence may be taken as a reasonable approximation, many grid spacings downstream of the grid. Thus, unlike Kolmogorov's theory of the small scales, this section is concerned with fairly idealized flows, of which grid turbulence is the archetypal example.

We saw earlier that statistical properties of turbulence in the final phase of viscous decay become self-similar.[8] It is natural to ask whether turbulence may also be self-similar before the final phase is reached, that is, while Re_L is still large and the turbulence has a wide continuum of scales and an active cascade. If one includes both energy-containing and dissipative scales in the proposed similarity, it is clear that the ratio, η/L, would need to remain constant during the decay, and this, in turn, requires a constant value of Re_L. Such complete statistical self-similarity is conjectured in the Von Karman similarity hypothesis and its consequences have been investigated in some detail (see Monin and Yaglom (G 1975, section 16.1)). However, it turns out to be in disagreement with experimental results, essentially because Re_L does not remain constant.

When Re_L is large, the asymptotic separation between large and small scales suggests that the large scales might be self-similar, even though the small ones do not partake of large-scale similarity because Re_L changes during the decay. Let us assume statistical self-similarity of the large scales, and examine the consequences. The large scales have velocity scale $u'(t)$ and length scale $L(t)$, and so statistical similarity means that the large-scale statistical properties of the scaled velocity field, $\mathbf{u}/u'(t)$, are solely a function of $\mathbf{x}/L(t)$. The dissipative scales remain free to evolve in a manner which does not share in this self-similarity, but is, instead, determined from Kolmogorov's theory by the mean energy flux, equal to $\bar{\varepsilon}(t)$, which is fixed by the large scales. Thus, according to the Kolmogorov theory, the small scales have their own self-similarity, described by (7.38). In what follows, we suppose that Re_L is sufficiently large that there is a clear separation of energy-containing and dissipative scales. In particular, we assume that direct viscous dissipation can be neglected for the large scales, which lose energy solely through transfer to higher wavenumbers, via the cascade.

Assuming large-scale self-similarity, the scaled correlation function, R/u'^2, is a function of r/L for $r = O(L)$. From (7.2), this implies a spectral form

$$E(k, t) = u'^2 L G(kL) \tag{7.52}$$

when $k = O(L^{-1})$. Likewise, the scaled cubic moment, S/u'^3 with S defined by (7.20), is a function of r/L, leading to

$$T(k, t) = u'^3 H(kL) \tag{7.53}$$

for $k = O(L^{-1})$, from (7.19). If these forms are introduced into the spectral evolution equation, (7.18), and the viscous dissipation term neglected, we obtain

$$\left[\frac{L}{u'^3}\frac{du'^2}{dt}\right] G(K) + \left[\frac{1}{u'}\frac{dL}{dt}\right]\frac{dKG(K)}{dK} = H(K) \tag{7.54}$$

where $K = kL$ is the spectral similarity variable. The justification for neglect of the dissipation term is that we are here concerned with the large scales, $k = O(L^{-1})$, for which direct viscous dissipation is small, thanks to the large Reynolds number. Such scales decay owing to transfer of energy to smaller scales by the cascade, represented

[8] Here, by statistically self-similar we mean that velocity and length scales can be found, which are solely functions of time, and that, if the scaled velocity is regarded as a function of the scaled position, then its statistical properties do not evolve with time.

by the right-hand side of (7.54). Thus, the self-similar behavior we consider here is *inviscid* and does not include the dissipative scales of turbulence, but the effects of dissipation appear via the term $H(K)$ in (7.54). We note that

$$\int_0^\infty G(K)dK = \frac{3}{2} \tag{7.55}$$

follows from (7.4) and (7.52).

The terms of (7.54) in square brackets are functions of t, whereas the rest are functions of K. By taking the partial derivative with respect to t, it can be shown that, either the bracketed terms are constant, or $G(K)$ is proportional to a power of K. The latter possibility gives spectra, (7.52), which are powers of k, and not in agreement with observations. To have an acceptable spectral form one must take the indicated terms as constant, leading to

$$\frac{du'^2}{dt} = -\alpha_1 \frac{u'^3}{L} \tag{7.56}$$

and

$$\frac{dL}{dt} = \alpha_2 u' \tag{7.57}$$

where α_1 and α_2 are nondimensional constants. When compared with (7.21), equation (7.56) provides an expression for $\bar{\varepsilon}$ that is consistent with the estimate (7.45). Since $\bar{\varepsilon}$ is positive, so is α_1, while α_2 is also observed to be positive. This leads to decaying $u'(t)$ and increasing $L(t)$. The reason why the correlation length increases is that, among the large scales, the smaller ones are shorter lived and tend to decay more quickly via transfer to still smaller scales, leaving the bigger ones to determine the correlation length at later times. Thus, the lowest wavenumbers are more persistent and provide most of the energy at later times. This suggests that there may be a connection between the behavior of the spectrum at the lowest wavenumbers and its evolution, a connection that will be investigated shortly.

Unlike u', the length scale, L, has not been completely specified (one could multiply the value of L by any constant number) and the values of the numerical constants α_1 and α_2 vary with the precise definition adopted for L. However, it can easily be shown, from (7.56) and (7.57), that the ratio, α_1/α_2, does not depend on the definition used. It follows that the essential properties of large-scale, self-similar turbulence can only depend on this ratio of constants. If desired, one may uniquely specify L by, for example, requiring that $\alpha_1 = 1$ in (7.56), or $\alpha_2 = 1$ in (7.57) (these particular normalizations require that $\alpha_1 \neq 0$ and $\alpha_2 \neq 0$ respectively), or adopt any other convenient normalization for L. Such normalization may be desirable, for instance, when one undertakes to solve similarity equations such as (7.54) using particular closure models for the nonlinear transfer term on the right-hand side. However, in this chapter, we do not use closures and prefer to maintain the freedom to alter the value of L to within a multiplicative constant.

By solving (7.56), (7.57), we obtain

$$u' \propto (t - t_0)^{-\alpha_1/(\alpha_1 + 2\alpha_2)} \tag{7.58}$$

and

$$L \propto (t - t_0)^{2\alpha_2/(\alpha_1+2\alpha_2)} \tag{7.59}$$

showing that u' and L should evolve according to power laws whose exponents are determined by the single parameter α_1/α_2. The size, $O(L)$, of the large scales of turbulence increase with time, while the velocity fluctuations decay. As discussed in the last chapter, the experimental results of Comte-Bellot and Corrsin (1966) on grid turbulence, among other authors, indeed show power laws like (7.58) and (7.59), which suggests that large-scale self-similarity occurs in such flows. The power-law exponents are found to vary depending upon the particular grid used, indicating that the properties of the large scales of turbulence are not universal, but vary from flow to flow. The exponents given in (6.173) and (6.174) of the previous chapter are consistent with (7.58) and (7.59) with $\alpha_1/\alpha_2 = 3.7$. Measurements of large-scale velocity correlations and spectra also seem to be consistent with large-scale self-similarity of grid turbulence at sufficient distances downstream of the grid. Although it is by no means certain that such similarity always occurs in homogeneous turbulence, we shall assume large-scale self-similarity here, and continue to work out its consequences.

At the beginning of Section 7.2, we discussed the time evolution of the low-wavenumber spectrum. A spectrum, for which $E \sim c_m k^m$ with $m < 4$ initially, persists at later times with a *constant* value of c_m. For this behavior to be consistent with the self-similar form (7.52) requires that $u'^2 L^{m+1}$ be constant, and hence, from (7.58) and (7.59), $\alpha_1/\alpha_2 = m + 1$, relates the small wavenumber spectral exponent to the constants, α_1 and α_2, which characterize self-similarity. In particular, turbulence whose spectrum has the low-wavenumber behavior (7.17) (i.e., $m = 2$) can only admit large-scale similarity with $\alpha_1/\alpha_2 = 3$, which implies that $u'^2 \propto (t - t_0)^{-6/5}$, $L \propto (t - t_0)^{2/5}$, according to (7.58), (7.59). If, on the other hand, the initial spectrum goes to zero like k^4 or faster as $k \to 0$, it should have the low-wavenumber form $E \sim c_4 k^4$ at later times, but with a coefficient, $c_4(t)$, that will generally vary with time. Since there is no requirement of constancy of c_4, the value of α_1/α_2 is not uniquely fixed by the low-wavenumber exponent, $m = 4$, unlike the case $m < 4$ considered above. Assuming large-scale similarity, we find that $c_4 \propto (t - t_0)^{2(5\alpha_2-\alpha_1)/(\alpha_1+2\alpha_2)}$, which is time varying unless $\alpha_2/\alpha_1 = 5$, a value that leads to $u'^2 \propto (t - t_0)^{-10/7}$, $L \propto (t - t_0)^{2/7}$. As mentioned earlier, it appears that grid turbulence has a k^4 spectrum at low wavenumbers and that different values of α_1/α_2 can occur, depending on the precise flow considered and corresponding to different self-similar behaviors. The observed values of α_1/α_2 are less than five, which implies that c_4 grows with time, that u'^2 decreases less rapidly than $(t - t_0)^{-10/7}$, and that L increases faster than $(t - t_0)^{2/7}$. Growth in c_4 is a particularly interesting result because it means that, although the spectrum is decaying overall, owing to transfer of energy to higher wavenumbers via the cascade, and dissipation at the highest wavenumbers, it *increases* with time at sufficiently low wavenumbers, due to energy transfer from the higher ones. Such a transfer of energy from higher to lower wavenumbers is in the opposite sense to the more familiar one induced by the cascade through the inertial range. It results from energy exchange *among* the large-scale components of turbulence, rather than the transfer of energy to smaller scales, which drives the cascade. Consequently, the transfer term, $T(k, t)$, in (7.18), whose behavior in the inertial and dissipative ranges was discussed earlier, should be positive at sufficiently low wavenumbers, becoming negative at higher ones, and positive again at still higher wavenumbers. The negative range, which includes the peak in $E(k,t)$, represents nonlinear extraction of energy from the corresponding wavenumber components, some of which goes to low wavenumbers, while the remainder is transferred to higher wavenumbers, via the cascade. In practice it is found that the region of positive $T(k,t)$ at low energy-containing wavenumbers is rather narrow and has a low peak value. The region of negative $T(k,t)$ and the positive range at high wavenumbers are considerably more prominent (see Figure 6.5).

Since $T(k,t)$ behaves like k^4 at low wavenumbers, the function $H(K)$ is proportional to K^4 as $K \to 0$. If such a form for $H(K)$ is used in (7.54), expressing the coefficients from (7.56) and (7.57), it can be shown that $G(K)$ must follow a power law with exponent $m \leq 4$. Low-wavenumber spectral forms other than powers are therefore inconsistent with large-scale similarity.

The Reynolds number, $\mathrm{Re}_L = u'L/\nu$, can be determined from (7.58) and (7.59) as

$$\mathrm{Re}_L \propto (t - t_0)^{(2\alpha_2 - \alpha_1)/(\alpha_1 + 2\alpha_2)} \tag{7.60}$$

and decreases with time, as observed, if $\alpha_1/\alpha_2 > 2$. The decreasing Reynolds number is a consequence of competition between decreasing u' and increasing L, in which u' wins, with the result that Re_L falls comparatively slowly. For instance, taking the value $\alpha_1/\alpha_2 = 3.7$, corresponding to the exponents of (6.173) and (6.174), gives $\mathrm{Re}_L \propto (t - t_0)^{-0.3}$. The dissipation rate can be calculated from (7.21) and (7.58) as

$$\overline{\varepsilon} \propto (t - t_0)^{-(3\alpha_1 + 2\alpha_2)/(\alpha_1 + 2\alpha_2)} \tag{7.61}$$

which may be used as input to the Kolmogorov model of the small scales, including the dissipative range, which is outside the scope of the similarity form, (7.52). We can determine the Taylor microscales via equations (6.A5) and (6.A6) of the appendix to Chapter 6. Both λ_f and λ_g are proportional to $(t - t_0)^{1/2}$, in agreement with the experimental result, (6.175). According to (7.37) and (7.61), the Kolmogorov scale evolves as

$$\eta \propto (t - t_0)^{(3\alpha_1 + 2\alpha_2)/(4\alpha_1 + 8\alpha_2)} \tag{7.62}$$

so that, using (7.59) and (7.60), we find $\eta/L \propto \mathrm{Re}_L^{-3/4}$, which is consistent with (7.46). Unless $\alpha_1/\alpha_2 = 2$, Re_L changes with time and the dissipative scales cannot share the self-similarity of the large ones. Indeed, the value, $\alpha_1/\alpha_2 = 2$, necessary for constancy of Re_L, leads to

$$\begin{aligned} u' &\propto (t - t_0)^{-1/2} \\ L &\propto (t - t_0)^{1/2} \end{aligned} \tag{7.63}$$

which are not in agreement with experiment.

If Re_L is sufficiently large that there is an inertial range, one may use (7.4), (7.21), and (7.56) to express $\overline{\varepsilon}$ in the Kolmogorov spectrum, (7.34), in terms of u' and L. Thus, one finds that the inertial-range spectrum has the self-similar form (7.52) with

$$G(K) \sim C\left(\frac{3}{2}\alpha_1\right)^{2/3} K^{-5/3} \tag{7.64}$$

which gives the large K limiting behavior of $G(K)$ necessary for consistency with the Kolmogorov theory. The fact that the Kolmogorov inertial-range spectrum is consistent with self-similarity of the large scales means that only the dissipative range lies outside the scope of (7.52). As noted above, according to the Kolmogorov theory, the dissipative range has its own self-similarity, (7.38), based on the length scale $\eta(t)$, similarity which is also shared by the inertial-range spectrum, thanks to (7.39) in the limit of small $k\eta$. Thus, the inertial range takes part in two self-similar forms and provides the bridge (or region of overlap if one thinks in matched asymptotic expansion terms) between them.

Self-similarity of the large scales presumably takes some time to develop when turbulence is started from some arbitrary initial state. The time scale for evolution of the large scales is $O(L/u')$ and so one must wait until $t \gg L_0/u'_0$, where L_0 and u'_0 are the initial correlation length and turbulent velocity, before self-similarity becomes apparent. The Reynolds number, supposed large initially, decreases slowly during the decay of the turbulence, due to competition between increasing L and more rapidly decreasing u'. When Re_L eventually becomes of O(1), the asymptotic separation between energy-containing and dissipative scales ceases to exist and, consequently, large-scale similarity no longer applies. The turbulence goes through an adjustment phase and emerges into the final period of passive decay, in which it is again self-similar, as we saw earlier. There is then no cascade or separation of scales, and the power laws, $u' \propto t^{-5/4}$ (from (7.29), assuming $m = 4$) and $L \propto t^{1/2}$, are different from (7.58), (7.59).

Von Karman suggested the following rough, semiempirical form for the spectrum in the energy-containing and inertial ranges:

$$E(k,t) = \frac{55\Gamma(5/6)}{9\Gamma(1/3)\pi^{1/2}} u'^2 L \frac{(kL)^4}{\left(1+(kL)^2\right)^{17/6}} \tag{7.65}$$

which has the self-similar form, (7.52). It has the low wavenumber limiting behavior, (7.16), and the Kolmogorov inertial range form, (7.7) with $n = 5/3$, at large kL. The coefficient in (7.65) has been chosen according to (7.4). Here, unusually for this book, the correlation length has been given a definite quantitative meaning, rather than being an order of magnitude. It can be shown that L, defined by (7.65), is related to the integral scales, (6.113) and (6.114), by

$$2L_n = L_p = \frac{\pi^{1/2}\Gamma(5/6)}{\Gamma(1/3)} L \approx 0.75L \tag{7.66}$$

To bring this section to a close, we remind the reader of the restriction to idealized homogeneous, isotropic turbulence without mean velocity gradients, with grid turbulence as the archetypal example. Unlike Kolmogorov's theory of the small scales, the above ideas concerning the properties of the large ones are not expected to apply to more general flows.

7.5 Beyond Kolmogorov's Original Theory

As we saw in Chapter 5, in shear flows such as boundary layers, jets, and wakes, motion of the frontier between turbulent and laminar fluid leads to intermittency. A fixed point in space finds itself sometimes inside, sometimes outside the turbulence as bulges in the frontier are convected past the point. The bulges engulf laminar fluid, which eventually becomes turbulent. Thus, the turbulent region spreads to include more and more fluid, but within the apparently turbulent flow there are regions of laminar fluid, which are more and more common toward the outskirts of the flow. In consequence, the turbulence becomes increasingly intermittent as one emerges from the shear flow. Intermittency can also occur when a laminar flow undergoes transition to turbulence as the Reynolds number increases, for instance, in the transitional zone of a boundary layer, or in a pipe. In such cases, turbulent and laminar flow alternate at a fixed point in space, producing intermittency of the large scales of turbulence within a statistically steady flow.

The above flows are inhomogeneous, but large-scale intermittency can also be envisaged in decaying, homogeneous turbulence without mean flow. Recalling that one regards a turbulent flow as an ensemble of different realizations of nominally identical flows, we are at liberty to choose the ensemble of initial conditions and consider an example that was introduced in Section 2.4. Suppose that there is no mean flow and that each realization consists of widely separated, statistically independent patches of turbulence in an infinite fluid, whose positions vary randomly from realization to realization with no preferred locations in space, so that the resulting ensemble is statistically homogeneous.[9] Although this flow is statistically homogeneous, the intensity of turbulence in a particular realization is highly nonuniform, which makes the example somewhat "pathological," but a good illustration of the fact that statistical homogeneity does not exclude extremely intermittent, patchy turbulence in which the patches move around from realization to realization to make the statistics uniform. However, by homogeneous turbulence, one usually has in mind flows for which, unlike this example, the properties of turbulence are roughly uniform in typical realizations (e.g., an initially Gaussian velocity field). In the above example, there is extreme large-scale intermittency: at a fixed point in space, there is no turbulence in most realizations, while occasionally the point finds itself inside a patch of turbulence. This means that average quantities, like u'^2 and $\bar{\varepsilon}$, are not representative of particular realizations of the flow, as discussed in Chapter 2. They overestimate the intensity of turbulence in most realizations and underestimate it for the rare cases where a patch is present at the given point in space. Thus, estimates, such as L/u' for the decay time of the turbulence, are inappropriate for grossly intermittent flows, like this one, which decay on a time scale shorter than L/u' because the turbulence is more intense than suggested by u'. More importantly, universality of the Kolmogorov constant, C, occurring in (7.34) and of the dissipative range spectral function, $F(\kappa)$, is called into question by this example, as we will show shortly.

Before discussing lack of universality for the above example, we should remind the reader that many experimental results, such as those shown in Figure 7.4, indicate similar inertial and dissipative range spectra for different flows using Kolmogorov scalings, and thus appear to support universality. Although the logarithmic scales employed in Figure 7.4 tend to suppress differences between the flows, one may conclude that departures from spectral universality are not dramatic for the given flows. Lack of Kolmogorov universality for general flows does not stop particular ones from having similar spectral forms, or at least approximately so, when

[9] It can also be made statistically isotropic by taking random orientations for the turbulent patches.

plotted using the Kolmogorov scalings. As we will see below, the example given above provides a case in which universality does not apply when the patches of turbulence become sparse enough. Such extreme large-scale intermittency can cause strong departures from Kolmogorov universality, with weaker departures in less extreme cases.

Let γ denote the average number of turbulent patches per unit volume. We imagine a sequence of different flows in which γ is allowed to decrease, while the properties of individual patches (e.g., the size and intensity of turbulence within a patch) are maintained the same. Mean values, such as u'^2 and $\bar{\varepsilon}$, of quantities which are effectively zero well away from any patches decrease proportional to γ. Likewise, the one-point velocity moments, $\overline{u_i u_j}$, which are of order u'^2, decrease with γ in the same way. The correlation functions, $R_{ij}(\mathbf{r})$, have the values $R_{ij} = \overline{u_i u_j}$ at $\mathbf{r} = 0$ and fall to zero over distances that are determined by the properties of turbulence within patches, since different patches are supposed uncorrelated, that is, the correlation length, L, remains constant as patches become rarer. Thus, we would expect $R_{ij}(\mathbf{r})$ to decrease in magnitude with γ, but $R_{ij}(\mathbf{r})/\gamma$ to remain unchanged, hence spectra, which are Fourier transforms of $R_{ij}(\mathbf{r})$, should also decrease in proportion to γ. Within the Kolmogorov inertial-range expression, (7.34), the left-hand side decreases like γ, whereas the right-hand side only decreases proportional to $\gamma^{2/3}$, owing to the factor $\bar{\varepsilon}^{2/3}$. Thus, the value of the supposedly universal constant C decreases as $\gamma^{1/3}$. The problem is that the spectrum is an average, as is $\bar{\varepsilon}$, and both are proportional to γ, but taking the 2/3 power of $\bar{\varepsilon}$ makes it scale differently with γ. This exemplifies the objection made by Landau to the universality assumptions of Kolmogorov's (1941) theory and indicates that C is not universal. Universality of the dissipative range using Kolmogorov scalings is also called into doubt by this example. For instance, the matching condition with the inertial range, (7.39), shows that $F(\kappa)$ cannot be universal unless C is. Furthermore, since the dissipation rate inside a patch is $O(\bar{\varepsilon}/\gamma)$, the size of the dissipative scales, which determines the range of dissipative wavenumbers for the spectrum, is $O(\gamma^{1/4}\eta)$, where we have employed (7.37) with $\bar{\varepsilon}/\gamma$ instead of $\bar{\varepsilon}$ to estimate dissipative scales within a patch. It follows that the wavenumber range at which viscosity becomes important is a factor of $\gamma^{-1/4}$ higher than suggested by Kolmogorov scalings and consequently that the spectrum is not universal using those scalings. By taking the limit $\gamma \to 0$, C can be made to decrease without limit, while the dissipative wavenumber range of the spectrum goes to infinity when expressed in terms of $k\eta$. Both effects show lack of Kolmogorov universality, although it is interesting to note the small exponents, $\gamma^{1/3}$ and $\gamma^{-1/4}$, involved. Since small exponents imply weak dependency on γ, this suggests that rather extreme large-scale intermittency may be required to produce strong changes in C and $F(\kappa)$ compared to the more usual cases in which the statistically homogeneous turbulence is approximately uniformly distributed in typical realizations, for example grid turbulence.

The large-scale intermittency in the above example comes from the intermittent initial conditions. The flow is a rather extreme example of intermittency of the large scales of turbulence and shows that such intermittency can invalidate Kolmogorov universality. Variability of C has also been found inside a jet flow by Kuznetsov, Praskovsky, and Sabelnikov (1992), where intermittency is due to the unsteady entrainment of laminar external fluid in individual realizations of the statistically steady flow. As noted at the beginning of this section, large-scale intermittency in such inhomogeneous shear flows increases towards their outskirts, leading to cross-stream variations of the Kolmogorov constant, C, which decreases as the intermittency rises, as we found for the homogeneous example discussed above. Interestingly, in the jet, Kuznetsov et al. (1992) found that $C \propto \gamma^{1/3}$, as for the above flow with patches, where γ represents the proportion of the time for which turbulence is present at the given point and therefore decreases the greater the intermittency. Such large-scale intermittency modulates the energy supply to the cascade, and hence the intensity of the small scales, but it may leave the internal dynamics of the cascade unchanged. If this is the case, one might expect that, although the value of the Kolmogorov constant can be changed by large-scale intermittency, the basic $k^{-5/3}$ spectral form would survive in the inertial range. Indeed, as noted earlier, numerous experimental studies of different flows have verified the $k^{-5/3}$ form, which is not seriously in doubt. However, from a theoretical point of view, variability of C implies that the cascade is affected by parameters other than $\bar{\varepsilon}$, for otherwise the dimensional analysis given earlier unambiguously leads to universality. Once one recognizes that Kolmogorov's assumption, that $\bar{\varepsilon}$ is the only parameter controlling the statistical properties of the cascade, is not always true, the theoretical justification for the $k^{-5/3}$ law is seriously weakened, and one may then ask why this power law is found to hold at all.

Following Frisch (G 1995), a reformulation of Kolmogorov's theory is possible that avoids the difficulties noted above, which we now explain. Let

$$\delta\mathbf{u}(\mathbf{r}) = \mathbf{u}(\mathbf{x}+\mathbf{r}, t) - \mathbf{u}(\mathbf{x}, t) \tag{7.67}$$

denote the velocity difference at separation $\mathbf{r}$, which, in principle, is a function of $\mathbf{x}$ and t, as well as $\mathbf{r}$. However, we will restrict attention to a particular point, $\mathbf{x}$, and time, t, so we do not explicitly show the dependence on $\mathbf{x}$ and t, and assume homogeneity, so that the statistics of $\delta\mathbf{u}(\mathbf{r})$ do not depend on the choice of $\mathbf{x}$. We also suppose isotropy, for simplicity sake, and a sufficiently high Reynolds number that there is a wide inertial range of scales, taking $r = |\mathbf{r}|$ to lie within that range. It may be recalled, from Chapter 3, that the reason for considering velocity differences, rather than $\mathbf{u}(\mathbf{x}, t)$ itself, is that such differencing emphasizes the "fur" in the velocity field and thus focuses attention on scales of size $O(r)$. In fact, one can consider $\delta\mathbf{u}(\mathbf{r})$ as a filtered version of $\mathbf{u}(\mathbf{x}, t)$, which suppresses the lower wavenumber (larger-scale) components. We will briefly discuss other possible filters toward the end of this section.

According to Kolmogorov's theory, the inertial-range statistics of $\delta\mathbf{u}$ depend only on $\bar{\varepsilon}$, which implies, via dimensional analysis, that $\overline{|\delta\mathbf{u}|^2} = (\Delta u_r)^2$ is a universal constant multiplying $(\bar{\varepsilon}r)^{2/3}$, as explicitly shown by (7.6) and (7.36). Furthermore, the nondimensional, scaled quantity $\mathbf{v}(\mathbf{r}) = \delta\mathbf{u}(\mathbf{r})/\Delta u_r$ should have universal statistical properties. It is easily shown that the probability distribution of $\delta\mathbf{u}(\mathbf{r})$ has the inertial-range form

$$P(\delta\mathbf{u}) = (\Delta u_r)^{-3}\Pi\left(\frac{|\delta\mathbf{u}|}{\Delta u_r}, \frac{\mathbf{r}.\delta\mathbf{u}}{r|\delta\mathbf{u}|}\right) \tag{7.68}$$

where isotropy has been used and Π is a universal function.[10] The first argument of Π shows the scaling of $|\delta\mathbf{u}|$ as $\Delta u_r \propto (\bar{\varepsilon} r)^{1/3}$, while the second expresses the dependence on the angle between $\delta\mathbf{u}(\mathbf{r})$ and $\mathbf{r}$. The multiplicative factor, $(\Delta u_r)^{-3}$, provides normalization, that is, that the integral of the probability distribution function, $P(\delta\mathbf{u})$, of $\delta\mathbf{u}$ over all values of the three components of $\delta\mathbf{u}$ should be unity.

As discussed above, the difficulty with Kolmogorov's original theory is the universality of functions such as Π in (7.68), which follows by dimensional analysis from the assumption of dependency on $\bar{\varepsilon}$ alone. However, if we drop the latter requirement, and hence the basis for dimensional analysis, how can we retain successful predictions, such as the $k^{-5/3}$ inertial-range spectrum? The $k^{-5/3}$ spectrum is a consequence of $\Delta u_r \propto r^{1/3}$ in the inertial range, which can be derived solely from *self-similarity* of the statistics of $\delta\mathbf{u}(\mathbf{r})$, as we now show. By statistical self-similarity, we mean that the scaled form, $\mathbf{v}(\mathbf{r}) = \delta\mathbf{u}(\mathbf{r})/\Delta u_r$, has the same statistical properties when viewed at different inertial-range scales within the given flow.[11] Such self-similarity follows from the earlier version of Kolmogorov's theory, but we now replace the original assumption, that the small scales are controlled by $\bar{\varepsilon}$, with the *weaker* hypothesis of inertial-range statistical self-similarity, a hypothesis that does not imply universal statistical properties. In particular, (7.68) gives the probability distribution of $\delta\mathbf{u}(\mathbf{r})$, but Π need not be universal now, nor do we assume a priori that $\Delta u_r \propto r^{1/3}$, but instead treat Δu_r as an unknown function of r and then show that, in fact, $\Delta u_r \propto r^{1/3}$.

To determine the form of Δu_r as a function of r, we introduce a result of Kolmogorov (1941c), namely the four-fifths law

$$\overline{\delta u_{||}^3(\mathbf{r})} = -\frac{4}{5}\bar{\varepsilon} r \tag{7.69}$$

where $\delta u_{||}$ is the component of $\delta\mathbf{u}$ parallel to $\mathbf{r}$,

$$\delta u_{||}(\mathbf{r}) = \frac{\mathbf{r}.\delta\mathbf{u}(\mathbf{r})}{r} \tag{7.70}$$

Equation (7.69) holds at inertial-range separations, asymptotically as $\mathrm{Re}_L \to \infty$, and is one of very few nontrivial exact results in the theory of turbulence. Its derivation is discussed at some length by Frisch (G 1995) and we give a relatively brief version in the appendix to this chapter.

The result, (7.69), may be interpreted in physical terms as follows. Taking the velocity difference $\delta\mathbf{u}(\mathbf{r})$ focuses on the dynamics of turbulence at inertial-range scales of size $O(r)$. An eddy of scale r may be thought of as having an associated energy of $O(\delta u^2)$ per unit mass to pass on to yet smaller scales via the cascade, where δu measures the velocity differences within the eddy. The eddy lifetime is of $O(r/\delta u)$, so that the eddy delivers energy to smaller scales at a rate $O(\delta u^3/r)$ per unit mass of the eddy. The average energy flux through scale r is thus of $O\left(\overline{\delta u^3}/r\right)$ per unit mass,

[10] More generally, one may consider N inertial-range separations, $\mathbf{r}_1, \ldots, \mathbf{r}_N$. Dimensional analysis then implies that the joint probability distribution of the random variables $\mathbf{v}_1 = \mathbf{v}(\mathbf{r}_1), \ldots, \mathbf{v}_N = \mathbf{v}(\mathbf{r}_N)$ is a universal function of $\mathbf{v}_1, \ldots, \mathbf{v}_N$ and of $\mathbf{r}_1/|\mathbf{r}_1|, \mathbf{r}_2/|\mathbf{r}_1|, \ldots, \mathbf{r}_N/|\mathbf{r}_1|$ (a function which is also invariant under rotations and reflections, by isotropy).

[11] This type of statistical self-similarity between different *spatial* scales should not be confused with that discussed in Section 7.4, in which we compared the statistics at different *times*.

and, since r is in the inertial range, $\bar{\varepsilon} = O\left(\overline{\delta u^3}/r\right)$, which corresponds to (7.69) if we take $\delta u = \delta u_{\|}$. Note that nothing in the above argument stops us using other measures of the velocity differences at scale r for δu, which suggests that any cubic moment of $\delta \mathbf{u}(\mathbf{r})$ will be proportional to $\bar{\varepsilon} r$. Thus, the *third-order* moments of velocity differences should be directly related to, and pinned down by, the mean energy flux through the inertial range, which is equal to the mean dissipation, as we saw earlier. The reasoning here is obviously fairly crude, being based on a number of unproven assumptions about the inertial-range cascade and giving only order-of-magnitude results, but is believed to provide an essentially correct interpretation of the appearance of cubic moments of velocity differences in energy flux expressions, such as (7.69).

With (7.69) in hand, we may evaluate the average, $\overline{\delta u_{\|}^3}$, using (7.70), as

$$\overline{\delta u_{\|}^3} = \int \left(\frac{\mathbf{r}.\delta \mathbf{u}}{r}\right)^3 P(\delta \mathbf{u}) d^3(\delta \mathbf{u}) \tag{7.71}$$

into which we substitute the assumed self-similar form (7.68) for $P(\delta \mathbf{u})$ to obtain

$$\overline{\delta u_{\|}^3} = 2\pi \Delta u_r^3 \int_0^{\infty} \int_{-1}^{1} \Pi(w, \xi) w^5 \xi^3 \, d\xi \, dw \tag{7.72}$$

where we have firstly adopted a spherical polar coordinate system for $\delta \mathbf{u}$, with $w = |\delta \mathbf{u}|/\Delta u_r$ as radial coordinate and θ as the angle between $\delta \mathbf{u}$ and $\mathbf{r}$, and secondly changed angular integration variable from θ to $\xi = \cos\theta$. Substituting (7.72) into (7.69), we can rewrite the result as

$$\Delta u_r^2 = \frac{9C\Gamma\left(\frac{1}{3}\right)}{5} \bar{\varepsilon}^{2/3} r^{2/3} \tag{7.73}$$

where the quantity

$$C = \frac{5}{9\Gamma\left(\frac{1}{3}\right)} \left[\frac{-2}{5\pi \int_0^{\infty} \int_{-1}^{1} \Pi(w, \xi) w^5 \xi^3 \, d\xi \, dw}\right]^{2/3} \tag{7.74}$$

has been introduced so that one obtains (7.36) from (7.73) using (7.6). Since the function $\Pi(w, \xi)$, describing the self-similar probability distribution of $\delta \mathbf{u}$, is fixed for a given flow, C is a constant, although it may vary from flow to flow, and, in general, with time for a given flow. Equation (7.73) shows that the velocity differences scale proportional to $r^{1/3}$ in the inertial range, while, using (7.6) one derives (7.36), which implies the Kolmogorov inertial range spectrum, (7.34) (recall the remarks following (7.10)). Thus, one recovers the results of Kolmogorov's theory without the assumption of dependence on $\bar{\varepsilon}$ alone, and consequent universality, but instead using the hypothesis of statistical self-similarity of $\delta \mathbf{u}$ at inertial-range separations. In particular, universality of the Kolmogorov constant, C, need no longer apply and it is free to vary from flow to flow. Observe that the average of any function of $\delta \mathbf{u}$ can be calculated from (7.68). For instance, the structure functions[12]

$$\chi_p(r) = \overline{|\delta \mathbf{u}|^p} \tag{7.75}$$

are found to be proportional to $r^{p/3}$, as in the original version of the theory.

[12] In general, the order, p, of $\chi_p(r)$ need not be an integer.

The lack of small-scale statistical universality, which has led us to reformulate Kolmogorov's theory, is a natural consequence of variability of the large-scale statistics from flow to flow in ways that are not simply described by $\bar{\varepsilon}$. Fluctuations in the energy flux from the large scales produce random modulations of the cascade and hence of the intensity of the small scales. Different flows yield differing statistics of the energy flux and hence of the small scales, in particular the distribution function $\Pi(w, \xi)$, which determines C via (7.74), may vary from flow to flow. For instance, in the example with sparse turbulent patches given earlier, outside of a patch there is no turbulence and so we would expect $\delta\mathbf{u}$ to be small there. Consequently, the probability distribution of $\delta\mathbf{u}(\mathbf{r})$ at both large and small scales should have a sharp peak at $\delta\mathbf{u} = 0$ in the three-dimensional space of $\delta\mathbf{u}$ in which that distribution is defined, surrounded by a much wider, lower skirt of the distribution function, whose width reflects the intensity of turbulence within a patch. This may be contrasted with a flow such as grid turbulence, for which the probability distribution of $\delta\mathbf{u}(\mathbf{r})$ consists of a broad hump about $\delta\mathbf{u} = 0$. The differing character of the distribution functions at small scales directly reflects the different large-scale statistics of the two flows, in particular, gross intermittency of the patchy flow.

Having replaced Kolmogorov's original assumption of small-scale universality, that is, dependence on $\bar{\varepsilon}$ alone, by one of inertial-range statistical self-similarity, interest centers on how self-similar the inertial-range statistics of $\delta\mathbf{u}(\mathbf{r})$ are in real turbulent flows. Departures from self-similarity are generally referred to as inertial-range intermittency, a topic which received considerable attention in the former Soviet Union (see Monin and Yaglom (G 1975), sections 25.2–25.5) and more recently in the West (see, e.g., Frisch (G 1995)). To clearly distinguish between lack of small-scale universality and lack of inertial-range similarity, it may be useful to draw an analogy between the inertial-range cascade and the propagation of signals along an electrical transmission line[13] in which we imagine probability distributions of $\delta\mathbf{u}(\mathbf{r})$ as being transmitted through the cascade to smaller scales. Thus, the distribution functions of $\delta\mathbf{u}(\mathbf{r})/\Delta u_r$ correspond to the electrical signal in the transmission line, while different distances along the line are analogous to different inertial-range scales, with the larger scales being represented by points closer to the source of the signals. The large scales are considered as providing the input of statistical information, in the form of probability distributions, like Π, to the top end of the inertial range, which is then passed down by the cascade to smaller scales. Here, it is important to note that, although the inertial-range statistics are determined by those of the large scales, the statistical properties of the velocity field at large scales are probably not a precise extrapolation of those in the inertial range, since the physical processes occurring differ significantly. For instance, the probability distribution of $\delta\mathbf{u}(\mathbf{r})/\Delta u_r$ at large scales is *not* necessarily the same as that at the top end of the inertial range, even though it is subsequently preserved through that range if self-similarity is respected. The probabilistic information that is introduced as input to the transmission line is that at the top end of the inertial range and not just the large-scale distribution functions. If statistical self-similarity holds good, this input is transmitted

[13] The reader may think in terms of, e.g., long-distance telephone lines, used to transmit signals. Electrical waveforms propagate along such a line, but undergo cumulative distortion with the distance propagated.

undistorted by the inertial-range cascade, or transmission line, in the sense that, for instance, the statistics of $\delta\mathbf{u}(\mathbf{r})/\Delta u_r$ are independent of inertial-range r. Lack of statistical similarity corresponds to distortion by the transmission line, that is, the probability distributions of $\delta\mathbf{u}(\mathbf{r})/\Delta u_r$ are cumulatively deformed by transmission and hence vary with r. On the other hand, nonuniversality results from passing *different* statistical information through the transmission line, due to the changing statistical input from the large scales, depending on the flow considered. It is also important to distinguish clearly between large-scale intermittency, reflected in the statistical input, and inertial-range intermittency, apparent as distortion in statistical transmission. Despite the similarity in nomenclature, the two are quite different in nature.

Inertial-range intermittency, that is, lack of statistical self-similarity of the inertial range (distortion in transmission), leads to departures from the reformulated version of Kolmogorov's theory, which predicts that $\chi_p(r) \propto r^{p/3}$, as we saw earlier. Experimentally, one may test the validity of this prediction and hence of the hypothesis of inertial-range self-similarity. Measurements suggest that $\chi_p(r)$ has the power-law form:

$$\chi_p(r) \propto r^{\zeta_p} \tag{7.76}$$

in the inertial range, where the value of the exponent ζ_p for a given p appears to be universal, that is, the same for all flows. As shown in Figure 7.5, the exponent, ζ_p, is close to the Kolmogorov value, $p/3$, for low values of p (up to about $p = 4$, say). From the definition, (7.75), we have $\zeta_0 = 0$, while observations indicate that $\zeta_3 = 1$. The reader will recall the qualitative argument given earlier, suggesting that cubic moments of $\delta\mathbf{u}$, such as $\chi_3(r)$, ought to be proportional to r (and hence $\zeta_3 = 1$), a result that may be derived quantitatively from (7.69) within most models of inertial-range intermittency, including the β-model considered later in this section, but that is

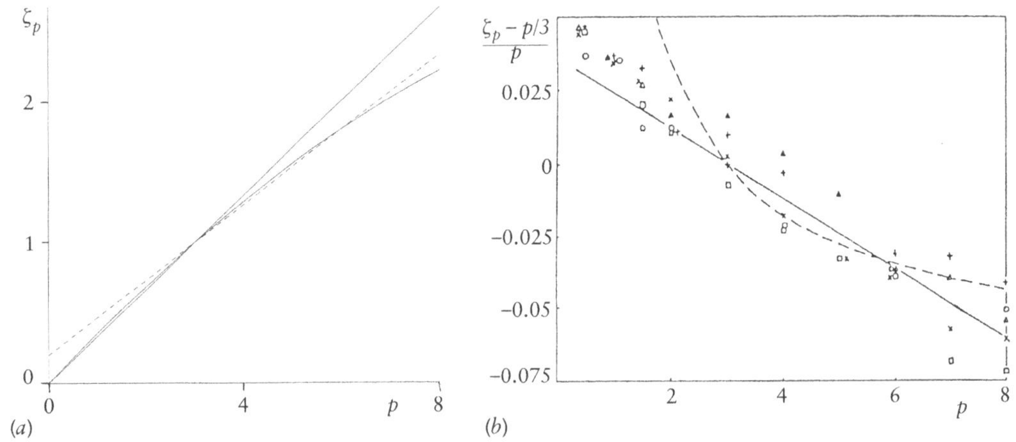

Figure 7.5. (a) Exponents, ζ_p, of the inertial-range moments of velocity differences, $\chi_p(r) \propto r^{\zeta_p}$, as a function of their order, p: the solid, straight line represents the Kolmogorov theory (no inertial-range intermittency); the dashed, straight line is for the β-model with $\mu = 0.2$; the solid curve summarizes experimental observations. (b) the quantity $(\zeta_p - p/3)/p$, measuring the departures of ζ_p from the Kolmogorov value: the solid line is a fit to the experimental data, equivalent to the solid curve of (a); the dashed curved represents the β-model with $\mu = 0.2$. ((b) Kuznetsov and Sabelnikov (1990), reproduced with permission.)

not an immediate consequence of (7.69) because we have used $\overline{|\delta \mathbf{u}|^3}$ rather than $\overline{\delta u_{\|}^3}$ to define $\chi_3(r)$. The measured value of ζ_p begins to depart from $p/3$ at larger p, as apparent in Figure 7.5. The curve in Figure 7.5a summarizes experimental data, while the solid line gives the Kolmogorov value, $\zeta_p = p/3$, and the dashed line is a fit to the data of the form $\zeta_p = \mu + (1 - \mu)p/3$ (as predicted by the β-model, considered later) with $\mu = 0.2$. Figure 7.5b shows the quantity $(\zeta_p - p/3)/p$, measuring departures from Kolmogorov's theory. Such departures reveal that the probability distribution of $\delta\mathbf{u}$ is not precisely self-similar through the inertial range, implying the presence of inertial-range intermittency.

It should be observed that accurate measurements of ζ_p are not easy, particularly for the large values of p at which departures from $p/3$ become apparent. One must first calculate $\chi_p(r)$ via averaging, as shown by (7.75), (7.76). The average may be written

$$\chi_p(r) = \int_0^\infty |\delta\mathbf{u}|^p P(|\delta\mathbf{u}|)d|\delta\mathbf{u}| \tag{7.77}$$

where $P(|\delta\mathbf{u}|)$ is the probability distribution function of $|\delta\mathbf{u}|$. If p is large and positive, the factor $|\delta\mathbf{u}|^p$ strongly weights high values of $|\delta\mathbf{u}|$, which lie in the tail of the probability distribution and represent rare events in which $|\delta\mathbf{u}|$ happens to be atypically large. In practice, $\chi_p(r)$ is usually determined by time averaging in a statistically steady flow. To obtain satisfactory time averages at large values of p, long time records must be used to encompass a sufficient number of the rare events which determine the average. The larger the value of p, the longer the record required to produce a converged average. At large enough p, one may imagine having to average over hours or days of data and questions can be asked as to whether the rare, large "spikes" in the measured $|\delta\mathbf{u}|$, which dominate the average, originate in the flow itself or are due to external perturbations (e.g., to be facetious, someone shutting the laboratory door). Thus, the measured $\chi_p(r)$ become less reliable the higher the value of p, which is why Figure 7.5 is limited to $p \leq 8$; better data should allow higher values of p.

It is possible, although unlikely, that the integral in (7.77) diverges at $|\delta\mathbf{u}| = \infty$ for p larger than some threshold value, so that higher order moments do not exist. Divergence of the integral at $|\delta\mathbf{u}| = 0$ is likely for *negative* values of p with $|p|$ greater than some threshold. As such a threshold is approached, the average is determined by rarer and rarer events and thus requires longer and longer time records for convergence of the time average.

Having measured $\chi_p(r)$ at different separations, r, one considers its variation with r to determine whether it obeys the inertial-range power law, (7.76), and if so what value is taken by the exponent ζ_p. This process is discussed in (G 1995), Section 8.3, who defines ζ_p using the longitudinal structure function, $\overline{\delta u_{\|}^p}$, rather than $\overline{|\delta\mathbf{u}|^p}$, as here. This is thought, however, not to affect the value of ζ_p (this can be shown to be the case under most intermittency models, for instance the β-model). The main difficulty in obtaining ζ_p given $\chi_p(r)$ is that, from the definition of the inertial range, the assumed power law, (7.76), is asymptotic in the limits $\mathrm{Re}_L \to \infty$, $r/L \to 0$, and $r/\eta \to \infty$. In practice, of course, the Reynolds number is finite, if large, and the inertial range is consequently of limited width. At energy-containing or dissipative scales, one would not expect a power law for $\chi_p(r)$ as a function of r. These scales border the inertial range and there is no clean division between, say, inertial and dissipative values of r. One may try to fit a power-law form to the observed $\chi_p(r)$, but a perfect fit is not possible and there is always leeway in the choice of ζ_p at finite Reynolds number. As discussed earlier, power laws with universal exponents are observed in the inertial range, although the experimental uncertainties are quite significant and should be borne in mind when considering theoretical models based on these results.

As apparent from Figure 7.5a, the departure of ζ_p from the Kolmogorov value, $p/3$, is small at low values of p, in particular $p = 2$, which corresponds to $\chi_2(r) = (\Delta u_r)^2 \propto r^{2/3}$ and leads to the $k^{-5/3}$ spectrum, as discussed earlier. If one instead uses $(\Delta u_r)^2 \propto r^{\zeta_2}$, rather than $r^{2/3}$, the spectrum is of the form k^{-n}, with $n = 1 + \zeta_2$, according to (7.6) and (7.10). From Figure 7.5b, one notes that $\zeta_2 - 2/3 \approx 3 \times 10^{-2}$, which implies that the spectrum is very slightly more rapidly decaying with increasing k than the $k^{-5/3}$ law would suggest. The predicted difference in the spectral exponent is only about 2% and is negligible for many purposes. Indeed, for the quantity, $k^{-\zeta_2+2/3}$, giving the ratio of the two spectral forms, to change by a factor of two requires that k vary by a factor of $2^{1/(\zeta_2-2/3)}$, which is of order 10^{10} if we take $\zeta_2 - 2/3 = 3 \times 10^{-2}$. To achieve such huge variations of k within the inertial range requires that it be extremely wide, that is, enormously large Re_L. More realistically, the value of k may change by, say, 100 across a more typical inertial range, giving variations in $k^{-\zeta_2+2/3}$ of only around 15%. Thus, as far as the spectrum is concerned, one may generally neglect departures from Kolmogorov's inertial-range theory, which is just as well given the considerable amount of experimental evidence confirming the $k^{-5/3}$ spectral form.

As apparent in Figure 7.5a, higher-order structure functions show larger departures from inertial-range statistical self-similarity, that is, inertial-range intermittency becomes more important for higher-order moments of $|\delta \mathbf{u}|$. As discussed above, high-order structure functions reflect the tail of the probability distribution of $|\delta \mathbf{u}|$, that is, rare, atypically large values of $|\delta \mathbf{u}|$, which become more common with decreasing r. The increasing frequency of such rare, high values of $|\delta \mathbf{u}|$ preferentially accentuates the high-order moments because they place greater weight on large $|\delta \mathbf{u}|$ than the lower-order ones, thus depressing the value of ζ_p at large p. The growing importance of the tail of the probability distribution of $|\delta \mathbf{u}|$ illustrates the progressive distortion of the distribution functions of $\delta \mathbf{u}/\Delta u_r$ by transmission through the cascade. Changes in the form of the main part of the scaled probability distribution, that is, where $\delta \mathbf{u}/\Delta u_r$ has a significant probability of being found, no doubt occur less rapidly with decreasing r, but considerable distortion of the main part of the distribution may nonetheless arise if the value of r/L is sufficiently small, corresponding to a long enough cascade. That is, large changes in r are required to produce strong departures from Kolmogorov similarity for the lower-order moments, so one would expect the main part of the distribution function to show only moderate changes unless quite small values of r/L (requiring correspondingly large Re_L) are considered. Regardless of the precise details of the changes that take place in the statistics of $\delta \mathbf{u}/\Delta u_r$ as r decreases through the inertial range, the important point is that such variations probably exist, as evidenced by the observed lack of proportionality of ζ_p

to p at the higher orders, even if very long inertial ranges are required to produce strong effects on the main parts of distribution functions.

The systematic changes in the inertial-range statistics of $\delta\mathbf{u}(\mathbf{r})/\Delta u_r$ with r are believed to be a consequence of randomness in the cascade process. The intensity of the large scales varies randomly with position, time, and from realization to realization, while small scales, whose intensity is also randomly variable, are formed inside the large ones. Within the small scales, yet smaller ones are randomly produced, and so on, down to dissipative scales. As a consequence of the random large scales and of the random cascade, the turbulent intensity at scale r varies randomly from place to place, realization to realization, and with time. In the nature of a random process, such as the cascade we envisage here, the turbulent activity at scale r will differ from place to place in a particular realization and, where it is more intense, there tends to be a greater energy flux to scales smaller than r. Thus, the turbulent intensity at scale r preconditions that of the smaller scales at the same location: the higher it is, the more active the resulting smaller scales tend to be. As a result, randomness introduced at each stage in the cascade is not forgotten, but is instead passed down and added to by each generation, leading to growing percentage fluctuations in turbulent intensity, with rare, atypically intense activity becoming more common at smaller scales. At the top end of the inertial range, randomness of turbulent intensity is principally due to that of the energy flux from the large scales, but is increasingly augmented by contributions from the random cascade. The tail of the probability distribution of quantities such as $|\delta\mathbf{u}|/\Delta u_r$ becomes more pronounced the smaller the scale considered, while the main part of the probability distribution may be significantly affected following a sufficient number of generations (assuming that the Reynolds number is large enough that one does not run into the dissipative range first). In summary, randomness of the process of formation of scales smaller than r is superimposed on that at scale r, leading to progressive changes in the statistical properties as the scale considered decreases.

To illustrate the effects of intermittency in the cascade, we introduce the β-model, which has its origins in work by Novikov and Stewart (1964), as reformulated by Frisch, Sulem, and Nelkin (1978). It is not particularly realistic as a model of inertial-range intermittency, and indeed, Frisch (G 1995) has described it as a "minimally complex toy model." However, it is relatively simple to explain and understand, and despite being unrealistic in detail, illustrates some general features of inertial-range intermittency.

According to the β-model, turbulent activity at any given inertial-range scale and time, and within any single realization of the flow is confined to only part of space, while there is no activity elsewhere at that scale. The cascade is conceived of as consisting of a series of discrete generations of eddies, with generation 0 eddies representing the large scales, $r = O(L)$, which are taken to fill space, thus excluding the possibility of large-scale intermittency. Eddies of generation m form inside those of generation $m-1$ and have scale $\alpha^m L$, occupying a fraction β^m of space. The parameters $0 < \alpha < 1$ and $0 < \beta < 1$ are constants of the model and are *not* random. However, despite occupying a fixed volume of space, the locations of the active eddies are allowed to vary randomly. The proportion of space occupied by eddies of size $r_m = \alpha^m L$ is $(r_m/L)^\mu$, where

$$\mu = \frac{\log \beta}{\log \alpha} \geq 0 \tag{7.78}$$

is the only really significant constant of the β-model. Note that, since the factors α and β are assumed to be the same at each generation of the cascade, the process of eddy formation is self-similar at different scales. Such self-similarity of the cascade *process* (rather than of the statistical properties of $\delta\mathbf{u}(\mathbf{r})$) is a common feature of inertial-range intermittency models and is what ultimately leads to power laws for quantities such as $\chi_p(r)$. Self-similar cascade processes can produce progressive changes in the small-scale probability distributions, and the resulting statistics of $\delta\mathbf{u}(\mathbf{r})$ are generally not themselves self-similar.

Having motivated the model, we may now forget the discrete generations of eddies and take the proportion of space in which there is turbulent activity at scale r to be $(r/L)^\mu$, while in the remaining fraction, $1-(r/L)^\mu$, there is no activity at scale r. This refers to a particular realization, whereas the regions of activity are supposed to move around randomly from realization to realization and we therefore interpret $(r/L)^\mu$ as the probability that a given point lies inside a region of activity at scale r. Thus, $\delta\mathbf{u}(\mathbf{r})$ is taken as zero with probability $1-(r/L)^\mu$. We further suppose that the statistical properties *within* active regions are self-similar, so that the distribution function of $\delta\mathbf{u}$ has the inertial-range form

$$P(\delta\mathbf{u}) = \underbrace{\left(1-\left(\frac{r}{L}\right)^\mu\right)\delta(\delta\mathbf{u})}_{\text{Inactive regions}} + \underbrace{\left(\frac{r}{L}\right)^\mu \lambda^{-3}(r)\Pi\left(\frac{|\delta\mathbf{u}|}{\lambda(r)}, \frac{\mathbf{r}.\delta\mathbf{u}}{r|\delta\mathbf{u}|}\right)}_{\text{Active regions}} \tag{7.79}$$

where, as indicated by the annotations, the Dirac function represents the inactive regions, while the active regions are described by the distribution function Π and their intensity by $\lambda(r)$. In the three-dimensional space of $\delta\mathbf{u}$, which the distribution function (7.79) describes, the first argument of Π gives dependency on the magnitude of $\delta\mathbf{u}$, while the second shows variations with its direction. The assumption of self-similarity of the active regions is an obvious generalization of the hypothesis of such similarity for the entire distribution of $\delta\mathbf{u}(\mathbf{r})$, used earlier in the revised version of Kolmogorov's theory.

Comparison of (7.68) and (7.79) shows their close connection and that one recovers (7.68) if $\mu = 0$, that is, if intermittency is absent, $\beta = 1$, and turbulent activity fills space at all scales. On the other hand, if $\mu > 0$ there is inertial-range intermittency, with the probability of $\delta\mathbf{u} = 0$, represented by the Dirac function in (7.79), increasing with decreasing r. That is, the regions of turbulent activity become more and more sparse as the scale considered decreases. The parameter μ is usually taken as $\mu = 0.2$, a small value implying weak intermittency. Given small μ, intermittency only becomes of prime importance when r/L is sufficiently small that $(r/L)^\mu$ differs significantly from 1, despite the small value of μ. From (7.79), it can be shown that Δu_r is proportional to $(r/L)^{\mu/2}\lambda$ and hence that the probability distribution of the scaled quantity $\mathbf{v} = \delta\mathbf{u}/\Delta u$ has the form

$$P(\mathbf{v}) = \underbrace{\left(1-\left(\frac{r}{L}\right)^\mu\right)\delta(\mathbf{v})}_{\text{Inactive regions}} + \underbrace{\left(\frac{r}{L}\right)^{5\mu/2}\Pi_v\left(\left(\frac{r}{L}\right)^{\mu/2}|\mathbf{v}|, \frac{\mathbf{r}.\mathbf{v}}{r|\mathbf{v}|}\right)}_{\text{Active regions}} \tag{7.80}$$

and depends on r, explicitly showing the lack of statistical self-similarity when $\mu > 0$. Thus, the cascade process alters the form of the distribution function, although acting quite slowly because $\mu = 0.2$ is small. As r decreases, the first term in (7.80) grows in importance, while the second extends to larger values of $|\mathbf{v}|$, thus both zero and large $|\mathbf{v}|$ are emphasized. However, it should be stressed that, the presence of the Dirac function at $\delta\mathbf{u} = 0$ (equivalently $\mathbf{v} = 0$) means that (7.79) and (7.80) do not provide very realistic expressions for the distribution functions. According to the β-model, at a given scale one is either inside an active region or else the turbulent intensity is zero. Real turbulence does not show such black-and-white behavior,[14] but rather has a continuous, *non-self-similar* probability distribution occupying $\delta\mathbf{u} \neq 0$. Furthermore, the assumption that the distribution function away from $\delta\mathbf{u} = 0$ is self-similar imposes unrealistic constraints on the statistics.

Regardless of the detailed realism of the model, one may compute the average in (7.69) using (7.79) to obtain $\lambda(r) \propto r^{(1-\mu)/3}$, just as we earlier determined Δu_r based on (7.68). The structure functions, (7.75), can then be calculated, yielding

$$\zeta_p = \mu + \frac{1}{3}(1 - \mu)p \tag{7.81}$$

for the exponents of order $p > 0$. Furthermore, it is easily shown that, as stated earlier, the moments of $\delta u_{\|}$ have the same exponents, ζ_p, as those of $|\delta\mathbf{u}|$, a result we expect to hold more generally than the β-model. If $\mu = 0$, intermittency is absent and we recover the Kolmogorov exponents, $\zeta_p = p/3$. The small value, $\mu = 0.2$, causes the exponents to diverge a little from the Kolmogorov form for moderate p and increasingly so at large p, as indicated by the dashed line in Figure 7.5a. The observed trend for ζ_p to fall away from $p/3$ as p increases above three is reproduced, although the straight-line form resulting from the model is probably incorrect. The value of ζ_2 given by the β-model with $\mu = 0.2$ is about 0.07 greater than the Kolmogorov value of 2/3. In consequence, the predicted inertial-range spectrum falls off more steeply than $k^{-5/3}$, having a spectral exponent about 4% higher than 5/3. The predicted departure of ζ_2, and consequently of the spectral exponent, from Kolmogorov's theory has the same sign, but is about a factor of two greater than that implied by the observations shown in Figure 7.5b. One may note a variety of unsatisfactory features due to the unrealistic Dirac function in (7.79), such as the divergence of $\chi_p(r)$ for all negative p and that the value of (7.81) at $p = 0$ is μ, whereas $\zeta_0 = 0$ in reality, as discussed earlier.

Under the β-model, as r decreases, turbulent activity becomes rarer, occurring in bursts of probability $(r/L)^\mu$, within which the turbulent intensity is characterized by $|\delta\mathbf{u}| = O(\lambda)$, according to (7.79). At the large scales, turbulent velocity fluctuations are of $O(u')$, allowing estimation of the constant of proportionality in $\lambda(r) \propto r^{(1-\mu)/3}$ to obtain $\lambda(r) \sim u'(r/L)^{(1-\mu)/3}$, which is larger than it would be in the absence of inertial-range intermittency (i.e., with $\mu = 0$) by a factor of $(L/r)^{\mu/3}$. It follows that, within an active region of inertial-range scale r, turbulence is more intense than without intermittency. This greater intensity compensates the cubic moments,

[14] Here we exclude gross *large-scale* intermittency, which may well produce regions in which turbulent activity is effectively zero, but is disallowed in the β-model by the assumption that large-scale turbulent activity fills space. At issue here is *inertial-range*, rather than large-scale intermittency.

which are related to the energy flux and therefore unaffected by inertial-range intermittency, for the increasing rarity of active regions at smaller r.

Despite being incorrect in detail, the β-model makes clear that a random cascade, here of unrealistic black-and-white type, *can* generate a non-self-similar probability distribution for $\delta\mathbf{u}(\mathbf{r})$, even though the cascade process is itself modeled in a self-similar way. This lack of statistical self-similarity is naturally reflected in the departure of ζ_p from $p/3$, discussed above. The reason for the term inertial-range intermittency is also elucidated: there are increasingly intense, but rare, peaks in turbulent activity as r decreases. However, these bursts of activity are separated by periods of *zero* activity within the model, extreme behavior that is not borne out in real turbulence. Furthermore, the model supposes the statistics within bursts to be self-similar at different scales, which is also unrealistic. In short, the model illustrates some of the effects of inertial-range intermittency, but should not be considered as giving a true account of turbulence.

It is interesting to consider the effect of inertial-range intermittency at the Kolmogorov scale, where viscosity becomes important. As noted above, within the active regions of scale r, we have $|\delta\mathbf{u}| \sim \lambda(r) \sim u'(r/L)^{(1-\mu)/3}$, so that one may consider the Reynolds number $|\delta\mathbf{u}|r/\nu \sim \mathrm{Re}_L(r/L)^{(4-\mu)/3}$. This becomes of $O(1)$, indicating the intervention of viscosity, when $r = O(\eta)$, where

$$\frac{\eta}{L} = O\left(\mathrm{Re}_L^{-3/(4-\mu)}\right) \tag{7.82}$$

gives an estimate of the Kolmogorov scale according to the β-model and reduces to (7.46) when $\mu = 0$. From (7.82), it appears that inertial-range intermittency somewhat decreases the scale at which viscous dissipation becomes significant because the turbulent intensity within active regions is higher than in the absence of intermittency.

There is another prediction of the β-model which is of particular interest, namely the fractal nature of the dissipation according to the model. To discuss this point, we first need to digress to introduce the notion of objects having fractional dimension, or *fractals*. Given an object, Θ, a collection of spheres is said to cover Θ if every point of Θ lies in at least one of the spheres. Question: How many spheres of a given size are needed to cover the object? In the case of a single point it is clear that one sphere suffices. In general, the number of spheres required will depend on the size of sphere used. Let us therefore suppose that the spheres are all of radius r and let $N(r)$ be the minimum number required to cover Θ. Readers may quickly convince themselves that, for a line $N(r)$ is of order r^{-1} as $r \to 0$. Similarly, it is clear that, for a surface $N(r) \propto r^{-2}$, while, for a volume $N(r) \propto r^{-3}$. In each case the exponent that occurs in the expression for $N(r)$ is equal to (minus) the dimension of the object considered: lines, surfaces, and volumes are generally described as having dimensions one, two, and three, respectively. Even the single point is included if we give it the dimension zero, since $N(r) = 1 = r^0$. A precise expression of this idea is the following: an object has Kolmogorov dimension D if the limit

$$D = -\lim_{r\to 0}\frac{\log N(r)}{\log r} \tag{7.83}$$

exists.

This definition of dimension agrees with the usual one for simple objects such as the line, surface, and volume considered above and in all these cases the value of D is an integer. However, there are objects for which D is not a whole number: such objects are called fractals, standing for "fractional dimension." An abstract example can be constructed iteratively as shown in Figure 7.6. Beginning with the object shown in Figure 7.6a, each line segment is replaced by a copy of the same figure reorientated and reduced in size by a factor of four, resulting in Figure 7.6b. Each line segment of Figure 7.6b is likewise replaced by a small, rotated version of the original object and the process continued ad infinitum. The object that is constructed in this way is a fractal of dimension 3/2. As is the case for this example, fractals are not smooth lines, surfaces, or volumes, because such classical geometrical objects have integer dimension. Instead a fractal object has a "furry" structure in which successive magnifications reveal more details at finer and finer scales.

The example we have constructed has no direct relevance to the study of turbulence and is intended solely for illustrative purposes. However, the energy cascade with its succession of smaller and smaller scales is obviously similar in character, provided we restrict attention to the inertial range of r. The object we consider is made up of those points within some finite volume, V, of space at which there is active energy dissipation at a given time and for a given realization. For the β-model, there is dissipation within the regions of turbulent activity, but not outside. The active regions occupy a proportion $(r/L)^{\mu}$ of the volume and therefore require a number, $N(r)$, given by $N(r)r^3 \sim (r/L)^{\mu} V$ of spheres of size r to cover them. It follows that

$$D = -\lim_{r\to 0} \frac{\log r^{\mu-3}}{\log r} = 3 - \mu \tag{7.84}$$

is the Kolmogorov dimension of the set of points with nonzero energy dissipation. In the absence of inertial-range intermittency ($\mu = 0$), the dimension is three and, as one might expect since $\beta = 1$, the set of nonzero dissipation fills space. With the value $\mu = 0.2$, which was quoted earlier, we obtain $D = 2.8$ as the inertial-range dimension of the set of nonzero dissipation according to the β-model. This dimension describes an object that does not fill a volume, since its dimension is less than (although quite close to) three. The set of nonzero dissipation gradually thins out as it is examined at smaller and smaller inertial-range scales. Of course, once the Kolmogorov scale is reached, the set is revealed to be a volume of space, rather than a fractal, but as the Reynolds number is increased the volume occupied by the dissipation shrinks to zero.

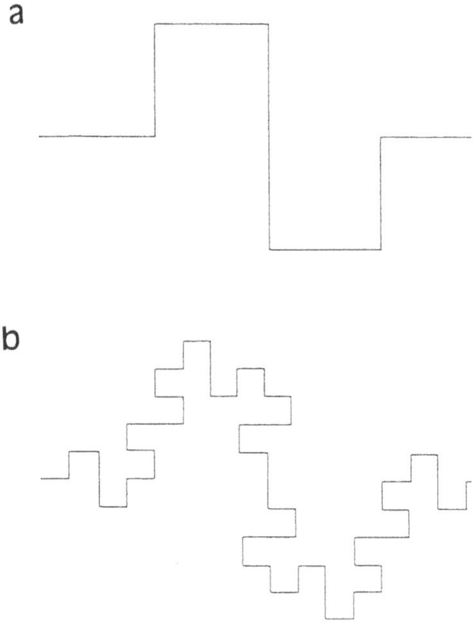

Figure 7.6. The first two stages in the iterative construction of a fractal.

Some general comments regarding fractals may be in order. Firstly, not all sets have a

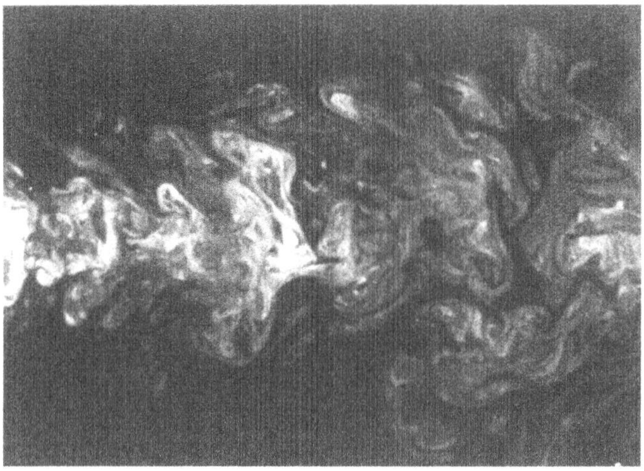

Figure 7.7. Image showing scalar concentration in a turbulent jet based on measurements using fluorescent dye. Data analysis indicates that contours of constant concentration are fractals. (Prasad and Sreenivasan (1990), reproduced with permission.)

Kolmogorov dimension, since it can happen that the limit (7.83) does not exist. This is one reason why mathematicians tend to prefer a different (more complicated) definition of dimension due to Hausdorff. Although the Hausdorff dimension has the virtue that it always exists, it is harder to use in practice because of its more complicated definition. The two definitions are closely related and concern the fine-scale structure ($r \to 0$) of an object.[15] The definition of the Kolmogorov dimension suggests a simple method for practical measurement of D. Given an object: cover it in spheres of size r and plot the logarithm of the minimum number required as a function of $\log r$. A straight line indicates the existence of a Kolmogorov dimension equal to minus the slope of the line. This method of finding D is known as the box-counting algorithm (note that covering with cubic boxes and spheres produces identical results for D). Needless to say, no object that derives from the real world is truly a fractal because it will always be smooth when looked at on a sufficiently fine scale[16] (at scale η in the present case). This is an inherent limitation of fractal models. Nonetheless, close approximations to fractals have been found to occur in other natural examples, including coastlines and cloud shapes. Figure 7.7 shows an experimental image of dye fluorescence in a turbulent jet, whose contours of constant fluorescence intensity, representing those of dye concentration, are found to yield fractals. This illustrates the fact that the convection of scalar quantities by turbulence tends to generate fractal isoscalar surfaces.

To bring this section to a close, we discuss the statistical properties of the dissipative range, beginning with the probability distribution of $\delta\mathbf{u}(\mathbf{r})$. The flatness factor of $\delta u_{||}$:

$$T = \frac{\overline{\delta u_{||}^4(\mathbf{r})}}{\left(\overline{\delta u_{||}^2(\mathbf{r})}\right)^2} \tag{7.85}$$

is one possible measure of the importance of the tails of the probability distribution (see Chapter 2) that would be independent of r if the distribution were self-similar (and equal to three if $\delta\mathbf{u}$ were Gaussian). Assuming that the moments of $\delta u_{||}$ are governed by the same power laws, (7.76), as those of $|\delta\mathbf{u}|$, we have $T \propto r^{\zeta_4 - 2\zeta_2}$ of which the exponent can be estimated from Figure 7.5b as about -0.1, representing a slow increase of

[15] Vassilicos and Hunt (1991) emphasize the difference between the two dimensions and suggest that the Kolmogorov dimension may be more appropriate to turbulence. The question is far from settled and we do not discuss the issue further.

[16] It follows that the statement that a physical object has fractional dimension D, ought also to give the range of scales for which this is true.

the flatness factor with decreasing inertial-range r due to intermittency. Observations indicate that T begins to increase more rapidly as the dissipative range is entered, indicating that the form of the distribution function of $\delta\mathbf{u}(\mathbf{r})$ departs from that in the inertial range, developing more pronounced tails. Thus, rare, large fluctuations become more common, a feature compounded at high Reynolds numbers by the intermittency inherited from the inertial range, which also tends to emphasize the tails of the distribution functions, as we discussed earlier. The value of T, and hence the form of the distribution function, varies across the dissipative range, until, at small enough separations, one may use the one-term Taylor's series

$$\delta u_i = \frac{\partial u_i}{\partial x_j} r_j \tag{7.86}$$

to relate $\delta\mathbf{u}(\mathbf{r})$ to the first derivatives of velocity, at which point it becomes statistically self-similar, but with different statistical properties than in the inertial range, properties that are determined by those of the velocity derivatives. Furthermore, the similarity scaling as $\delta\mathbf{u} \propto r$ is different from that, $\delta\mathbf{u} \propto r^{1/3}$, which would be implied by inertial-range similarity.

Since dissipative scales arise from the cascade, their statistical properties are determined by, though different from, those of the small scales of the inertial range. In consequence, the dissipative-range statistics depend on both the statistical input from the large scales into the top end of the cascade, which may vary with the flow considered, and the distortion of the statistics resulting from transmission across the inertial range, which varies with the width of that range, and hence with Re_L, albeit only slowly for the main parts of the probability distribution functions. Thus, changes in either the large-scale statistical input or in Re_L may alter the statistics expressed using Kolmogorov dissipative-range scalings, and hence invalidate universality in that range. The intensity of the inertial range fluctuates owing to random variations in the large scales and cascade, producing random changes in the size and intensity of the dissipative range scales. The higher the intensity of the inertial range, the higher the effective Reynolds number at any given scale, and hence the smaller the scale at which viscosity eventually becomes important. This is similar in nature to the reduction in the Kolmogorov scale, (7.37), when $\bar{\varepsilon}$ is increased, although $\bar{\varepsilon}$ measures the *average* inertial-range intensity, rather than the random fluctuations we have in mind here. It is as if the Kolmogorov scale, characterizing the size of the dissipative scales, becomes a random variable, along with the intensity of such scales.

In the far dissipative range, consisting of sufficiently small scales that the velocity field is smooth in all but very occasional realizations, one may use (7.86) to relate velocity differences and derivatives, at least for the main parts of their probability distributions. Since the turbulent dissipation rate, ε, is determined by the velocity derivatives (see equation (4.26)), its statistical properties can be derived from those of the derivatives. As noted above, the distribution functions in the dissipative range, including those of $\delta\mathbf{u}$, tend to have more pronounced tails than in the inertial range. This is reflected by more frequent, though still rare, large fluctuations in, for instance, the velocity derivatives and dissipation rate.

A striking variety of small-scale intermittency is observed if, rather than differencing the velocity field to obtain $\delta\mathbf{u}(\mathbf{r})$, one uses other filters to suppress more strongly the large-scale, low-frequency components. Historically, this was the first form of small-scale intermittency to be recognized. For instance, Batchelor and Townsend (1949) measured the velocity at a fixed point in grid turbulence and applied successive time derivatives. The velocity itself fluctuates randomly, but shows no obvious periods of inactivity (see Figure 3.1a). Taking the derivative of the velocity emphasizes the small-scale "fur," as discussed in Chapter 3, and the results are dominated by the dissipative scales of turbulence (see Figure 3.1b), but again show no obvious periods of calm. However, as the order of the derivative is increased, periods of activity, separated by intervals of calm, become clearly evident. The process of taking derivatives filters the velocity field, suppressing low-frequency Fourier components in favor of high-frequency components. The higher the derivative that is taken, the higher the frequencies that remain after the filtering operation. A filter can be considered either as a convolution with some filter function in the time domain or, equivalently, as multiplication of the Fourier transform of the time signal by a function of frequency called the transfer function of the filter. As usual, one may adopt the Taylor hypothesis to translate results into spatial terms, thus interpreting time signals as spatial line cuts through the flow. Taking the velocity difference, $\delta\mathbf{u}(\mathbf{r})$, in physical space is equivalent to multiplying the spatial Fourier transform by $e^{i\mathbf{k}.\mathbf{r}} - 1$, which goes to zero at $\mathbf{k} = 0$ and therefore suppresses the large scales, as does the derivative $\partial/\partial x_i$, which has the transfer function ik_i and may be obtained from $\delta\mathbf{u}(\mathbf{r})$ via the limit $\mathbf{r} \to 0$. A single derivative suppresses wavenumbers below the dissipative range, while high derivatives correspond to high powers of wavenumber and emphasize far-dissipative wavenumbers at the expense of lower ones. A high-pass filter, which suppresses low wavenumbers completely, may be constructed by taking a transfer function that is 0 for $|\mathbf{k}| < k_c$ and 1 for $|\mathbf{k}| > k_c$, where k_c is a cutoff wavenumber, whose value can be chosen freely. Increasing k_c through the dissipative range of wavenumbers, the output of a high-pass filter is found to become more and more obviously intermittent, showing isolated bursts of activity, just as when taking derivatives of increasing orders. Both processes strongly emphasize high dissipative wavenumbers, and we conclude that the higher the dissipative wavenumber considered, the more sporadically it appears in the flow.

Using a high-pass filter with k_c in the far dissipative range, or taking a high derivative, is thought to focus on rare events for which the dissipative scale happens to be exceptionally small because the intensity of the inertial range is atypically large. That this occurs only infrequently is the reason why the filter output shows isolated bursts of activity. It may be useful to think of this process in spectral terms. The effect of a filter is to multiply the spectrum by the squared modulus of its transfer function. Thus, a high-pass filter sets the spectrum at wavenumbers below k_c to zero, while high derivatives multiply the spectrum by correspondingly high powers of the wavenumber. Since the raw spectrum probably has exponential decay in the far dissipative range, either filter yields an output with exponentially small energy, which is concentrated at high dissipative wavenumbers, that is, very small scales. The spectrum is defined in terms of the average properties of the turbulence, namely the velocity correlation functions, and is not a random variable, but, qualitatively, one may think of the effects of fluctuations in the energy flux to the dissipative scales as random variations of the location of the exponential falloff in the spectrum (the "knee" in any one of the spectra of Figure 7.1). Such variations produce very large percentage fluctuations of the spectral level in the far dissipative range, owing to exponential decay there. After filtering to suppress lower wavenumbers, these large fluctuations appear as highly sporadic output. The spectral interpretation of observed intermittency following dissipative-range filtering is due to Kraichnan (1967).

7.6 Conclusions

In this chapter we have covered quite a wide range of important topics. In particular, using the spectral methods of the previous chapter, assuming high-Reynolds-number, developed turbulence, we have recognized as many as three asymptotic ranges of wavenumbers and corresponding scales in physical space: large energy-containing scales at small wavenumbers, small dissipative scales at high wavenumbers and, if Re_L is sufficiently large, an intermediate zone, called the inertial range. The smallest scales are of size $O(\eta)$, where η is the Kolmogorov scale, which is fixed by the energy supply from the large eddies, and the viscosity. Thus, the smallest scales adjust their size so as to dissipate the energy delivered to them by the cascade. Kolmogorov's theory supposes that the dissipative and inertial-range scales have universal statistical properties when suitably scaled. In particular, the scaled spectrum takes on a universal form when the nondimensional wavenumber, $k\eta$, is used. If there is an inertial range, one obtains the famous $k^{-5/3}$ energy spectrum there, corresponding to a $r^{1/3}$ law for velocity differences. The asymptotic separation of energy-containing and dissipative scales is expressed by $L/\eta = O(\mathrm{Re}_L^{3/4})$.

According to Kolmogorov's theory, provided the Reynolds number is high enough, the small scales are insensitive to the particular turbulent flow in which they find themselves, apart from the value of $\bar{\varepsilon}$. However, the large scales have an intimate relationship with the overall flow and vary considerably in their properties according to the specific flow considered. For high-Reynolds-number homogeneous, isotropic turbulence without mean flow, in particular grid turbulence, we found that the large scales probably approach self-similarity, but that the small ones do not generally participate in that similarity. As the turbulence decays, the value of Re_L, and hence the separation of scales between the large and smallest, decreases. Eventually, the Reynolds number is no longer sufficiently large that there is a clear separation, the turbulence goes through an adjustment phase, emerging into its final phase of decay, again self-similar, in which there is no large separation of scales and no cascade, and the turbulence decays passively under the effects of viscosity (the turbulence is effectively dead).

Although Kolmogorov's original theory is good enough for many purposes, we have seen that the assumption on which it is based, that the properties of the small scales are determined solely by $\bar{\varepsilon}$ (and ν in the case of the dissipative range), is not strictly valid. As a result, universality of the scaled statistical properties of the small scales, and, in particular, of the inertial-range spectral constant, C, and dissipative range spectral form do not hold true for general flows. Such lack of universality reflects differences in the large-scale statistics of different flows. One may recover the inertial-range power laws for the structure functions, and in consequence the $k^{-5/3}$ spectrum, using the weaker hypothesis of inertial-range statistical self-similarity, but universality of C and of the dissipative-range spectrum are lost. This does not mean that there need be huge variations of these quantities between flows, of course, and indeed many observations indicate similar values for the Kolmogorov constant and scaled dissipative-range spectrum in different flows. Such modest variability may be connected with the small exponents occurring in expressions such as $C \propto \gamma^{1/3}$, derived earlier for the flow with sparse turbulent patches, with γ the fraction of the time the flow is turbulent. The small exponent suggests that changes with the large-scale properties of the flow should be mild, as do the small exponents occurring in the expressions for the Kolmogorov dissipative-range scalings as functions of $\bar{\varepsilon}$.

Since the revised version of Kolmogorov's theory is based on inertial-range statistical self-similarity, departures from similarity will generally invalidate the theory. Such departures are referred to as inertial-range intermittency and measurements of the structure functions, $\chi_p(r)$, indicate that they obey inertial-range power laws as a function of r, with exponents, ζ_p, which do indeed appear to depart from the Kolmogorov value, $\zeta_p = p/3$, at the high orders. However, the discrepancies are rather small at low orders. In particular, the velocity correlations and spectra ($p = 2$) show departures which are negligible for most purposes, thus leaving the well-verified $k^{-5/3}$ law essentially intact, although there may be intermittency corrections to the spectral exponent of the order of a few percent.

The larger departures from $\zeta_p = p/3$ at higher orders suggest that the scaled statistical properties of $\delta\mathbf{u}(\mathbf{r})$ are systematically modified by transmission through the inertial-range cascade, in particular the tails of the probability distribution functions, which become more pronounced at smaller scales. For sufficiently long inertial ranges, requiring very large Reynolds numbers, the main parts of the probability distributions may also be modified by their transmission through many generations of the cascade.

Appendix: Kolmogorov's Four-Fifths Law

Using (7.67) to express $\delta\mathbf{u}(\mathbf{r})$ in (7.70) gives

$$\delta u_{\parallel} = \frac{r_i}{r}\{u_i(\mathbf{x}+\mathbf{r}) - u_i(\mathbf{x})\} \tag{7.A1}$$

whose cube is taken and averaged to obtain

$$\begin{aligned}\overline{\delta u_{\parallel}^3} = \frac{r_i r_l r_m}{r^3}\{&\overline{u_i(\mathbf{x}+\mathbf{r})u_l(\mathbf{x}+\mathbf{r})u_m(\mathbf{x}+\mathbf{r})} - \overline{u_i(\mathbf{x})u_l(\mathbf{x})u_m(\mathbf{x})} \\ &+ 3\overline{u_i(\mathbf{x}+\mathbf{r})u_l(\mathbf{x})u_m(\mathbf{x})} - 3\overline{u_i(\mathbf{x})u_l(\mathbf{x}+\mathbf{r})u_m(\mathbf{x}+\mathbf{r})}\}\end{aligned} \tag{7.A2}$$

of which the first two terms in brackets cancel by homogeneity (they are, in any case, both zero by isotropy). The others can be determined using homogeneity and (6.134), leading to

$$\overline{\delta u_{\parallel}^3} = 3\frac{r_i r_l r_m}{r^3}\{Q_{ilm}(\mathbf{r}) - Q_{ilm}(-\mathbf{r})\} \tag{7.A3}$$

which can be further simplified from (6.135) and (6.140) as

$$\overline{\delta u_{\parallel}^3} = 6u'^3 K(r) \tag{7.A4}$$

The quantity $K(r)$, appearing in (7.A4), is related to the spectral transfer via (6.149), which we write as

$$T(k) = \frac{u'^3}{\pi} \int_0^\infty \Lambda(r)\, kr \, \sin kr \, dr \tag{7.A5}$$

where

$$\Lambda(r) = \frac{1}{r^2} \frac{\partial}{\partial r} \left(\frac{1}{r} \frac{\partial}{\partial r} \left(r^4 K \right) \right) \tag{7.A6}$$

The inverse of the transform (7.A5) is

$$\Lambda(r) = \frac{2}{u'^3} \int_0^\infty T(k) \frac{\sin kr}{kr} dk \tag{7.A7}$$

and it may be observed that the transform pair, (7.A5), (7.A7), are exactly analogous to (6.116), (6.117).[17] We next evaluate (7.A7) asymptotically for large Reynolds number and inertial-range r.

When r is in the inertial range, the function $(\sin kr/kr)$ in (7.A7) is close to 1 for energy-containing wavenumbers, $k = O\left(L^{-1}\right)$, while it oscillates and decays as k increases through $k = O\left(r^{-1}\right)$, to become negligible at dissipative wavenumbers. It will be recalled (see Figure 6.5) that $T(k)$ has a strong negative peak in the energy-containing range, which dominates the integral in (7.A7). Indeed, from (7.31) and (7.33) it is apparent that the integral of $T(k)$ from $k = 0$ up to any point in the inertial range is $-\bar{\varepsilon}$, being dominated by energy-containing-range contributions. Thus, approximating $(\sin kr/kr)$ by 1 over the energy-containing range of k, the integral in (7.A7) has the inertial-range value $-\bar{\varepsilon}$, leading to

$$\Lambda(r) \sim -\frac{2\bar{\varepsilon}}{u'^3} \tag{7.A8}$$

at inertial range r. Substituting (7.A8) into (7.A6), one may integrate to find

$$K(r) \sim -\frac{2\bar{\varepsilon}\, r}{15u'^3} \tag{7.A9}$$

where we have ignored possible terms of the form r^{-2} and r^{-4}, which are solutions of the homogeneous version of the equation for $K(r)$, on the grounds that such singular behavior as $r \to 0$ is inadmissible. Finally, we employ (7.A4) to derive (7.69).

Note that Kolmogorov's original hypothesis that the small scales are controlled by $\bar{\varepsilon}$ has not been employed in deriving (7.69). Indeed, no universality assumption is required for Kolmogorov's four-fifths law to hold.

References

Abramowitz, M., Stegun, I. A., 1970. *Handbook of Mathematical Functions*. Dover, New York.

Batchelor, G. K., Proudman, I., 1956. The large-scale structure of homogeneous turbulence. *Phil. Trans. Roy. Soc. Lond.*, **A248**, 369–405.

[17] It will be recalled from Chapter 6 that such isotropic transform pairs are obtained from the usual three-dimensional Fourier transform pairs using (6.104).

Batchelor, G. K., Townsend, A. A., 1949. The nature of turbulent motion at large wavenumbers. *Proc. Roy. Soc. Lond.*, **A199**, 238–55.

Comte-Bellot, G., Corrsin, S., 1966. The use of a contraction to improve the isotropy of grid-generated turbulence. *J. Fluid Mech.*, **25**(4), 657–82.

Frisch, U., Sulem, P.-L., Nelkin, M., 1978. A simple dynamical model of intermittent fully developed turbulence. *J. Fluid Mech.*, **87**, 719–36.

Gibson, C. H., Schwarz, W. H., 1963. The universal equilibrium spectra of turbulent velocity and scalar fields. *J. Fluid Mech.*, **16**, 365–84.

Grant, H. L., Stewart, R. W., Moilliet, A., 1962. Turbulence spectra from a tidal channel. *J. Fluid Mech.*, **12**, 241–68.

Kolmogorov, A. N., 1941a. The local structure of turbulence in incompressible viscous fluid at very high Reynolds number. *Dokl. Akad. Nauk SSSR*, **30**, 301–5.

Kolmogorov, A. N., 1941b. On decay of isotropic turbulence in an incompressible viscous liquid. *Dokl. Akad. Nauk SSSR*, **31**, 538–41.

Kolmogorov, A. N., 1941c. Energy dissipation in locally isotropic turbulence. *Dokl. Akad. Nauk SSSR*, **32**, 16–18.

Kolmogorov anniversary edition, 1991. *Proc. Roy. Soc. Lond.*, **A434**.

Kraichnan, R. H., 1967. Intermittency in the very small scales of turbulence. *Phys. Fluids*, **10**, 2080–2.

Kuznetsov, V. R., Praskovsky, A. A., Sabelnikov, V. A., 1992. Fine-scale turbulence structure of intermittent shear flows. *J. Fluid Mech.*, **243**, 595–622.

Kuznetsov, V. R., Sabelnikov, V. A., 1990. *Turbulence and Combustion*. Hemisphere Publishing Corp., New York.

Novikov, E. A., Stewart, R. W., 1964. The intermittency of turbulence and the spectrum of energy dissipation. *Izv. Akad. Nauk. SSSR, Ser. Geoffiz.*, 3, 408–13.

Pond, S., Stewart, R. W., Burling, R. W., 1963. Turbulence spectra in the wind over waves. *J. Atmos. Sci.*, 20, 319–24.

Prasad, R. R., Sreenivasan, K. R., 1990. The measurement and interpretation of fractal dimensions of surfaces in turbulent flows. *Phys. Fluids A*, **2**, 792–807.

Saffman, P. G., 1967. The large scale structure of homogeneous turbulence. *J. Fluid Mech.*, **27**, 581–93.

Sandborn, V. A., Marshall, R. D., 1965. Local isotropy in wind tunnel turbulence. Technical report, Fluid Dynamics and Diffusion Laboratory, Colorado State University.

Stewart, R. W., Townsend, A. A., 1951. Similarity and self-preservation in isotropic turbulence. *Phil. Trans. Roy. Soc.*, **A243**, 359–86.

Vassilicos, J. C., Hunt, J. C. R., 1991. Fractal dimensions and spectra of interfaces with application to turbulence. *Proc. Roy. Soc. Lond.*, **A435**, 505–34.

CHAPTER EIGHT

Numerical Simulation of Turbulent Flows

The ability of modern computers to store and rapidly manipulate large quantities of data has allowed the development of flow simulations that complement, and sometimes even replace, traditional laboratory experiments. In industry, the considerable expense of constructing and exploiting physical models motivates the search for reliable simulation techniques to reduce the cost of the design and development phases. Full, detailed coverage of numerical methods for turbulent flows goes well beyond the scope of this book and would require a volume to itself. This chapter is therefore limited to a relatively brief overview of the three main types of numerical techniques, represented by the following three sections, in which the reader desiring further details of numerical methods will find references to more specialized works.

8.1 Direct Numerical Simulation (DNS)

The first, and most obvious, approach to the numerical simulation of turbulent flows treats them like any others, solving the unsteady, three-dimensional Navier–Stokes equations numerically without further ado. Such methods are referred to as direct simulations, in contrast with those based on the statistically averaged equations considered in Section 8.3, which involve closure approximations, but are much less computationally costly than DNS at high Reynolds numbers. Direct simulation allows numerical experiments to be carried out that are often difficult or impossible to realize in the laboratory, and yields detailed information concerning the flow field in individual realizations and its statistical properties that again often surpasses what can be measured in real flows. Furthermore, unlike other simulation techniques for turbulent flows, DNS does not involve approximations, other than those due to discretization, which are inherent in any numerical solution of differential equations, and so provides benchmarks against which less costly simulation methods may be evaluated and parameterized, just as they can by using the results of physical experiments. However, the continuum of different spatial (and temporal) scales intrinsic to turbulence, a continuum whose width increases with the turbulent Reynolds number, Re_L, leads to computer time and storage requirements that increase rapidly with Re_L. In consequence, DNS is limited to relatively low Reynolds numbers in practice and it is currently very expensive to conduct a DNS calculation at even moderately high Reynolds number. To undertake such simulations at the high Reynolds numbers typical of most industrial and naturally occurring flows is not feasible at present, nor, owing to the rapid increase of required computer resources with Re_L, is it likely to be in the near future. Thus, as has been emphasized by Moin and Mahesh (1998) in their recent and highly readable review of the field, to which the reader is encouraged to refer, DNS is a research tool, rather than an aid to engineering design.

The incompressible Navier–Stokes equations

$$\frac{\partial U_i}{\partial t} + U_j \frac{\partial U_i}{\partial x_j} = -\frac{1}{\rho}\frac{\partial P}{\partial x_i} + \nu \frac{\partial^2 U_i}{\partial x_j \partial x_j} \tag{8.1}$$

and

$$\frac{\partial U_i}{\partial x_i} = 0 \tag{8.2}$$

may be treated numerically in a variety of ways. At any instant of time, the flow field, characterized by $\mathbf{U}(\mathbf{x}, t)$ and $P(\mathbf{x}, t)$ has, in principle, an infinite number of degrees of freedom. However, only a finite set of variables can be handled numerically and numerical schemes therefore involve an approximate representation of the flow in terms of a finite number of unknowns, a procedure referred to as spatial discretization. For instance, one might use the values of $\mathbf{U}(\mathbf{x}, t)$ and $P(\mathbf{x}, t)$ at a discrete set of grid points or a truncated Fourier series to describe the flow. Obviously, spatial discretization is an approximation and, for it to be acceptable, it needs to be able to represent the real flow with reasonable accuracy. In particular, it should be capable of resolving the smallest dynamically important scales of turbulence, that is, the dissipative scales of size $O(\eta)$, where η denotes the Kolmogorov scale. Since, for a given flow, the size of the dissipative scales decreases as the Reynolds number is increased, the spatial resolution of the scheme must be improved at larger Reynolds numbers, for instance the grid should be made finer in grid-based schemes. This leads to the rapidly increasing memory and calculation requirements referred to earlier and limits the Reynolds numbers attainable by DNS.

Based on an accurate representation of the spatial structure of the flow in terms of a discrete set of variables, a numerical scheme provides a set of equations describing the time evolution of those variables, thus allowing the flow to be followed forwards in time from some prescribed initial velocity field. Time, which is, in principle, a continuous variable, is usually divided into discrete time steps, with the numerical scheme providing a means for advancing from time t to time $t + \Delta t$, where Δt is the time step. In summary, a numerical scheme consists of a spatial representation of the flow in terms of a finite number of variables and a prescription for marching forward, one time step after another, from any specified initial conditions.

The sheer number of different schemes that have been proposed as approximate representations of the governing equations (8.1), (8.2), together with appropriate boundary conditions, if boundaries there are, precludes any attempt at exhaustive coverage here. The reader is referred to the specialist works by Canuto et al. (1987), Hirsch (1988, 1990), and Ferziger and Peric (1996), while Press et al. (1992) give a highly readable account of numerical techniques in general, with an overview (chapter 19) of the solution of partial differential equations. Schemes are often divided into general classes, each of which have their individual particularities, advantages, and disadvantages, such as spectral, finite-difference, finite-volume, and finite-element methods. For instance, finite-difference schemes represent the flow by its values at a discrete grid of points and approximate the spatial derivatives in (8.1), (8.2) by differencing. The time derivative in (8.1) may also be expressed using a finite-difference approximation, or more elaborate methods, such as predictor–corrector techniques, can be employed.

Note that, whereas (8.1) involves a time derivative and describes the dynamics of the velocity field, the incompressibility condition (8.2) is a kinematic constraint on the flow to be enforced at each time step for the determination of $P(\mathbf{x}, t)$. The requirement that the velocity field satisfy the discrete version of (8.2) constrains the pressure gradient term in discretized (8.1), allowing solution for the pressure field. This is the numerical equivalent of the discussion in Section 4.3, where it was shown that the pressure is instantaneously determined by the velocity field via the Poisson equation (4.46). Here, in keeping with the declared policy of this book, we are restricting attention to incompressible fluids of uniform density. If compressibility is allowed for, the equivalent of (8.2) involves a time derivative of the density and no longer provides a purely kinematic constraint, leading to propagation of pressure disturbances as acoustic waves of finite speed. In the incompressible limit, the sound speed goes to infinity, and the limit is nonuniform. Numerical schemes designed for compressible flows (e.g., gas flows at significant Mach numbers) are qualitatively different in character from those used for incompressible ones because they must resolve the rapidly moving sound waves and are generally not well adapted to the treatment of even weakly compressible ones, requiring smaller and smaller time steps to maintain numerical stability as the incompressible limit is approached. Weakly compressible flows are one of the main objects of study of aeroacoustics and are usually analyzed theoretically via "acoustic analogies" in which the Laplacian on the left-hand side of (4.46) is replaced by an acoustic wave operator, rendering the incompressible limit uniform. The theory of aeroacoustics, as well as the design of numerical schemes for aeroacoustic applications, are specialized areas that lie beyond the scope of this book (see Goldstein (1984) and Wells and Renaut (1997) for further information), as are compressible flows in general. Thus, we shall restrict attention to incompressible flows in what follows.

Although the design of numerical schemes to solve (8.1), (8.2), together with appropriate boundary conditions, if boundaries there are, often appears as a black art based on individual taste and experience, there are certain important requirements. The fundamental need is for the numerical solution to approach the true solution as the continuous limit is approached, that is, as the time step and spatial resolution tend to zero, a property referred to as convergence. Thus, if the initial conditions are given and the scheme used to obtain the solution at some fixed later time, the results should converge to the true flow in the continuous limit.[1] The main obstacle to convergence is numerical instability. Scales of the flow that are much larger than the smallest spatial resolution of a numerical scheme, Δx, are usually well described, but scales comparable with Δx are not accurately simulated, and may be spuriously amplified, resulting in a numerical solution containing fine scales that have little to do with the real flow. Apparently reasonable looking schemes can exhibit such numerical instabilities and are effectively unusable. Even among those

[1] Solution of the system of equations resulting from the scheme may be done iteratively, in which case there is a further convergence issue associated with the iteration that is not considered here.

that are in common use, stability often places constraints on the numerical parameters of the scheme, typically an upper bound on the time step which decreases with the spatial resolution. In that case, one cannot simply take the continuous limit Δx, $\Delta t \to 0$ indiscriminately and obtain convergence, but should do so while respecting the stability criterion. The usual technique for analyzing the stability of a scheme is Von Neumann's method (see, e.g., Hirsch (1988), chapter 8), which should certainly be invoked when developing a new scheme, but is not foolproof since it involves linearization and the replacement of variable coefficients by constants. Furthermore, boundaries are not allowed for, whereas numerical instabilities can appear at the sides or corners of a flow, invalidating the numerical solution. In the end, it is often a question of trial and error.

Presuming convergence of the numerical scheme, the question remains how rapidly the real solution is approached as Δx, $\Delta t \to 0$. The difference between the numerical and true solutions is referred to as the discretization error, which represents the cumulative effects of discretization errors committed at past time steps and whose magnitude determines the precision of the scheme. It is usually possible to estimate the error incurred by a single time step in the limit Δx, $\Delta t \to 0$, for instance, using a Taylor's expansion of the flow field to examine the error introduced by finite-differencing derivatives. This local error estimate generally has the form $\phi(\Delta x, \Delta t)\Delta t$, where ϕ is a polynomial function of Δx and Δt, determining the order of the scheme. For example, if $\phi = O(\Delta x^2, \Delta t^2)$ as $\Delta x, \Delta t \to 0$, the scheme is said to be second order in space and time. The current trend appears to be towards the use of higher-order schemes, which are, in principle, more precise when Δx and Δt are small enough, but are more complicated to implement and require more calculations for a given Δx and Δt. The hope is that, for a particular discretization error, higher-order schemes allow larger values of Δx and Δt to be taken. However, it is not obvious that, for a given accuracy, a higher-order scheme will involve less computer time, which is the bottom line.

Uniformity of convergence of numerical schemes can be a significant issue. That is, the discretization error may be larger in certain regions of the flow, for instance, near boundaries or at later times. Spatial nonuniformities can often be dealt with by refining the spatial resolution (and perhaps the time step) of the scheme in the regions concerned. Temporal nonuniformity is unavoidable for turbulent flows owing to their extreme sensitivity to perturbations. Thus, as discussed in Chapter 1, small changes in the initial conditions are rapidly amplified, leading to effective unpredictability of the details of the flow in particular realizations, although one hopes and expects that its statistical properties are well defined. The discretization error of a numerical scheme can be thought of as a perturbation (the local error discussed above) applied to the flow at each time step and subsequently amplified, leading to growing departures from the exact solution and hence nonuniformity of convergence in time. Moreover, even if the scheme had zero discretization error, the finite precision of calculations on a computer means that small perturbations are continuously introduced and amplified by the flow. In consequence, due to the fundamental nature of turbulence, and just as in real flows, one is led to consider the statistical properties of numerical solutions, which one hopes will uniformly converge to those of the true flow as the continuous limit is approached. Questions of convergence and precision then need to be recast in terms of the flow statistics and, although it seems likely that a scheme that describes individual realizations with reasonable accuracy will give a good account of the statistics, to the authors' knowledge, little work has been done on quantifying the statistical convergence and precision of different schemes. For instance, it could be that, in a statistical sense, a low-order scheme produces better results than a higher-order one, at least for some statistical measures, or even that some apparently reasonable schemes fail to model the Navier–Stokes statistics at all.

While we are on the important subject of the statistics of numerical solutions, a number of further points should be touched on, many of which were discussed in the context of real flows in Chapter 2. In principle, an ensemble of different initial conditions should be used for any given turbulent flow, with the number of realizations being sufficiently large that converged statistics can be obtained. Given the high cost of DNS, this is very expensive to achieve in practice. In many cases, one expects that the flow eventually forgets its initial conditions and settles down to a statistically steady state with statistical independence at wide temporal separations. In that case, a single, sufficiently long numerical simulation yields the flow statistics, since it effectively contains many independent simulations, as discussed in Chapter 2 for real flows. One must nonetheless wait long enough that the flow has attained the steady state, and subsequently continue the simulation over many correlation times to obtain converged statistics. Statistical steadiness and independence at large temporal separations are assumptions that need to be verified in particular cases and, as described in Chapter 2, some flows exhibit periodic oscillations or statistical variations on a variety of disparate time scales that can make the choice of appropriate statistical formulation for a particular flow less than obvious. Use of statistics conditioned on some phase-locked physical variable in periodic flows provides a relatively simple example.

The widely studied case of homogeneous turbulence that approaches statistical independence at large spatial separations also allows calculation of flow statistics without explicitly performing an ensemble of different calculations, now using volume, rather than time averaging. The flow volume must be large enough compared with the correlation length that averaging encompasses a sufficient number of effectively independent samples.

If homogeneous turbulence is allowed to decay without mean flow, rather than being maintained by random forcing, the choice of initial conditions can play an important role, whereas if random forcing is used to produce a statistically steady state, the result is obviously dependent on the statistics of the forcing field. This is in contrast with, for instance, wake, jet, or channel flows in which turbulence is generated spontaneously when the Reynolds number is high enough and the flow presumably forgets its initial conditions, taking on whatever statistical properties it has naturally. The homogeneous case is idealized and artificial, but, as we have seen in earlier chapters, allows analysis to be taken further, yielding insights into the dynamics of turbulence that may hold true of more general flows. Although DNS is obviously capable of handling inhomogeneous flows, and has indeed been widely used for such cases, homogeneous turbulence has provided one of the main objects of study, particularly in the early works on the subject. This is partly due to the general importance of homogeneous flows in the theory of turbulence, but also to the numerical efficiency and precision of the Fourier-spectral schemes that are usually employed in the homogeneous case. For the above reasons, as well as its relative simplicity and close relationship with the spectral theory developed in Chapter 6, we want to examine the Fourier-spectral approach in some detail. In so doing, we will have in mind homogeneous turbulence without mean flow.

Fourier-spectral schemes apply to spatially periodic flows without boundaries, of which the simplest case, and one considered here, consists of those that are periodic with respect to each of the coordinates x_1, x_2, and x_3 separately. Thanks to periodicity, the velocity and pressure fields can be expressed as the Fourier series

$$U_i(\mathbf{x}, t) = \sum_{\mathbf{k}} a_i(\mathbf{k}, t) e^{i\mathbf{k}.\mathbf{x}}$$
$$P(\mathbf{x}, t) = \sum_{\mathbf{k}} b(\mathbf{k}, t) e^{i\mathbf{k}.\mathbf{x}} \tag{8.3}$$

where the sums are taken over the discrete set of wavenumbers

$$\mathbf{k} = 2\pi\left(\frac{n_1}{\ell_1}, \frac{n_2}{\ell_2}, \frac{n_3}{\ell_3}\right)$$

with ℓ_1, ℓ_2, and ℓ_3 representing the periods in the x_1-, x_2-, and x_3-directions and n_1, n_2, and n_3 running over integer values. Periodicity means that the flow repeats itself throughout space, which can be thought of as filled with rectangular boxes of sides ℓ_1, ℓ_2, and ℓ_3, in each of which the flow is identical. The periods ℓ_1, ℓ_2, and ℓ_3 are usually taken to have comparable magnitudes, $O(\ell)$; indeed, they are often chosen as equal. Since turbulence is not really periodic, the spatial periods ought, in principle, to be allowed to tend to infinity, but should, in any case, be large compared with the large scales, $O(L)$, of turbulence. That is, one hopes and expects that the dynamics of the flow will not notice its periodicity provided that $\ell \gg L$. Observe that, as $\ell \to \infty$, the distance between neighboring wavenumbers, $\Delta k = O(\ell^{-1})$, in the sum of (8.3) goes to zero. Thus, in the large-period limit, the discrete **k**, **x**-periodic, representation used here attempts to approach the continuous **k**, aperiodic description of Chapter 6. In particular, (8.3) is the discrete **k** equivalent of the Fourier integral (6.17).

Assuming a periodic flow in what follows, we introduce (8.3) into (8.2), leading to the incompressibility condition $k_i a_i = 0$, equivalent to (6.47), while (8.1) yields

$$\frac{da_i}{dt} + ik_j a_i \otimes a_j = -\frac{i}{\rho} k_i b - \nu k^2 a_i \tag{8.4}$$

where

$$(a_i \otimes a_j)(\mathbf{k}, t) = \sum_{\mathbf{k}'} a_i(\mathbf{k}', t) a_j(\mathbf{k} - \mathbf{k}', t) \tag{8.5}$$

is a discrete **k** convolution. Multiplying (8.4) by k_i and employing the incompressibility condition $k_i a_i = 0$, we have

$$b = -\rho \frac{k_i k_j}{k^2} a_i \otimes a_j \tag{8.6}$$

as the Fourier equivalent of the Poisson equation, (4.46), for the pressure. Using (8.6) in (8.4) gives the final evolution equation

$$\frac{da_i}{dt} = -ik_m \Delta_{il}(\mathbf{k})\, a_l \otimes a_m - \nu k^2 a_i \tag{8.7}$$

for the Fourier coefficients of the velocity field, where

$$\Delta_{il}(\mathbf{k}) = \delta_{il} - \frac{k_i k_l}{k^2} \tag{8.8}$$

is the usual projection tensor in **k**-space. It may be observed that (8.4), (8.6), and (8.7) are the discrete equivalents of (6.48)–(6.50), and indeed, the above reasoning closely parallels that given in Chapter 6 for the continuous wavenumber case. If desired, spatially periodic volume forcing can be easily incorporated via an additional term on the right of (8.1), whose Fourier coefficient, projected perpendicular to the vector **k** by $\Delta_{il}(\mathbf{k})$, then appears in (8.7).

The above formulation is exact for the class of periodic flows without boundaries to which it applies. However, as things stand, an infinity of coefficients, $a_i(\mathbf{k}, t)$, are needed to describe the flow. Since numerical methods are only capable of handling a finite number of unknowns, the Fourier series (8.3) are truncated, that is, a_i and b are taken as zero for

$$\mathbf{k} = 2\pi\left(\frac{n_1}{\ell_1}, \frac{n_2}{\ell_2}, \frac{n_3}{\ell_3}\right)$$

when $|n_1| > N_1$, $|n_2| > N_2$ or $|n_3| > N_3$, leaving a finite sum over $-N_1 \le n_1 \le N_1$, $-N_2 \le n_2 \le N_2$, and $-N_3 \le n_3 \le N_3$, where we have in mind large, but comparable, values for N_1, N_2, and N_3. Truncation of the Fourier series means that the flow resulting from (8.3) cannot exactly satisfy the Navier–Stokes equations because, using truncated a_i, the convolution (8.5) yields nonzero values beyond the truncation range, leading to nonzero a_i outside that range if the full system, (8.7), of equations is employed. A number of standard prescriptions, for instance, Galerkin, collocation, and τ-methods, are available for deriving approximate evolution equations for the nonzero a_i following truncation (Gottlieb and Orszag (1977) describe the general theory with examples). In the present context, the most obvious approach, adopted here and equivalent to a Galerkin method, is to simply drop all equations of the infinite system (8.7) corresponding to a_i that have been set to zero, yielding a finite system of evolution equations for the nonzero $a_i(\mathbf{k}, t)$. Truncation is, of course, an approximation, since, as noted above, the a_i that have been set to zero would become nonzero if the full system (8.7) were used to describe their time evolution.

In truncating the Fourier series (8.3), one effectively neglects the high wavenumbers, that is, the small scales, which limits the size of the smallest flow structures that can be accurately represented by the scheme. Wavenumbers with $|k_i| \le 2\pi N_i/\ell_i$ are included, forming a rectangular box in **k**-space. Let

$$k_c = 2\pi \min\left(\frac{N_1}{\ell_1}, \frac{N_2}{\ell_2}, \frac{N_3}{\ell_3}\right)$$

then $|\mathbf{k}| \le k_c$ is the largest sphere that lies wholly inside the box, while the smallest neglected wavenumber lies just above $|\mathbf{k}| = k_c$. The truncated version of (8.3) cannot accurately resolve flow structures of size $O(k_c^{-1})$ and below. Thus, for truncation to be acceptable, the spatial resolution,[2] $\Delta x = k_c^{-1}$, of the scheme should be much less than the smallest scales of the flow. In the case of turbulence, one expects that the smallest significant scales are of size $O(\eta)$, so that Δx must be smaller than $O(\eta)$. Recalling from earlier chapters that $k = O(\eta^{-1})$ describes the dissipative range of wavenumbers, we see that the truncation in **k**-space should be chosen to lie above the dissipative range, where the spectrum falls off rapidly with increasing k. That is, the wavenumbers that are retained by the truncation ought to include those responsible for the dissipation. Given the observed exponential decay of the spectrum with wavenumber above the dissipative range, one would expect that the energy associated with such wavenumbers, which is a measure of the truncation error, to be exponentially small as $\Delta x \to 0$. This suggests that the spatial precision of the scheme is better than any power of Δx, which represents a significant advantage of the Fourier-spectral approach compared with, say, finite-difference methods, for the restricted class of periodic flows without boundaries to which it applies. Indeed, it is found that one can generally get away with using a smaller number of variables to describe homogeneous turbulence using a Fourier-spectral scheme, which, combined with the relative simplest of the evolution equations (8.7), implies greater numerical efficiency and explains why schemes of the above type are usually the method of choice in the homogeneous case.

Having truncated the a_i and the system of equations (8.7) to a finite number, while being careful to include dissipative wavenumbers, one is faced with solving a set of ordinary differential equations to determine the time evolution of the flow from given initial conditions. Many techniques are available, including Runge–Kutta and predictor–corrector methods of different orders of precision in the time step Δt, which must be chosen small enough that the time scales associated with the smallest structures of the flow are resolved and the scheme is stable. Whichever method is employed to integrate forwards in time,[3] the most time-consuming part of the calculation is the evaluation of the convolution in (8.7). This is best achieved using fast Fourier transforms (FFTs), rather than directly from (8.5). That is, one employs three-dimensional, inverse FFTs to evaluate the velocity field on a rectangular grid of points in **x**-space, performs the multiplication of U_l and U_m at each point, and returns to **k**-space, again using FFTs (see Press et al. (1992), chapters 12 and 13 for details of the FFT algorithm and its application to the calculation of convolutions). This procedure involves fewer arithmetic operations than direct use of (8.5) and is therefore more efficient, as well as incurring smaller rounding errors due to the finite arithmetic precision of the computer. However, care needs to be taken to avoid what are referred to as aliasing errors (see, e.g., Press et al. (1992) for a discussion of aliasing in general and Orszag (1971) on aliasing errors in Fourier-spectral schemes for fluid flows).

The convolution in (8.7) arises from the quadratic term in the Navier–Stokes equation (8.1). Given Fourier components of wavenumbers $\mathbf{k}'$ and $\mathbf{k}''$, their product has wavenumber $\mathbf{k}' + \mathbf{k}''$. Thus, if the a_i are truncated to zero at wavenumbers above $2\pi N_1/\ell_1$ in the x_1-direction, the convolution (8.5) will yield nonzero values up to

[2] Note that Δx should be thought of as an order-of-magnitude scale.

[3] Given the high spatial precision of the Fourier-spectral approach, it may be desirable to employ high-order temporal schemes.

wavenumber $4\pi N_1/\ell_1$, twice that used for the truncation. This is of no consequence if (8.5) is employed directly to calculate the convolution: one simply does not bother to calculate the convolution for wavenumbers beyond the truncation. However, if FFTs are used for reasons of numerical efficiency, the higher wavenumbers generated by nonlinearity corrupt the convolution within the range of retained wavenumbers unless that range is extended prior to performing the FFTs. It can be shown that, to avoid such errors, the a_i should be explicitly extended with zeroes to wavenumbers of at least $\pm 3\pi N_1/\ell_1$ in the x_1-direction before taking FFTs, and likewise in the other coordinate directions. That is, the top one-third of the wavenumbers used in the FFT, of both signs and all three directions, should have zero a_i. In practice, one chooses a certain length of FFT, usually a power of two since that simplifies the FFT algorithm, fills the lowest two-thirds of wavenumbers with the nonzero a_i retained by the truncation and zeroes the remaining one-third. Inverse FFTs are then performed, followed by multiplication in physical space, and FFTs to return to spectral space, at which point the top one-third of wavenumbers is corrupted by aliasing errors, but is discarded, since they lie beyond the a_i retained by the truncation. The net result is the same (to within rounding errors) as direct use of (8.5). The fact that $\mathbf{U}(\mathbf{x}, t)$ is real, or equivalently that $a_i(-\mathbf{k}, t) = a_i^*(\mathbf{k}, t)$ (where the star denotes complex conjugation), is usually employed in practical implementations to reduce the storage and calculation requirements of the FFTs (see Press et al. (1992) for details). The general issue of aliasing errors due to wavenumbers beyond the spatial resolution of a scheme being mistaken for lower wavenumbers is of wider scope than the technical issues involved in the use of FFTs for Fourier-spectral schemes that are addressed here and can even be important in the analysis of schemes that do not explicitly use Fourier components at all (see the references to aliasing given above).

Having developed an efficient and accurate procedure for advancing the truncated $a_i(\mathbf{k}, t)$ from one time step to the next, one needs a set of initial values. These initial values should be zero mean, while satisfying the incompressibility condition $k_i a_i = 0$ and the requirement $a_i(-\mathbf{k}) = a_i^*(\mathbf{k})$, which, as noted above, reflects the fact that $\mathbf{U}(\mathbf{x}, t)$ is real (these properties are preserved by time evolution according to (8.7)). A turbulent flow involves random initial $a_i(\mathbf{k})$, whose statistics could obviously be chosen in many ways, reflecting those of the velocity field they represent. Recalling that we have in mind homogeneous turbulence without mean flow, one method of generating the initial $a_i(\mathbf{k})$ given an initial spectral tensor $\Phi_{ij}(\mathbf{k})$ is as follows. The $a_i(\mathbf{k})$ at different wavenumbers are taken to be statistically independent random variables (apart from the constraint $a_i(-\mathbf{k}) = a_i^*(\mathbf{k})$) of the form $a_i = c_i e^{i\phi}$, where ϕ is uniformly distributed over the range $0 \leq \phi \leq 2\pi$ and independent of the complex random vector c_i, which satisfies $k_i c_i = 0$ and has statistics such that the discrete approximation

$$\Phi_{ij}(\mathbf{k}, t) = \frac{\ell_1 \ell_2 \ell_3}{8\pi^3} \overline{a_i(\mathbf{k}, t) a_j^*(\mathbf{k}, t)} \tag{8.9}$$

to the spectral tensor (cf. (6.24)) has the prescribed initial values. The random phase factor ϕ leads to homogeneity of the flow described by (8.3) and, together with initial statistical independence of different wavenumbers, generates approximately Gaussian statistics for the initial velocity field. It is perhaps unfortunate that, since it is considerably easier to take different initial wavenumbers as statistically independent,[4] rather than employ more complicated initial conditions, most simulations of homogeneous turbulence have implicitly used similar statistics, thus missing the chance to exploit the flexibility of DNS to examine the results of more general initial statistics.

Given the initial $a_i(\mathbf{k})$, one integrates the truncated system (8.7) forward in discrete time to determine $a_i(\mathbf{k}, t)$, and hence the velocity field, at later times. In principle, a series of initial realizations should then be used to calculate the flow statistics, such as the spectral tensor (8.9), although the high computational cost of even a single DNS realization at moderately large Reynolds number means that converged statistics are rarely, if ever, obtained by this method. The results for a single realization are difficult to interpret in general because they contain fluctuations specific to that realization. In particular, as discussed in Chapter 6, the spectral density computed from the right-hand side of (8.9) without averaging has fluctuations from one wavenumber to the next whose magnitude is comparable to the average, $\Phi_{ij}(\mathbf{k}, t)$ (recall Figure 6.2). In the isotropic case, the energy density can be averaged over thin spherical shells in $\mathbf{k}$-space to smooth out the fluctuations and thus obtain what is presumably an approximation to $E(k, t)$ from a single realization, an approach that is not, however, available for anisotropic turbulence. In the anisotropic case, a local moving average over $\mathbf{k}$ can be performed to smooth the spectral tensor, although this may significantly degrade the wavenumber resolution. The need to obtain properly converged statistics from DNS is an important issue that adds to the already heavy cost of such calculations. Notice that here the term statistical convergence refers to increasing the number of realizations in order to obtain reliable statistical information. This is quite distinct from the earlier notion of convergence of the statistics as the spatial and temporal resolutions of the scheme are refined. In our view, both questions need to be given more attention than is traditionally the case.

Homogeneous turbulence decays under viscous action in the absence of mean flow and the resulting statistical unsteadiness disallows use of time averaging to calculate the flow statistics. However, homogeneous turbulence can be maintained by artificial random body forcing, appearing as an additional term on the right of (8.7). If statistically steady random forcing is used, the flow itself may eventually settle down to a statistically steady state, whose properties depend on those of the forcing. Since the forcing is a random function of both time and spatial location, there is considerable scope in the choice of its statistics. Spatially, one may use methods similar to those described above for the initial $a_i(\mathbf{k})$ to generate statistically homogeneous random forcing terms for (8.7) at each time step, while the choice of statistical relationships between different times adds to the already wide range of possibilities. The forcing is usually concentrated at low wavenumbers, corresponding to the large scales of turbulence, with the smaller scales appearing naturally via the cascade. The role of the forcing is thus to maintain the large scales, despite their loss of energy to smaller ones.

Although, as we saw above, statistically homogeneous flows of the form (8.3) can easily be constructed, spatial periodicity of individual realizations means that they

[4] Homogeneity constrains wavenumber pairs, other than $\pm\mathbf{k}$, to be uncorrelated, but not necessarily statistically independent.

are not statistically independent at wide separations, which is one of the properties one usually has in mind for homogeneous turbulence. On the other hand, provided that the size of the large scales of turbulence, $O(L)$, is much less than the spatial period, one would expect there to be only a small effect on the flow dynamics. Thus, as noted earlier, the spatial periods should be chosen sufficiently large compared with the correlation length that the large eddies do not notice the fact that the flow repeats itself throughout space. This condition should ensure that the large scales are properly handled, while, as also discussed earlier, at the other extreme, the scheme must be capable of resolving the smallest, dissipative scales. In spectral terms, the spacing between discrete wavenumbers is $\Delta k = O(\ell^{-1})$, where ℓ represents the spatial period, assumed comparable in the three coordinate directions, so the above requirement that $\ell \gg L$ is equivalent to $\Delta k \ll L^{-1}$. In other words, the spectral discretization, Δk, should be sufficiently fine that the energy-containing part of the spectrum, $k = O(L^{-1})$, representing the large scales, is well resolved. At the same time, as we saw earlier, proper description of the small scales requires that the spectral cutoff due to truncation lie above the dissipative range of wavenumbers, including the bulk of the viscous dissipation within the wavenumbers handled by the scheme. In addition to these small and large-scale spatial constraints on the scheme, the time step needs to be small enough that the shortest temporal scales are well resolved.

Notice that, as discussed in the last chapter, the correlation scale, $L(t)$, of freely decaying homogeneous turbulence increases with time, eventually becoming comparable with the spatial period, $O(\ell)$. This places an upper limit on the times for which a given simulation provides an accurate account of decaying turbulence.

The Fourier-spectral approach to DNS described above can be extended to some more general flows. In particular, mean flows having uniform velocity gradients acting on homogeneous turbulence can be treated by methods very similar to those discussed above. We do not give details here, but the interested reader may refer to Rogers and Moin (1987) for examples of such DNS calculations. Some inhomogeneous flows are amenable to treatment using truncated Fourier-spectral schemes of the kind discussed above, an example being a localized patch of turbulence, in a nominally unbounded flow volume. As before, care should be taken to ensure that the periodicity introduced by the scheme does not significantly influence the dynamics of the flow. For instance, a finite patch of turbulence described by Fourier series will be treated as if it had periodic "images" throughout space and the spatial period must be sufficiently large compared with the patch size that one patch is not affected by the others.

Fourier-spectral schemes are generally not well adapted to flows with boundaries, and alternatives are usually sought in such cases. Developed flow in an infinite, plane channel, $0 < x_2 < 2D$, driven by a mean pressure gradient in the x_1-direction, was discussed in Section 4.5 (see "Two-Dimensional Channel Flow"), and provides another widely studied case to which spectral DNS methods are applicable. Unboundedness and homogeneity of the flow in planes parallel to the channel walls suggest the continued use of Fourier series in the coordinates, x_1 and x_3, of which x_1 represents the streamwise direction. However, due to the flow boundaries at $x_2 = 0$ and $x_2 = 2D$, Fourier modes are not the best choice in the x_2-direction and it is usual to employ Chebyshev polynomials instead. Like Fourier modes for periodic functions, Chebyshev polynomials form a complete set, allowing expansion of an arbitrary function on the interval between -1 and $+1$, to which $0 < x_2 < 2D$ is first transformed using the variable $(x_2 - D)/D$. Thus, the channel flow is taken as periodic, with a rectangular unit cell, in the x_1 and x_3 directions and is expanded using products of appropriate Fourier components in x_1, x_3 and Chebyshev polynomials in $(x_2 - D)/D$. The expansion is truncated and equations describing the time evolution of the expansion coefficients developed, usually by Galerkin or collocation techniques. These equations are then integrated to follow the development of the flow, which is expected to eventually become statistically steady. General approaches using spectral expansions to solve partial differential equations numerically are described in Gottlieb and Orzag (1977), while the specific application to channel flow is the subject of Kim, Moin, and Moser (1987). As always, it is important to make sure that the scheme is capable of resolving the smallest structures of the flow, both in space and time, and that the periods in x_1 and x_3 are big enough that the large scales are correctly handled.

Spectral techniques, based on expansion of the flow variables in terms of a complete set of smooth functions over the entire flow volume, are difficult to devise and implement for more complicated boundary geometries than channel flow and it is usual to employ other numerical methods, like finite-difference schemes, in such cases. Space precludes going into details of such methods here and the interested reader should consult the general references to numerical techniques given earlier. The issue of appropriate conditions to apply at inflow boundaries where the turbulence is inhomogeneous is particularly delicate (see Moin and Mahesh (1998), section 2.3, for a discussion). However, before bringing this section to a close, we want to examine the general issue of the dependence of computer time and storage requirements of DNS on the Reynolds number.

For a given large-scale flow, the size of the smallest scales, $O(\eta)$, is proportional to $\mathrm{Re}_L^{-3/4}$ at large Reynolds numbers, as we saw in Chapter 7. Thus, there are $O(\mathrm{Re}_L^{9/4})$ spatial degrees of freedom that must be represented no matter what numerical scheme is used for DNS, for instance, by the a_i of the Fourier-spectral scheme described earlier or by the grid-point values of a finite-difference scheme. These degrees of freedom of the flow require storage proportional to[5] $\mathrm{Re}_L^{9/4}$. Each degree of freedom needs to be updated at every time step, requiring at least $O(\mathrm{Re}_L^{9/4})$ arithmetic operations, and perhaps more. For example, a Fourier-spectral scheme implemented using FFTs involves $O(\mathrm{Re}_L^{9/4} \log \mathrm{Re}_L)$ operations per time step.

As discussed in Chapter 3, the shortest Eulerian time scale of a turbulent flow is determined, not by the Kolmogorov time t_η, which describes the dynamics of dissipative eddies and is a Lagrangian quantity, but by convection by the large scales and mean velocities of the smallest spatial scales past a fixed point. Since numerical schemes usually employ an Eulerian representation, the time step must be small compared with $\eta/\max(u', |\overline{\mathrm{U}}|)$ and scales proportional to η, that is, like $\mathrm{Re}_L^{-3/4}$ as a function of Reynolds number. The number of time steps required to reach a given time is thus proportional to $\mathrm{Re}_L^{3/4}$, which we multiply by the number of arithmetic

[5] The turbulent Reynolds number, Re_L, is usually smaller than, but proportional to, the overall Reynolds number of the flow, based on, for instance, the free-stream velocity and body size of an obstacle placed in a flow. Thus, in such flows, one can replace Re_L by the overall Reynolds number in the power-law estimates derived here.

operations, at least $O(\mathrm{Re}_L^{9/4})$, per step, to obtain $O(\mathrm{Re}_L^3)$ as an estimate for the dependence of computer time on Reynolds number. In summary, at large Reynolds numbers, we expect DNS storage requirements to increase proportional to $\mathrm{Re}_L^{9/4}$ and arithmetic operations at least as fast as Re_L^3. The relatively large exponents in these power laws indicate rapid increase of required computer resources with Reynolds number. The constants of proportionality depend, of course, on the particular flow, numerical scheme, and precision used.

Empirically, it is found that computer speed doubles roughly every 18 months, an observation known as Moore's law and which provides a remarkably good description of the long-term trend over, say, the last forty years. If this continues to hold true in the future, it indicates a factor of ten increase in attainable Reynolds number about every fifteen years. Thus, one can foresee steady, but slow, progress towards higher Reynolds number flows with homogeneous turbulence in the vanguard, owing to the greater efficiency of the associated numerical schemes, and complex geometries attaining a given Reynolds number later. Bearing in mind the wide gap between currently tractable Reynolds numbers and, say, the flow in the wake of an automobile, such cases are not likely to be susceptible to DNS in the near future. Furthermore, here we are talking of simulations taking significant time on the fastest available machines, rather than standard PCs, so we are even further from the day when DNS might be used in routine industrial design work for such flows, rather than as a research tool.

As has been emphasized by Moin and Mahesh (1998, section 2.5), if one is only interested in calculating properties of the flow that are determined by the large scales, such as the mean velocity or turbulent velocity moments $\overline{u_i u_j}$, it may not be necessary to attain the Reynolds numbers of the real flow. For example, in the case of homogeneous turbulence, once the Reynolds number is high enough that there is a clear distinction between the large and dissipative scales, further increases in Reynolds number are thought to modify the Kolmogorov scale, but not the large-scale properties, which approach limiting values as Re_L is raised. More general flows may contain regions in which the correct treatment of the small scales is vital to a description of the overall dynamics, for instance near walls. Thus, one is led to consider a continuously varying viscosity, taking its true value in regions where accurate resolution of the smallest scales is required, but having larger values in the bulk of the flow. Such an approach may significantly reduce the required computer resources, concentrating them where needed, and is similar in philosophy to large-eddy simulation, which forms the subject of the next section.[6] On the other hand, if one is interested in the small scales of the flow, say, to study their intermittency properties, the viscosity must obviously be small enough that it does not affect those scales unduly and the numerical scheme should resolve them. The question of how high a Reynolds number is attainable by simulation thus becomes one of which quantities one is interested in, of the use of numerical methods that are well-adapted to the flow, and of the computer resources one is willing and able to devote to the problem.

[6] So as not to muddy the waters, it should be made clear that the name DNS is usually reserved for simulations which resolve all scales of the real flow.

8.2 Large-Eddy Simulation (LES)

As discussed above, DNS, in which, by definition, all scales of the flow are properly resolved, is limited in Reynolds number by available computer power. As its name suggests, large-eddy simulation has the less ambitious goal of describing the larger scales of the flow by numerical simulation. A numerical representation of the flow and of the Navier–Stokes equations is used, as in DNS, but the Reynolds number is such that the spatial resolution of the scheme is insufficient to describe the smallest scales, at least in some parts of the flow. Thus, LES generates an approximation to the real flow in which scales below a certain size are missing. However, all scales of turbulence are dynamically significant and so the lack of scales below a certain size must somehow be corrected for. This correction is applied via additional terms in the equations of motion, known as subgrid terms, which only come into play at the smaller of the scales resolved by the LES. That is, the subgrid terms are not supposed to effect the larger scales of the flow, which are handled as in DNS, but are there to improve the description of the smaller scales. The art of LES lies in the appropriate choice of subgrid terms, matched to the particular flow and numerical scheme used, a process known as subgrid modeling. Such an approach is obviously approximate, and intrinsically incapable of yielding information about the smallest scales of the real flow, but allows higher Reynolds numbers to be achieved than DNS at a given computational cost. Given the lack of small scales, LES can only be used to calculate flow statistics which are essentially determined by the larger scales, such as the mean velocity and second-order velocity moments, but these are often what is wanted in practice. The reader is referred to Galperin and Orszag (1993), Ferziger and Peric (1996), and Lesieur and Métais (1996) for further reading material, in addition to the detailed references given in the course of this section.

The turbulent energy cascade generates smaller and smaller scales and is terminated by viscosity in a real flow. If a numerical simulation is carried out (without subgrid terms) for which the spatial resolution is insufficient to discern the dissipative scales, the cascade transfers energy from the large scales down to those of size comparable with the resolution of the scheme. The latter are subject to severe discretization errors and are thus not treated correctly by the simulation. The result depends on the details of the particular numerical scheme employed, that is, discretization errors come to the fore whose properties differ from scheme to scheme. For instance, the truncated Fourier-spectral scheme discussed in Section 8.1, which is based on the evolution equations (8.7), conserves energy in the absence of viscosity. In that case, energy arriving at the smallest resolved scales via the cascade piles up there.[7] Of course, the viscosity is not really zero, but if the cutoff wavenumber lies well below the dissipative range, which is the grossly underresolved case we are interested in here, viscous dissipation is insufficient to balance the energy flux via the cascade, and hence reach statistical equilibrium, until the small scales have accumulated considerably more energy than they ought to. Other schemes may behave differently, with discretization errors yielding sufficient effective dissipation that, at equilibrium, the energy in the smallest scales is less than in the real flow or, at

[7] The end result of a truncated Fourier-spectral scheme without viscosity is probably equipartition of energy among Fourier modes (see Lesieur (G 1990), chapter 10), which places most of the energy at the smallest resolved scales.

the other extreme, produce explosive growth in the small scales, that is, numerical instability. Presuming equilibration rather than instability, Figure 8.1 sketches developed spectra (here we have in mind homogeneous, isotropic turbulence without mean flow and a scheme of spatial resolution of $O(k_c^{-1})$) resulting from a resolved DNS calculation (Figure 8.1a) and three underresolved ones for which the numerical scheme is (b) under-dissipative, (c) over-dissipative, and (d) "just right." Since raw DNS schemes do not usually yield case (d), the objective of LES is to approach that case by the addition of subgrid terms to the governing equations. Viewed in this light, subgrid terms should correct the natural behavior of the underlying numerical scheme, yielding an overall LES scheme which dissipates the energy flux from the cascade in the right way and at the right rate that statistical equilibrium leads to a realistic small-scale spectrum. Notice that, since the subgrid model is a correction to a given numerical scheme, in general one would expect it to depend on the particular scheme used, as well as on the physical properties of the small scales of turbulence. It may not always be possible to devise a subgrid model with the desired properties. For instance, the case shown in Figure 8.1c, in which the raw numerical scheme is already too dissipative, is probably a bad candidate for LES because the subgrid terms would have to inject energy at the smallest resolved scales, a recipe for potential numerical instabilities. This indicates that the choice of numerical scheme, to which the subgrid model is applied, can also be important.

Most current LES schemes are based on an eddy-viscosity model for the subgrid terms, that is, (8.1) is modified to yield the LES equation

$$\frac{\partial U_i}{\partial t} + U_j \frac{\partial U_i}{\partial x_j} = -\frac{1}{\rho}\frac{\partial P}{\partial x_i} + \frac{\partial}{\partial x_j}\left((\nu + \nu_s)\left(\frac{\partial U_i}{\partial x_j} + \frac{\partial U_j}{\partial x_i}\right)\right) \tag{8.10}$$

while (8.2) remains unchanged. Here, U_i and P denote the LES approximation to the flow, which is missing the smaller scales, and ν_s is the subgrid viscosity, which in general can vary with both $\mathbf{x}$ and t and adapts to the flow according to the subgrid model employed within the class considered here. As discussed in Section 4.1 (see "The Eddy-Viscosity Approximation and One-Point Modeling"), the idea of an eddy viscosity to express the effects of turbulence on the *mean flow* has a long history, forming, for instance, the basis of the k–ε model of turbulence examined in Section 8.3. However, the subgrid viscosity, ν_s, concerns, not the entire influence of turbulence on the mean flow, as in traditional eddy-viscosity models, but corrections to the behavior of the small scales in *individual realizations*. The subgrid viscosity is small compared with

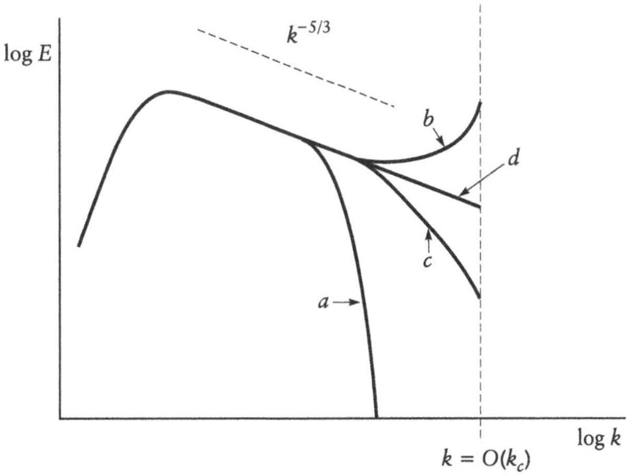

Figure 8.1. Sketch of spectra resulting from possible numerical simulations of the Navier–Stokes equations with spatial resolution of $O(k_c^{-1})$. (a) Properly resolved DNS; (b), (c), and (d) illustrate under-resolved simulations having: (b) too little, (c) too much numerical energy dissipation, and (d) ideal LES.

the eddy viscosity used in k–ε, so its effects are indeed confined to the small scales, leaving larger ones free to be simulated by the numerics. Addition of the subgrid viscosity to form the total LES viscosity $\nu + \nu_s$ in (8.10) can be roughly thought of as decreasing the effective Reynolds number of the flow so that the smallest scales of the LES field, U_i, P, are comparable in size or larger than the resolution of the numerical scheme employed to solve the LES equations (8.10), (8.2). Thus, the value used for ν_s should adapt itself to the flow and numerical resolution. If the resolution of the scheme is insufficient for the dissipative scales of the real flow, ν_s ought to be larger than ν, dominating the real viscosity in the grossly underresolved cases one usually has in mind for LES and becoming comparable to ν when the resolution approaches the smallest scales of the flow. On the other hand, if the dissipative scales are well-resolved, the subgrid viscosity is unnecessary and ν_s should be small compared with ν so that LES reduces to DNS. For instance, in wall-bounded flows, although the turbulent Reynolds number may be high away from walls, and the resolution insufficient to resolve the dissipative range, indicating a need for LES there, the turbulent Reynolds number decreases near walls, where the scheme may resolve the smallest scales of the flow and become a DNS. The distinction between LES and DNS is thus blurred near the walls.

It is important to recognize that it is not in fact (8.10) with (8.2), but *a discretized version of these equations that are solved numerically*. This distinction can be very important because, in underresolved regions, the subgrid viscosity is often chosen such that the LES field contains scales comparable with the resolution of the scheme, in which case discretization effects are highly significant and vary with the scheme used. In that case, the model employed to adapt ν_s to the small-scale properties of the flow should take into account the properties of the particular numerical scheme. It is, of course, possible to choose ν_s sufficiently large that the LES field only contains scales much larger than the numerical resolution, in which case the effects of discretization are small. This has the advantage that the subgrid model no longer need depend on the numerical scheme, simplifying the modeling, but the full resolution potential of the scheme is not then exploited. If one is willing to discard numerical resolution in this way, the precise choice of ν_s probably does not matter all that much, provided the resulting scheme is sufficiently dissipative that it avoids numerical instability and excessive build-up of energy at the smallest resolved scales, but not so dissipative that it affects scales that are too large: one simply discards the smaller scales, accepting that they are inaccurate. To quote Kraichnan (1976): "The need for a faithful representation of subgrid-scale effects on resolved scales may be strong only when computing economy does not permit the luxury of a throw-away inaccurate octave of resolved scales." However, in our view, the goal of LES should be to reproduce the behavior of the small scales down to as near as possible to the numerical resolution, aiming for a spectrum like Figure 8.1d, rather than Figure 8.1c or 8.1a. This implies tackling the awkward problem of devising subgrid models matched to particular numerical schemes, a problem which, in our opinion, has not been given enough attention to date.

The determination of an appropriate distribution, $\nu_s(\mathbf{x}, t)$, of subgrid viscosity for a given scheme and flow is a difficult problem. To examine the question in more detail, we specialize to the case of homogeneous, isotropic, developed turbulence without mean flow, using the Fourier-spectral scheme (8.3), (8.7) with truncation of the Fourier modes. For homogeneous turbulence, whose statistics are independent of position, it is natural to choose a subgrid viscosity, $\nu_s(t)$, dependent only on time, leading to (8.7) with ν replaced by $\nu + \nu_s(t)$. Let us further assume that Re_L is sufficiently large that there is a well-defined inertial range and that the spectral truncation occurs within that range, making the scheme grossly underresolved. Since viscosity is negligible in the inertial range, we can forget the real viscosity for all resolved scales, but must allow for the subgrid viscosity, $\nu_s(t)$, which acts on the smallest resolved scales, mimicking their loss of energy to subgrid scales via the cascade. The upshot is that ν is replaced by $\nu_s(t)$ in (8.7).

If we were to use the spectral truncation described in Section 8.1, setting the a_i to zero outside a rectangular box in $\mathbf{k}$-space, the smallest resolved scales would be anisotropic, reflecting anisotropy of the truncation, even though the larger ones are not because we are supposing isotropic turbulence. In that case, one ought really to use an anisotropic subgrid model, rather than the isotropic one which results from a simple subgrid viscosity. This exemplifies the need to adapt the subgrid model to the numerical scheme used, compensating for its unrealistic features so that the smallest resolved scales have reasonable properties. In any case, here we adopt a different spectral truncation, better suited to isotropic turbulence, in which wavenumbers $|\mathbf{k}| > k_c$ are dropped from (8.3) and (8.7), where k_c is the inertial-range cutoff wavenumber of the scheme.

In the absence of viscosity, the truncated version of (8.7) conserves energy, as we noted earlier. Thus, the energy cascade would lead to a pile-up of energy at the smallest resolved scales, as sketched in Figure 8.1b. The subgrid viscosity is introduced to dissipate this energy appropriately, hopefully leading to the spectrum shown in Figure 8.1d. The energy equation resulting from (8.7), with ν replaced by $\nu_s(t)$, has the form

$$\frac{d}{dt}\left[\sum_{\mathbf{k}} a_i a_i^*\right] = -2\nu_s \sum_{\mathbf{k}} k^2 a_i a_i^* \tag{8.11}$$

of which the right-hand side represents dissipation by the subgrid viscosity. Taking the average of (8.11) and assuming ν_s to be the same in all flow realizations, which places an important additional constraint on our choice of subgrid model, we obtain

$$\frac{d}{dt}\left[\sum_{\mathbf{k}} \overline{a_i a_i^*}\right] = -2\nu_s \sum_{\mathbf{k}} k^2 \overline{a_i a_i^*} \tag{8.12}$$

describing the decay of the averaged total energy.

As discussed in Section 8.1, correct treatment of the large scales of turbulence requires that the spatial periods, $O(\ell)$, of the Fourier representation are much larger than the correlation length. As we also saw in Section 8.1, this condition makes the spectral discretization, $\Delta k = O(\ell^{-1})$, small compared to the width of the energy-containing peak, $O(L^{-1})$, of the spectrum and hence the latter is well resolved. Thus, using (8.9) to express $\overline{a_i a_i^*}$ in terms of the spectral tensor, we can approximate the sums in (8.12) as integrals, replacing $\sum_{\mathbf{k}}$ by

$$\frac{\ell_1 \ell_2 \ell_3}{8\pi^3} \int d^3\mathbf{k}$$

Isotropy can then be employed to write $\Phi_{ii} = E(k,t)/(2\pi k^2)$ (recall (6.103)), leading to

$$\frac{d}{dt}\left[\int_0^{k_c} Edk\right] = -2\nu_s \int_0^{k_c} k^2 Edk \tag{8.13}$$

where $E(k,t)$ is the energy spectrum arising from the LES calculation. This result expresses, in an averaged sense, the fact that the energy of the resolved scales, $k < k_c$, decreases due to dissipation by the subgrid viscosity acting on those scales, which is intended to mimic energy transfer to smaller scales in the real flow.

Notice that, in deriving (8.13), the subgrid model has been made rather specific. Firstly, a subgrid viscosity has been used, and secondly ν_s has been taken both independent of **x** and of the particular flow realization. The choice of subgrid model is ultimately up to us, although it may perform better or worse at its assigned task of yielding reasonably realistic behavior at the smallest resolved scales depending on what is built into the model, the flow considered, and numerical scheme used. The class of subgrid models considered here is perhaps the simplest, but is far from being the only possible one, even for homogeneous, isotropic flows using Fourier-spectral schemes with isotropic truncation.

Since k_c lies well above the energy-containing range, the integral on the left of (8.13) should be a close approximation to the turbulent energy of the real flow, whose time derivative is (minus) the energy flux delivered by the large scales into the inertial range. This should equal $-\bar{\varepsilon}$ for developed turbulence, where $\bar{\varepsilon}$ is the average dissipation rate of the real turbulent flow. Thus (8.13) yields

$$2\nu_s \int_0^{k_c} k^2 Edk = \bar{\varepsilon} \tag{8.14}$$

expressing developed equilibrium of the subgrid viscous dissipation, which is dominated by the smallest resolved scales, $k = O(k_c)$, with the energy flux arriving at such scales via the cascade. For a given large-scale flow, depending on ν_s, different equilibrium LES spectral forms $E(k)$ are obtained for $k = O(k_c)$, each of which satisfies (8.14), and our aim is to adjust ν_s to obtain a spectrum like that shown in Figure 8.1d. According to Kolmogorov's theory, the inertial-range spectrum of the real flow is $E = C\bar{\varepsilon}^{2/3}k^{-5/3}$, which one would expect to be an automatic consequence of the simulation for inertial-range wavenumbers below $k = O(k_c)$. We suppose that a reasonable approximation to the Kolmogorov spectrum in $k = O(k_c)$ is possible by appropriate choice of ν_s and attempt to determine the corresponding ν_s using (8.14). Thus, the desired spectrum, $E = C\bar{\varepsilon}^{2/3}k^{-5/3}$, is introduced into the left-hand side of (8.14), leading to

$$\nu_s = \frac{2\bar{\varepsilon}^{1/3}}{3Ck_c^{4/3}} \tag{8.15}$$

as the subgrid viscosity which produces dissipation at the correct rate to mop up the energy supply, $\bar{\varepsilon}$, from the large scales with the desired energy spectrum, $E = C\bar{\varepsilon}^{2/3}k^{-5/3}$. One may also rewrite (8.15) using $E(k_c) = C\bar{\varepsilon}^{2/3}k_c^{-5/3}$ to express $\bar{\varepsilon}$, giving

$$\nu_s = \frac{2}{3C^{3/2}} \left[\frac{E(k_c, t)}{k_c} \right]^{1/2} \tag{8.16}$$

which is the expression for the subgrid viscosity proposed by Leslie and Quarini (1979).

Equation (8.16) can be used to determine the subgrid viscosity for an LES computation based on the isotropically truncated version of (8.7) with ν replaced by $\nu + \nu_s$. The quantity $E(k_c, t)$ needs to be calculated at each time step and, in principle, involves an average over an ensemble of realizations and use of

$$E = \frac{k^2 \ell_1 \ell_2 \ell_3}{4\pi^2} \overline{a_i(\mathbf{k}, t) a_i^*(\mathbf{k}, t)} \tag{8.17}$$

which follows from (8.9) and the isotropic identity $E = 2\pi k^2 \Phi_{ii}$. Thus, one might conduct a large number of LES calculations in parallel, each with different initial conditions, and calculate the ensemble average in (8.17). However, for the isotropic flows considered here, $E(k, t)$ is usually estimated for single realizations by calculating the average over a spherical shell in **k**-space. In any case, an LES calculation based on the given subgrid model consists of solving the isotropically truncated version of (8.7) in discrete time, with ν replaced by ν_s and suitable random initial conditions for the a_i. From a computational point of view, this is much the same as a Fourier-spectral DNS and the choice of time step, initial conditions, and the use of FFTs to compute the convolutions are essentially as discussed in Section 8.1. In particular, the time step should be small compared with the shortest time scale associated with the resolved spatial scales, which is determined by the size of the smallest resolved scales and the large-scale flow velocities. Having performed a simulation to obtain the LES velocity field, U_i, one can study its statistical properties, which should approximate those of the real flow in the case of statistics, such as $\overline{u'^2}$, which are determined by the large scales. For instance, one may compute $E(k, t)$ and integrate over wavenumber to obtain $3\overline{u'^2}/2$, perhaps extrapolating the LES spectrum via assumed $k^{-5/3}$ and dissipative-range spectral forms to improve the accuracy of the result.

If, during the simulation, the spectral energy, $E(k_c, t)$, at the cutoff wavenumber were to rise, (8.16) would increase the subgrid viscosity, and hence the subgrid dissipation at the smallest resolved scales, bringing the spectrum back down. Hopefully, the result is an equilibrium that is close to the desired Kolmogorov spectrum in the inertial range of resolved wavenumbers. However, there is no a priori guarantee that this will be the case for $k = O(k_c)$ because the subgrid model we have adopted might not be compatible with a Kolmogorov spectrum there. In practice, it is found that the results of LES using (8.16) are reasonably good, but the Kolmogorov spectrum is never precisely achieved all the way up to the cutoff wavenumber, calling into question the analysis leading to (8.16), which assumed $E = C\bar{\varepsilon}^{2/3}k^{-5/3}$ for $k = O(k_c)$. Since the Kolmogorov spectrum is thus not exactly attainable for the given subgrid model (and numerical scheme), one can envisage modifying the numerical constant in (8.16), while maintaining the $(E(k_c, t)/k_c)^{1/2}$ proportionality of ν_s. Changing the constant alters the detailed form of the spectrum in $k = O(k_c)$ and can be used to optimize the spectrum there according to personal taste, in particular by comparison with the expected $k^{-5/3}$ spectrum of the real flow.

For a given value of the constant, Kolmogorov universality indicates that the form of the inertial-range spectrum as a function of k/k_c should not depend on the flow considered or the choice of inertial-range cutoff wavenumber, k_c.

Kraichnan (1976) considered a more general subgrid model than that used above, again for the case of homogeneous, isotropic, high-Reynolds-number turbulence and an isotropically truncated Fourier-spectral scheme. As before, the truncated version of (8.7) is modified by replacing ν by ν_s, where ν_s does not vary from realization to realization of the flow, but is now a function of k as well as t, providing more degrees of freedom in the choice of ν_s to optimize the spectrum in $k = O(k_c)$. The dependence of ν_s on k means that the subgrid model is more complicated than a simple eddy viscosity, that is, the LES equation is no longer (8.10) for Kraichnan's model, but is instead derived from (8.7) with $\nu_s(k, t)$ instead of ν. The function $\nu_s(k, t)$ is determined using a triadic spectral closure, such as EDQNM, to express the average energy transfer, $T(k, t)$ in (6.128), to wavenumber k from all other wavenumbers as an integral over nonlinearly interacting wavenumber triads such that $\mathbf{k} + \mathbf{k}' + \mathbf{k}'' = 0$. Taking $k < k_c$, the integral can be split into two components as $T = T_< + T_>$, where $T_<$ consists of the integral over triads for which $|\mathbf{k}'| < k_c$ and $|\mathbf{k}''| < k_c$, and $T_>$ is the remainder, representing transfers involving wavenumbers above k_c, which are not resolved by the scheme. It is supposed that $T_<$ is already accounted for in the LES via the convolution term in the truncated version of (8.7), while $T_>$ is not and should be mimicked by the subgrid energy dissipation, $-2\nu_s k^2 E(k, t)$. Equating $T_>$ and $-2\nu_s k^2 E(k, t)$, one obtains

$$\nu_s = -\frac{T_>(k, t)}{2k^2 E(k, t)} \tag{8.18}$$

for $\nu_s(k, t)$. Using the Kolmogorov spectrum, $E = C\bar{\varepsilon}^{2/3} k^{-5/3}$, $T_>(k, t)$ can be evaluated for any given spectral closure, and hence $\nu_s(k, t)$ calculated from (8.18). Expressing $\bar{\varepsilon}$ in terms of $E(k_c, t)$, the final result has the form

$$\nu_s = C^{-3/2} \left[\frac{E(k_c, t)}{k_c} \right]^{1/2} f\left(\frac{k}{k_c} \right) \tag{8.19}$$

for $k < k_c$, where f is a purely numerical function dependent on the closure used and which satisfies

$$\int_0^1 x^{1/3} f(x) dx = \frac{1}{2} \tag{8.20}$$

in order that the overall energy balance, (8.14), holds with $\nu_s(k, t)$ taken inside the integral, since it is now a function of k. Note that, if one insisted that f be constant, that is, ν_s be independent of k, as was the case with the earlier subgrid model, (8.20) yields $f = 2/3$ and we recover (8.16) from (8.19). However, the function $f(k/k_c)$ determined from spectral models is not constant and instead is found to increase from $k = 0$ to $k = k_c$, indicating that the $\nu_s(k, t)$ should be allowed to rise and the subgrid model become more dissipative as the cutoff wavenumber is approached, in order that the detailed spectral energy transfers across wavenumber k_c be respected.

Although LES calculations based on (8.19) probably yield a more realistic description of the smallest resolved scales than (8.16), it is found that they do not give a perfect Kolmogorov spectrum up to $k = k_c$. This should not be too surprising since, among other things, it is implicitly supposed in deriving (8.18) that the spectral closure, which is designed as an approximation to real turbulence, describes transfers among the resolved wavenumbers of the LES resulting from (8.19). This is a rather strong assumption, at best approximately true. In practice, it is far from clear that the more complicated subgrid model resulting from (8.19) is worth the additional effort compared with the simple subgrid viscosity model using (8.16) and an appropriately chosen constant.

As discussed in Section 8.1, the initial conditions of homogeneous DNS computations are usually chosen to have statistically independent Fourier components with randomly chosen phases and the same is true of LES, leading to an approximately Gaussian initial velocity field. The resulting turbulent flow is similar to grid turbulence and does not display gross large-scale intermittency of the type discussed at the beginning of Section 7.5, which can invalidate Kolmogorov universality of the small scales. Both (8.16) and the spectral closures leading to (8.19) are based on such universality, which implies that the resulting subgrid models may have to be modified for more exotic initial conditions. For instance, the constant multiplying $(E(k_c, t)/k_c)^{1/2}$ in (8.16) could well need adjustment depending on the flow considered, reflecting changes in the Kolmogorov constant from flow to flow and perhaps with time for decaying turbulence. Since, as we saw in Section 7.5, the $k^{-5/3}$ inertial-range spectrum survives the loss of universality, one can still employ such a power-law form as the criterion for success of subgrid models.

Until now we have managed to avoid introducing filtering, which is often the starting point of presentations of LES, so let us briefly put that right, even though, in our view, filtering does not ultimately bring very much to the party. Given some flow quantity, such as $U_i(\mathbf{x}, t)$, one may perform Fourier analysis, discard the higher wavenumbers by truncation, as discussed earlier for Fourier-spectral schemes, followed by Fourier synthesis to reconstitute the filtered field, denoted $\hat{U}_i(\mathbf{x}, t)$. Thus, filtering discards the high wavenumbers, representing small scales, leaving only the larger scales of the flow. Such truncation in **k**-space yields one special class of filters, but more general ones can be devised. Truncation is equivalent to multiplying the Fourier transform by a function of **k** that is either 0 or 1, but a general filter involves multiplication by an arbitrary function of **k** that is close to 1 at energy-containing and lower wavenumbers and goes to zero sufficiently rapidly at large k. Since multiplication of Fourier transforms translates to convolution in **x**-space, a filter has the effect of convolving with a fixed function in physical space, for instance

$$\hat{U}_i(\mathbf{x}, t) = \int G(\mathbf{x} - \mathbf{x}')U_i(\mathbf{x}', t)d^3\mathbf{x}' \tag{8.21}$$

represents a general filter applied to the velocity field. The filter function $G(\mathbf{r})$ integrates to 1 and decays to zero outside a range $|\mathbf{r}| = O(\Delta)$, where Δ is referred to as the filter width. Roughly speaking, filtering, as in (8.21), corresponds to taking an average over a region of size Δ around the point considered. This averaging operation removes scales smaller than $O(\Delta)$, but leaves larger ones intact, producing a filtered field which contains only scales of size $O(\Delta)$ and above. One often has in mind that Δ is comparable with the spatial resolution of the numerical scheme which is subsequently used to implement the LES.

Following the standard formalism, the filtered flow, which contains only the larger scales, is treated as the object of study of LES and so one looks for equations describing its evolution. Since filtering commutes with space and time derivatives, applying it to (8.1) and (8.2) gives

$$\frac{\partial \hat{U}_i}{\partial t} + \hat{U}_j \frac{\partial \hat{U}_i}{\partial x_j} = -\frac{1}{\rho}\frac{\partial \hat{P}}{\partial x_i} + \frac{\partial \tau_{ij}}{\partial x_j} + \nu \frac{\partial^2 \hat{U}_i}{\partial x_j \partial x_j} \tag{8.22}$$

and

$$\frac{\partial \hat{U}_i}{\partial x_i} = 0 \tag{8.23}$$

where

$$\tau_{ij} = \hat{U}_i \hat{U}_j - \widehat{U_i U_j} \tag{8.24}$$

is known as the subgrid stress tensor and $\widehat{U_i U_j}$ denotes the result of filtering applied to $U_i U_j$. Equation (8.23) tells us that the filtered flow satisfies the usual incompressibility condition, while (8.22) involves an additional subgrid stress term that implies lack of closure of the filtered equations, since $\widehat{U_i U_j}$ and hence τ_{ij}, is not exactly expressible in terms of $\hat{U}_i$. Note that the subgrid stresses and filtering play roles in equations (8.22) and (8.23) similar to those of the Reynolds stresses and averaging in the mean-flow ones, (4.9) and (4.5).

A number of subgrid closures, that is, approximations for τ_{ij} in terms of the filtered flow variables, have been proposed, of which the most commonly used are of the eddy-viscosity type

$$\tau_{ij} = \frac{1}{3}\tau_{kk}\delta_{ij} + \nu_s \left(\frac{\partial \hat{U}_i}{\partial x_j} + \frac{\partial \hat{U}_j}{\partial x_i} \right) \tag{8.25}$$

The result of introducing this closure into (8.22) is (8.10) with $\hat{U}_i$ in place of U_i and $\hat{P} - \rho\tau_{kk}/3$ instead of P, i.e. the LES equations with a subgrid viscosity are

$$\frac{\partial \hat{U}_i}{\partial t} + \hat{U}_j \frac{\partial \hat{U}_i}{\partial x_j} = -\frac{\partial \Gamma}{\partial x_i} + \frac{\partial}{\partial x_j}\left((\nu + \nu_s) \left(\frac{\partial \hat{U}_i}{\partial x_j} + \frac{\partial \hat{U}_j}{\partial x_i} \right) \right) \tag{8.26}$$

and (8.23), where $\Gamma = \hat{P}/\rho - \tau_{kk}/3$. Note that, if one forgets that $\hat{U}_i$ is supposed to represent the filtered velocity field, the filter plays absolutely no role in the final LES equations, (8.23) and (8.26), which can be understood in terms of adding a subgrid viscosity term to the Navier–Stokes equations, without introducing the notion of filtering. Nor does the filtering formalism help with the central problem of LES, namely the construction of subgrid models, which amounts to an appropriate choice of expression for ν_s in terms of the LES field if an eddy-viscosity model is used.

Whether filtering is introduced or not, the LES equations with an eddy-viscosity model, that is, (8.2) and (8.10), are solved numerically for the time evolution of the LES field. This involves discretization in space and time, which introduces differences between the differential equations and their numerical equivalents. If the LES field contains scales comparable with the resolution of the numerical scheme, as is generally the case in practice, these differences are important and one is not really solving the LES differential equations, but a numerical version of them. This does not mean that the resulting simulation is somehow a poor approximation of some ideal LES based on (8.10) with (8.2), rather that LES is not really expressed by the differential equations, but instead by their numerical implementation. Indeed, as we will see in the following, rather than being an undesirable artifact of the numerical scheme, numerical "errors" due to discretization are, in fact, vital to a successful LES description of the smallest resolved scales.

In the case of an eddy-viscosity subgrid model and *in the absence of discretization*, one expects the size of the smallest resolved scales of the nondiscretized LES field resulting from solution of (8.10) with (8.2) to be determined by the dissipative range resulting from the total effective viscosity $\nu + \nu_s$. Supposing high-Reynolds-number, developed turbulence, Kolmogorov's theory suggests that this size will be of $O(\eta_{LES})$, where $\eta_{LES} = (\nu + \nu_s)^{3/4}/\bar{\varepsilon}^{1/4}$ is the "Kolmogorov scale" calculated using the effective viscosity and energy supply rate, $\bar{\varepsilon}$, from the large scales, allowing one to estimate the smallest scales of the developed LES field at high Reynolds number without discretization effects. The real flow has a Kolmogorov scale $\eta_K = \nu^{3/4}/\bar{\varepsilon}^{1/4}$, leading to three length scales in all: the spatial resolution of the numerical scheme, η_K and η_{LES}, with $\eta_K \leq \eta_{LES}$ reflecting the fact that the smallest scales of LES are larger than those of the real flow, owing to the addition of the subgrid viscosity. One can distinguish three cases. Firstly, if the numerical resolution is small compared with η_K, the scheme is capable of resolving the real flow and LES is unnecessary, which is often the case near walls, as noted earlier. Addition of a significant subgrid viscosity is then not only unnecessary, but undesirable, since it modifies the flow predictions. In other words, the problem becomes one of ensuring that the subgrid model switches itself off in regions of the flow where it is not needed. The second case arises when the numerical resolution is large compared with η_{LES}. One is then relying on the discretization errors of the scheme alone at the smallest resolved scales, with the both real and subgrid viscosities producing only small effects on the resolved field. While, depending on the properties of the numerical scheme used, this might lead to acceptable results, the purpose of LES is to correct for deficiencies of under-resolved schemes and it is not being exploited here.

The third possibility, in which the resolution scale lies somewhere between $O(\eta_K)$ and $O(\eta_{LES})$ is the one we usually have in mind for LES, and can be subdivided into two cases. If the resolution is small compared with η_{LES}, the scheme can resolve the LES field well and discretization effects are small. However, scales intermediate between $O(\eta_{LES})$ and the spatial resolution lie above the dissipative range of wavenumbers of the effective viscosity and have energies below those of the real flow. Thus, such scales are not accurately simulated and must be discarded, despite the fact that they lie within the potential for resolution of the scheme. On the other hand, when the numerical resolution is $O(\eta_{LES})$, the scheme is being fully exploited, but discretization effects are significant and should be taken into account in analyzing the properties of LES at the smallest resolved scales, $O(\eta_{LES})$. In this case, there is a delicate interplay between the subgrid model and discretization effects at the smallest scales, the result of which determines the quality of the final LES. Without discretization, the LES spectrum would be as in Figure 8.1c, rather than having the desired form of Figure 8.1d, since k_c lies in the dissipative range of the nondiscretized LES field. Discretization effects may bring the spectral level up to that shown in Figure

8.1d if the subgrid model and scheme are appropriately matched. Thus, as stated above, discretization effects are essential to obtaining a successful LES description of the smallest resolved scales.

We saw earlier that Fourier-spectral LES of high-Reynolds-number, homogeneous, isotropic turbulence, based on (8.7) with suitably chosen ν_s, has good spectral properties up to the cutoff wavenumber. This implies that the LES field contains scales down to the resolution of the scheme and hence significant discretization errors at the smallest resolved scales. Discretization places a spectral boundary at $k = k_c$ and energy tends to pile up there as shown in Figure 8.1b. The subgrid viscosity dissipates this energy, pulling the net spectrum towards Figure 8.1d. This again illustrates the fact that good LES results require discretization effects and involve subtle interactions between the subgrid model and numerical scheme. Note that schemes other than the one considered here might be sufficiently dissipative that they generate a spectrum such as Figure 8.1c even in the absence of subgrid viscosity, in which case one would not expect it to be possible to recover Figure 8.1d using a subgrid viscosity.[8] It is important to choose schemes with appropriate properties, in particular ones which are not too dissipative, and to match the subgrid model to the given scheme and flow.

Although (8.16), with perhaps a different numerical constant, provides a suitable expression for ν_s in the case of homogeneous, isotropic turbulence using an isotropically truncated Fourier-spectral scheme, the question remains as to what might be an appropriate expression for the subgrid viscosity for more general flows and other schemes. Even in the case of anisotropic, homogeneous turbulence without mean flow, for which the standard truncated Fourier-spectral scheme continues to be suitable, it is not obvious how to proceed: What is the equivalent of $E(k_c, t)$ in (8.16) when the distribution of energy in **k**-space is anisotropic? Should one take other properties of the spectral tensor Φ_{ij} than the energy distribution $\Phi_{ii}/2$ into account? Is an isotropic truncation of the spectral scheme the best choice? Would a different subgrid model, for instance one with a tensorial viscosity, be more appropriate, and if so how does one fix its precise form? In practice, one usually assumes, in Kolmogorov style, that the turbulence approaches isotropy at small enough scales and that it is sufficiently close to isotropy by the time the cutoff wavenumber of the scheme is reached that one can treat it as if it were isotropic at the smallest resolved scales. Thus, one may continue to use an isotropically truncated Fourier-spectral scheme and a subgrid viscosity given by (8.16), with a possibly different constant and $E(k_c, t)$ determined by an average of the spectral energy density over spherical shells in **k**-space.

Things become more complicated when the flow is inhomogeneous because ν_s should be allowed to vary with position and spectral analysis is no longer available to define $E(k_c, t)$ in (8.16), except perhaps in an approximate, local sense. Furthermore, Fourier-spectral schemes are not appropriate for LES of inhomogeneous flows, although a Fourier series description may still be used in directions of homogeneity, such as parallel to the walls in the channel flow discussed in Section 8.1. More complex geometries are usually treated by nonspectral methods, such as finite-

[8] One might consider using a negative subgrid viscosity in such cases, but that would probably lead to numerical instability.

difference schemes. Whatever numerical technique is used, a subgrid model adapted to the local flow and scheme is needed, which, in the case of eddy-viscosity models, amounts to an expression for ν_s.

The standard nonspectral expression for ν_s is due to Smagorinsky (1963), who proposed it for meteorological simulations some time before the mainstream turbulence community began to investigate the possibilities of LES. The Smagorinsky subgrid viscosity has the form

$$\nu_s = (C_s\delta)^2 \left\{ \left(\frac{\partial U_i}{\partial x_j} + \frac{\partial U_j}{\partial x_i} \right) \left(\frac{\partial U_i}{\partial x_j} + \frac{\partial U_j}{\partial x_i} \right) \right\}^{1/2} \tag{8.27}$$

where U_i is the LES velocity field and $C_s\delta$ is a length scale that is usually chosen to be of the order of the numerical resolution. As we now show, in the absence of discretization effects and assuming high-Reynolds-number, developed turbulence, (8.27) leads to a nondiscretized LES field whose smallest scales are of order $\max(C_s\delta, \eta_K)$, an estimate for η_{LES}. If U_i were the real velocity field, rather than the LES field governed by (8.2) and (8.10), the quantity inside the curly brackets of (8.27) is related to the viscous energy dissipation via (1.3), which is dominated by the turbulent part, ε, of the dissipation since the Reynolds number is high. One can estimate the mean dissipation rate in the LES flow using (1.3) with $\nu + \nu_s$ in place of ν and taking averages with U_i interpreted as the LES velocity field. Once the turbulence has developed, one expects the average dissipation to balance the mean energy supply rate from the large scales in both the real and LES flows, and so to be the same in the two flows, equal to $\bar{\varepsilon}$. Thus, one estimates the right-hand side of (8.27) as $(C_s\delta)^2(\bar{\varepsilon}/(\nu + \nu_s))^{1/2}$, which equals $(\nu + \nu_s)(C_s\delta/\eta_{LES})^2$ according to the definition, $\eta_{LES} = (\nu + \nu_s)^{3/4}/\bar{\varepsilon}^{1/4}$, of η_{LES}. In consequence, (8.27) yields

$$\frac{\eta_{LES}}{C_s\delta} = O\left(\left(\frac{\nu + \nu_s}{\nu_s} \right)^{1/2} \right) \tag{8.28}$$

If $\nu_s \geq O(\nu)$, the right-hand side of (8.28) is $O(1)$ and we obtain $\eta_{LES} = O(C_s\delta)$. On the other hand, when $\nu_s \ll \nu$, the subgrid viscosity can be neglected and the nondiscretized LES field is essentially the same as the real flow, having smallest scales $\eta_{LES} = O(\eta_K)$ that are larger than $O(C_s\delta)$, according to (8.28). In other words, η_{LES} is of order $\max(C_s\delta, \eta_K)$, as stated above.

Suppose that the length scale $C_s\delta$ in (8.27) is chosen to be of the same order as the numerical resolution. If the resolution is sufficient to describe the real flow, $C_s\delta$ is smaller than $O(\eta_K)$ and, according to the above discussion, $\nu_s \ll \nu$, that is, the subgrid viscosity becomes negligible in well-resolved cases. In contrast, if the flow is underresolved, the subgrid viscosity resulting from (8.27) is significant at the smallest resolved scales and adjusts itself so that $\eta_{LES} = O(C_s\delta)$ is comparable to the resolution of the scheme. This means that discretization effects are important at such scales, including those involved in the derivatives in (8.27) Thus, (8.27) has the general properties one might hope of a subgrid viscosity.

In the underresolved case, although the requirement that $C_s\delta$ be comparable with the resolution places one in the right ballpark, its precise value must still be tuned to optimize the characteristics of the smallest resolved scales and hence obtain a good LES. The length scale δ is usually chosen according to the local numerical resolution of the scheme, while the nominally O(1) parameter C_s is employed for empirical optimization of the results of LES depending on the details of the local flow and numerical scheme. For instance, finite-difference schemes often use a grid whose geometrical form varies from place to place, in addition to changes in the mesh size that are presumably allowed for by δ, and one would expect that this would require letting the value of C_s depend on position to compensate. The absence of spectral analysis for inhomogeneous flows means that the criterion for success of LES is not as clearly defined as for homogeneous ones. Nonetheless, the $k^{-5/3}$ spectral law can still be used if there are directions of homogeneity, or, in more general flows, one might employ locally defined spectra or the $r^{2/3}$ law for mean-squared velocity differences at inertial-range separation r. Note that the $k^{-5/3}$ and $r^{2/3}$ criteria are not identical. For instance, a truncated $k^{-5/3}$ inertial-range spectrum, usually regarded as optimal in the homogeneous, isotropic case with a Fourier-spectral scheme, does not yield $r^{2/3}$ mean-squared velocity differences at separations comparable to or below the spectral truncation, so criteria based on these two measures of success may lead to somewhat different LES schemes.

The analysis of the Smagorinsky subgrid model given above assumes a large turbulent Reynolds number, which will not be the case near walls, for instance. As discussed earlier, LES in general is not well-adapted to such regions and the scheme should be capable of fully resolving the flow in near-wall regions and reduce to DNS there. The problem is one of suppressing the effects of the subgrid model, which, left to itself, may be too diffusive. This is often achieved with the Smagorinsky model by using empirical wall functions for the parameter C_s, involving functions of the scaled wall distance y_+ (defined by (5.197)), which was introduced in Chapter 5 to describe the viscous layer at the wall. These wall functions tend to zero at the boundary, reducing C_s and hence lessening the subgrid viscosity in the near-wall region. Other cases, such as strong density stratification or rotation, have been found in which C_s needs to be modified even when the turbulent Reynolds number is large.

Despite the difficulties associated with determining an appropriate C_s, the Smagorinsky subgrid model has been successfully used to carry out LES calculations in a number of different flows. A technique, known as the dynamic model, has been developed to automatically adapt the value of C_s to the flow, and so avoid at least some of the problems. This involves the application of two filters to the LES velocity field, from the results of which one infers an appropriate C_s. We do not want to describe the model in detail here, but instead refer the reader to Germano et al. (1991) and Ghosal et al. (1995). Overall, the dynamic model appears to do quite well in adapting C_s to the flow. For instance, it automatically reduces its value near walls and in the presence of strong stratification or rotation. However, it is not without its difficulties, generating values for C_s that fluctuate rapidly with position and leading to negative subgrid viscosities, which can destabilize the numerics. For these reasons, the original model is generally modified using one of a variety of averaging methods. Furthermore, in common with other subgrid formulations, the model does not allow for possible variability from one numerical scheme to another, nor across the flow due to, for instance, changes in the form of a finite-difference grid. It would be comforting, at least to the present authors, if sounder theoretical foundations for subgrid modeling were developed, allowing less reliance to be placed on empiricism. However, analysis of the small scales of turbulence, allowing for discretization effects, although a central question of LES, is obviously a very difficult problem and currently the dynamic model appears to be the best general-purpose option. Finally, we should mention that, at least in principle, the approach used in the dynamic model provides a general technique for adaptively determining the parameters of subgrid models. Thus, it is not limited to the Smagorinsky form, (8.27), of eddy viscosity and may even allow the parameterization of non-eddy-viscosity subgrid descriptions.

To summarize the discussion of this section, successful LES allows the simulation of the large scales of flows whose Reynolds number is too high to permit DNS at reasonable computational cost. It amounts to an underresolved DNS together with a subgrid model that corrects the smallest resolved scales by dissipating the energy that would otherwise accumulate at such scales via the cascade. The underlying numerical scheme should not be too dissipative, and, given the subtle interaction between numerical discretization effects and subgrid model, the latter needs to be adapted to both the flow and scheme considered. Subgrid models are usually based on the idea of an eddy viscosity and their theoretical basis is relatively well established for homogeneous, isotropic, developed turbulence at high Reynolds numbers using a truncated Fourier-spectral scheme. Less idealized flows and other numerical schemes present more of a challenge, with significant fundamental questions remaining unanswered. Development of subgrid models is an area of continuing research, but the dynamic model with a Smagorinsky subgrid viscosity appears to give acceptable results in many cases.

8.3 One-Point Statistical Modeling

Although industrial applications of LES are envisaged, like DNS it is currently too computationally expensive to allow routine simulations. More traditional approaches work with the evolution equations for the velocity moments, or their spectral equivalents, but this introduces a closure problem, resolved by introducing additional hypotheses that can, at best, be only approximately true. As we saw in Chapter 6, spectral analysis allows the treatment of multipoint moments in a relatively straightforward manner for homogeneous turbulence. However, if one restricts attention to one-point moments, nonlocality leads to additional closure problems, as discussed in Chapter 4. In going from DNS to LES to multipoint, then one-point, statistical models, the computational cost of the simulations decreases, but the degree of approximation increases. Thus, there is a compromise to be made between the cost and accuracy of simulations.

Given the lengthy description of spectral closures to be found in a second volume of this book, we do not discuss such models here. It is nonetheless important to note that, since they provide a quantitative representation of the different scales of turbulence and the energy transfers between them, multipoint models potentially capture much more of the physics of turbulent flows than the one-point models examined in this section. The restriction of spectral analysis to homogeneous turbulence is a serious drawback of classical spectral models for most practical applications, but some work is in progress in an attempt to extend them to inhomogeneous flows. For example, Parpais and Bertoglio (1996) have developed an inhomogeneous model based on isotropic spectral calculations[9] that shows promise and that is intermediate in approximation and cost between one-point and full two-point statistical modeling.

One-point moments and their evolution equations formed one of the main topics of Chapter 4 and are the basis of one-point statistical modeling. Following Reynolds decomposition into mean and fluctuating parts, the mean-flow equations are the averaged momentum equation

$$\frac{\partial \overline{U}_i}{\partial t} + \overline{U}_j \frac{\partial \overline{U}_i}{\partial x_j} = -\frac{1}{\rho}\frac{\partial \overline{P}}{\partial x_i} + \nu \frac{\partial^2 \overline{U}_i}{\partial x_j \partial x_j} - \frac{\partial \overline{u_i u_j}}{\partial x_j} \tag{8.29}$$

and

$$\frac{\partial \overline{U}_i}{\partial x_i} = 0 \tag{8.30}$$

expressing incompressibility of the mean flow. The Reynolds tensor, $-\overline{u_i u_j}$, represents the mean flux of momentum due to turbulent fluctuations and its divergence appears as volumetric forcing of the mean flow in (8.29).

In the k–ε model, which is undoubtedly the most widely used technique for the practical simulation of turbulent flows and forms the principal topic of this section, an eddy-viscosity approximation is used to write

$$-\overline{u_i u_j} = -\frac{2}{3} k \delta_{ij} + \nu_T \left(\frac{\partial \overline{U}_i}{\partial x_j} + \frac{\partial \overline{U}_j}{\partial x_i} \right) \tag{8.31}$$

where $k = \overline{u_i u_i}/2$ is the mean turbulent kinetic energy per unit mass, not to be confused with wavenumber, and $\nu_T(\mathbf{x}, t)$ is the eddy viscosity. The eddy-viscosity approximation has a long history dating back to the earliest days of turbulence research. It was discussed in Section 4.1 (see "The Eddy-Viscosity Approximation and One-Point Modeling"), to which the reader should refer, and was used in Chapter 5 to model the behavior of jets and wakes. It is also worth observing the similarity of (8.29)–(8.31) with the filtered equations, (8.22) and (8.23), and subgrid eddy-viscosity model, (8.25), although their physical interpretation is rather different, since the eddy viscosity introduced here is supposed to represent the influence of turbulence on the mean flow, whereas the subgrid viscosity expresses that of subgrid scales on resolved scales in particular flow realizations.

Having adopted (8.31) for the Reynolds tensor, one still needs to have a definite expression for $\nu_T(\mathbf{x}, t)$ to carry out numerical simulation using (8.29) with (8.30). One possibility is to suppose that ν_T can be directly expressed in terms of derivatives of the mean velocity field, as in classical mixing-length theories (see, e.g., Hinze (G 1975); Wilcox (1993); Schiestel (1993)). Using such an expression for ν_T leads directly to closure of the mean-flow equations. The next step up in sophistication is to introduce an evolution equation for some quantity, for instance, $k = \overline{u_i u_i}/2$, characterizing the turbulence, with ν_T expressed in terms of that quantity and the mean velocity field

[9] Isotropic spectral calculations are now computationally cheap, but anisotropic ones still require considerable computer resources.

(see, e.g., Wilcox (1993); Schiestel (1993)). The mean-flow equations are thus supplemented by an extra evolution equation, which models the turbulence, hence the name one-equation model often used to describe such an approach. By analogy, zero-equation modeling refers to the mixing-length type formulations noted above, which do not allow for additional degrees of freedom associated with the turbulence and hence have no need for extra evolution equations. Zero- and one-equation models are still used for some applications, but they have been largely superseded for general-purpose flow simulation by the more sophisticated two-equation models, of which k–ε is the commonest example.[10] Although less popular than k–ε, we should also mention the k–ω two-equation model, which is similar in philosophy, but may lead to superior results in some flows (see, e.g., Wilcox (1993), sections 4.3.1 and 4.11, among others, for further information).

As is perhaps implied by its name, the k–ε model describes turbulence using two parameters, namely k and ε, both of which have model evolution equations. Here, the reader should note that, in this section alone, and in conformity with conventional usage in the context of one-point modeling, ε denotes the *average* turbulent dissipation rate, despite the absence of an overbar. Thus, the model uses the variables $\overline{U_i}$, k, and ε to represent the total flow field and consists of evolution equations for each of these variables. In the case of homogeneous, isotropic turbulence without mean flow, it is evident from equations (7.4) and (7.5), which give k and ε in terms of integrals over the spectrum[11] $E(k, t)$, that one could obtain the same values of k and ε with many quite different spectra, which would no doubt evolve in different ways. However, when applying the k–ε model, one has in mind that the flow is of high turbulent Reynolds number and that the turbulence has been allowed to develop for sufficient time that, in the case of homogeneous, isotropic turbulence, it has become self-similar in the sense of Section 7.4. Waiting for large-scale similarity to appear makes it plausible that the large scales of turbulence can be described using just the two parameters k and ε, at least in the homogeneous, isotropic case.

The quantity ε represents the average viscous dissipation, which occurs at the smallest scales. However, for the high-Reynolds-number, developed turbulence implicitly assumed by the model, dissipation is determined by the rate of energy supply to the cascade by the large scales, which can be estimated in the usual fashion as $\varepsilon = O(u'^3/L) = O(k^{3/2}/L)$, leading to $L = O(k^{3/2}/\varepsilon)$ as the order of magnitude of the correlation scale in terms of the variables k and ε. Thus, the quantity $k^{3/2}/\varepsilon$ may be interpreted as representing the size of the large scales of turbulence, while $k = \overline{u_i u_i}/2$ measures their intensity. For homogeneous, isotropic turbulence, one can imagine a plot of the self-similar energy spectrum, $E(k)$, using the wavenumber scaling $\varepsilon/k^{3/2}$ and height scaling $k^{5/2}/\varepsilon$, parameterizing the large-scale part of the spectrum in terms of k and ε. In this way, k and ε can be considered as representing properties of the large scales, despite the small-scale origin of the dissipation ε. If one really wanted to describe the details of the dissipative scales, a third parameter, such as a turbulent Reynolds number, would be needed, because the position of the dissipative range of wavenumbers is not otherwise fixed in relation to the large-scale spectral peak, whose location and height already consume the two parameters k and ε. Thus, despite appearances,

[10] See, e.g., Wilcox (1993) and Schiestel (1993) for further reading on two-equation models, including k–ε.

[11] Recall the distinction between k representing mean kinetic energy and wavenumber.

the k–ε model of turbulence should be interpreted as a description of the large scales. More generally, spectral analysis of anisotropic turbulence introduces a tensor Φ_{ij}, which is a function of the direction, as well as the magnitude, of the wavenumber, rather than the single scalar quantity $E(k)$, suggesting that more than the two parameters k and ε would be needed to parameterize the large-scale spectral properties. This points the way towards models having more turbulence parameters (for instance the R_{ij}–ε models discussed briefly later). The relationships between spectral analysis and one-point turbulence modeling are explored in more detail in Cambon and Scott (1999).

Assuming the turbulence is described by k and ε, the stronger hypothesis is made that the eddy viscosity in (8.31) is determined solely by the values of k and ε *at the same point and time*. This is one of many closure hypotheses (including (8.31) itself) making up the model and, if one accepts it, leads to

$$\nu_T = C_\mu \frac{k^2}{\varepsilon} \tag{8.32}$$

by dimensional analysis, where C_μ is the first of a number of constant numerical parameters of the model. Equations (8.29)–(8.32) describe the mean flow, but (8.32) contains the unknown quantities k and ε, representing the turbulence. The other components of the model are evolution equations for k and ε, which we first examine in the special case of homogeneous turbulence.

The turbulent kinetic energy, k, satisfies the evolution equation (4.37), which we write in the symbolic form

$$\frac{dk}{dt} = \Pi - \varepsilon \tag{8.33}$$

where

$$\Pi = -\overline{u_i u_j}\frac{\partial \overline{U}_i}{\partial x_j} \tag{8.34}$$

is the turbulent energy production and is approximated using the eddy-viscosity hypothesis (8.31) as

$$\Pi = \frac{1}{2}\nu_T\left(\frac{\partial \overline{U}_i}{\partial x_j} + \frac{\partial \overline{U}_j}{\partial x_i}\right)\left(\frac{\partial \overline{U}_i}{\partial x_j} + \frac{\partial \overline{U}_j}{\partial x_i}\right) \tag{8.35}$$

from which it is apparent that the turbulent energy production is directly determined by the mean-flow gradients and is always positive according to the eddy-viscosity model. Equation (8.33), with (8.32) and (8.35) for Π, provides the evolution equation for k in homogeneous turbulence.

The ε equation is more problematic. One can write down the exact evolution equation for ε, which is, of course, not closed, and analyze the different terms, but one is obliged to make the crucial closure assumptions in the end. For homogeneous turbulence, (4.68) with (4.42) give the equation for ε. As discussed in Chapter 4, following equation (4.68), the two terms on the right dominate those on the left of (4.68). Thus, to a first approximation, the equation for ε expresses approximate equilibrium between vortex stretching and viscous effects at the smallest scales, rather than giving an evolution equation for ε. To obtain (8.36) by this route, one introduces closure assumptions for the entire right-hand side of the equation, that is, the small difference between the stretching and viscous terms. In our view, it is this closure step that is crucial, rather than the order-of-magnitude arguments surrounding the inhomogeneous equivalent of the ε equation that occupy pride of place in many presentations. It seems to us simpler and less opaque to go directly to the result, ignoring the exact equation for ε. Thus, by analogy with (8.33), we suppose that $d\varepsilon/dt$ consists of two terms, one proportional to Π, which takes into account the effect of the mean-flow gradient via (8.35), the other dependent on k and ε alone. Thus, dimensional analysis leads to

$$\frac{d\varepsilon}{dt} = C_{\varepsilon 1}\frac{\varepsilon}{k}\Pi - C_{\varepsilon 2}\frac{\varepsilon^2}{k} \tag{8.36}$$

as the evolution equation for ε, where $C_{\varepsilon 1}$ and $C_{\varepsilon 2}$ are two more numerical parameters of the model. Clearly, the basis for (8.36) is weaker than that of (8.33), but together they form the standard k–ε model for homogeneous turbulence.

Let us examine the consequences of the model in the absence of mean flow, that is, setting $\Pi = 0$ in (8.33) and (8.36). The solution of the equations makes k proportional to $(t - t_0)^{-1/(C_{\varepsilon 2}-1)}$, while $\varepsilon = -dk/dt$ is proportional to $(t - t_0)^{-C_{\varepsilon 2}/(C_{\varepsilon 2}-1)}$. These power-law dependencies on time are consistent with the large-scale self-similar behavior examined in Section 7.4, and comparison with the exponent in (7.58) shows that

$$C_{\varepsilon 2} = \frac{3}{2} + \frac{\alpha_2}{\alpha_1} \tag{8.37}$$

should relate $C_{\varepsilon 2}$ to the self-similarity parameter, α_1/α_2, used in Chapter 7. As discussed in Section 6.5, the experimentally observed values of the time exponent of k vary somewhat from experiment to experiment,[12] corresponding to different α_1/α_2. The value $C_{\varepsilon 2} = 1.92$ is often used in k–ε calculations, which corresponds to $k \propto (t - t_0)^{-1.09}$, a somewhat slower decay than given by (6.173), which would yield the nonstandard value $C_{\varepsilon 2} = 1.77$. This illustrates how experimental results for simple flows may be used to determine the parameters of one-point models.

For inhomogeneous flows, the exact turbulent energy equation (4.35) can be written as

$$\frac{\partial k}{\partial t} + \overline{U}_i\frac{\partial k}{\partial x_i} = \Pi - \varepsilon - \frac{\partial I_i}{\partial x_i} \tag{8.38}$$

where I_i is the turbulent energy flux labeled "Diffusion" in (4.35), which is modeled as if it really represented diffusion of the scalar quantity k with diffusivity determined by k and ε being proportional, like ν_T, to k^2/ε. That is, $I_i = -(\nu_T/\sigma_k)(\partial k/\partial x_i)$ is used in (8.38), where σ_k is a further constant numerical parameter, to obtain

$$\frac{\partial k}{\partial t} + \overline{U}_i\frac{\partial k}{\partial x_i} = \Pi - \varepsilon + \frac{\partial}{\partial x_i}\left(\frac{\nu_T}{\sigma_k}\frac{\partial k}{\partial x_i}\right) \tag{8.39}$$

[12] Recall that, as discussed in earlier chapters, the homogeneous case is generally realized experimentally using grid turbulence and a frame of reference moving downstream with the mean flow.

which, with expressions (8.32) and (8.35) for ν_T and Π, is the final evolution equation for k. Compared with (8.33), inhomogeneity has resulted in two modifications. Firstly, the time derivative has become a mean-flow convective derivative and, secondly, a diffusive term has been added. Both additional terms are zero in the homogeneous case. Similar changes are made to (8.36), leading to

$$\frac{\partial \varepsilon}{\partial t} + \overline{U}_i \frac{\partial \varepsilon}{\partial x_i} = C_{\varepsilon 1} \frac{\varepsilon}{k} \Pi - C_{\varepsilon 2} \frac{\varepsilon^2}{k} + \frac{\partial}{\partial x_i} \left(\frac{\nu_T}{\sigma_\varepsilon} \frac{\partial \varepsilon}{\partial x_i} \right) \tag{8.40}$$

as the evolution equation for ε. A final numerical parameter, σ_ε, has been introduced here, making five in all: C_μ, $C_{\varepsilon 1}$, $C_{\varepsilon 2}$, σ_k, and σ_ε. Standard values of these parameters, obtained by empirical fitting of the predictions of the overall k–ε model, are $C_\mu = 0.09$, $C_{\varepsilon 1} = 1.44$, $C_{\varepsilon 2} = 1.92$, $\sigma_k = 1.0$, and $\sigma_\varepsilon = 1.3$.

In summary, the full set of k–ε equations consists of (8.39) and (8.40) for the turbulence, and (8.29), (8.30) with (8.31), i.e.,

$$\frac{\partial \overline{U}_i}{\partial t} + \overline{U}_j \frac{\partial \overline{U}_i}{\partial x_j} = -\frac{\partial \Gamma}{\partial x_i} + \frac{\partial}{\partial x_j} \left((\nu + \nu_T) \left(\frac{\partial \overline{U}_i}{\partial x_j} + \frac{\partial \overline{U}_j}{\partial x_i} \right) \right) \tag{8.41}$$

and

$$\frac{\partial \overline{U}_i}{\partial x_i} = 0 \tag{8.42}$$

for the mean flow, where $\Gamma = (P/\rho) + 2k/3$. Equations (8.39)–(8.42) are completed by (8.32) and (8.35) for ν_T and Π, yielding a closed set of evolution equations for the variables $\overline{U}_i$, k, and ε, with the pressure-like variable Γ, needed to satisfy the mean-flow incompressibility constraint (8.42), playing a secondary role, similar to that of the pressure in the Navier–Stokes equations.

All three evolution equations, (8.39)–(8.41), of the k–ε model contain a diffusive term involving the eddy viscosity, ν_T. The model is restricted to high turbulent Reynolds numbers, which means that ν_T is considerably larger than ν. This has several consequences, one of which is that $\nu + \nu_T$ can as well be replaced by ν_T in (8.41), leading to essentially the same form for the diffusive terms in the turbulence equations, (8.39) and (8.40), on the one hand, and the mean-flow equation, (8.41), on the other. One could also symmetrize the equations by using $\nu + \nu_T$ instead of ν_T in (8.39), (8.40). The difference between the two formulations should be unimportant in the regime of high turbulent Reynolds numbers for which they are valid. Dropping ν in (8.41) makes the model independent of viscosity, although as we shall see later, the boundary conditions at solid surfaces reintroduce the viscosity into the problem, making the "inviscid" k–ε model Reynolds-number dependent after all. As discussed towards the beginning of Section 4.1, for high-Reynolds-number turbulence, the Reynolds stress term in the mean-flow equation, represented by ν_T in (8.41), dominates over the genuine viscous term, that is, $\nu_T \gg \nu$ as stated above, except for thin viscous layers near walls (and, of course, parts of the flow in which turbulence is absent). The turbulent Reynolds number is not high enough for validity of the k–ε model in the viscous layers and it must be modified if a realistic description of those layers is required. That is, viscosity becomes important in the viscous layers, making the difference between $\nu + \nu_T$ and ν_T significant, but the model must then be changed in more ways than simply retaining ν in (8.41) and replacing ν_T by $\nu + \nu_T$ in (8.39), (8.40). The issue of a proper description of viscous layers is of importance to the flow as a whole because, to get to a solid wall one must pass through such a layer, so the imposition of boundary conditions involves such considerations, a point we will return to shortly.

Another important consequence of $\nu_T \gg \nu$ for turbulence away from walls is considerably greater diffusion in k–ε compared to the Navier–Stokes equations at the high turbulent Reynolds numbers for which the model applies. This in turn makes the smallest spatial structures of $\overline{U_i}$, k, and ε larger than those of the velocity field in individual flow realizations, which is reasonable, since they are averaged quantities that typically vary on scales determined by the overall geometry of the flow, rather than showing the small, dissipative structures of individual realizations.[13] In consequence, the spatial resolution required for k–ε simulations is considerably less stringent than for DNS or LES, which is, of course, the reason they are computationally cheaper. The downside is the approximate nature of the model due to the closure assumptions used in its construction, whose effects on the accuracy of the results are hard to predict. Near walls, large gradients in statistical properties such as $\overline{U_i}$ and ε may require refinement of the resolution needed for k–ε normal to the wall, becoming comparable to that for DNS in the viscous layer, where the required resolution is determined by the layer thickness. However, even in the viscous layer, where the relatively low turbulent Reynolds number implies modification of the standard k–ε model, large gradients in the one-point statistics occur solely in the direction normal to the wall and so, in principle, it is only in that direction that the numerical resolution needs refinement, unlike DNS, which must resolve the small scales in all directions.

The difficulty in applying boundary conditions at walls, namely that the basic k–ε equations given above do not apply within the viscous layer, can be tackled in at least two different ways. Firstly, one may modify the model so that it describes the viscous layer, leading to "low-Reynolds-number" models,[14] of which a considerable number have been proposed (see Patel, Rodi, and Scheuerer (1985), Rodi and Mansour (1993), Wilcox (1993), for further information). Low-Reynolds models attempt to reproduce the experimentally observed viscous-layer scaling laws, based on ν and the turbulent friction velocity, u_*, which were discussed in Section 5.5. These scaling laws apply a priori to flat-plate, zero-pressure-gradient boundary layers, but, as discussed in Chapter 5, may be relatively robust to moderate departures from this ideal case. However, one should not expect them to apply near separation, for instance.

If a low Reynolds number k–ε is used, the wall boundary conditions can be applied without further ado. Since $U_i = 0$ at the surface, we have the exact relations

$$\overline{U_i} = 0$$

$$k = 0$$

[13] The "effective Reynolds number" of the model, based on ν_T, u', and L, is always of the same order of magnitude, as follows from (8.32), $k = 3u'^2/2$ and $\varepsilon = O(u'^3/L)$.

[14] Here, low does not mean small Reynolds number and a more descriptive name might be near-wall modeling.

$$\varepsilon = \nu\frac{\partial^2 k}{\partial y^2} = 2\nu\left(\frac{\partial k^{1/2}}{\partial y}\right)^2 \tag{8.43}$$

at $y = 0$, where, as in Chapter 5, y is distance from the wall. The first two equations of (8.43) follow directly from $\overline{U_i} = u_i = 0$ and the definition of k, while the first equality of the third equation can be derived from the identity

$$\nu\frac{\partial^2 k}{\partial x_i \partial x_i} = \varepsilon + \nu\left\{\overline{u_j\frac{\partial^2 u_j}{\partial x_i \partial x_i}} - \overline{\frac{\partial u_i}{\partial x_j}\frac{\partial u_j}{\partial x_i}}\right\} \tag{8.44}$$

which stems from taking the Laplacian of $k = \overline{u_j u_j}/2$ and the definition, (4.27), of ε. Evaluating (8.44) at the surface, the term in brackets is zero because $u_i|_{y=0} = 0$ and the only nonzero components of $(\partial u_i/\partial x_j)|_{y=0}$ are $\partial u_x/\partial y$ and $\partial u_z/\partial y$. The second equality of the third equation is derived from the first using the Taylor's series expansion

$$k \sim \frac{1}{2}y^2\frac{\partial^2 k}{\partial y^2}\bigg|_{y=0}$$

This version of the ε boundary condition has the numerical advantage that it does not involve second derivatives. Equation (8.43) provides five scalar boundary conditions for the unknowns of the model, namely the three components of $\overline{U_i}$, k, and ε, giving sufficient wall conditions for a low-Reynolds k–ε calculation. Use of a low-Reynolds-number k–ε model introduces additional numerical parameters.

In the second approach to wall boundary conditions, such conditions are applied, not at the wall $y = 0$, but instead, near the wall, outside the viscous layer, where the basic k–ε model is supposed to hold. Numerical simulation using k–ε is then restricted to the flow external to the viscous layer, where viscosity may be neglected in the mean-flow equation, (8.41), leading to an inviscid set of model equations, namely (8.32), (8.35), and (8.39)–(8.42), with ν set to zero in (8.41). In effect, an informal matched asymptotic expansion philosophy is adopted, in which the viscous layer is described using empirical scaling laws, while inviscid k–ε is applied to simulate the remainder of the flow.[15] The wall boundary conditions for k–ε result from matching of the two regions. It will be recalled from the discussion of boundary layers in Chapter 5 that a turbulent boundary layer consists of the viscous layer at the wall, much thinner than the overall boundary layer thickness and in which the experimentally observed universal scaling law (5.198) applies, and an outer region, while between the two lies the inertial layer in which the mean velocity profile is observed to have the characteristic logarithmic form

$$\overline{U_x} = u_*\left(\frac{1}{\kappa}\log y_+ + a\right) \tag{8.45}$$

[15] Of course, viscosity still plays a role outside the viscous layer via the energy dissipation and other intrinsically dissipative-scale phenomena, but not in the mean-flow momentum equation.

where $u_*(x)$ is wall friction velocity, x is streamwise distance along the wall, $y_+ = u_*y/\nu$ is distance from the wall, scaled appropriately for the viscous layer, and $\kappa \approx 0.4$, $a \approx 4$ are universal numerical constants. Notice that, in using the viscous-layer scaling law (5.198) and its large y_+ limiting form, (8.45), we suppose a well-defined streamwise direction, represented by x, for the mean velocity near the wall. This is the case for two-dimensional flows, to which, for simplicity sake, we specialize in what follows.

An exact solution of the k–ε equations given earlier, with ν dropped from (8.41), is obtained by taking Γ constant, $\overline{U_2} = \overline{U_3} = 0$, and

$$\overline{U_1} = u_*\left(\frac{1}{\kappa}\log x_2 + A\right)$$

$$k = \frac{u_*^2}{C_\mu^{1/2}} \tag{8.46}$$

$$\varepsilon = \frac{u_*^3}{\kappa x_2}$$

where u_* and A are arbitrary constants, while

$$\kappa^2 = (C_{\varepsilon 2} - C_{\varepsilon 1})\sigma_\varepsilon C_\mu^{1/2} \tag{8.47}$$

This solution of the k–ε equation is interpreted as representing a flat-plate inertial layer in which $x = x_1$ is the streamwise coordinate and $y = x_2$ the coordinate normal to the wall. Equation (8.47) yields the value of the Von Karman constant, $\kappa = 0.43$, according to the k–ε model with the standard parameter values given earlier. Good agreement with the experimental value of κ is no coincidence, since the boundary-layer log law was one of the empirical inputs used to fix the standard model parameters. As implied by the notation, the constant u_* occurring in the above solution represents the wall friction velocity, since

$$-\overline{u_1 u_2} = \nu_T \frac{\partial \overline{U_1}}{\partial x_2} = u_*^2 \tag{8.48}$$

of which the first equality follows from (8.31) and the second from (8.32) and equations (8.46). As discussed in Chapter 5, within the inertial layer, the turbulent shear stress, $-\rho\overline{u_x u_y}$, is equal to the wall friction, ρu_*^2. Thus, (8.48) implies that the constant, u_*, in the above solution should indeed be set equal to the wall friction velocity, a requirement that can be regarded as a consequence of informal matching to the viscous layer. Likewise, the constant A can be determined by matching $\overline{U_1}$ from (8.46) with the experimental result, (8.45). Applying empirical results directly in this way avoids the need to explicitly model the viscous layer.

Assuming that the above solution represents the near-wall behavior of more general flows modeled by k–ε in the near-wall limit, we have

$$
\begin{aligned}
\overline{U_x} &\sim u_*\left(\frac{1}{\kappa}\log\frac{u_* y}{\nu} + a\right)\\
\overline{U_y} &\to 0\\
k &\to \frac{u_*^2}{C_\mu^{1/2}}\\
\varepsilon &\sim \frac{u_*^3}{\kappa y}
\end{aligned}
\tag{8.49}
$$

for the limiting behavior as $y \to 0$ of a k–ε solution with ν dropped from (8.41). Equations (8.49) provide boundary conditions to apply to such an inviscid model, which is to be solved for the mean velocity, k, ε, and the unknown wall-friction velocity, $u_*(x)$. Observe that viscosity only enters the problem via the first equation in (8.49), where it expresses the residual influence of viscosity on the mean-flow outside the viscous layer. That is, k–ε simulations using the above formulation are dependent on Reynolds number solely via one of the boundary conditions at walls (and then with only a rather weak, logarithmic dependency). The parameters of such a simulation are those coming from the k–ε model, which determine κ from (8.47), together with a, which appears, along with the viscosity, in the first of the boundary conditions (8.49) and is given its experimentally determined value. Notice that, being based on flat-plate inertial-range scaling laws, the above formulation will not properly describe flows near separation, where simulations usually lead to relatively poor results for separated flows. Indeed, separation is not generally well-handled by low-Reynolds-number k–ε models either, being one of the Achilles heels of k–ε as a simulation technique.

Flows, such as the wake of an obstacle placed in a uniform stream, in which there is a frontier between turbulent and laminar flow lead to a sharp interface[16] between the two according to inviscid k–ε, and a thin boundary for low-Reynolds-number k–ε models (for more details, see Wilcox (1993), section 7.2.2). This phenomenon is unphysical, since in reality the location of the turbulent/laminar interface fluctuates and averaged quantities are smooth functions of position, illustrating a further deficiency of the model (which also manifests itself in other models). Note that, under inviscid k–ε, one should set $k = \varepsilon = 0$ on the laminar side of the frontier and interpret the indefinite quotient in (8.32) as zero.

Whether one employs a low-Reynolds-number k–ε model or the "inviscid" version, the choice of numerical method is much as for DNS or LES. As discussed earlier, the spatial resolution requirements are less stringent for k–ε because the averaged variables do not contain the fine scales of individual realizations. However, care must be taken near walls to ensure that the numerical scheme is capable of handling the singular behavior of ε and $\overline{U_x}$ implied by (8.49), which persists all the way to the wall for "inviscid" k–ε or until the viscous layer is entered for low-Reynolds-number models (whose resolution of the viscous layer itself must also be sufficiently fine, as noted earlier). Furthermore, attention should be paid

[16] On the turbulent side of the interface, the quantities k and ε go to zero like distance to the power of 10/7, while the difference between the tangential component of the mean velocity and its value at the frontier behaves in the same way. Thus, these quantities are continuous, but have singular second derivatives.

to correct numerical treatment of any laminar/turbulent interface that is present. In addition to the wall boundary conditions described above, conditions are required at any inlet and outlet boundaries of the flow domain. For instance, one might specify the values of the mean velocity and the turbulent quantities k and ε there.

Although, because the time scale for evolution of the averaged quantities $\overline{U_i}$, k, and ε is longer than that of the smallest scales of individual realizations, bigger time steps can, in principle, be taken in k–ε calculations than DNS or LES, one obviously needs to take care to ensure that the scheme is stable. According to (8.49), the quantity k/ε, which provides a natural time scale for the k–ε equations and corresponds physically to a large-scale eddy lifetime, goes to zero at a wall. This means that the solution can evolve quickly in the near-wall zone and hence ought to equilibrate rapidly, subsequently responding passively to changes in the rest of the flow. However, in consequence, the k–ε equations have the potential for generating rapid changes, a property known as stiffness, and which requires considerable care to avoid numerical instability. Furthermore, low-Reynolds-number models typically become stiffer still as the wall is approached through the viscous layer. Overall, the construction of satisfactory numerical schemes for k–ε calculations is, as for flow simulation in general, but perhaps more so, something of an art. Nonetheless, it goes without saying that the usual numerical disciplines of verifying the correctness of the results against standard test cases and their convergence with respect to all numerical parameters (e.g., grid sizes and time steps), need to be applied.

One often wants to simulate statistically steady turbulent flows. Such flows are not steady in individual realizations and so DNS treats them like any other flows. However, using statistical models, one can exploit statistical steadiness by looking for steady solutions of, for instance, the k–ε equations. The usual way of obtaining steady solutions amounts to integrating the time-dependent model equations forward in time until the flow variables settle down to a steady state. The fact that one does not really care about the accuracy of the time integration, only the final state, allows acceleration of the convergence using large time steps and the addition of terms to the equations that damp out instabilities, but go to zero once the steady state is attained. In a similar vein, one may suppress problems of stiffness near the surface by altering the relative rates of change at different locations. Interestingly, in some cases, it is found that there are several possible steady solutions of the k–ε equations, which one is attained depending on the initial conditions. Whether such nonuniqueness of the steady k–ε model reflects different statistically steady states of the real turbulent flow is unclear, but it provides a warning to the unwary modeler. Given multiple equilibria, some or all of which may be unstable, and with dynamical systems theory in mind, one may ask if the time-dependent model (without convergence acceleration) might have oscillatory or chaotic solutions, which would again pose problems of physical interpretation.

Given the sweeping closure approximations involved, the k–ε model provides only a rather crude representation of the physics of turbulence. However, it contains five basic parameters, plus those introduced by low-Reynolds-number modifications, or through the "inviscid" boundary conditions, all of which are adjustable to fit experimental data. Standard values of the basic parameters were given earlier, but better results can usually be obtained by tuning the parameters to particular flows, or classes of flows. That is, although the physical basis of the modeling is crude, the flexibility provided by the parameters is sufficient that, once the model has been parameterized for one member of a class of similar flows, it may be capable of predicting the others with reasonable precision. This may be just what is needed in many design and development applications, where the basic form of the flow geometry is fixed by other considerations and one wants to optimize the details. Thus, k–ε is in the best traditions of pragmatic engineering, meeting a pressing practical need for flow predictions with a relatively modest computational cost allowing routine calculations. What is perhaps surprising is not that k–ε is inaccurate in some circumstances, notably in the presence of separation, boundary layers with adverse pressure gradients, recirculating zones, secondary flows, and strong swirl or streamline curvature, but that it does remarkably well at predicting many flows and better if its numerical parameters are tuned. The main practical difficulty lies in not knowing a priori when new physical effects, potentially leading to modeling deficiencies, may appear as the geometry or other physical parameters of the flow are changed within a given class of flows.

The widespread adoption of the k–ε model for practical simulations of turbulence has spurred attempts to remedy its defects. These include the R_{ij}–ε models, based on the evolution equation for $R_{ij} = \overline{u_i u_j}$, and which we briefly outline below. For further information, we refer the reader to the seminal papers by Hanjalic and Launder (1972), Launder, Reece, and Rodi (1975), whose model set the standard, and Lumley (1975), as well as the review articles by Launder (1989) and Hanjalic (1994), and the book by Schiestel (1993), while Cambon and Scott (1999) discuss, among other things, R_{ij}-based modeling in the light of anisotropic spectral analysis.

Like k–ε, R_{ij}–ε models employ $\overline{U_i}$ to describe the mean flow, but R_{ij} and ε, in place of $k = R_{ii}/2$ and ε, to represent the turbulence. As with the k–ε model, one has in mind developed turbulence at high Reynolds numbers, for which ε can be thought of as providing limited information concerning the spatial structure of turbulence (e.g., a single correlation length scale $k^{3/2}/\varepsilon$) to supplement the rather more detailed velocity data provided by R_{ij}. No closure is needed for the Reynolds stress term in (8.29), that is, the mean-flow equations are retained as is. On the turbulence side, R_{ij} evolves according to (4.32), which can be written as

$$\frac{\partial R_{ij}}{\partial t} + \overline{U_k}\frac{\partial R_{ij}}{\partial x_k} = \underbrace{-\frac{\partial \overline{U_j}}{\partial x_k}R_{ik} - \frac{\partial \overline{U_i}}{\partial x_k}R_{jk}}_{\text{Production}} + \underbrace{\Psi_{ij}}_{\substack{\text{Pressure}\\ \text{term}}} - \underbrace{\varepsilon_{ij}}_{\text{Dissipation}} - \underbrace{\frac{\partial D_{ijk}}{\partial x_k}}_{\text{Diffusion}} \tag{8.50}$$

where

$$\Psi_{ij} = \overline{\frac{1}{\rho}p\left(\frac{\partial u_i}{\partial x_j} + \frac{\partial u_j}{\partial x_i}\right)} \tag{8.51}$$

$$\varepsilon_{ij} = 2\nu\overline{\frac{\partial u_i}{\partial x_k}\frac{\partial u_j}{\partial x_k}} \tag{8.52}$$

and

$$D_{ijk} = \overline{u_i u_j u_k} + \frac{1}{\rho}\overline{p(u_i\delta_{jk} + u_j\delta_{ik})} - \nu\frac{\partial R_{ij}}{\partial x_k} \tag{8.53}$$

All three of the final terms of (8.50), that is, (8.51)–(8.53), need closure, since, with the exception of the viscous term of (8.53) (which is anyway negligible at high Reynolds numbers, outside viscous layers), they are not exactly expressible in terms of the basic variables, $\overline{U}_i$, R_{ij}, and ε of the model.

We do not want to detail the closure approximations used, which vary from author to author, with those of Launder et al. (1975) as the prototype. As discussed in Section 4.3, the pressure term, Ψ_{ij}, is generally decomposed into "slow" (nonlinear), "rapid" (linear), and wall components, based on different parts of the solution of the Poisson equation, (4.48), for the fluctuating pressure. Each component is modeled separately; for instance, Rotta's closure, (4.55), may be used for the slow part, while, being retained in linear theory, the rapid component can be calibrated using rapid distortion theory. The dissipative term, ε_{ij}, is often closed by assuming isotropy of the dissipative scales that dominate the derivatives in (8.52), that is, (4.45) is used for this term. As well as (8.29), and the closed version of (8.50), an evolution equation for ε is needed, which is often obtained by heuristically modifying that, (8.40), of the k–ε model. As for the earlier model, wall boundary conditions should be carefully formulated, either using an "inviscid" R_{ij}–ε model and the near-wall semi-empirical forms, or by tailoring the model to describe the viscous layer.

Description of the turbulent velocities using the tensor quantity R_{ij}, rather than the scalar $k = R_{ii}/2$, gives more degrees of freedom than k–ε (and more adjustable parameters) to express variability from flow to flow and with position in a given flow, although it is clear, by considering the case of homogeneous turbulence, that one is still very far from the infinite number of degrees of freedom of the spectral tensor $\Phi_{ij}(\mathbf{k})$, representing two-point moments, let alone a full statistical description of turbulence. In consequence, models based on R_{ij} have more potential for describing turbulent flows than k–ε, but as for k–ε models, the many closure assumptions used in constructing Reynolds-stress models mean that they are rather gross approximations, whose numerous adjustable parameters allow empirical fitting to match some classes of flows. The models have been found to perform better than k–ε in some cases, but not for others. The cynically inclined might attribute the improved result with some flows to the greater number of parameters, but this is probably too extreme a view. Compared with k–ε, the more sophisticated R_{ij}-based models are more complicated, involve more variables, and appear to have more stringent numerical stability constraints. The result is heavier computer requirements, though still considerably less than DNS or LES at the high turbulent Reynolds numbers to which the models apply. From a practical standpoint, one may ask whether the increased computing requirements justify adoption of the more complicated modeling, the answer to which may depend on the particular type of flow one wants to simulate.

Rodi (1976) proposed the use of models derived from the R_{ij}–ε formalism by making certain equilibrium assumptions to approximate differential equations as algebraic relation. Such algebraic-stress models yield equations to solve for R_{ij} in terms of k, ε, and derivatives of the mean-velocity field (see e.g., the discussion in Gatski and Speziale (1993)). If the equations are solved for R_{ij} and the result used in (8.29), an evolution equation for the mean flow is obtained and the model completed by equations for k and ε. Although it is perhaps more systematic, being derived from R_{ij}-based modeling, this technique is similar to, and has revived interest in, nonlinear[17] k–ε models. In nonlinear k–ε, the usual Newtonian expression, (8.31), for the Reynolds stresses is modified to depend on the mean flow in a nonlinear fashion. This approach treats turbulence like an anisotropic, nonlinear material and is intended as a more realistic representation of R_{ij} than (8.31). As for standard k–ε models, the advantage over R_{ij}-based modeling is reduced computational cost owing to having fewer evolution equations with less stringent numerical stability constraints. The reader is referred to Shih (1996) for a review of nonlinear k–ε modeling.

8.4 Conclusions

Despite more than a century of work on the theory of turbulence, and a number of important insights, many basic issues remain unresolved. For instance, although some of the sequence of instabilities leading to transition have been elucidated for certain flows, a fully satisfactory theory of transition is elusive. Furthermore, even the description of the small scales of developed turbulence in the high-Reynolds-number limit, where Kolmogorov's theory and its derivatives provide glimpses of a general theory, is currently based on rather strong, heuristic hypotheses, rather than deduced from the Navier–Stokes equations. Add fundamental questions about developing turbulence, large-scale self-similarity of homogeneous flows, the log law near walls, to name but a few, and it is apparent that much remains to be fully explained. Trying to find a succinct characterization of turbulence, capable of distinguishing it from a complicated laminar flow, might also help clarify ideas. Turbulence theory is a notoriously difficult subject area and, given its history, it is perhaps too much to expect a miraculous solution of the "turbulence problem" (whatever that may mean) in the near future. However, work towards answering some of the outstanding fundamental questions will no doubt lead to further advances in our understanding in the years to come.

Imaginatively designed numerical experiments based on DNS, or even LES, allowing "thought experiments" and the determination of quantities that are difficult to measure in the laboratory, will have a role to play alongside more traditional laboratory experiments and analytical theory, particularly as the steady advance in computer technology extends the range of accessible Reynolds numbers. At the same time, the pressing practical need for reliable, routine simulations of industrial and naturally occurring turbulent flows will be met by increasingly sophisticated models, and perhaps eventually by LES or one of its successors.

[17] The name nonlinear k–ε is perhaps unfortunate, since the standard k–ε equations are themselves nonlinear. What is specifically nonlinear here is the assumed dependence of the Reynolds stresses on the mean velocity field.

References

Cambon, C., Scott, J. F., 1999. Linear and nonlinear models of anisotropic turbulence. *Ann. Rev. Fluid Mech.*, **31**, 1–53.

Canuto, C., Hussaini, M. Y., Quateroni, A., Zang, T. A., 1987. *Spectral Methods in Fluid Mechanics*. Springer, Berlin.

Ferziger, J. H., Peric, M., 1996. *Computational Methods for Fluid Dynamics*. Springer, Berlin.

Galperin, B., Orszag, S.A. (eds), 1993. *Large Eddy Simulation of Complex Engineering and Geophysical Flows*. Cambridge University Press, Cambridge.

Gatski, T. B., Speziale, C. G., 1993. On explicit algebraic stress models for complex turbulent flows. *J. Fluid Mech.*, **254**, 59–78.

Germano, M., Ugo, P., Moin, P., Cabot, W. H., 1991. A dynamic subgrid-scale eddy viscosity model. *Phys. Fluids*, **A3**, 1760–5.

Ghosal, S., Lund, T. S., Moin, P., Akselvoll, K., 1995. A dynamic localization model for large-eddy simulation of turbulent flows. *J. Fluid Mech.*, **286**, 229–55.

Goldstein, M. E., 1984. Aeroacoustics of turbulent shear flows. *Ann. Rev. Fluid Mech.*, **16**, 263–85.

Gottlieb, D., Orszag, S. A., 1977. *Numerical Analysis of Spectral Methods: Theory and Applications*. SIAM, Philadelphia.

Hanjalic, K., 1994. Advanced turbulence closure models: review of current status and future prospects. *Int. J. Heat Fluid Flow*, **15**, 178–203.

Hanjalic, K., Launder, B. E., 1972. A Reynolds stress model of turbulence and its application to thin shear flow. *J. Fluid Mech.*, **52**, 609–38.

Hirsch, C., 1988 (vol. 1), 1990 (vol. 2). *Numerical Computation of Internal and External Flows*. Wiley, New York.

Kim, J., Moin, P., Moser, R. D., 1987. Turbulence statistics in fully developed channel flow at low Reynolds number. *J. Fluid Mech.*, **177**, 133–66.

Kraichnan, R. H., 1976. Eddy viscosity in two and three dimensions. *J. Atmos. Sci.*, **33**, 1521–36.

Launder, B. E., 1989. Second-moment closure and its use in modelling turbulent industrial flows. *Int. J. Numer. Methods Fluids*, **9**, 963–85.

Launder, B. E., Reece, G. J., Rodi, W., 1975. Progress in the development of a Reynolds-stress turbulence closure. *J. Fluid Mech.*, **68**, 537–66.

Lesieur, M., Métais, O., 1996. New trends in large-eddy simulations of turbulence. *Ann. Rev. Fluid Mech.*, **28**, 45–82.

Leslie, D. C., Quarini, G. L., 1979. The application of turbulence theory to the formulation of sub-grid modelling procedures. *J. Fluid Mech.*, **91**, 65–91.

Lumley, J. L., 1975. Prediction methods in turbulent flows. Von Karman Institute, Lecture Series 76.

Moin, P., Mahesh, K., 1998. Direct numerical simulation: a tool in turbulence research. *Ann. Rev. Fluid Mech.*, **30**, 539–78.

Orszag, S. A., 1971. Numerical simulation of incompressible flows within simple boundaries: accuracy. *J. Fluid Mech.*, **49**, 75–112.

Parpais, S., Bertoglio, J.-P., 1996. A spectral closure for inhomogeneous turbulence applied to turbulent confined flow. In *Proc. 6th Eur. Turbul. Conf.*, Lausanne, Switzerland, pp. 75–6.

Patel, V. C., Rodi, W., Scheuerer, G., 1985. Turbulence models for near-wall and low-Reynolds number flows: a review. *AIAA J.*, **23**, 1308–19.

Press, W. H., Teukolsky, S. A., Vetterling, W. T., Flannery, B. P., 1992. *Numerical Recipes in Fortran: The Art of Scientific Computing*. Cambridge University Press, Cambridge.

Rodi, W., 1976. A new algebraic relation for calculating the Reynolds stresses. *Z. angew. Math. Mech.*, **56**, T219–21.

Rodi, W., Mansour, N. N., 1993. Low Reynolds number k–ε modelling with the aid of direct simulation data. *J. Fluid Mech.*, **250**, 509–29.

Rogers, M. M., Moin, P., 1987. The structure of the vorticity field in homogeneous turbulent flows. *J. Fluid Mech.*, **176**, 33–66.

Schiestel, R., 1993. *Modélisation et Simulation des Ecoulements Turbulents.* Hermès, Paris.

Shih, T. H., 1996. Developments in computational modeling of turbulent flows. Contractor's report 198458, ICOMP-96-04, CMOTT-96-03, NASA.

Smagorinsky, J., 1963. General circulation experiments with the primitive equations, part I: the basic experiment. *Monthly Weather Rev.*, **91**, 99–164.

Wells, V. L., Renaut, R. A., 1997. Computing aerodynamically generated noise. *Ann. Rev. Fluid Mech.*, **29**, 161–99.

Wilcox, D. C., 1993. *Turbulence Modeling for CFD.* DCW Industries Inc., La Cañada, CA.

Index

Lightning Source UK Ltd.
Milton Keynes UK
UKOW07f2338190615

253793UK00005B/146/P

9 780521 775380